VOLCANOES AND SOCIETY

David Chester

University of Liverpool

Edward Arnold
A division of Hodder & Stoughton
LONDON MELBOURNE AUCKLAND

To Jennifer and Olivia

First published in Great Britain 1993
Distributed in the USA by Routledge, Chapman and Hall, Inc.
29 West 35th Street, New York, NY 10001

British Library Cataloguing in Publication Data

Chester, D. K.
Volcanoes and Society
I. Title
551.2

ISBN 0–340–51761–1

Library of Congress Cataloging-in-Publication Data

Chester, David K.
Volcanoes and society / David K. Chester.
p. cm.
Includes bibliographical references and index.
ISBN 0–340–51761–1 : $25.00
1. Volcanoes. 2. Volcanic hazard analysis. I. Title.
QE522.C46 1994 93–1620
551.2′—dc20 CIP

Typeset in 10/11pt Sabon by Wearset, Boldon, Tyne & Wear.
Printed in Great Britain for Edward Arnold, a division of Hodder and Stoughton Limited, Mill Road, Dunton Green, Sevenoaks, Kent TN13 2YA by St. Edmundsbury Press, Bury St. Edmunds, Suffolk and bound by Hartnolls Ltd, Bodmin, Cornwall.

Cover photograph

'Eruption of Vesuvius on 10th August 1767' by Peter Fabris, from *Campi Phlegraei: Observations on the Volcanoes of the Two Sicilies*, by Sir William Hamilton (1776), reproduced by kind permission of Liverpool University Library.

Preface

For several years I have taught a course on pure and applied volcanology to senior undergraduate students. During this time it has become apparent that, although some excellent introductory texts are available, those of a more advanced character specialize in either the processes responsible for producing volcanic activity of various kinds, or else are essays in human adjustments to volcanic risk. To my knowledge no current work examines all aspects of pure and applied volcanology at a level suitable for this group of students. The primary aim of the book is to remedy this deficiency and, in addition, provide an introductory text for graduate and professional audiences. I am conscious that students who are interested in volcanoes and volcanic hazards are drawn from two very different traditions. One cohort is already grounded in the fundamentals of earth science, but has a very sketchy knowledge of social theory, whereas the other group has a contrasting pedigree. The book, therefore, includes some explanations of basic concepts in the earth and social sciences.

Many organizations and individuals have helped directly and indirectly in the production of this work. In addition to those acknowledged for agreeing to the use of copyright materials, I wish to thank my friends and colleagues for their encouragement, support, assistance and, in some cases, comments on draft chapters. Particular thanks are due to Angus Duncan for first awakening my interest in volcanoes and John Guest and Chris Kilburn for helping in its development. No writer of a general work can do other than acknowledge the debt he owes to the published work of a large number of scientists. I note in particular the constant stimulation I have received from the writings of: R. J. Blong, A. R. McBirney, D. W. Peterson, R. S. J. Sparks, R. I. Tilling, G. P. L. Walker, G. F. White and L. Wilson. The late Professor W. B. Fisher of the University of Durham was responsible for the early career development of many geographers, and I am pleased to admit the debt I owe to him. Stimulating research supervision by Chalmers Clapperton and David Sugden was also very positive and I trust that they are not too disappointed that the book is not on glacial geomorphology. In Liverpool I have been fortunate to work in a stimulating academic environment and I wish to record the debt I owe to colleagues past and present, especially: Ian Bradbury, Andy Charlesworth, John Dickenson, Adrian Harvey, Ann Henderson-Sellers, Peter James, Dick Lawton, Phil Lister, Peter Lloyd, Andy Morse, Andy Plater, Geoff Thomas and many final year students. Special thanks are due to Sandra Mather, who has so expertly drafted the figures, and to Ian Qualtrough and Suzanne Yee for photographic assistance. Laura McKelvie has been a most supportive editor.

Research for the book began in 1988 during study leave at the United States Geological Survey, Reston, Virginia. I place on record my gratitude to all, in particular Mr Frank A. Taylor, who made my sojourn so productive. Over the years the author's research has been supported by the Leverhulme Trust, the Natural Environment Research Council, the North Atlantic Treaty Organisation, the Royal Society and the University of Liverpool, to all I am most grateful.

Finally, special mention should be made of Jennifer and Olivia, the two people to whom this work is dedicated.

David Chester
October 1992

Acknowledgements

The publishers would like to thank the following for permission to use copyright material:

Chapman and Hall for Greeley, R. 1987: *Planetary Landscapes*. London: Allen and Unwin, fig. 8.1, page 190 (Figure 1.2); The American Geophysical Union for Howard, K.A. *et al.* 1974: Lunar basin formation and highland stratigraphy. *Reviews of Geophysics and Space Physics*, 12, fig. 14, page 322 © by the American Geophysical Union (Figure 1.3); The American Geophysical Union for Trask, N.J. and Guest, J.E. 1975: Preliminary geologic terrain map of Mercury. *Journal of Geophysical Research* 80, plate 1, pages 2472–3 © by the American Geophysical Union (Figure 1.6); The American Geophysical Union for Masursky, H., Eliason, E., Ford, P.G., McGill, G.E., Pettengill, G.H., Schaber, G.G. and Schubert, G. 1980: Pioneer Venus radar results: geology from images and altimetry. *Journal of Geophysical Research* 85, fig. 4, page 8237 © by the American Geophysical Union (Figure 1.10); reprinted from Condie, K.C., *Plate Tectonics and Crustal Evolution*, © 1989, fig. 10.27, page 381, with permission from Pergamon Press Ltd, Headington Hill Hall, Oxford OX3 0BW, UK (Figure 1.12); Cambridge University Press for Glass, B.P. 1982: *Planetary Geology*, fig. 3.14, page 59 (Figure 2.1b); The AAAS for ARCYANA 1975: Transform fault and rift valley from bathyscaph and diving saucer. *Science* 190, fig. 3, page 112, © 1975 by the AAAS (Figure 2.4a); Macdonald, K.C. and Luyendyk, B.P. 1982: The crust of the East Pacific Rise. In *Volcanoes and the Earth's Interior*, by Robert Decker and Barbara Decker © 1982 by W.H. Freeman. Reprinted by permission of W.H. Freeman and Co. (Figure 2.4b); Longman Group UK Limited for Summerfield, M.F. 1991: *Global Geomorphology* London: Longman, fig. 4.5, page 88 and fig. 4.12, page 94 (Figure 2.6); the authors for Henyey, T.L. and Lee, T.C. 1976: Heat flow in Lake Tahoe, California-Nevada and the Sierra Nevada-Basin and Range transition. *Geological Society of America Bulletin* 87, fig. 8, page 1185 (Figure 2.7); the Geological Society Publishing House and the authors for Le Bas, M.J. and Streckeisen, A.L. 1991: The IUGS systematics of igneous rocks. *Journal of the Geological Society of London* 148, fig. 5, page 830 (Figure 3.2); The American Geophysical Union for Ringwood, A.E. 1969: Composition and evolution of the upper mantle. *American Geophysical Union Monographs* 13, fig. 5, page 12 © by the American Geophysical Union (Figure 3.4); the authors and the AAAS for Fiske, R.S. and Kinoshita, W.T. 1969: Inflation of Kilauea volcano prior to its 1967–1968 eruption. *Science* 165, fig. 8b, page 347 © 1969 by the AAAS (Figure 3.5); New Scientist for Morgan, N.: The fires that cracked a continent. *New Scientist* 8.6.91, diagram page 44 (Figure 4.2); Chapman and Hall for Holmes, A. 1965: *Principles of Physical Geology*. London: Chapman and Hall, fig. 63, page 101 (Figure 4.3a); reprinted with permission from the author and *Nature*, Cox, K.G. 1989: The role of mantle plumes in the development of continental drainage patterns. *Nature* 342, fig. 2, page 874 © Macmillan Magazines Limited (Figure 4.5a); Elsevier Science Publishers BV and the authors for Peterson, D.W. and Tilling, R.I. 1980: Transition of basaltic lava from *pahoehoe* to *aa*, Kilauea volcano, Hawaii: Field observations and key factors. *Journal of Volcanology and Geothermal Research* 7(3/4), fig. 9, page 285 (Figure 4.7); Elsevier Science Publishers BV and the authors for Rowland, S.K. and Walker, G.P.L. 1987: Toothpaste lava: Characteristics and origin of a lava structural type transitional between pahoehoe and aa. *Journal of Volcanology and Goethermal Research* 49, figs 3a and 3b, page 633 (Figure 4.9); Chapman and Hall for Chester, D.K. *et al.* 1985: *Mount Etna: The Anatomy of a Volcano*. London: Chapman and Hall, fig. 4.19, page 162 (Figure 4.10); Springer Verlag and the authors for Blake, S. 1990: Viscoplastic models of lava domes. In Fink, J.H. (ed.) 1990: *Lava flows and domes*. IAVCEI Proceedings in Volcanology 2, Berlin, Springer Verlag, fig. 1, page 90 (Figure 4.11b); Elsevier Science Publishers BV for Furnes, H., Friedleifsson, I.B., and Atkins, F.B. 1980: Subglacial volcanics—on the formation of acid hyaloclastites. *Journal of Volcanology and Geothermal Research* 8, pages 95–110 (Figure 4.15); Springer Verlag for Pichler, H. 1965: Acid Hyaloclastites. *Bulletin Volcanologique* 28, pages 293–310 (Figure 4.16); Geologische Rundshau for Walker, G.P.L. 1973: Explosive volcanic eruptions—a new classification system. *Geologische Rundshau* 62, fig. 5, page 439 (Figure 5.1); Springer-Verlag for Fisher R.V. and Schmincke, H.U. 1984: *Pyroclastic Rocks*. Berlin: Springer-Verlag, figs 1.3, page 8; 1.5, page 9 and 1.6, page 10 (Figure 5.2); Springer-Verlag for Carey, S. and Sparks, R.S.J. 1986: Quantitative models of the fallout and dispersal of tephra from volcanic eruption columns. *Bulletin of Volcanology* 48, fig. 1, page 111 (Figure 5.4); Springer-Verlag for Fisher R.V. and Schmincke, H.U. 1984: *Pyroclastic Rocks*. Berlin: Springer-Verlag, fig. 4.5, page 65 (Figure 5.5); Prentice Hall for Gordon A. Macdonald, *Volcanoes*, © 1972, fig. 8.2, page 149. Reprinted by permission of Prentice Hall, Englewood Cliffs, New Jersey (Figure 5.6); Chapman and Hall for Cas, R.A.F. and Wright, J.V. 1987: *Volcanic Successions*. London: Allen and Unwin, fig. 8.13, page 233 (Figure 5.7); Chapman and Hall for Case, R.A.F. and Wright, J.V. 1987: *Volcanic Successions*. London: Allen and Unwin, fig. 8.15, page 234 (Figure 5.8); Lipman, P.W. and Mullineaux, D.R.: The 1980 eruptions of Mount St Helens, Washington. *United States Geological Survey Professional Paper* 1250, fig. 219, page 381 (Figure 5.9); Elsevier Science Publishers BV and the author for Wohletz, K.H. 1983: Mechanisms of hydrovolcanic pyroclast formation: grain-size, scanning electron microscopy, and experimental studies. *Journal of Volcanology and Geothermal Research* 17, fig. 1, page 38 (Figure 5.11); Chapman and Hall for Cas, R.A.F. and Wright, J.V. 1987: *Volcanic Successions*. London: Allen and Unwin, fig. 7.33, page 204 (Figure 5.12); Springer-Verlag for Kokelaar, P. 1986: Magma-water interactions in subaqueous and emergent basaltic volcanism. *Bulletin of Volcanology* 48, fig. 4, page 282 (Figure 5.13); Chapman and Hall for Chester, D.K. *et al.* 1985: *Mount Etna: The Anatomy of a Volcano*. London: Chapman and Hall, fig. 4.6, page 133 (Figure 5.14); Research and Exploration for Sigurdsson, H. *et al.* 1985: The eruption of Vesuvius A.D. 79. *Research and Exploration* 1(3), fig. 18, page 349 (Figure 5.15); Chapman and Hall for Cas, R.A.F. and Wright, J.V. 1987: *Volcanic Successions*. London: Allen and Unwin, fig. 13.17, page 377 (Figure 5.16); Springer-Verlag for Fisher, R.V. and Schmincke, H.U. 1984: *Pyroclastic Rocks*. Berlin: Springer-

Verlag, fig. 8.23, page 204 (Figure 5.18a); the authors for Sparks, R.S.J., Self, S. and Walker, G.P.L., 1983: *Geology* 1, pages 115–18 (Figure 5.18b); Kluwer Academic Publishers for Wright, J.V. *et al.* 1981: Towards a facies model for ignimbrite-forming eruptions. In Self, S. and Sparks, R.S.J. 1981 *Tephra Studies.* D. Reidel: Amsterdam, fig. 3, page 437 (Figure 5.18c); Springer-Verlag for Guest, J.E. *et al.* 1988: Mount Etna Volcano (Basilicata, Italy): An analysis of morphology and volcaniclastic facies. *Bulletin of Volcanology* 50, fig. 4, page 250 (Figure 5.19); Longman Group UK Ltd for Summerfield, M.A. 1991: *Global Geomorphology*, fig. 10.3, page 238 (Figure 5.21a); W.H. Freeman and Co for *Volcanoes Revised and Updated*, by Robert Decker and Barbara Decker © 1989 by W.H. Freeman and Company. Reprinted by permission by W.H. Freeman and Company (Figure 6.2); Elsevier Science Publishers BV for Walker, G.P.L. 1981a: *Journal of Volcanology and Geothermal Research* 11, fig. 3, page 89 (Figure 6.4); the Royal Meteorological Society for Durbin C.S. and Henderson-Sellers, A. 1981: *Weather* 36(10), fig. 1, page 285 (Figure 6.5); Blackwell Scientific Publications for Booth, B. 1979: *Journal of the Geological Society of London* 136, fig. 4, page 335 (Figure 7.3); the United Nations Department of Humanitarian Affairs, Geneva for United Nations 1977: *Disaster Prevention and Mitigations, Vol. 1 Volcanological Aspects*, fig. 1, page 2 (Figure 7.4); reprinted from *Journal of Geodynamics* 3, Tilling, R.I. and Bailey, R.A. 1984: Volcanic hazards program in the United States, page 430 © 1985, with permission from Pergamon Press Ltd, Headington Hill Hall, Oxford OX3 0BW, UK (Figure 7.5); Lipman, P.W. and Mullineaux, D.R. 1981, *The 1980 Eruptions of Mount St Helens, Washington*, Washington D.C., United States Geological Survey 1250, fig. 455, page 796 (Figure 7.7); The Geographical Journal for Duncan, A.M. *et al.* 1981: Mount Etna Volcano: Environmental impact and problems of volcanic prediction. *The Geographical Journal* 147, fig. 6, page 176 (Figure 7.8); Springer-Verlag for Latter, J.H. (ed.): *Volcanic Hazards*, IAVCEI Proceedings in Volcanology, Berlin, Springer-Verlag, fig. 1, page 59 (Figure 7.9); Blackwell Scientific Publications for Baker, P.E. 1985: Volcanic hazards on St Kitts and Montserrat, West Indies. *Journal of the Geographical Society of London* 142, fig. 11, page 293 (Figure 7.10); Unger, J.D. 1974: *Earthquake Information Bulletin* 6, page 7, United States Geological Survey (Figure 7.13); Springer-Verlag for Decker, R.W. 1973: State of the art in volcano forecasting. *Bulletin Volcanologique* 37(4), fig. 16, page 388 (Figure 7.14); Chapman and Hall for Chester, D.K. *et al.* 1985: *Mount Etna: The Anatomy of a Volcano*. London: Chapman and Hall, fig. 8.4, page 304 (Figure 7.15); Chapman and Hall for Chester, D.K. *et al.* 1985: *Mount Etna: The Anatomy of a Volcano*. London: Chapman and Hall, fig. 9.1, page 338 (Figure 8.1); Routledge for Hewitt, K. ed., 1983: *Interpretations of Calamity.* London: Allen and Unwin, fig. 14.3, page 279 (Figure 8.3); Routledge for Dickenson, J.P. *et al.* 1983: *A Geography of the Third World.* London: Methuen & Co, fig. 1.2, page 2 and fig. 1.3, page 3 (Figures 8.4a and 8.4b respectively); Basil Blackwell for Ollier, C.D. 1988: *Volcanoes.* Oxford: Blackwell, fig. 12.4, page 188 (Figure 8.4d); Foxworthy, B.L. and Hill, M. 1982: *Volcanic eruptions of 1980 at Mount St. Helens: The first 100 days.* Washington D.C., United States Geological Survey Professional Paper 1249, fig. 13, page 24 (Figure 9.1); BBC Enterprises for Jones, D.K.C. 1974: Japan under strain. *Geographical Magazine* 47(3), page 186 (Figure 9.2a); BBC Enterprises for Coates, B.E. 1974: Giant of the East. *Geographical Magazine* 47(3), page 172 (Figure 9.2b); reprinted from *Journal of Geodynamics*, Vol. 3, Dibble, R.R. *et al.* 1985: Volcanic hazards of North Island,

New Zealand—overview. Page 371, © 1985, with permission from Pergamon Press Ltd, Headington Hill Hall, Oxford OX3 0BW, UK (Figure 9.3); Elsevier Science Publishers BV for Tazieff, H. and Sabroux, J.C.: *Forecasting volcanic eruptions*. Amsterdam: Elsevier, figs 15.1 and 15.2, page 194 (Figures 9.4a and 9.4b); Williams, R.S. and Moore, J.G. 1983: *Man against volcano: The eruption on Heimaey, Vestmannaeyjar, Iceland*. Washington D.C., United States Geological Survey publication 1983–381–618/103, page 6 (Figure 9.5); Tilling, R.I. 1982: The 1982 eruption of El Chichón volcano, Southeastern Mexico. *Earthquake Information Bulletin* 14(5), page 172, United States Geographical Survey (Figure 9.7); Elsevier Science Publishers for Zen, T. 1983: Mitigating volcanic disasters in Indonesia. In Tazieff, H. and Sabroux, J.C. 1983: *Forecasting volcanic events*. Amsterdam: Elsevier, fig. 17.1, page 221 (Figure 9.10); IAVCEI for Nuemann van Padang, M. 1963b: *Catalogue of the active volcanoes and solfatara fields of Arabia and the Indian Ocean*. Rome: International Association of Volcanology (Figure 9.11); Episodes for Sudradjat, A. and Tilling, R. 1984: Volcanic hazards in Indonesia: the 1982–1983 eruption of Galunggung. *Episodes* 7(2), fig. 3, page 14 (Figure 9.12a); Academic Press Inc. for Blong, R. 1984: *Volcanic Hazards*. Australia: Academic Press, fig. 4.7, page 169 (Figure 9.12b); Elsevier Science Publishers BV for Parra, E. and Cepeda, H. 1990: Volcanic hazard maps of the Nevado del Ruiz volcano, Colombia. *Journal of Volcanology and Geothermal Research* 42, fig. 1, page 118 (Figure 9.13); BBC Enterprises for Nolan, M.L. and Nolan, S. 1979: The five towns of Parícutin. *The Geographical Magazine* 51(5), page 342 (Figure 9.15); Springer-Verlag for Barberi, F. *et al.* 1984: Phlegraean fields 1982–1984: Brief chronicle of a volcano emergency in a densely populated area. *Bulletin Volcanologique* 47(2), fig. 2, page 177 and fig. 3, page 178 (Figure 9.16); W.H. Freeman and Co for Bolt, B.A. 1982: *Inside the Earth*, W.H. Freeman and Co., San Francisco, table 4.1, page 79 (Table 2.1); Springer-Verlag for Fisher, R.V. and Schmincke, H.-U., 1984: *Pyroclastic Rocks*. Springer-Verlag, fig. 2.2, page 13 (Table 2.3); Kluwer Academic Publishers for Wright, J.V. *et al.* 1981: A terminology for pyroclastic deposits. In Self, S. and Sparks, R.S.J.: *Tephra Studies*. Amsterdam: D. Reidel, table 5, page 461 (Table 5.1); Kluwer Academic Publishers for Wright, J.V. *et al.* 1981: A terminology for pyroclastic deposits. In Self, S. and Sparks, R.S.J.: *Tephra Studies*. Amsterdam: D. Reidel, table 6, page 462 (Table 5.2); Springer-Verlag for Kokelaar, P. 1986: Magma-water interactions in subaqueous and emergent basaltic volcanism. *Bulletin of Volcanology* 48, pages 279–81 (Table 5.7); Springer-Verlag for Fisher, R.V. and Schmincke, H.U. 1984: *Pyroclastic Rocks*. Berlin: Springer-Verlag, table 6.1, page 126 (Table 5.8); Elsevier Science Publishers BV for Sheridan, M.F. and Wohletz, K.H. 1983: Hydrovolcanism: Basic considerations and review. *Journal of Volcanology and Geothermal Research* 17, pages 18–19 (Table 5.19); Prentice Hall for Macdonald, Gordon, A.: *Volcanoes*, © 1972, p. 171. Adapted by permission of Prentice Hall, Englewood Cliffs, New Jersey (Table 5.12); Newhall, C.G. and Self, S. 1982: The volcanic explosivity index (VEI): An estimate of the explosive magnitude for historical volcanism. *Journal of Geophysical Research* 87C, table 1, page 1232 © by the American Geophysical Union (Table 6.4); Springer-Verlag for Decker, R. 1973: *Bulletin Volcanologique* 37(4), table 2, page 377 (Table 7.3); Chapman and Hall for Chester, D.K. *et al.* 1985: *Mount Etna: The Anatomy of a Volcano*, table 9.1, pages 340–1 (Table 8.2); Chapman and Hall for Chester, D.K. *et al.* 1985: *Mount Etna: The Anatomy of a Volcano*, summary of text on page 344 (Table 8.3); Chapman and Hall for Chester, D.K. *et al.* 1985 *Mount Etna: The Anatomy of a Volcano*, table 9.2, page 346 (Table 8.4); Routledge for Susman, P. *et al.* 1983: Global disasters, a radical interpretation. In Hewitt, K. (ed.): *Interpretation of Calamity*. London: Allen and Unwin, pages 279–80 (Table 8.5).

Every effort has been made to trace copyright holders of material reproduced in this book. Any rights not acknowledged here will be acknowledged in subsequent printings if notice is given to the publisher.

Contents

1

Planetary volcanism

If we are to understand the Earth, we must have a comprehensive knowledge of the other planets (Carl Sagan 1975).

1.1 Introduction

According to Ron Greeley the distinguished American planetary geologist, the 'mid-1960s witnessed two fundamental revolutions' (Greeley 1987, p. 1): the first was plate tectonics; the second the new insights which emerged through 'exploration' of the Solar System by spacecraft. It is easy to generalize and claim from a volcanological perspective that the principal impact of plate tectonics has been to improve understanding of spatial patterns of volcanic activity on Earth (Chapter 2), whereas planetary geology has provided an extended time frame into which eruptions on Earth may be placed. This is to over-simplify, because plate tectonics includes models which show how volcanic activity on Earth has changed through time, whilst planetary volcanology has been concerned to open windows and shed light on many themes, including the vital one of why the Earth has such a wide variety of volcanic processes and landscapes. Linked to both 'revolutions' are even more crucial concerns about the role of volcanism in atmospheric evolution and, not least, in the development of life on Earth (see Chapter 6, section 6.4.2.2).

Despite the fact that the pace of planetary research is less frenetic than it was 20 or even 15 years ago—when well-funded research programmes reflected geopolitical rivalries between the governments of the United States of America and the former Soviet Union—in recent years many hundreds of individual pieces of research have been published. Today both planetary science in general and planetary geology in particular are distinctive disciplines and it is impossible in one short chapter to do full justice to them. What follows is a summary, centred around those aspects of planetary science which are relevant to volcanic activity on Earth. The chapter will take the form of a comparison between volcanic activity on Earth and that occurring on the other 'terrestrial' planets. These are the inner planets of the Solar System—Earth, Mercury, Venus and Mars—to which the Moon is normally added because of its size. These are comparatively small bodies, have high densities and, like the Earth, are stony. In contrast and with the possible exception of Pluto—a planet some 5886×10^6 km away from the Sun about which relatively little is known—the 'giant' outer or *Jovian* planets are gaseous and much less dense. Often described as being 'sun-like', they are less well studied and not so relevant to a comparative study of volcanoes on Earth. Their volcanic activity is summarized below.

1.2 The volcanological evolution of the giant planets

The Earth has one satellite, the Moon. Mars has two, but the giant planets have over 30, some of which are important volcanologically. *Io*, a satellite of Jupiter, is known to have extant volcanic activity (see Fig. 1.1), whilst others have been in the past and remain capable of eruption. In recent years the giant planets and their satellites have been observed by spacecraft of the United States Pioneer and later Voyager missions, but the volcanological interpretation of the data obtained is difficult. As Michael Carr (1987, p. 137, my emphasis) has noted perspicaciously 'as we travel outward in the Solar System we ... encounter stranger and stranger forms of volcanism. Most volcanologists would feel they were on familiar ground as they traversed the younger lava plains of Mars. . . . Even the sulfur (*sic*) eruptions of *Io* and the ice-filled fissures of *Europa* have some remote terrestrial equivalents. But in the Saturnian and Uranian systems we find truly alien worlds and our terrestrially biased geologic intuition is of little use in attempting to understand them'. Because it is the best studied and has the closest parallels to volcanoes on Earth, the principal features of Jupiter and its inner moons are summarized in Table 1.1 and Fig. 1.2.

The moons (or satellites) of Saturn and Uranus are generally very small bodies, a few hundred kilometres or less in diameter, and before the Voyager missions were assumed to be volcanologically inactive. Their small sizes, it was assumed, would have caused any heat produced by the decay of radioactive elements to be dissipated early in their histories and size would again have precluded melt being generated by the tidal influences of other planetary bodies (see Chapter 3, section 3.3.1). It was argued further, that because these satellites were composed mainly

Fig. 1.1 A volcanic plume from the vent Prometheus on Io. The plume is ~100 km high and ~300 km across (NASA Voyager I image 1637748 supplied by Professor J.E. Guest, Regional Space Imagery Library, University of London Observatory).

Table 1.1: The principal geological and volcanological attributes of Jupiter and its four large, or *Galilean*, moons. Compiled from a number of sources

Planet/moon	Principal attributes
Jupiter	This is the largest gaseous planet and its volume is some 1300 times larger that that of the Earth. It is some 775×10^6 km from the Sun. Astronomical inferences and remote sensing from spacecraft, imply that it is composed of hydrogen, helium, ammonia, methane and may have a small rocky core. If it were slightly larger it would have attained a temperature sufficiently high for nuclear reactions to have been established. It would then have been a star, rather than a planet. Principal geological and volcanological interest has centred on its moons, information on which was obtained by Pioneer and Voyager spacecraft. Its four major moons are discussed below in order of increasing distance from Jupiter.
Io	Io's colour implies a sulphur-rich surface and the Voyager 1 mission of 1979 revealed a gigantic volcano in eruption and a young uncratered surface; so implying rapid resurfacing. It is thought that volcanic activity may be caused by tidal stresses due to the gravitational fields of Jupiter and Europa (see Chapter 3, section 3.3.1; and Peale *et al.* 1979). Unlike most eruptions on Earth, plume colours imply that much of the erupted material is sulphurous, although silicate melts may be of some importance (Carr 1987).
Europa	Impact craters are lacking, suggesting rapid resurfacing and there is some evidence of volcanism. The density of Europa implies that it is of a silicate composition, covered by ice tens of kilometres thick. Dark linear markings may be eruptions of water and/or icy materials. The process of melt production may be reflective of tidal movements but, because of the greater distances involved, these are less than those of Io.
Ganymede	There are no signs of extant volcanism, though it may have been active in the past. Much of its surface is densely cratered, implying great age, but other areas have ridges and valleys and are less heavily cratered. The formation of a silicate core may have initiated some ice eruptions like those of Europa, and caused disruption of the surface.
Callisto	Callisto is pock-marked with impact craters and shows no evidence of volcanism.

of water and ice, there could be 'no plausible combination of radionuclide content and tidal heating that could raise interior temperature to the melting point of water after the first billion years of the planet's history' (Carr *op. cit.*, pp. 134–135).

When results from the Voyager missions were analysed they caused a flutter of excitement because several moons of Saturn and Uranus showed clear evidence that their surfaces were not heavily cratered and, therefore ancient, but had at some time been resurfaced—presumably by volcanic materials—which had caused impact craters to be buried. *Enceladus*, a satellite of Saturn, for instance, is 500 km in diameter and is so sparsely cratered that its volcanoes must have been active in the last 2 billion years. The five largest moons of Uranus show signs of

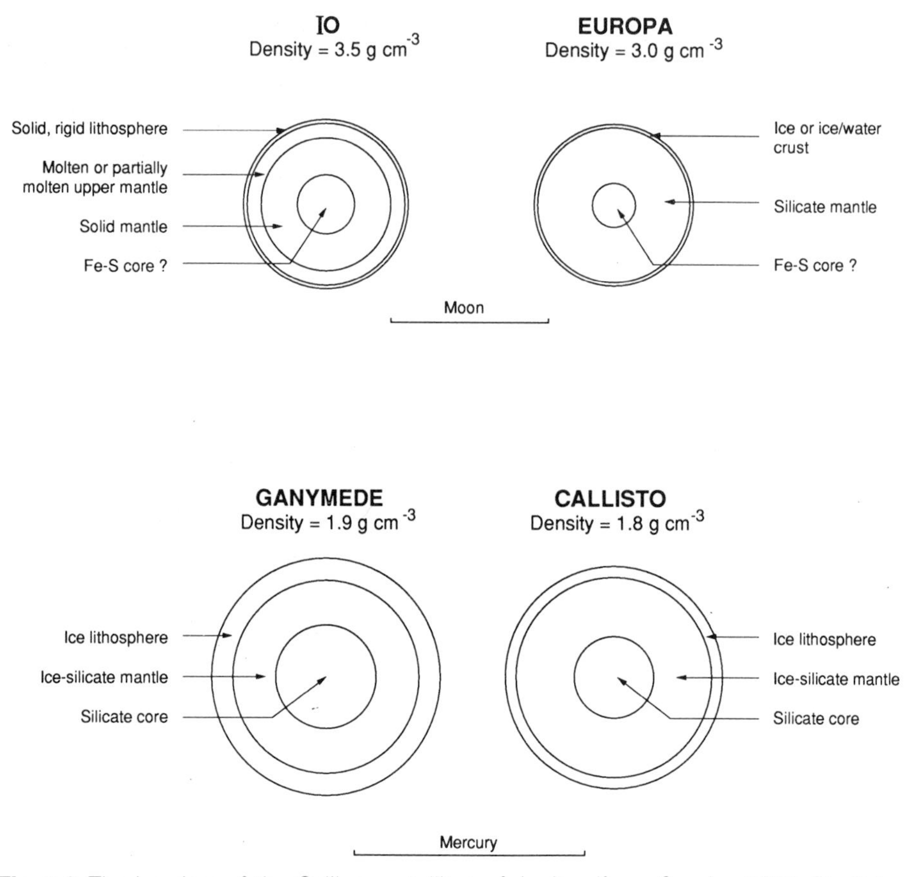

Fig. 1.2 The interiors of the Galilean satellites of Jupiter (from Greeley 1987, Fig. 8.1, p. 190).

resurfacing (Greeley 1987, p. 251) and on *Miranda* this seems to have occurred by a combination of eruptions and tectonism. It could well be that the material that has been erupted is not silicate melt as is the case on the terrestrial planets, or sulphur as on Io, or even water and ice, but rather a mixture of ammonia and water. An ammonia/water mixture could have lowered melting temperatures to as low as 173°K—much lower than that of water, although an alternative possibility is that a methane/water mixture was involved.

In August 1989 Voyage 2 reached Neptune and study of its largest moon, *Triton*, revealed a complex and fascinating set of features. Before the mission it had been assumed that Triton had been geologically inactive for a very long time and that its surface would be heavily cratered. Remote sensing imagery revealed, however, evidence of resurfacing and the clear implication that volcanic activity was not so ancient as had been assumed. The reasons why volcanic activity has occurred during Triton's geological history is still debated, but it is generally thought that it may be related to peculiarities in the satellite's orbit around the planet, which produces melt by tidal energy.

Further information on the volcanology and geological evolution of the giant planets may be obtained from the following references: Peale *et al.* (1979); Ingersoll (1981); Glass (1982); Soderblom and Johnson (1982); Smith *et al.* (1986); Allison and Clifford (1987); Carr (1987); Greeley (1987); Moore and Malin (1988); Lunine (1989); Binzel (1990); McKinnon (1991); and Summerfield (1991).

1.3 The volcanological development of the terrestrial planets

For centuries astronomers have observed the terrestrial planets, but in recent years space programmes have produced massive amounts of data and allowed many new insights into planetary volcanic processes to emerge. An initial difficulty in reviewing the geological and volcanological evolution of the terrestrial planets is that data vary in coverage and quality. The Earth and to a lesser extent the Moon have been studied remotely from spacecraft and by direct observation, whereas the other terrestrial planets have been analysed by remote sensing and in some cases by unmanned landings. It follows that models of planetary evolution are never so well founded as those developed to explain phenomena on Earth and differences in scientific opinion occur frequently. Despite this, and not least because of the ingenuity of researchers, a consensus has emerged that there are sufficient similarities in the evolution of the terrestrial planets to make comparison worthwhile.

The terrestrial planets are made of rock, their densities and probably their compositions are similar and their dimensions, masses and distances from the Sun have been calculated with reasonable degrees of precision (Table 1.2). It is far from certain how the Solar System in general, and specifically the terrestrial planets, formed. The origin of the Solar System is a topic which has fascinated scientists for centuries, but controversy reigns and to quote from one well-used textbook 'it is a rare geological or astronomical congress that does not witness a fresh debate triggered by the latest experimental data or the newest theoretical advance' (Press and Siever 1986, p. 6). Not only is the origin of the Solar System a

Table 1.2: Selected properties of the terrestrial planets. Based on information in: Murray *et al.* (1981); Glass (1982); and Condie (1989). Note, some figures vary between sources, and where this occurs, a majority opinion has been followed

	Mass (Earth) (=1)	Density (gm/cm^2)	Distance from the Sun (10^6 km)	Ratio of planet's area to its mass (Earth) (=1)	Ratio of planet's core to its mantle volume (Earth) (=1)
Mercury	0.055	5.44	58	2.5	~12
Venus	0.815	5.27	107	1.1	0.9
Earth	1	5.52	149	1	1
Mars	0.107	3.95	227	2.5	0.8
Moon	0.012	3.34	0.38[1]	6.1	0.12

[1] From the Earth.

Table 1.3: Summary of the stages involved in forming the planets. It should be noted that this is one view of the sequence of events and variations upon it are frequently encountered in the literature (based on several sources including Murray *et al.* 1981; and Glass 1982)

Stages	Events
1	The solar nebula from which the planets were formed, was probably similar in composition to the present-day Sun (i.e. hydrogen, helium, with heavier elements representing only a small percentage of the total). It is hypothesized that as the nebula contracted, its rate of rotation increased due to the law of *conservation of angular momentum*, and it became disc shaped. With continued contraction, temperature and density increased (i.e. temperature possible $\sim 10^7$ °K) and, at the centre of the disc, processes of hydrogen fusion caused the Sun to flare as a star.
2	For reasons which are not fully understood, the other material in the disc began to form small accretions known as *planetesimals* and around these the present planets started to form, by means of particle collision and the capture of material by gravity.
3	Planetesimals, and later the planets which evolved from them, varied in composition with distance from the Sun; this being caused by a marked temperature gradient. The gradient controlled the composition of materials from which the planets grew. Near to the Sun only metal oxides and silicates were found, due to their high melting temperatures. These formed the terrestrial planets, which are rich in iron and silica. Planets closest to the Sun (e.g. Mercury) have the highest densities, whereas those at distance (e.g. Mars) are less dense (Table 1.2). Because of greater distances, temperatures in the vicinity of the giant planets were lower and, in addition to metal oxides and silicates, lighter materials such as ices of: water; methane; ammonia; and carbon dioxide, were stable. The giant planets grew to their large sizes because 'ices' are more abundant in the Solar System than iron and silicates.
4	Jupiter and Saturn in particular, but also Uranus and Neptune, grew so large that their gravitational fields were able to attract hydrogen, helium, neon and argon and this accounts for their Sun-like character.

controversial topic, it is also vast, complex and peripheral to the primary concern of the present volume. Detailed treatments may be found in Glass (1982) and Condie (1989) and a summary of one possible model is reproduced in Table 1.3. Three points, however, need to be made about Table 1.3.

First, in the era before space flight it was held that the material forming the planets was initially uniform or *homogeneous* in composition across the Solar System, although some variations may have been caused by distance from the Sun. The Earth was assumed to have been cold when it formed and to have been heated by radio-isotopes early in its history producing a differentiation into core, mantle and crustal regions. From the early 1960s new data were acquired and the favoured model became one in which accretion occurred under hot conditions; this being the situation shown in Table 1.3. Planets may have been initially molten, with chemical differentiation occurring as they formed (Murray *et al.* 1981). The history of planetary evolution is, therefore, one of progressive cooling, at least over most of geological time. A number of processes contributed to early heating (see Clapperton 1977) and are discussed in Chapter 3, section 3.3.1.

Secondly, Table 1.3 depicts a model of *heterogeneous* accretion, in which the solar nebula was not uniform in composition. Iron and silica were common near to the Sun and lighter elements increased in importance at greater distances from it. It should be noted that certain writers favour a model involving a hot *homogeneous* nebula (see Condie 1989).

The age of the planets is the third point. The oldest dated rocks on Earth are around $3.8\text{–}3.9 \times 10^9$ years old, although there are some indications that small amounts of crust may be around 4.2×10^9 years old. Samples of the oldest lunar rocks yield maximum ages of 4.5×10^9 years and those of some meteorites, which are thought to be fragments of asteroids and comets dating from earliest stages of Solar System development, ages of $\sim 4.55 \times 10^9$ years. The geological and volcanological evolution of the terrestrial planets has occurred, therefore, over some $\sim 4.6 \times 10^9$ years at the most.

Before volcanic activity on Earth may be compared with that of the other planets, it is necessary to say something about the other terrestrial planets, since these have certain styles of activity which are not found on Earth.

1.3.1 The Moon

With the exception of Earth, the Moon is the best known object in the Solar System. Being close to Earth the Moon has been observed by telescope for hundreds of years and landings by spacecraft, both manned and unmanned, have afforded the opportunity to add to the data provided by remote sensing (Moore *et al.* 1980; Greeley 1987). Samples of rock collected on the Moon have been analysed and in some cases dated.

Landforms on the Moon may be divided into two categories: cratered highlands and flat, much smoother *maria*. In earlier centuries it was thought that the maria were seas, hence the name. Following exploration by spacecraft, a more detailed subdivision can now be sustained (see Howard *et al.* 1974; Figs 1.3 and 1.4) and, in addition to maria which are now known to be basaltic lava flows, four distinctive landscapes may be distinguished.

Many details of the geological history and volcanological evolution of the Moon are still subjects for debate, but five time-stratigraphic units are commonly recognized (Guest and Greeley 1977; Wilhelms 1980). The earliest—the *Pre-Nectarian*—began around 4.6×10^9 years ago and is notable for its impact craters. Craters were produced by 'bolides': a collective noun which includes meteoroids, asteroids and comets. On Earth impact craters are rare, many bolides burn up in the atmosphere, others are slowly eliminated by processes of degradation and evidence of ancient impacts is often not preserved because of crustal recycling through the operation of plate tectonics. On the Moon none of these conditions exist and impact-cratering has been an important process throughout its history. Craters are an important stratigraphical tool, for in general the higher the density of impact craters, the older the surface. Impacts can occur at velocities as high as $150{,}000\ \text{kmh}^{-1}$ and rapid conversion of kinetic energy to heat takes place as the bolide penetrates the surface, explodes and vaporizes. Large craters are partly a function of bolide size, but velocity is more important as kinetic energy is proportional to the square of velocity (Summerfield 1991, p. 485). In the Pre-Nectarian heat was generated by bolide impact, possibly supplemented by radioactivity, and these agencies are thought to have been responsible for melting the outer few hundred kilometres of the planet, so

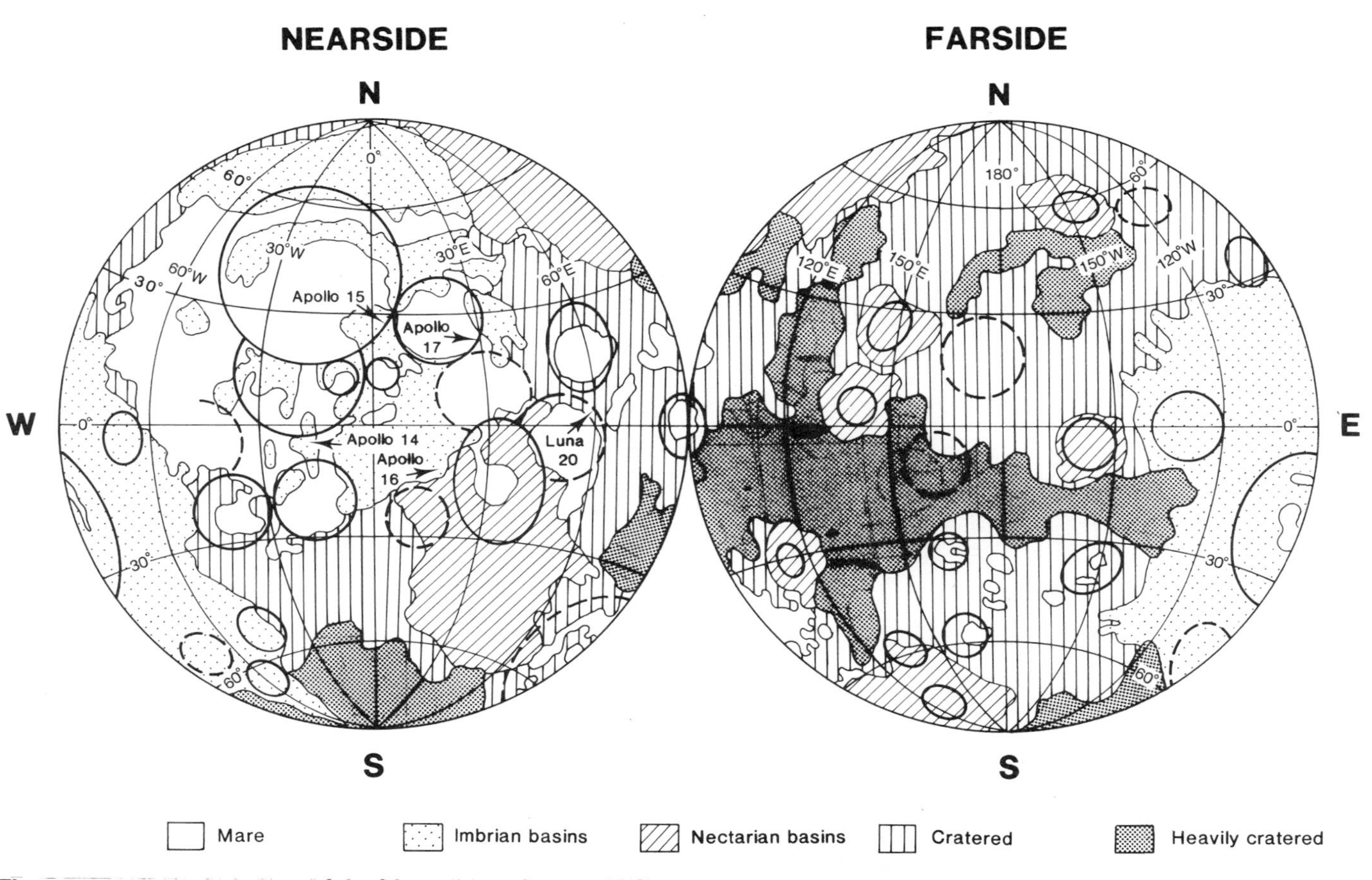

Fig. 1.3 Morphological maps of the Moon (from Howard *et al.* 1974, Fig. 14, p. 322).

Fig. 1.4 The surface features of the Moon, showing a predominance of heavily cratered uplands and darker mare areas (NASA image supplied by Professor J.E. Guest, Regional Space Imagery Library, University of London Observatory).

forming a magmatic ocean. Subsequently, processes of differentiation caused a crust to develop. More specifically this involved plagioclase-rich material floating, producing an anorthositic crust, and olivines and pyroxenes sinking—forming an upper-mantle rich in incompatible elements (see Condie 1989; and Chapter 3, section 3.2). By 3.9×10^9 years the crust was sufficiently rigid for impact craters to be preserved and subsequent volcanic activity was caused by partial melting of the upper-mantle. The Pre-Nectarian is represented by older heavily cratered terrain which pre-dates the Nectaris Basin (Fig. 1.3).

The *Nectarian* is the second unit and began with the formation of the large impact basin known as Nectaris. Other basins such as Crisium and Humorum were formed around the same time and volcanic activity involved the eruption of a suite of lavas rich in potassium, rare earth elements and phosphorus (Greeley 1987). The Pre-Nectarian and Nectarian represent a period of about 600 million years (Guest and Greeley 1977) when the planet was subjected to intense cratering. This episode is called *The Heavy Bombardment Period.* Some writers claim that heavy bombardment lasted for the whole 600 million years, but others argue that it occurred around 4.0×10^9 years. Evidence of a Heavy Bombardment

Period is found on other terrestrial planets and acts as an important time reference.

Following the Nectarian, the *Imbrian* began with the formation of the Imbrium Basin some 3.9×10^9 years ago. The Imbrian is associated with the formation of Orientale Basin and volcanic activity on a truly massive scale, involving the eruption of basalts which now comprise about two-thirds of the volume of visible maria. Lavas, which form a secondary crust, were produced by processes operating within the Moon and cover some 17% of the lunar surface (Head 1976). Many flows partially fill earlier basins where they are up to 4 km thick, but much thinner flows are, however, more common.

The next unit—the *Eratosthenian*—sees the formation of more craters and eruption of the remaining one-third of maria basalts. Finally, the *Copernican* unit ends the story. This is distinguished by fresh, undegraded craters and a low density of impacts. From the interpretation of photographs there is some tentative evidence that limited volcanic activity may have continued until around 2.0×10^9 years (Greeley 1987).

Flows of lunar basalt differ in several important respects from those found on Earth. Geochemically they contain less silica, aluminium, sodium and potassium and are closer to tholeiites than alkalic basalts in composition (see Chapter 3, section 3.2). Another distinguishing feature is that there are variations in titanium content, which has led to the division of mare basalts into groups (Middlemost 1985). In general titanium, magnesium and iron are more abundant in lunar basalts. 'The mineralogy of mare basalts is essentially similar to that of the basalts of the Earth' (Middlemost *op. cit.*, p. 109), but a lack of water means there are no hydrous minerals and the total mineralogical range is reduced.

The morphology of lunar basalts shows major differences, when compared to chemically similar lava flows on Earth (see Chapter 4). Early eruptions had high effusion rates, were rich in titanium, possessed high mobility and were able partially to fill low-lying basins. Laboratory experiments using simulated lunar flows imply that viscosity (see Chapter 4, section 4.2.2) was low, while studies of conductivity are suggestive of minimal heat losses during transport (Greeley 1987). Both attributes would have contributed to high mobility. Formed as vast lava lakes, these early flows probably took hundreds of years to cool and have virtually no surface features (Basaltic Volcanism Study Project 1981). One exception is the occurrence of mare (or wrinkle) ridges. These are 'irregular, discontinuous, sinuous ridges several kilometres in width and up to 100 km in length [which] form concentric rings in the circular mare basins and tend to be located preferentially along mare boundaries' (Glass 1982, p. 186). Some mare ridges were formed by vertical faulting and are related to highland scarps (Lucchitta 1976), but the origin of others is uncertain. Several authors have ascribed them to folding and faulting, others believe that they were caused by intrusion of magma (Hodges 1973), or by the extrusion of lava along ring-shaped fractures (Hartmann and Wood 1971). It is possible that some may be buried features. Mare ridges remain enigmatic and it is possible that they were produced by several processes (see Guest and Greeley 1977).

Later eruptions had lower titanium contents and are associated with distinctive landforms known as *sinuous rilles* (see Fig. 1.5). Again controversy surrounds their origin. These channels, which are also found on Mars, are around 50 km in length and some appear to be collapsed lava tubes (see Table 4.4; and Oberbeck *et al.* 1969). An alternative view (Hulme 1973; and Carr 1974) is that rilles were

Fig. 1.5 Two different types of terrain on the Moon: heavily cratered highlands and smoother, more sparsely cratered mare. The channels shown on the photograph are *sinuous rilles* and are thought to be of volcanic origin (see text). Differences in crater frequency are used to infer the relative ages of surfaces. The area shown is ~150 km across (NASA Apollo 15 image AS15 2484, supplied by Professor J.E. Guest, Regional Space Imagery Library, University of London Observatory).

cut by lava flows which melted existing surfaces. In Chapter 4 (section 4.2.2) flowage of lava is discussed and on Earth flows are virtually always *laminar*. Precambrian komatiites are an exception and are thought to have been *turbulent*; melting the ground over which they flowed. It is argued that a similar situation may have occurred on the Moon, where eruptions were notable for their very high eruption rates (Hulme 1973) of up to 10^9 kgs^{-1}, with eruptions lasting several months and having discharges up to 1000 km^3 per event (L. Wilson *et al.* 1987). Changes from laminar to turbulent flow may also be related to differences in ground slope-angle (Hulme and Fielder 1977). The final lavas to be erupted were again titanium-rich, but much smaller volumes of material were involved.

Regardless of age, some features which are common on lava flows on Earth are very rare on the Moon. These include flow-fronts and margins and it has been argued that lavas were either too fluid to preserve these features, or were eroded by subsequent impact cratering (Greeley 1987). Domes are found on flows of lunar basalt and these range from less than 1 to more than 20 km in diameter (Head and Gifford 1980). Some are flat-topped, have summit craters and are thought to have been formed during episodes when eruption rates were lower. Higher viscosity lava is another possibility. Some domes may be high-standing 'islands' which have not been covered by mare basalts (Greeley 1987).

1.3.2 Mercury

Mercury is a small planet with a high density (Table 1.2) and it is argued that this implies an iron/nickel core, which comprises some 66% of planetary volume (Condie 1989). Of all the terrestrial planets Mercury is closest to the Sun, has a negligible atmosphere and, until the results of the Mariner 10 mission were analysed in 1974–75, virtually nothing was known about its volcanic and wider geological history. It should be noted that, although 'volcanism is repeatedly called upon to explain various features on Mercury and is inferred to have occurred throughout its history. . . . Very few features, such as domes, cones and shield volcanoes, can be unequivocally identified as volcanic' (Greeley 1987, p. 129). Notwithstanding this note of caution, virtually all writers have accepted that volcanism has been important, indeed crucial, in the geological development of Mercury.

Remote sensing images are only available for about 50% of the surface, but Trask and Guest (1975) were able to use them to produce a map of terrain units (Fig. 1.6). Superficially Mercury appears to be very similar to the Moon, with heavily impacted crater regions and large impact basins, but when examined in more detail it becomes clear that major differences are apparent. Mercury has major scarps and those with an arcuate form are interpreted as high-angle reverse faults, which were formed during global contraction (see Fig. 1.7). This may have been related to cooling of the interior and formation of a core (Strom *et al.* 1975; Solomon 1976). In addition lobate scarps are found on areas of elevated terrain and have been ascribed a tectonic origin by Dzurisin (1978), though other authors have claimed them as lava flows (see Greeley 1987, p. 124). The smooth plains of Mercury ('plains' on Fig. 1.6) are considered to be relatively young and, although some writers have argued for an impact origin (e.g. Wilhelms 1976), the general consensus is that they are volcanic and may equate to lunar maria. They certainly possess features which are similar to mare ridges (Cintala *et al.* 1977). On Fig. 1.6, 'hilly and lineated terrain' is defined. Often termed 'weird terrain', it consists of a chaotic assemblage of hills 5–10 km in length and from 100 m to 1.8 km in height. Its location may be important for it occurs in an antipodal position with respect to the large Caloris Basin. It may have been formed by the focusing of seismic energy when the Caloris Basin was formed by a large impact (Schultz and Gault 1975). In the absence of radiometric dates the chronology of Mercury is relative rather than absolute but, using the Moon as an analogy, it is probable that major volcanic events occurred before the large basins were formed around 3.9×10^9 years ago (Condie 1989).

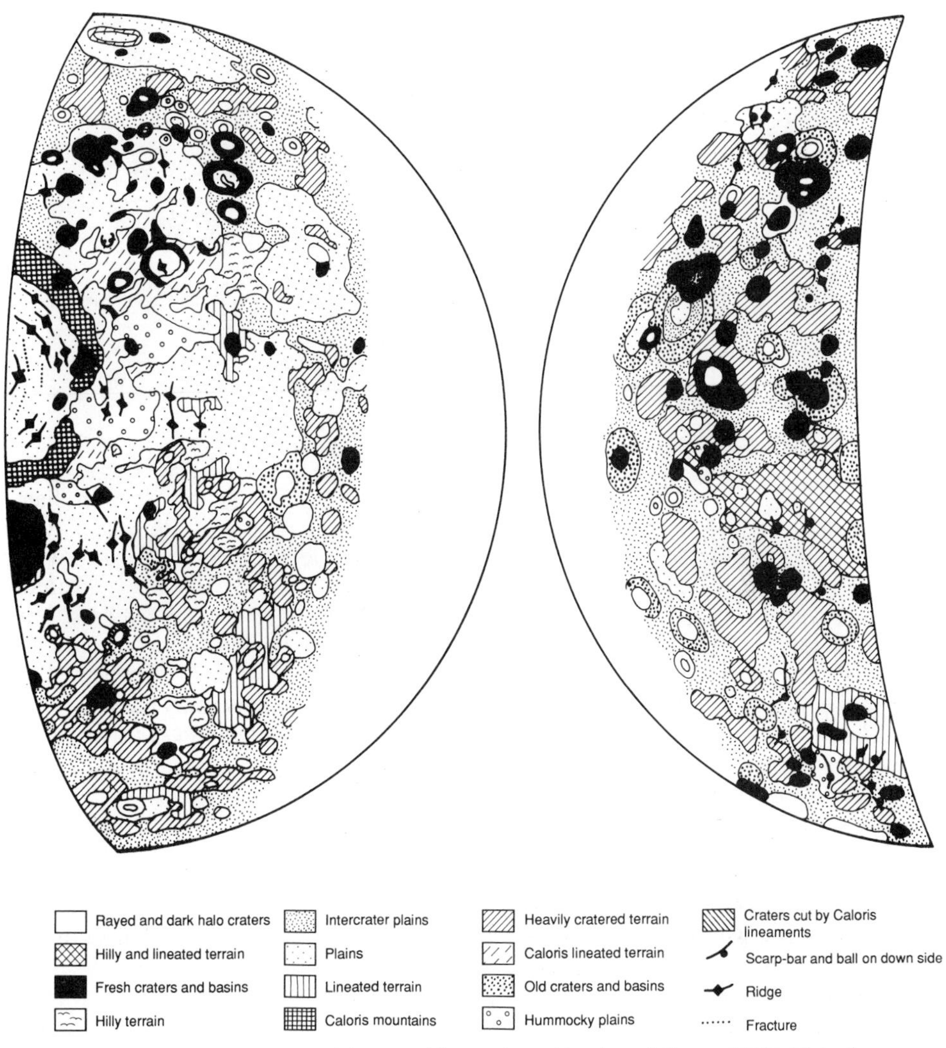

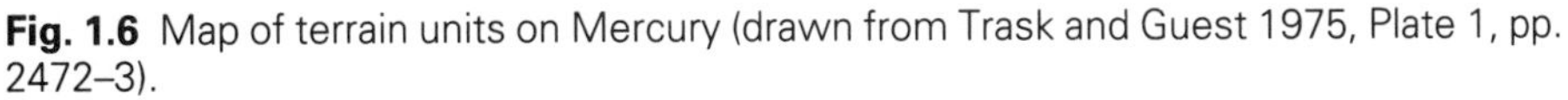

Fig. 1.6 Map of terrain units on Mercury (drawn from Trask and Guest 1975, Plate 1, pp. 2472–3).

1.3.3 Mars

Although over half its area is covered by cratered regions and there are some large basins, Mars differs significantly from Mercury and the Moon. Remote sensing shows aeolian deposits, large canyons thought to have been cut by water, together with polar caps of water and solid CO_2. There are abundant signs of fracture zones and rifts. The geological history of Mars is dominated by volcanic activity. This has included the eruption of material to form spectacular shield volcanoes but, as Greeley (1987) notes, these occupy only ~1% of the surface area, whereas volcanic plains account for more than 35%. In contrast to Mercury volcanism on

Fig. 1.7 Image showing an arcuate scarp on the surface of Mercury. Also shown is cratered terrain of different ages. The view is ~700 km across (NASA Mariner 10 image 10 FDS 166842, supplied by Professor J.E. Guest, Regional Space Imagery Library, University of London Observatory).

Mars (see Fig. 1.8) has been verified by analysis of samples at landing sites, these showing that rocks are 'mafic' (i.e. basic) to 'ultramafic' (i.e. ultrabasic) in composition (see Chapter 3, section 3.2).

Volcanic landforms may be divided into two groups: those found on plains and those related to central volcanoes. The largest plains appear to have been formed by flood eruptions of basaltic composition, in a manner similar to the early mare basalts on the Moon and the smooth plains of Mercury. Other plains possess features that bear witness to the wide range of volcanic styles which have occurred during the planet's evolution. These are summarized in Table 1.4 (see Greeley and Spudis 1981).

Amongst the central volcanoes, many edifices are similar to Hawaiian shields, but are much larger. Olympus Mons, for instance, stands 26 km above its surroundings and is 650 km in diameter (see Fig. 1.9). The large size of Martian shield volcanoes, in comparison with those on Earth, may be due to two factors (Carr 1973). The first is the total volume of magma erupted, which for Olympus Mons was greater than that of the whole Hawaiian/Emperor chain system in the Pacific Ocean. The second factor is the lack of plate motions on Mars. In Chapter 2 (section 2.3.1) there is a discussion of oceanic intraplate-volcanism, which is typified on Earth by Hawaii. In the Hawaiian province plate movements have occurred over a stationary magma source, or 'hot spot', producing several

Fig. 1.8 The surface of Mars. The image shows cratered terrain in the southern hemisphere and also dendritic valleys and fracturing. The image is more than 200 km across (NASA Viking Orbiter image 63A09, supplied by Professor J.E. Guest, Regional Space Imagery Library, University of London Observatory).

volcanoes of different ages and modest size. Because Mars does not have plate motions, the interior regions of the planet can erupt vast amounts of lava through the same vents and so produce far larger shields. Other central volcanoes are not analogous to Hawaiian shields. Some are called *paterae*: low-angle forms, of great areal extent and showing gentle relief. The Alba Petera is ~1500 km in extent and only 10 km in height (Carr 1987). On its flanks are extensive lava flows, which must have had very low viscosities to enable them to move over such gentle slopes (see Plescia and Saunders 1979). In contrast are *tholii* (or dome volcanoes) which are smaller and have steeper slopes. Higher slope angles may have been caused by more viscous lavas, intercalations of pyroclastic materials, lower eruption rates or all these factors in varying combinations (Greeley 1987).

The large Martian shield volcanoes are concentrated into an area known as the Tharis region. This is a crustal bulge and is thought to have formed $\sim 3.9 \times 10^9$ years ago. Processes of magma generation were probably similar to those producing 'hot spots' on Earth (see Chapter 3, section 3.3).

Because of a lack of radiometric dates, constraints on the geological history of Mars are fairly weak and it is not surprising that several chronological schemes

Table 1.4: Possible additional examples of volcanic activity which has occurred on the Martian plains (based on the references cited)

Type of 'volcanic' plain	Principal features and possible causal processes
Ash plains	It is generally thought that for much of Martian history water was available at the surface and as groundwater. According to Scott (1982) the interaction of water and magma could have generated explosive hydrovolcanic eruptions and these were responsible for the fine ashes (see Chapter 5, section 5.3.6.2). Other writers dispute this hydrovolcanic origin.
Northern plains with cones	Hydrovolcanism on a small scale is invoked by Greeley and Theilig (1978) to explain small constructs with summit craters on the northern plains. In Iceland similar craters are small pyroclastic cones which form on lavas flowing over waterlogged ground.
Hummocky plains	Some plains have hummocky surfaces and small channels. Greeley (1987) claims a volcanic origin.

have been proposed (e.g. Mutch *et al.* 1976; Carr 1981), none of which has been accepted in all its particulars. All schemes use the density of impact cratering as a stratigraphical tool and supplement this by inferences drawn from other planets. A possible sequence of events is presented in Table 1.5. Two points should be noted when reading Table 1.5 (see Glass 1982). The first is that fluvial and volcanic activity has occurred throughout most of Martian history, and the second that, depending on the values placed on the energy delivered by impacts, Martian geological and volcanic evolution ended either $\sim 1\text{–}2 \times 10^9$ years ago, or continued until $\sim 100 \times 10^6$ years ago (Condie 1989). Whatever the exact timing, the consensus opinion is that Mars is 'a dead or dying planet' (Glass 1982, p. 289).

1.3.4 Venus

Until very recently much less was known about Venus than the other terrestrial planets. Because of its thick atmosphere, remote sensing had to rely on low-resolution Earth-based and spacecraft radar images and, whilst these allowed major landforms to be picked out, they lacked fine detail. It was established that Venus is about the same size as Earth, has a similar density (Table 1.2) and roughly equal amounts of N_2 and CO_2. For these reasons astronomers referred to the planet as 'Earth's twin', but space missions in the 1970s and 1980s demonstrated that, whereas on Earth much CO_2 is held in sediments and used in photosynthesis (see Chapter 6, section 6.4.2.2), Venus has an atmosphere which is rich in CO_2 and nine times as dense. Atmospheric pressure at the surface is equivalent to an ocean depth of ~1 km on Earth, which has had significant effects on many surface processes. Combine this with a marked 'greenhouse effect', which sustains surface temperatures at between 420 and 485°C, and it can be seen that Venus is quite different from Earth; not least in being hostile to life.

Much of the literature on Venus is based on radar data acquired during the

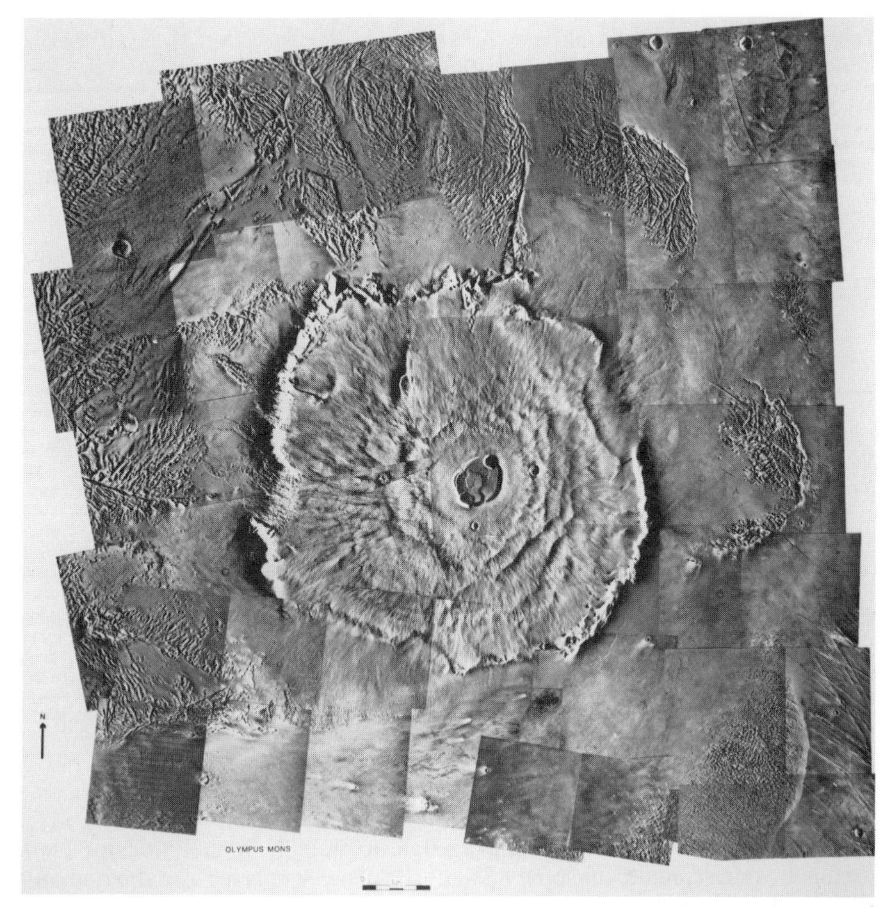

Fig. 1.9 Olympus Mons, a large shield volcano in the Tharsis region of Mars (NASA image, supplied by Professor J.E. Guest, Regional Space Imagery Library, University of London Observatory).

American Pioneer and Russian Venera missions, that took place between 1967 and 1985 (Greeley 1987). Because the radar imagery was of low resolution, the conclusions that could be drawn were both vague and ambiguous. In 1992 the first major scientific reports of the United States Magellan mission were published in a special issue of *Journal of Geophysical Research* (vol. 97E, 1992). Magellan, which mapped the surface of the planet in September 1990, carried a more sophisticated radar scanner and mapped 84% of the surface at resolutions up to ~300 m. The new insights flowing from Magellan will take some years to be fully digested by the scientific community, but they are incorporated where appropriate in the paragraphs which follow.

Radar images from the Pioneer and Venera missions showed the surface of Venus to be flat overall, but with marked relief in some regions. Mapping by Masursky *et al.* (1980) allowed three large-scale landscape types to be identified (Fig. 1.10):

Table 1.5: Possible stages in the evolution of Mars (based on Mutch *et al.* 1976; Glass 1982; Condie 1989; and several other sources)

Stage	Events
1	The early history of Mars was very similar to the Moon and Mercury. Differentiation of a magmatic ocean into: crust, mantle and core was probably complete by $\sim 4.4 \times 10^9$ years and heavy bombardment formed heavily cratered terrain. The first stage ended with the formation of large basins.
2	The Tharsis region was uplifted and rifting of the crust allowed widespread volcanic activity to occur. An early, dense atmosphere was rich in CO_2 and H_2O and extensive fluvial erosion was initiated under 'greenhouse' conditions. Losses from the atmosphere were greater than replacement by volcanic volatiles and a drop in temperature followed. Ground ice formed. Volcanic activity caused localized melting of ice and some fluvial activity was initiated.
3	Formation of the volcanic plains and continued fluvial activity.
4	The Tharsis uplift continued, forming radial faults and releasing ground water. There was continued cutting of channels by water.
5	Central volcanoes (including shields) were erupted in the Tharis region. The northern lowlands were mantled by pyroclastic and/or windblown materials.
6	Cooling of the planet became a major controlling factor. The relative importance of windblown activity increased and continues to the present day.

1) upland rolling plains (plateaux), up to 2 km in height;
2) highlands, over 2 km; and
3) lowlands, below the 0 km datum.

Calculations made following the Magellan mission show that volcanic plains, generally lowlands, make up some 85% of the planet's surface and the remaining 15% is higher (Saunders *et al.* 1992). Highlands are dominated by ridge terrain and plateaux have compressional tectonic features. Many surface features seem to have been formed by cratering, weathering, aeolian and channel-forming processes (Table 1.6).

'The similarity of Earth and Venus in mass, density and diameter would suggest that volcanism and tectonism should be as important on Venus as on earth. Although ... definitive evidence for volcanism is lacking, radar images reveal various landforms that resemble volcanoes' (Greeley 1987, p. 142). Since these words were written the Magellan mission has produced very strong evidence for volcanic activity and it is now recognized as being 'the most widespread and important geologic process' (Saunders *et al.* 1992, p. 13,068). In addition to radar images, evidence of igneous processes has been obtained from a number of Venera spacecraft landings on the planet's surface and these are suggestive of basaltic volcanism. Many lowlands are covered by lava flows of a flood-field type (Surkov *et al.* 1983), possibly analogous to lunar maria. In 1979 the volcanologist Charles Wood (see Greeley 1987) posited that for a given eruption rate, volume and composition, a lava will flow further on Venus than on Earth because of reduced heat losses. In addition to flood-fields there are over 550 groups of shields, 156 edifices in excess of 100 km, over 80 calderas, around 50 sinuous lava channels and many other volcanic features. Most landforms are compatible

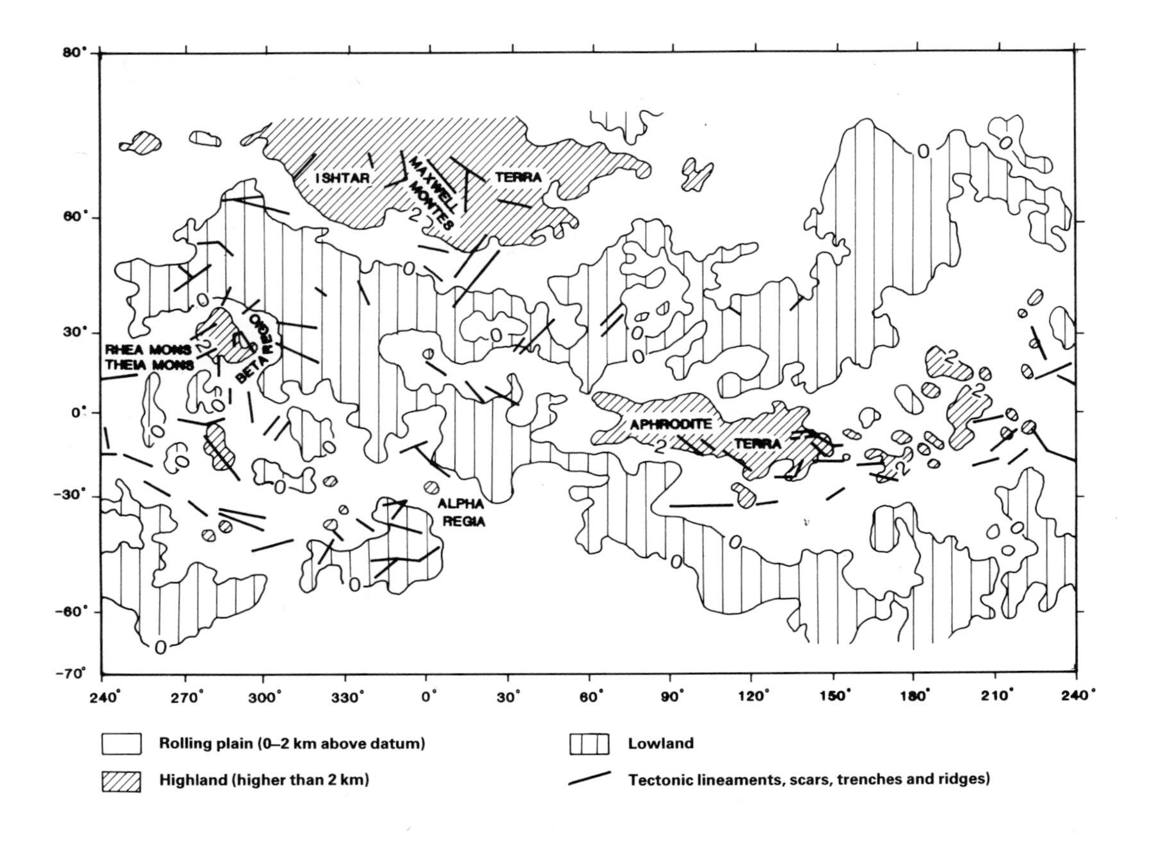

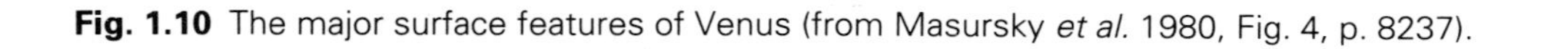

Fig. 1.10 The major surface features of Venus (from Masursky *et al.* 1980, Fig. 4, p. 8237).

Table 1.6: Features produced by cratering, weathering, aeolian and channel-forming processes on Venus (based on the references cited)

Process	Details and landforms
Cratering	On Venus there are many circular features which were probably formed by bolide impact (Campbell and Burns 1980; Cutts *et al.* 1981). According to Schaber *et al.* (1992), the surface of Venus is not modified rapidly, since impact craters are not significantly weathered. Craters range from 2–275 km in diameter and their estimated ages from 200–700 × 10^6 years. The 'youthfulness' of craters may represent a period of major resurfacing of the plant by volcanoes, with relatively little subsequent activity. The Venusian atmosphere is thought to have had a major influence on cratering (Tauber and Kirk 1976), in particular flight through the dense atmosphere would have caused fracturing and reduction in the efficiency of the cratering process.
Weathering, aeolian and fluvial? processes	Radar images show evidence of loose materials at the surface. These were probably produced by weathering (Arvidson *et al.* 1992) and are transported by aeolian processes; wind streaks and dunes being common landforms (Greeley *et al.* 1992). The nature of weathering under the conditions of the Venusian surface are much debated, are likely to have been complex and probably involved a mixture of chemical and physical processes (see Nozette and Lewis 1982). Channels are common on Venus and their cutting is commonly ascribed to silicate melts (i.e. lavas) of varying composition (V.R. Baker *et al.* 1992). A more remote possibility is that they were cut by fluvial action (Donahue *et al.* 1982).

with basaltic volcanism; though ultramafic melts may have been erupted in the planet's history (see Chapter 3, section 3.2; and Head *et al.* 1992).

James Head and Lionel Wilson (Wilson and Head 1983; Head and Wilson 1986) have considered some of the controls on volcanic activity. Whereas on Earth pyroclastic materials are usually formed by gases fragmenting ascending batches of magma (see Chapter 5), Wilson and Head argue that on the Venusian lowlands atmospheric pressure is too high for this to occur. Even if pyroclastic materials were erupted during the planet's evolution, column heights and the dispersal of materials would have been very restricted. The scarcity of pyroclastic materials has been confirmed by the Magellan mission (Head *et al.* 1992).

In 1987, Michael Carr wrote that 'like virtually everything on this fascinating planet, the style and timing of volcanic activity ... are poorly understood' (Carr 1987, p. 129). Today this opinion cannot be sustained and, following the Magellan mission, severe constraints have been imposed on models of volcanic, thermal and wider tectonic evolution. In the 1980s, three groups of models were popular (Phillips and Malin 1983). The first held that the dominant process in the evolution of Venus was progressive heat loss through conduction. As Condie (1989, p. 380) points out, whilst this model provide a reasonable explanation of the volcanic geology of the Moon, Mercury and Mars, the comparative youthfulness of the Venusian surface rules it out (Table 1.6). Mantle plumes (see Chapter 3, section 3.3) provided an alternative model, with the planet losing heat by this process throughout geological time. Plumes seemed able to explain observed volcanic activity, associated uplift and some tectonic features. The third group of models invoked plate tectonics and many such models were proposed, using

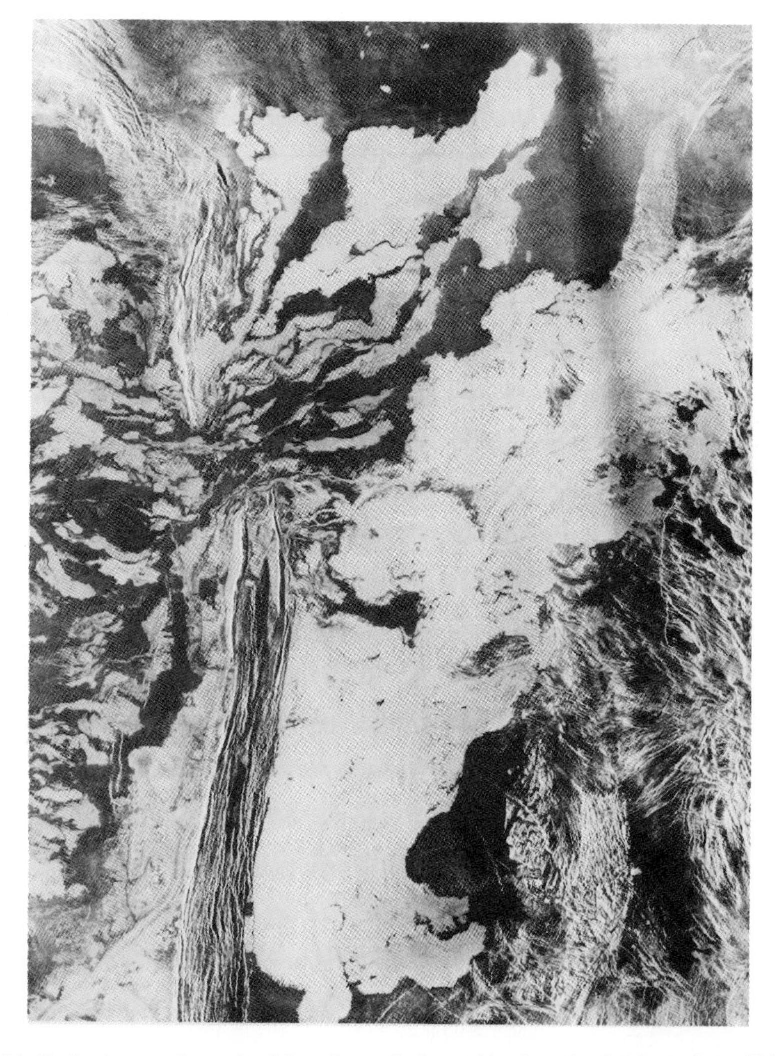

Fig. 1.11 Radar image from the Magellan mission, showing the Lada region of Venus. The area is ~550 km wide by ~630 km long (north is at the top of the image), and the image shows radar-bright and radar-dark lava flows moving east and breaching a belt of elongated ridges running north–south. The lavas are derived from the Ammavaru caldera, which is located 300 km to the west (NASA image, Jet Propulsion Laboratory (JPL 400-406C 9/91), supplied by Professor J.E. Guest, Regional Space Imagery Library, University of London Observatory).

evidence which was at the same time superficially convincing, yet circumstantial (see McGill *et al.* 1983 for a review). Evidence included:
1) the putative existence of major rifts;
2) the possible association between upland plains and lowlands on Venus and continents and ocean basins on Earth; and
3) the suggestion that fold mountains may occur on Venus.

There were equally strong arguments against plate tectonics. One hinged on a theoretical calculation by Anderson (1981) which cast doubt on whether plate motions were possible because of high surface temperatures, while another hung on the atmospheric content of the isotope ^{40}Ar. This isotope is found in planetary atmospheres and is a guide to the effectiveness of tectonic and volcanic processes (Condie 1989). ^{40}Ar is produced within planets by radioactive decay and escapes through the operation of tectonic movements and volcanic eruptions. The amount in the Venusian atmosphere is only one-third that found in the Earth's atmosphere.

Results from Magellan have produced many important data germane to tectonic and volcanological evolution. These show, *inter alia*, a surface deformed with zones of compression and extension, slopes of high angle maintained over tens of kilometres (i.e. fault scarps?) and a correlation of large shield volcanoes and positive surface height anomalies, suggestive of mantle upwelling (Solomon

Table 1.7: Some of the factors affecting the thermal evolution of the terrestrial planets (based on Condie 1989, with additional information from other sources)

Factor causing heat loss	Mechanisms involved
Size	The greater the area/mass ratio, the more rapidly will a planet lose heat. The Moon, Mercury and Mars, Venus and Earth stand in a descending sequence (see Table 1.2).
Position within the Solar System	Position may have determined the sequence of condensation of elements within the Solar nebula.
Abundance of radionuclides	Linked to the above, the position in the nebula may have affected the occurrence and abundance of isotopes of Al, U, Th, and K. These probably affected processes of partial melting (see text). Limited evidence from Lunar and Martian landings, implies that these planets are relatively depleted in radionuclides with respect to Earth.
Estimated size of the iron core	It is hypothesized (see text) that much heat was added to planets during their early histories, through the formation of iron cores. Taking the core/mantle ratio for Earth as unity (Table 1.2), figures for the other planets are: Venus (0.9); Mars (0.8); and the Moon (0.12). Mercury probably has a far larger core/mantle ratio than the other planets, but because it has a low total mass, it probably lost this heat very quickly.
Volatile contents	These are not well known, but the relative abundance of volatiles would probably have affected the amount and viscosity of magmas. The Earth is known to be volatile-rich.
Convection	Planets lose heat by convection. It is generally believed that Mars has a very thick lithosphere and this may have inhibited convection.

et al. 1992; see Fig. 1.11). Tectonism seems to have been a feature of Venus throughout its history. In a well argued paper, McKenzie *et al.* (1992) are strongly in favour of a plate tectonics model, interpreting some features as subduction zones and others as transform faults. Also volcanic activity appears to be driven by mantle plumes, possibly by processes of adiabatic decompression (see Chapter 3, section 3.3.1). However, there is other evidence which implies that if plate tectonics has occurred it is not the same as on Earth. Deformation is distributed across wide zones, whereas on Earth it is concentrated spatially into the narrow regions around plate boundaries, and hot spots on Venus do not describe the same patterns as they do in an intraplate situation such as Hawaiian volcanic province (see Chapter 2, section 2.3.1).

1.3.5 Comparative evolution of the terrestrial planets

For the last three decades, planetary scientists and volcanologists have attempted to answer the fundamental question of why the Earth is as it is. As mentioned at the start of section 1.3, it is generally accepted that the terrestrial planets formed when hot, increased in temperature during their earliest stages of evolution and gradually cooled towards a state of 'tectonic stability and quiescence' (Condie 1989, p. 374). Writers have focused on the relative size (i.e. mass and area/mass ratio) of planets as being the single most important factor in controlling rates of heat loss and, thus, variations in tectonic and volcanological development.

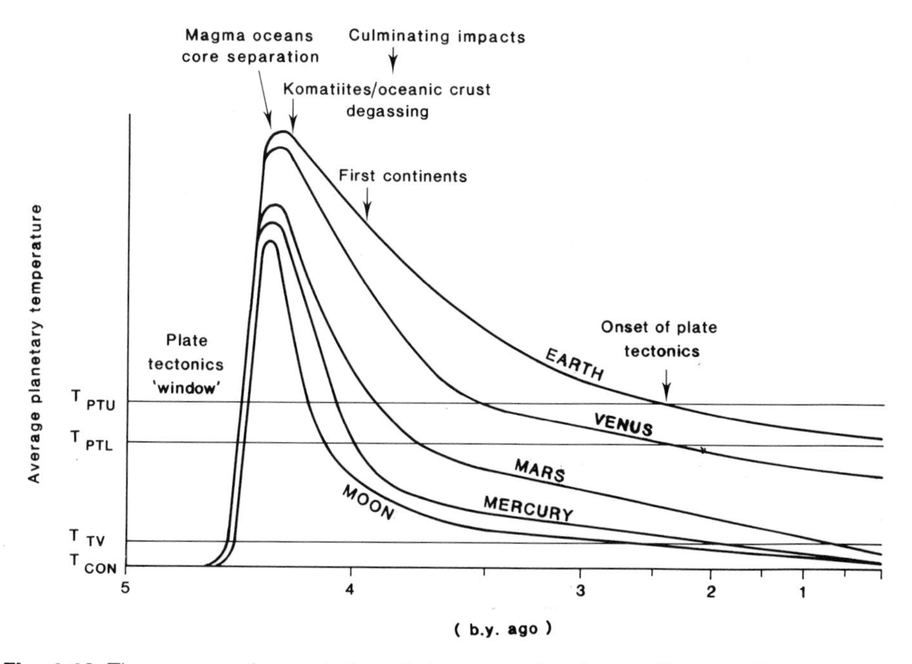

Fig. 1.12 The comparative evolution of the terrestrial planets. T_{PTU} and T_{PTL} represent, respectively, the upper and lower temperature limits for Earth-like plate tectonics. T_{TV} is the terminal volcanic temperature and T_{CON} the terminal convection temperature. Information currently available does not allow the vertical axis to be scaled (from Condie 1989, Fig. 10.27, p. 381, with some modifications).

Although mass may well be the principal factor, others are also of importance (see Lowman 1976; Schubert 1979; Condie 1989) and are listed in Table 1.7.

Several models have been suggested which place the evolution of the terrestrial planets within comparative contexts (e.g. Lowman 1976; Soderblom and Johnson 1982). The American geologist Kent Condie has published one such model which is remarkable for its simplicity (Condie 1989). As Fig. 1.12 shows and following heating in the earliest stages of evolution, all the planets cool but at different rates. The Moon and Mercury cool so rapidly and pass through the 'plate tectonics window' so quickly, that plate motions are not initiated, a thick lithosphere develops and volcanic activity ceases around $1–2 \times 10^9$ years ago. Earth and Venus, in contrast, cool more slowly and experience a greater variety of tectonic and volcanic processes. Both volcanism and tectonism are active for billions of years. Mars lies between the two extremes. It is interesting to note that, according to Fig. 1.12, volcanic activity on Earth will cease in $\sim 500 \times 10^6$ years.

It should be noted that it is not possible to place the giant planets and their satellites within the model shown in Fig. 1.12 because, in addition to mass and the other factors listed in Table 1.7, composition—specifically whether they are 'icy' or 'rocky'—is a critical consideration (see section 1.2). However, using a three-dimensional plot of thermal history, composition and size, Soderblom and Johnson (1982) were able to place all bodies in the Solar System into one evolutionary scheme.

As more data are collected it is highly probable that models like those of Condie and Soderblom and Johnson will either be modified or discarded. At present they enable volcanoes on Earth to be placed in wider contexts and offer interim answers to the question: Why the Earth is so unique?

2

Volcanoes and plate tectonics

The same regions do not remain always sea or always land, but all change their condition in the course of time (Aristotle 384–322 BC).

2.1 Introduction

The observation that volcanoes do not describe random patterns on the Earth's surface has fascinated people for thousands of years. It was commented upon by Plato, Strabo and Seneca, held as an 'article of faith' by the founding fathers of modern geology in the nineteenth century (Zittel 1901; Macdonald 1972) and is now universally accepted. Today the theory of plate tectonics is a well supported model and one which is able successfully to describe the global distribution of volcanoes of different types at the surface of the Earth; both at the present time and, more tentatively, in the geological past (Windley 1977; Brown and Mussett 1981; Condie 1989). The present chapter is essentially descriptive because, although the model of plate tectonics casts some light on the processes responsible for the generation, movement to the surface and eruption of magma, other mechanisms operating deep within the Earth are very important and are dealt with in Chapter 3.

2.2 The plate tectonics model

It took more than 50 years for the idea that the major structural and tectonic units of the Earth had moved in their relative positions to be accepted. Although Alfred Wegener first put forward the theory of continental drift before the First World War (Wegener 1912), it was not until the late 1960s that plate tectonics became part of the accepted geological canon. The story of the fierce resistance to and often grudging acceptance of the model has been told many times (e.g. Muir-Wood 1985), but from the 1950s the accumulation of increasing amounts of mutually supporting evidence from investigations into polar wandering, the nature and significance of ocean trenches and palaeomagnetism of ocean floors, led to the synthesis of earlier notions of continental drift with those of sea-floor spreading to produce the theory of plate tectonics (e.g. Hess 1962; Vine and Matthews 1963; J. T. Wilson 1965).

Plate tectonics is embraced by the vast majority of earth scientists, but an occasional dissident voice is still to be heard (e.g. Lyttleton 1982). It must be stressed that plate tectonics is a model; in other words 'an idealized representation of reality in order to demonstrate some of its properties' (Haggett 1965,

p. 19): it is also a theory—the best available at the moment—but a theory nevertheless. In some parts of the world plate tectonics does not as yet fully explain all field observations and is still being developed. The volcanic activity of the central Mediterranean and its relationship to plate tectonics is, for instance, very complex and models proposed by Dewey *et al.* (1973) and Biju-Duval *et al.* (1977) both fail adequately to account for the varied volcanic activity both now and in the past (see Chester *et al.* 1985). Science advances as prevailing theories are falsified by new evidence which cannot be explained by them. This causes theories to be modified and, thereby, improved (Popper 1959). Presumably this will happen as more data on the volcanology of the central Mediterranean are collected and published. Any model is based on assumptions and in the case of plate tectonics two are fundamental (Summerfield 1991):

1) that the surface area of the Earth has not changed with respect to the generation of crust at spreading centres, otherwise on an expanding Earth spreading could occur without subduction; and
2) there has been relatively little deformation within plates compared to the movements between them.

According to Summerfield (1991, p. 47) both assumptions are justified at least for the last 500 million years.

Detailed discussion of the evidence supporting the model of plate tectonics and what it means in terms of the structural make up of the planet are enticing topics, yet peripheral to the study of volcanoes. What follows is a simplification of the major elements of the model, sufficient to provide a framework in which volcanic activity may be examined within different structural and tectonic regions. Readers who are already familiar with plate tectonics can safely skip the rest of this section, while those requiring a more detailed exposition should make reference to the following: Hallam (1975); Brown and Mussett (1981); Condie (1982, 1989); Boss (1983); Van Andel (1984); Cox and Hart (1986); Achache (1987); Jurdy (1987); Hamilton (1988); and Le Grand (1988).

Over a century of seismological and other geophysical findings and their interpretation, has allowed the Earth to be divided into a number of vertical layers or 'shells' (Bolt 1982). It is conventional to consider the crust to be thin, composed of rocks of relatively low density and separated by a seismic discontinuity from a more dense mantle, comprising materials rich in iron and magnesium (Table 2.1). This discontinuity, first recognized by Yugoslavian geophysicist A. Mohorovičić, is known as the *Moho*. The crust makes up less than 0.1% of the total volume of the Earth and is composed of two types of material: *sima*, the predominant material of the ocean floors; and *sial*, occurring on the continents and itself overlying a zone of sima. Both terms are acronyms of their principal constituents: sima (Si = silica and Ma = magnesium) and sial (Si = silica and Al = aluminium). Separating the mantle and the outer core is another transition 'shell' (i.e. D″, Table 2.1) known as the Gutenberg, formerly Oldham, discontinuity. On the basis of the propagation of seismic waves, the outer core of the Earth is considered to be liquid and the inner core solid. Both are composed predominantly of iron.

In the plate tectonics model the scheme of vertical zoning of the Earth, given above and based primarily on compositional differences, is modified. The crust/mantle distinction is retained, but it is recognized that it is sensible to divide the upper ~640 km of the Earth's interior on other criteria. First there is the *lithosphere*. This comprises the crust and the upper part of the upper mantle (i.e.

Table 2.1: The vertical structure of the Earth (based on Bolt 1982, with some modifications)

Shell	Name	Depth range (km)	Physical attributes
A	Crust (Crustal) (Lithosphere)	0–5 (oceans) 0–50 or more (continents)	Solid and liquid
B	Upper mantle (Non-crustal) (Lithosphere)	'Moho' to 100	Solid
C	Upper mantle (asthenosphere)	100–640[1]	Solid (upper parts near melting)
D′	Lower mantle	640–2780	Solid
D″	Transition shell	2780–2885	Solid (lower velocities)
E	Outer core	2885–4590	Liquid
F	Transition shell	4590–5155	Liquid
G	Inner core	5155–6371	Solid

[1] This figure is often given as 650 km, or even 670 km.

shells A and B—Table 2.1). Together these 'shells' have the property of a relatively rigid solid. The layer below it (i.e. shell C) is marked by lower seismic wave velocities and was first recognized by Beno Gutenberg (Bolt 1982). It is termed the *asthenosphere*. The asthenosphere is both semi-molten and semi-plastic (Greeley 1987) and its base at a depth of ~640 km is not only very close to the maximum depths of earthquake foci, but also coincides with further zone of seismic wave discontinuities.

Over 60 years ago the British geologist Arthur Holmes first suggested that the decay of radioactive materials could have an effect on earth movements (Holmes 1928) and today it is generally accepted that radioactive and other heat sources are able to generate convection cells within the asthenosphere (see Chapter 3 for more details). Where upward tending cells converge this may cause arching of the lithosphere and concentrate heat at and near to the surface (Fig 2.1b). Sometimes heat is localized and expressed as 'hot spots', which produce volcanoes such as those of the Hawaiian Islands (Fig. 2.1a; see section 2.3), but in other situations thermal plumes fracture and pull the crustal rocks apart. Volcanism is more widespread and involves large volumes of magma moving to the surface, so creating new oceanic floor and mid-ocean ridges (Fig. 2.1a; see section 2.4). Lateral movements of the asthenosphere drag segments of lithosphere away from these zones of rifting so that the oceanic floor is at its youngest nearest to ridges and increases in age with increasing distance from them. Magnetic 'stripes' occur roughly parallel to ridge axes and represent reversals in magnetic polarity during the time the ocean floor was being formed (Vine and Matthews 1963). It is hypothesized that heat sources are not of equal size and so convection cells are of differing dimensions. This in turn causes segments of the lithosphere, known as *plates*, to vary in surface area. Plates are in constant movement and rates average 70 $mm a^{-1}$ and may reach 100 $mm a^{-1}$ (Summerfield 1991). The zones where plates come into contact are known as *plate boundaries* and these and adjacent

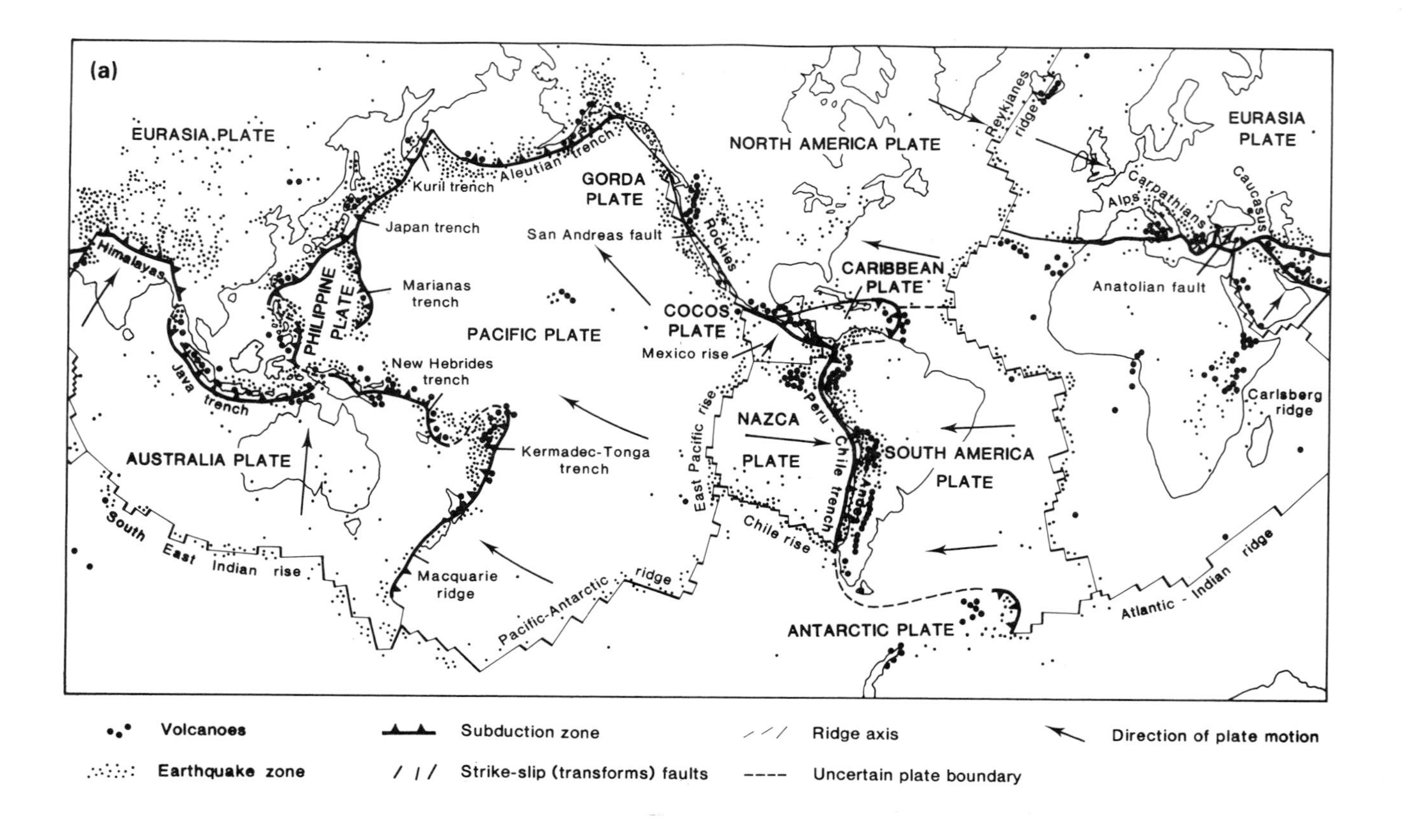
(a)
EURASIA PLATE
Himalayas
Kuril trench
Aleutian trench
Japan trench
Marianas trench
PHILIPPINE PLATE
New Hebrides trench
Java trench
AUSTRALIA PLATE
South East Indian rise
Kermadec-Tonga trench
Macquarie ridge
Pacific-Antarctic ridge
PACIFIC PLATE
GORDA PLATE
San Andreas fault
Rockies
COCOS PLATE
Mexico rise
East Pacific rise
NAZCA PLATE
Chile rise
Peru - Chile trench
Andes
NORTH AMERICA PLATE
CARIBBEAN PLATE
SOUTH AMERICA PLATE
ANTARCTIC PLATE
Reykjanes ridge
EURASIA PLATE
Alps
Carpathians
Caucasus
Anatolian fault
Carlsberg ridge
Atlantic - Indian ridge
Volcanoes
Earthquake zone
Subduction zone
Strike-slip (transforms) faults
Ridge axis
Uncertain plate boundary
Direction of plate motion

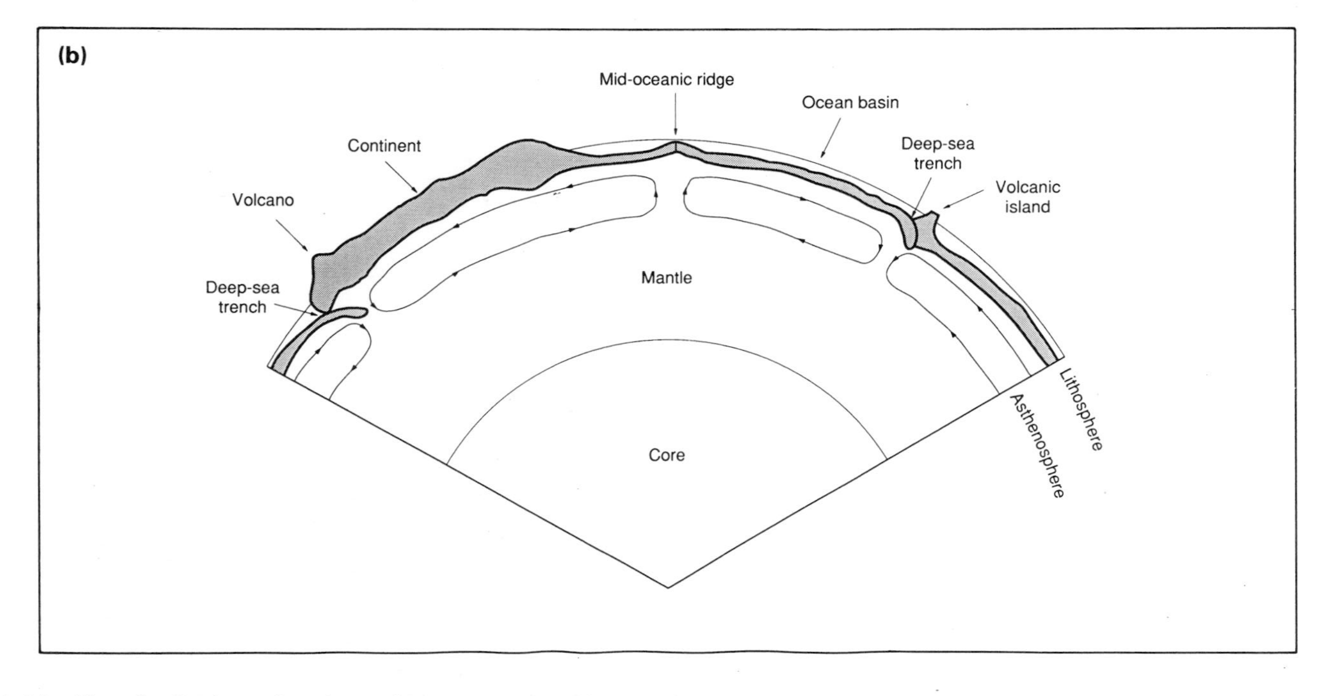

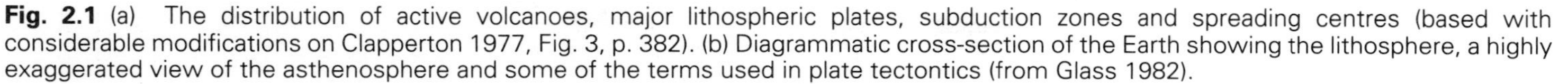
Fig. 2.1 (a) The distribution of active volcanoes, major lithospheric plates, subduction zones and spreading centres (based with considerable modifications on Clapperton 1977, Fig. 3, p. 382). (b) Diagrammatic cross-section of the Earth showing the lithosphere, a highly exaggerated view of the asthenosphere and some of the terms used in plate tectontics (from Glass 1982).

regions as *plate margins*. Three plates coming together create what is called a *triple junction* (Burke and Dewey 1973).

As Fig. 2.1a shows most but by no means all the world's earthquakes and volcanoes are related to plate boundaries and these may be classified under three heads: *divergent*; *transform*; and *convergent*. *Divergent* boundaries are mid-ocean ridges where new crust is being formed and continental rifts like those of East Africa. *Transform* boundaries are typified by the infamous San Andreas Fault in California and represent areas in which plates merely move past each other, without any divergence or convergence. They are not associated with volcanic activity, though earthquakes are common and may be devastating. Finally, there are *convergent* boundaries. These are very complex geological regions and 'it is the processes at these junctions that ultimately result in the linear mountain chains and metamorphic belts of the world' (Carr 1984, p. 85). Convective cells within the asthenosphere trend downwards as well as upwards (Fig. 2.1b) and this has the effect of drawing lithospheric plates together, involving the *subduction*, or slipping down, of one plate below another. Whereas material is added to the crust at divergent boundaries, it is reduced at convergent boundaries; thereby maintaining a global equilibrium. In detail there are a number of quite distinct styles of plate collision and these produce contrasting forms of volcanic activity (section 2.5).

2.3 Intraplate volcanic activity

Intraplate volcanic activity is driven by mantle plumes which create volcanoes at 'hot spots' on the surface of the Earth. Activity may occur within both oceanic and continental lithospheric plates.

2.3.1 Oceanic intraplate volcanism

Lines of active and inactive volcanic islands and seamounts are features of intra-oceanic plates in the Indian, Atlantic and Pacific Oceans. Examples include the Maldive Islands in the Indian Ocean and the Azores in the Atlantic Ocean (Clapperton 1977). It is, however, in the Pacific that this type of activity is best developed and closest studied (Fig. 2.2). Plume-related islands are overwhelmingly basaltic. They are not, however, exclusively tholeiitic like the ocean floors, but are more varied with alkaline basalts and limited volumes of more silica-rich materials being commonly erupted (see Chapter 3, section 3.3; and Frey and Clague 1984). Cas and Wright (1987) building on work by Moore and Fiske (1969) show that these volcanic lines[1] are built of pillow lavas and hyaloclastites (Chapter 4, section 4.4), which may have thicknesses measured in kilometres. Hydrovolcanic materials often become important as a volcano builds and approaches the surface of the ocean (Chapter 5, section 5.5) and, where conditions are appropriate, limestone and other organic detritus may become part of the succession—sometimes forming atolls. It has long been suspected from studies on land that voluminous submarine lava flows will be erupted from

---

[1] In this volume the term 'volcanic line' will be adopted, but in the literature these islands and seamounts are sometimes known as 'chains'. The use of the 'chain' will follow Clapperton's (1977) usage and will refer to volcanoes arranged in straight lines within mountain belts (section 2.5.1).

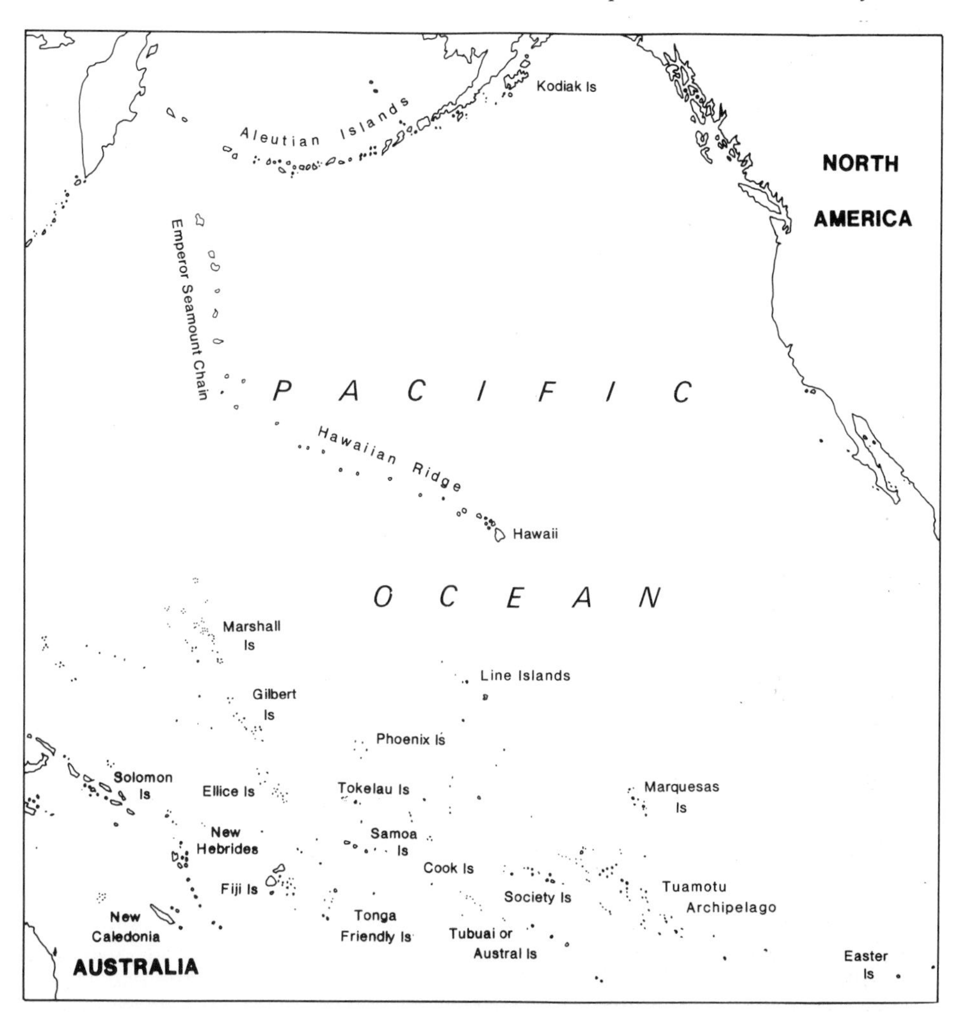

Fig. 2.2 Volcanic lines (sometimes called 'chains') of active and inactive volcanoes and seamounts in the Pacific Ocean (from Williams and McBirney 1979, Fig. 12.4, p. 284, and based originally on work by Dalrymple *et al.* 1973).

intraplate oceanic islands. For Hawaii this has now been confirmed by a programme of long-range sonar imaging around the east and south of the islands (Holcomb *et al.* 1988).

More than a century ago the pioneer American geologist and volcanologist J. D. Dana noted with commendable insight that the volcanoes of the Hawaiian Islands became progressively older towards the northwest (Dana 1890) and it is now known that this is the direction of plate movement. J. T. Wilson (1963a and b) and Burke and Wilson (1976) showed that this pattern is also evident at other volcanic lines and suggested that it is consistent with the formation of a series of islands as a moving plate passes over a stationary hot spot (Fig. 2.3). As Fig. 2.2 shows there is a change of direction along the Hawaiian volcanic line, with the

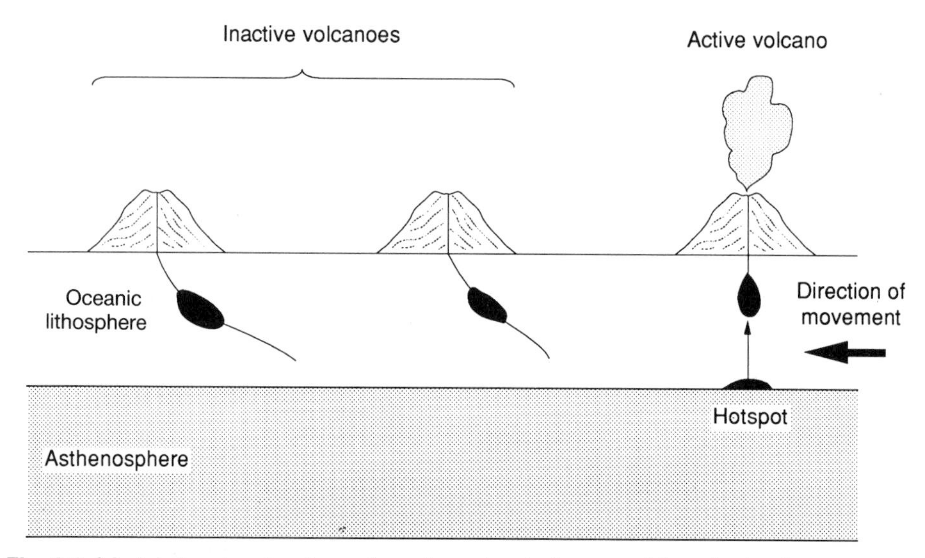

Fig. 2.3 Model showing the formation of a volcanic line as a lithospheric plate moves over a stationary 'hot spot' (based on information in Williams and McBirney 1979, Fig. 12.5, p. 285).

Hawaiian ridge being on a different alignment from the Emperor Seamount Chain. According to Morgan (1972) this represents an abrupt change in direction of the Pacific plate, during the formation of the volcanic line.

2.3.2 Continental intraplate volcanism

Textbook reviews of continental intraplate volcanism have tended to be terse[2]. This does not reflect the rigour with which authors have tackled the topic, but merely that during the 1980s and especially since 1988 the whole subject has been given new direction and impetus by the publication of a number of seminal papers (e.g. Cox 1989; White and McKenzie 1989). The principal expression of continental intraplate volcanism is the eruption of vast areas of flood basalt in regions such as the Snake River Plain (USA), Ethiopia, the Karoo region of South Africa and the Deccan region of India and, although it has been accepted for many years that activity is due to mantle plumes, here the similarity with oceanic intraplate activity ends. A discussion of the ideas of Cox (1989), White and McKenzie (1989) and other writers, together with a description of flood basalts, is contained in Chapter 4 (section 4.3.1.1), to which reference should be made.

2.4 Volcanic activity at divergent plate margins

Before discussing activity at divergent margins, it is necessary to refer to some recent proposed modifications of the theory of plate tectonics since research has

[2] Compare, for example, accounts in two of the most stimulating and scholarly recent studies by Fisher and Schmincke (1984, p. 14) and Cas and Wright (1987, pp. 452–53).

led to a new model being proposed known as the *Wilson Cycle*[3]. As its name implies, this model originated in the writings of the eminent earth scientist J. T. Wilson (1966) and has been developed by him (i.e. J. T. Wilson 1988) and by a number of other writers who include: Gurnis (1988); Nance *et al.* (1988); and Murphy and Nance (1991, 1992).

From the time of Alfred Wegener it has been held by advocates of continental drift and later of plate tectonics, that until about 180 million years ago the continents now bordering the Atlantic Ocean were part of a single landmass known as *Pangaea*, its breakup being viewed as a unique event. More recently John Tuzo Wilson, Brendan Murphy, Damian Nance and others have argued that Pangaea was not a unique entity, but rather that there has been a regular cycle of supercontinental building and dispersal throughout geological history (Table 2.2). They note that over geological time major episodes of mountain building, or orogenesis, have had return intervals of $\sim 500 \times 10^6$ years and they explain this in terms of their suggested cycle.

Ideas about supercontinental breakup and aggregation are inchoate, still much debated and not accepted fully by all earth scientists, but there is little doubt that they are already having effects on the ways in which global tectonics, orogenies—both pre- and post-Pangaea—and the geomorphology of large continental land

Table 2.2: The principal elements of and processes involved in the Wilson Cycle (based on references in the text and information in Summerfield 1991)

Stage	Processes and principal events
1	It is hypothesized that two complementary processes may be involved in the breakup of Pangaea, or any supercontinent. The first makes use of ideas first posited by Anderson (1982) and is based on the observation that continental lithosphere is not so good a heat conductor as the much thinner, basaltic oceanic lithosphere. Hence, over long periods of time heat will build up under continents and after about 80 million years this will lead to uplift, rifting and fragmentation. A second hypothesis attributes the breakup to the rotation of the Earth. Because it is elevated, a supercontinent makes the planet lopsided. This means that supercontinents have high angular momentum, causing long-term stresses which lead to continental rifting. Both the 'heat loss' and the 'angular momentum' models may be of importance.
2	New ocean basins (so called *interior oceans* like the present-day Atlantic, as opposed to *exterior oceans* like the present-day Pacific) are formed and these become progressively wider. The continental margins of expanding interior oceans show little tectonic or volcanic activity (e.g. the Atlantic today), whereas subduction occurs on the margins of exterior oceans.
3	The dispersal of continents does not continue indefinitely. Oceanic crust becomes cooler and denser with distance from mid-ocean ridges, subduction becomes established and there is a change from continental dispersal to aggregation. This leads eventually to the creation of a new supercontinent. The whole cycle is claimed to take around 500 million years to complete.

[3] The Wilson Cycle is also known as the *supercontinent model.*

masses are being perceived. The geomorphologist Michael Summerfield (1991, p. 102) quotes the case, for example, of the new light which has been cast on the old problem of average continental elevation. The breakup of Pangaea probably involved what is now Africa remaining in approximately the same position and the other continents moving away from it. In other words it is still close to the area of heating and uplift and has, therefore, the highest mean elevation, though it lacks major orogenic belts. Australia, in contrast, which was on the edge of Pangaea, has moved far from its original location and has the lowest mean continental elevation.

From a volcanological perspective the Wilson Cycle is already having a profound impact on the conception of plate-boundary volcanism, particularly over long time scales. The nature of this impact will be discussed in this section with reference to divergent margins and later when subduction zone activity is considered (section 2.5).

Volcanic activity is associated with two classes of divergent plate boundary. Mid-ocean ridge activity is the best known, but broadly similar activity occurs within continental lithospheric plates in the form of rift valleys. As Table 2.3 shows, 61% of all intrusive and 73% of all extrusive rocks are generated at divergent margins, most at mid-ocean ridges.

2.4.1 Mid-ocean ridge (open ocean) volcanism

As might be expected mid-ocean ridges (MOR) are not only noteworthy because of their high geothermal heat flux, but are also the locations where most of the igneous rocks on Earth are formed. The first and most fundamental question raised by the study of MOR volcanism is its relationship to sea-floor spreading. Since the pioneer work of Vine and Matthews (1963) the striped configuration of alternating magnetic anomalies running roughly parallel to ridge axes have been interpreted as representing the progressive movement of new material away from ridges, but the mechanisms involved in producing such patterns remain unclear. As Williams and McBirney (1979, p. 282) have noted with considerable perspicacity 'the magnetic properties of rapidly quenched basalts can account for at least a large component of the observed magnetic intensities, . . . but the steep magnetic gradients require individual bodies with vertical dimensions of hundred of meters (*sic*) and steep regular margins. They could not be caused by overlapping lava flows; . . . the random lengths of flows would produce irregular and indistinct boundaries between polarity strips'. They conclude that the

Table 2.3: Rates of magma production in different tectonic situations (based on information in Fisher and Schmincke 1984)

Tectonic situation	Intrusive km^3/year	(%)	Extrusive km^3/year	(%)
Intraplate				
Oceanic	2.0	(7)	0.4	(10)
Continental	1.5	(5)	0.1	(2)
Divergent	18.0	(61)	3.0	(73)
Convergent	8.0	(27)	0.6	(15)
Totals	29.5		4.1	

magnetic pattern is probably caused by dyke intrusion. In other words intrusive igneous activity is very important indeed (Table 2.3; and White and McKenzie 1989).

Information about the nature of MOR volcanic activity is not easy to collect, but in the last few years information has been made available from a number of sources. These include the 'direct' study of ridges by submersibles, instruments towed at depth beneath ships, sonar, bathymetric surveys and photography. In addition geological data have been collected by 'indirect' methods. Noteworthy amongst these has been the study of *ophiolites*; uplifted areas of oceanic crust (Gass 1982). Traditionally ophiolites have been interpreted as ocean-floor materials formed at mid-ocean ridges but several recent reports have cast doubt on this hypothesis, favouring instead mechanisms related to island-arc magmatism (e.g. Pearce *et al.* 1984; Shervais and Kimbrough 1985). At the present time interest in MOR is stimulated by commercial demands, for it has been found first on the East Pacific Rise (Francheteau *et al.* 1979; Ballard *et al.* 1984) and later more generally (Edmond and von Damm 1983; Cronan 1985) that ridges are associated with vents depositing sulphide compounds named 'black smokers'. Percolating sea water is thought to react with the hot rocks, metals are leached and hot hydrothermal jets rise through the sea floor as 'black smokers'. The black smoke comprises fine particles of metallic and other minerals which precipitate as the 'smoker' cools.

Much of what is known about volcanic activity on and around ocean ridges is derived from research which has been carried out since the early 1970s (Ballard and van Andel 1977; Luyendyk and Macdonald 1977; Macdonald and Luyendyk 1977, 1982; Macdonald 1982; Hékinian 1984). It is now established that volcanic activity is related to fissures set within central rifts (Ballard and van Andel 1977) and that these produce: massive and pillow lava flows (see Chapter 4); large accumulations of debris at the base of fault scarps; numerous dykes and much hydrothermal activity (Fig. 2.4a and b). The type of basalt produced at a MOR is known as tholeiite and its petrology and petrogenesis is discussed in Chapter 3 (section 3.2). It is also evident that the detailed morphology of central rifts is variable and depends critically on the rate of spreading, which alters the stresses within the rift (Macdonald 1982). As Table 2.4 shows three situations may occur depending on whether spreading is slow, medium or fast.

According to the putative Wilson Cycle (Table 2.2), volcanic activity at mid-ocean ridges is restricted in time. As oceanic crust moves away from a ridge axis it becomes older and more dense. Subsidence, which has been observed in the Atlantic, occurs at increasing distance from ridges (Murphy and Nance 1992) and gradually as subduction becomes dominant the ocean will begin to close. It is argued that the whole cycle from ocean formation to eventual closure takes around 500 million years.

2.4.2 Continental rift volcanism

It is conventional to make a distinction between narrow, linear rifts like those of East Africa and much wider ones, typified by the Basin and Range Province of the USA.

Narrow rifts (Fig. 2.5) are not located randomly within continental plates, but represent long-established zones of crustal weakness. Although the current phase

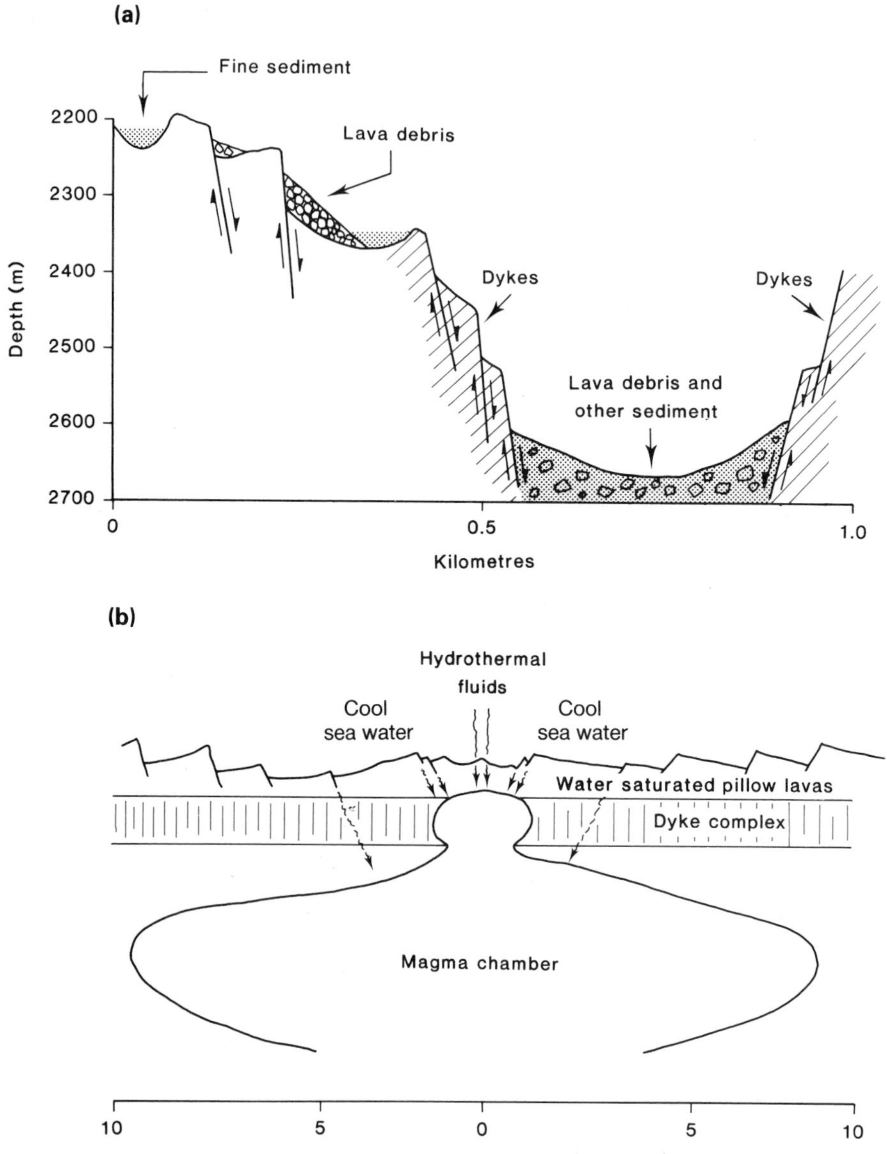

Fig. 2.4 Aspects of mid-ocean ridge volcanic activity.
(a) The morphology of the central rift of the mid-Atlantic ridge to the southwest of the Azores. Most volcanic activity is associated with fissures within this narrow valley and complex faulting has created depressions where lava debris and other sediments may accumulate (based on ARCYANA 1975, Fig. 3, p. 112).
(b) Diagrammatic section showing the possible structure of the East Pacific Rise based on the results of geophysical investigations. The results indicate a 'cupola' above the main magma chamber, and its episodic existence may explain the episodic nature of hydrothermal activity, which usually follows intense volcanic activity and spreading. The layering of the ocean floor, the faulted central rift and the water-saturated pillow layers should be noted (based on Macdonald and Luyendyk 1982, p. 26).

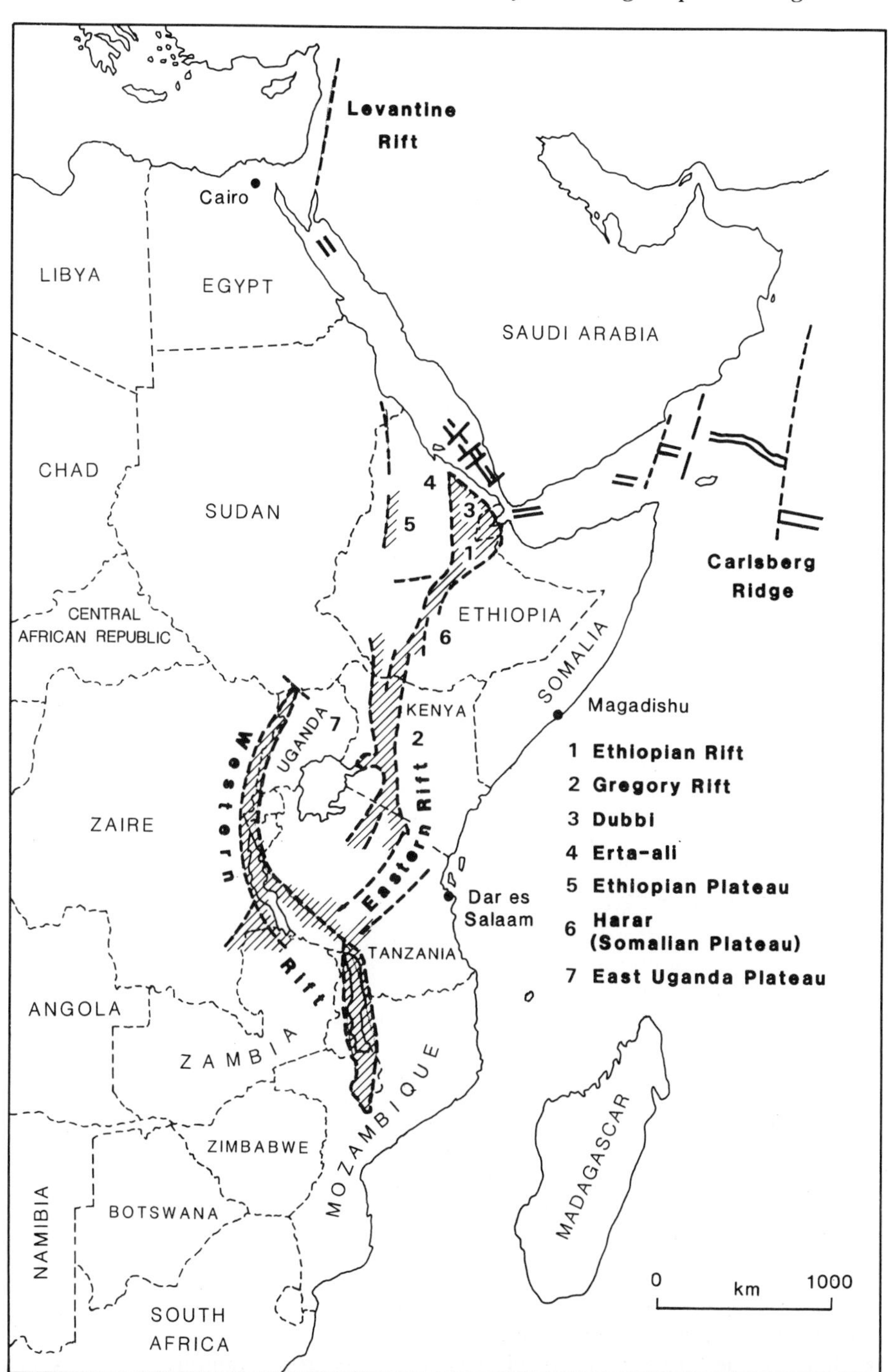

Fig. 2.5 The East African Rift, showing major structural and volcanic provinces. The numbers refer to volcanic provinces, named above. The diagram is a composite and based on B.H. Baker *et al.* (1972); Mohr and Wood (1976); and several other sources.

Table 2.4: The influence of spreading rate on the nature of volcanic activity at mid-ocean ridges (based on information in Macdonald 1982)

Spreading rate	Nature of volcanic activity
Slow	Volcanoes occupy a distinctive and well-defined rift valley and activity occurs as a discontinuous set of central volcanoes, aligned parallel to the ridge axis. Pillow lavas are dominant and edifices have basal diameters of 1–4 km and heights up to 250 m.
Intermediate	As the rate of spreading increases the central rift valley is not so prominent a feature and volcanoes are fed by fissures. Sheet flows of pahoehoe morphology (see Chapter 4) are very common and edifices rarely exceed 50 m in height.
Fast	The central rift valley is not well defined and edifices resemble shield volcanoes on land, except that they are elongated along fissures. Massive pillow lavas are common.

of activity began ~25–35 million years ago, the East African Rift runs parallel to much older orogenic belts and follows the margins of Precambrian shields (cratons). Stretching from the Red Sea in the north to Zimbabwe in the south, the East African Rift Valley is one of the principal landforms of the planet, is geologically complex and has been studied intensely, particularly over the last 25 years. Today its structural, volcanological and geomorphological evolution is known in detail and many excellent research reports and reviews have been published, which have revolutionized understanding of this and other rifted regions (e.g. B. H. Baker *et al.* 1972; Di Paola 1972; Mohr and Wood 1976; Gass *et al.* 1978; Ramberg and Neumann 1978; Mohr 1983; Summerfield 1985, 1991; Bonatti 1987; Buck *et al.* 1988; White and McKenzie 1989; Morley *et al.* 1992).

Ever since the plate tectonics model was accepted, it has been recognized that ruptures like those in East Africa represent areas where old plates are starting to separate and that rifts 'form along a zone of incipient divergence and are characterized by updoming, rifting and volcanism' (Carr 1984, p. 85). It is evident that rifts may be formed by two processes (Summerfield 1991, p. 93). The first known as *active rifting* (Fig. 2.6a) is a response to tensions set up in the crust as a result of uplift caused by welling up of material from the asthenosphere. The second process—*passive rifting*—is permissive because upwelling occurs because of previous extension and thinning of the lithosphere (Fig. 2.6b; Summerfield 1991, p. 93). In active rifting the expected sequence of events is: volcanic activity, uplift, and rifting; whereas in passive rifting the anticipated sequence is: rifting, volcanism, and uplift. In East Africa it is widely held that uplift preceded rifting, and hence the whole system is active. The question, however, is by no means settled, the evidence is equivocal and some workers argue that rifting may have occurred before uplift, implying a passive system.

Volcanic products generated at linear rifts are much more variable than those of mid-ocean ridges. Compositions range from basalts to silica-rich trachytes and rhyolites and are generally more alkalic and contain greater amounts of rare earth elements (see Chapter 3). Products include lava flows and variety of pyroclastic materials, including ignimbrites (see Chapter 5, section 5.3.4). It has long been held that the principal reason why magmas produced at continental rifts are so

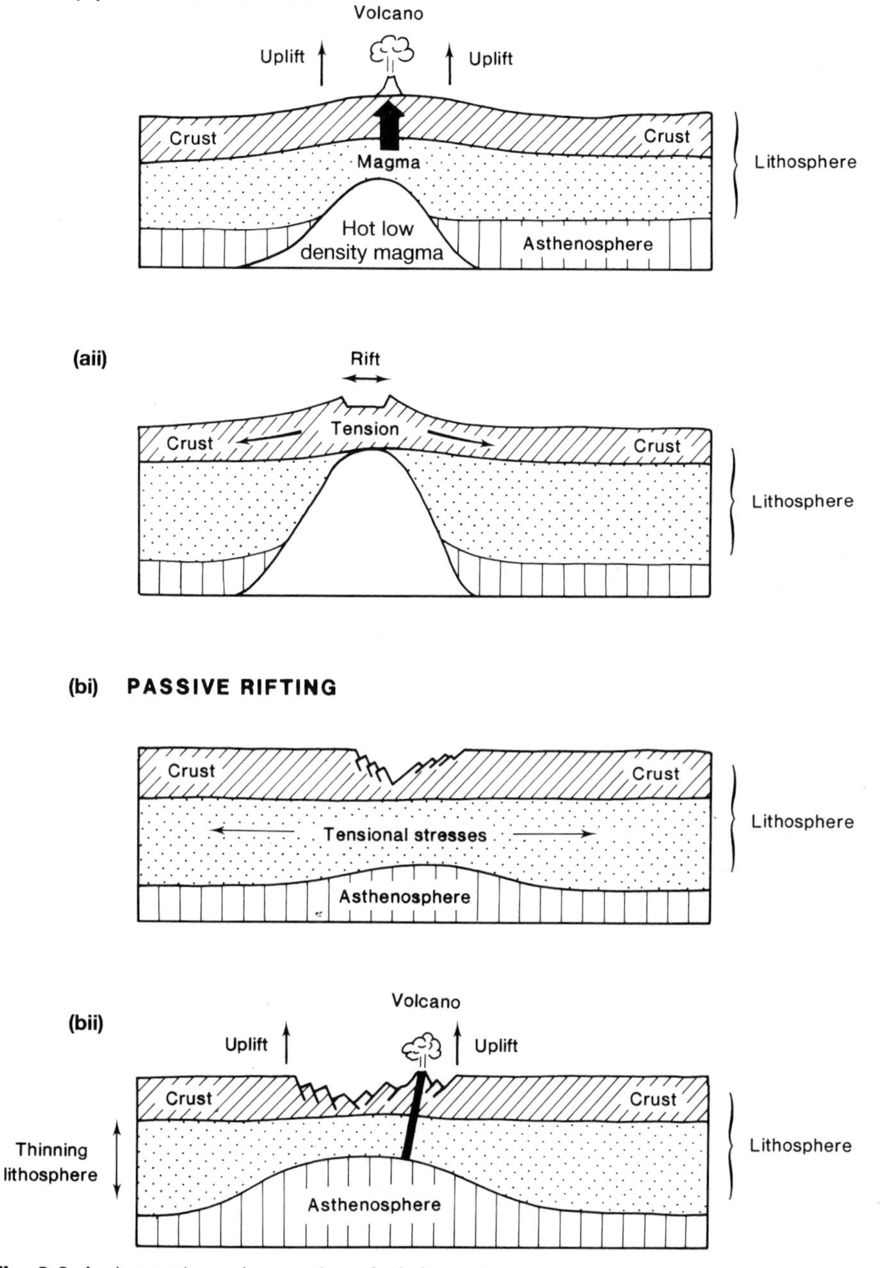

Fig. 2.6 Active and passive modes of rift formation.
(a) Active rifting: volcanism and uplift followed by rifting.
(b) Passive rifting: rifting followed by volcanism and uplift (based with some modifications on Summerfield 1991, Figs 4.5 and 4.12, pp. 88 and 94).

varied, in comparison with MOR basalts, is because of their differing depths of origin (see Chapter 3, section 3.3; and Green 1972). It has also been argued (Green and Ringwood 1968) that the wide variety of volcanic products is due to selective removal of minerals such as olivine, pyroxene and garnet from these melts, although contamination by crustal rocks may also be significant (Carr 1984). Because of the varied interaction of volcanic and other subaerial, crustal and subcrustal processes at different stages in the evolution of a rift, vertical successions and spatial patterns of activity are complicated. For instance, early stages often go hand in hand with normal faulting and successions showing mixtures of volcanic facies, alluvial fans, river and lake sediments. Later as rifts widen and subside associations change. As in the Red Sea and Gulf of California the sea may flow in, a 'mid-ocean' ridge may develop and products will become more tholeiitic. They are frequently interbedded with marine sediments (Cas and Wright 1987).

In contrast to continental rifts such as the East African which are narrow, the Basin and Range Province in the western USA and northern Mexico, is much wider (several hundred kilometres) and volcanically diverse (Eaton 1982, 1984). It is typified by: a thin continental crust (less than 30 km) (Summerfield 1991); a high mean elevation; a marked geothermal heat flux; a distinctive topography of basins and dissected uplands and strong seismic, as well as volcanic, activity. Petrologically and volcanologically the province is varied with rock types ranging from tholeiitic and alkalic basalt to silica-rich calc-alkaline rocks, the latter being associated with ignimbrites and large calderas. Much controversy surrounds the geological interpretation of this region, though it seems certain that its origin does not involve processes similar to those operating in East Africa, but expressed on a larger canvas. Menard (1964) posited that the region was formed by the subduction of an active mid-ocean ridge under the western USA, while later models such as those of Scholz *et al.* (1971), Henyey and Lee (1976) and Eaton (1984) have argued that the province is the terrestrial equivalent of an oceanic back-arc basin (see section 2.5.2.3). Fig. 2.7 summarizes one model which has been popular in recent years.

2.5 Volcanic activity at convergent plate margins

It is estimated that 80% of the volcanoes currently active occur above subduction zones (Clapperton 1977). In the past the circum-Pacific volcanic belt was known as 'the ring of fire', reflecting the high magnitude and frequent devastation wrought by eruptions, with the imaginary boundary between it and the inner Pacific being termed the 'andesite line'; emphasizing the supposed composition of magmas (see Holmes 1965, pp. 1013–14). Whilst useful these memorable expressions are misleading because not all eruptions are violent and, although andesite (~57–63% SiO_2) and basaltic andesite (~52–57% SiO_2) are the most abundant constituents in certain parts of the circum-Pacific, magmas range from basalt (~45–52% SiO_2), which is very common in some island arcs, to dacite (~63–70% SiO_2) and rhyolite (over ~70% SiO_2). Ringwood's (1974) term 'orogenic volcanic series' is both more accurate and emphasizes that magma generation is but part of a wider story of mountain-building, or orogenesis. Before plate tectonics was established the 'orogenic volcanic series' together with

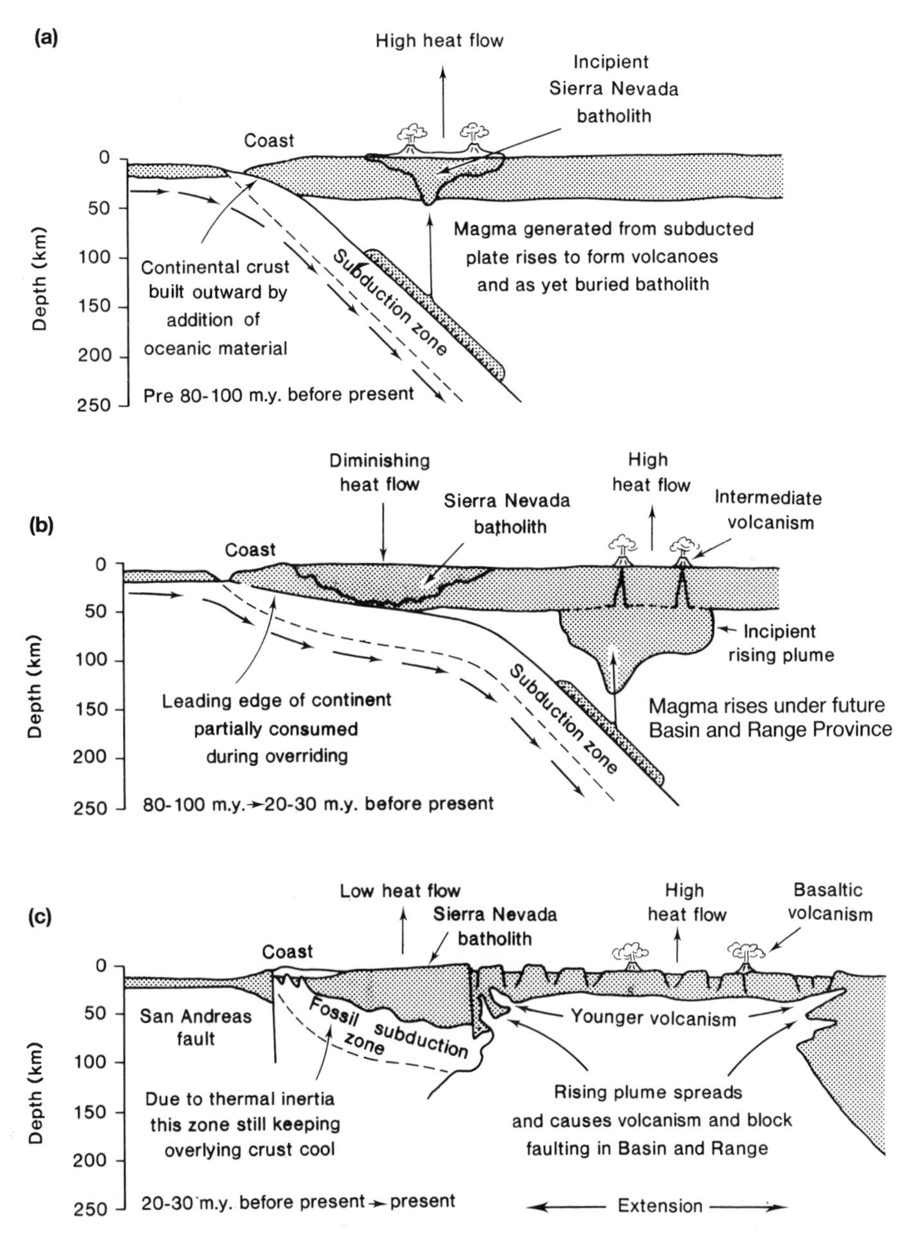

Fig. 2.7 The model of Henyey and Lee describing the long-term effects of subduction under North America and, more specifically, the origin of the Basin and Range Province.
(a) The continent grows by accretion of oceanic material.
(b) The continent 'overrides' and magma generation shifts to the future Basin and Range Province.
(c) Subduction stops, the plate boundary becomes transform and the heated, extended and thinned crust under the Basin and Range Province produces block faulting and volcanic activity (based on Henyey and Lee 1976, Fig. 8, p. 1185).

its plutonic[4] equivalents, such as granite, were explained by the melting and selective fusion of crustal rocks (*anatexis*) at the base of thick mountain piles (Clapperton 1977). Today the 'orogenic volcanic series' is placed within the context of plate subduction.

The 'ring of fire' is known not only for its volcanoes but also for its devastating earthquakes. Recognition of subduction zones came about largely through the interpretation of seismic evidence and 'without exception every region of plate convergence that has active "andesitic" volcanoes also has earthquakes deeper than 100 km. Similarly, there are no major regions of deep seismicity, with the possible exception of the eastern Himalayan chain, that are devoid of active volcanoes. It is rare in geology to find such impressive correlations' (Williams and McBirney 1979, p. 289). The implications of these correlations for the prediction of eruptions is considered in Chapter 7.

Processes of subduction vary according to the nature of plate collision and these are discussed later in this chapter (sections 2.5.1 and 2.5.2.), but common features do occur and are summarized in Table 2.5. From reading Table 2.5 it might be supposed that the generation of magma at subduction zones is a simple matter and in need of no further explanation. This is unfortunately not the case and production of magma is complicated, much debated and not fully understood. The spatial patterns described by subduction zone volcanoes, however, hold valuable clues to the importance of depth and, therefore, temperature and pressure in the generation of magmas belonging to the 'orogenic volcanic series'. Pioneer work by Dickinson (1973) showed that volcanoes do not occupy a surface zone more than 300 km in width, that there is a gap of 125 ± 50 km between ocean trenches and the volcanic front at island arcs and 225 ± 50 km at continental margins. Although these figures have been modified by later writers (e.g. Sakuyama and Nesbitt 1986; Condie 1989), their implications have been accepted, namely, that there is a specific range of depths along subduction zones where magmas are generated (Table 2.5). This range is 80 to 150 km according to Dickinson (1973) and 125 to ~200 km according to Sakuyama and Nesbitt (1986). It has also been observed that the height difference between subduction zones and the ground surface is related to the chemical composition of the magma produced and that the volume of magma diminishes as the height above subduction zones increases. The significance of these spatial patterns as constraints on possible mechanisms of magma generation is discussed in Chapter 3 (section 3.3).

Some features at subduction zones provide strong evidence in support of the Wilson Cycle. Hamilton (1988 p. 1507) notes 'that the Pacific Ocean is becoming smaller with time as flanking continents and marginal sea plates advance trenchward over ocean-floor plates'.

Volcanoes found at convergent plate margins may be placed into two categories:
1) those occurring on thick crustal rocks at continental margins such as the Andes; and
2) those related to island arcs, where both the overriding and subducting plates are oceanic.

---

[4] Plutonic rocks are igneous rocks formed at depth. They have different textures from effusive rocks (lavas and pyroclastic materials), but may have similar chemical compositions. They are discussed in greater detail in Chapter 3, section 3.2.

Table 2.5: Summary of the mechanisms involved in the subduction of the lithosphere. This is the most straightforward example and is based on subduction under the Island of Honshu (Japan) (based on information in Toksöz (1980); with amendments from numerous sources)

Stage	Mechanisms and attributes
Initial	Two plates converge. The oceanic plate flexes and is pushed under the thicker continental plate. An ocean trench may mark the site of subduction. Seismic profiling shows that the descending slab is curved and that an initial low angle of dip steepens as the slab descends into the asthenosphere.
Heating	Studies from a number of research fields (e.g. experimental petrology, seismology, geothermal studies and physical and chemical modelling) are suggestive of heating from a number of sources. Of first-order importance are: 1) Conduction of heat to the cool descending oceanic slab from the warmer mantle. This is a 'positive feedback process', since the output of the system—the warmed slab— reinforces the input because as temperatures rise so the efficiency of conduction improves. At depth this is a very efficient process. 2) Increased pressure is associated with increased depth. Heat is produced by energy released as minerals change to denser phases and/or more compact crystals as they experience higher pressures during descent. Increased pressure on the lithosphere as it descends also introduces heat. Of secondary importance are: 3) The descending slab is heated by the decay of radioactive isotopes of uranium, thorium and potassium. These are present throughout the Earth's crust and add heat at a constant rate. 4) Heat is generated by friction and shear stresses.
Slab assimilation	The temperature of the descending slab increases with depth and changes its nature. Theoretical modelling implies that at a maximum of ~700 km depth it is assimilated into the mantle. This also coincides with the maximum depth at which earthquakes can be traced along subduction zones. A descending slab does not always reach 700 km. If it moves slowly then it reaches thermal equilibrium at a shallower depth. If subduction ends, then it is estimated that the slab will lose its identity after ~60 million years.

Arcs include the Kurile, Marianas, Aleutian and Tonga Islands of the western Pacific and the South Sandwich Islands of the western Atlantic (see Fig. 2.1a). They occur, thus, 'on the western sides of oceans and are underlain by a subducting oceanic plate moving westward from an axis of spreading to the east' (Williams and McBirney 1979, p. 289). On continental margins the volcanoes of South, Central and North America face westward and do not show the same marked curvature. According to Williams and McBirney (*op cit.*) 'this asymmetry of east-facing island arcs and west-facing active continental margins on opposite sides of spreading axes suggests that their structure is in some way influenced by directional forces, such as those derived from the earth's rotation', however Toksöz (1980, p. 130) argues that the Andes are 'merely an overgrown island arc' and simply reflective of differences in geometry, age and scale of subduction. This is an important point because in examining plate tectonics and volcanoes emphasis is placed on present-day spatial arrangements where valid distinctions may be drawn between, say, island arcs of different types. This can give a misleading impression of stability, yet over the wider canvas of geological time 'many arcs are continuous across a wide range of crustal types . . . (and) belong to a continuum. Arcs are not steady-state tectonic systems but instead evolve and change complexly and rapidly, and different parts of a single, continuous arc can have grossly different histories and characteristics' (Hamilton 1988, p. 1503; see also Honza 1983).

2.5.1 Continental margin volcanism

The principal features of this class of volcanism are shown on Fig. 2.8. Some features are either self-explanatory or have been mentioned before, but others require elaboration. Deposits from ocean floors, portions of oceanic lithosphere—*ophiolites*—and sediments from the land are forced against continental margins to form complexes which are folded and metamorphosed (i.e. changed in texture and mineralogy) under conditions of high pressure and low temperature. The French word *mélange*, meaning to mix or blend, is used to describe these rock suites and, because continental margins are typified by mountain building and crustal buoyancy, it is common for *mélange* to be uplifted many metres above sea level even though metamorphism occurred at depths of up to 30 km (Silver and Beutner 1980). Inland from *mélange*, continental margin volcanism is dominated by magma ascending from subduction zones. Melts are generated at depths of between ~80 to 200 km and produce not only andesite predominant volcanic

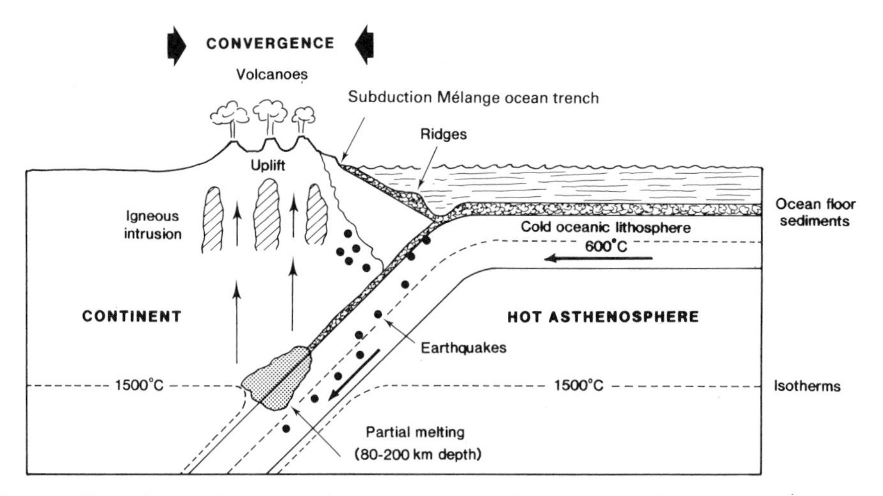

Fig. 2.8 The principal features of continental margin volcanism. The diagram shows a cross-section through this type of subduction zone. The diagram is not to scale. (Based on several sources including Clapperton 1977; Press and Siever 1986; and Achache 1987.)

products but also massive granitic intrusions. Crustal rocks are metamorphosed under high temperature, but low pressure, conditions and lifted up to form mountain belts. Igneous intrusions contribute to thickening of the crust and in parts of the Andes the crust is estimated to be ~70 km, heavily faulted and folded (Summerfield 1991). The presence of such thick crusts along continental margins may well have important implications in understanding the generation of andesitic melts (Davidson *et al.* 1990; see Chapter 3, sections 3.3 and 3.4).

In any marginal continental region the number of volcanoes, their spacing and activity may be explained in part by differences in the geometry of subduction (see Guffanti and Weaver 1988). In the Central Andes de Silva and Francis (1991) were able to show that, whereas in southern Peru volcanoes define a narrow belt some ~250–300 km east of the ocean trench—implying a deeply dipping subduction zone—further south the zone is wider and may suggest a shallower dip. It is not only the geometry of subduction which is of importance, and de Silva and Francis (1991) for the central Andes and Guffanti and Weaver (1988) for the Cascades, argue cogently that crustal structure and regional tectonic setting may also be relevant in accounting for spatial and temporal patterns of activity. Unravelling the volcanic evolution of continental margins is difficult, because of the complexity of processes involved: it requires study at a detailed scale and the application of considerable geological aptitude (e.g. de Silva 1989).

2.5.2 Island-arc volcanism

A generalized section through a typical island arc is shown in Fig. 2.9. Writers

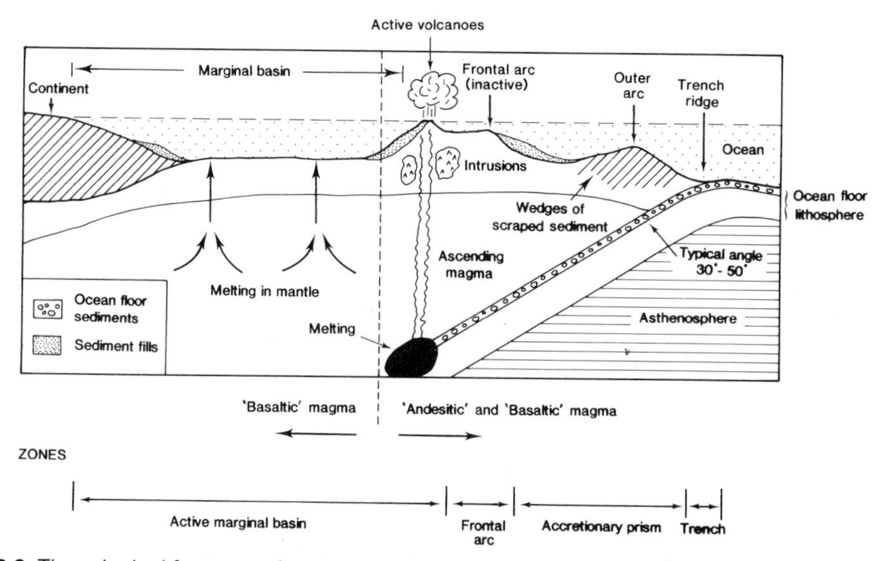

Fig. 2.9 The principal features of and terminology used to describe island-arc volcanism. The diagram is not to scale, but it should be noted that situation may be more complicated because to the landward of active marginal basins there may be 'remnant arc zones' and beyond these 'inactive (or retro-arc) basins' (based on numerous sources).

have used a varied assortment of terms to describe arcs, but adopting those employed by Karig (1974), it is possible to conceive of arcs as consisting of four zones. From seaward to landward they are: *ocean trenches*; *accretionary prisms*; *frontal arcs*; and *active marginal basins* (Table 2.6). As Fig. 2.9 implies arcs represent a sequence of splitting, rifting and oceanward migration of frontal arcs, following small scale sea-floor spreading within active marginal basins. In certain arcs two additional zones may be recognized: *remnant arcs* and *inactive marginal basins*. These are relict features again formed by splitting of the arc and are found ~500 km inland from ocean trenches.

Three subtypes of island-arc may be recognized, based on tectonic and volcanological differences, but all belonging to a continuum and showing similarities to the general model depicted in Table 2.6 and Fig. 2.9.

Table 2.6: The zones commonly found in island arcs and their principal attributes (compiled from various sources)

Zones	Attributes
Ocean trench	This is normally a well-defined feature, the Marianas trench being up to 11 km in depth.
Accretionary prism	This may be either well developed, not developed in full or virtually absent. Normally two subzones are recognized. The first immediately inland from the trench, consists of sediment scraped off the descending oceanic slab and ophiolite, which is commonly metamorphosed to form a tectonic melange. Frequently it produces a ridge (outer arc ridge) which is below sea level, though there are cases such as Barbados (West Indies) where parts protrude above the ocean surface. The second zone, known as a fore-arc basin, occurs between the outer and frontal arcs and may be partly filled with sediment.
Frontal arc	This is an inactive volcanic zone, which is formed of both granitic intrusions and effusive volcanic products. It represents an earlier phase of volcanic activity, before rifting occurred, within the active marginal basin. It may be below sea level.
Active marginal basin	This comprises two subzones. Immediately inland from the frontal arc, is a zone of active volcanoes, only some of which rise above sea level. The type of magma is variable, but in some arcs is dominated by andesite and basaltic andesite (see text). The second zone is an active basin; sometimes known as a back-arc basin. This is an area of little sediment cover, high heat flow and an elevated centre. It is probably formed through rifting by a process known as back-arc spreading, which is analogous to oceanic sea-floor spreading.
Remnant (or third arc)	This zone is commonly absent. As its name implies, it is inactive and represents older volcanic activity. It is detached from the frontal arc because of back-arc spreading. It is a submarine feature.
Inactive marginal basin	Sometimes found to the landward of the remnant arc, this submarine feature is also known as a retro-arc.

2.5.2.1 Young arcs

Typified by the Tonga-Kermadec and Mariana Islands in the Pacific, young arcs have similar characteristics to those shown in Table 2.6 and Fig. 2.9 and comprise: ocean trenches; often, but not always, accretionary prisms; frontal arcs; active marginal basins and, sometimes, remnant arcs. The zone of active volcanism is narrow (i.e. ~50 km), many volcanic edifices are submarine and there is a marked parallelism with ocean trenches, which may cause surface patterns to range from straight to arcuate. Because this type of arc is geologically young and the plates involved oceanic, melts do not have to pass through either continental crust or 'thick terrigenous sedimentary rocks proxying for such crust' and melts are predominantly basaltic and of restricted volume (Hamilton 1988, p. 1518). They are similar to ocean-floor tholeiites, but differ from them in their detailed geochemistry and mineralogy (Gill *et al.* 1984). As well as basaltic materials, a magnesian-rich andesite (boninite) and even more silica-rich melts are sometimes erupted.

2.5.2.2 Micro-continental arcs

The main difference between young arc volcanism and that found in areas such as Japan, Indonesia and New Zealand is related to the nature of magmas produced. Whereas in the former case they are predominantly silica-poor, in the latter silica-rich andesites and dacites are more common (Cole 1979, 1984). In Chapter 3 it will be shown that andesites and dacites are associated with violent explosive volcanism and have profound effects on both the prediction of (Chapter 7) and responses to hazards (Chapter 9). The fundamental reason why andesites and dacites are common is because of the 'maturity' of micro-continental arcs. These are older features and as such there is more material above the subducting slab, through which melt has to pass (Hamilton 1988). Magmas are variable both through time and over space. For instance, just as the thickness of 'crustal' materials increases inland from ocean trenches (Fig. 2.9), so also the character of magmas changes from tholeiite to andesite and dacite. Also the active volcanic belt may change its relative position above the trench as an arc develops. Finally, different parts of an arc may be at different stages in their evolution. The East Indonesian arc, for example, is less 'mature' and more basaltic than the arc near to Java and Sumatra (Honza 1983).

2.5.2.3 Marginal sea back-arc and inter-arc basins

The processes involved in producing these features have already been discussed in section 2.5.2 (see Fig. 2.9 and Table 2.6) and will be expanded upon in Chapter 3. The products of marginal basins are basalts, which are similar to mid-ocean ridge tholeiites. However, they are more accurately described as being transitional between mid-ocean ridge and island-arc tholeiites.

2.6 Conclusion

This brief account provides a summary of the relationships between magma generation and plate tectonics. In preparing the chapter the author was struck by how convincing the model of plate tectonics is in explaining spatial patterns of volcanic activity and how few are the areas, both scientific and geographical, where anomalies still remain to be explained. Whilst not agreeing with those who reject the conventional wisdom (e.g. Lyttleton 1982) or even major parts of it (e.g.

Carey 1988), it is important to reiterate that plate tectonics is a model and as such may be modified in the future. It is sobering to recall that just over 40 years ago it was stated with confidence that 'everything that is now known concerning the configuration and structure of the floors of the oceans proves conclusively that Wegener's hypothesis of continental drift is wholly untenable' (Bucher 1950, p. 40).

3

The generation of magma and the eruption of volcanoes

It is useful to be assured that the heavings of the earth are not the work of angry deities. These phenomena have causes of their own (Seneca 4 BC–65 AD).

3.1 Introduction

Generation of magma, its movement to the surface and eruption are themes which lie at the heart of volcanology. Study of magma not only provides a 'window' through which processes operating within the Earth may be viewed, but magma also controls directly and in combination with other factors the geomorphology of volcanic regions and the hazardousness of place. Whereas it is possible to argue a case in favour of the proposition that the history of research into factors which control the geomorphology of volcanoes, whether they are for instance *shields* or *strato* constructs, has been marked by steady progress as more observational data have become available so allowing existing models to be refined and/or discarded, a similar case cannot be sustained for the other themes of this chapter. Indeed, generally accepted views about the generation of magma, its movement to the surface and eruption have been subject to frequent changes.

Until the end of the last century the interior of the Earth was *terra incognito*, inaccessible to study, inducing speculation and providing a theme for writers of science fiction including Jules Verne whose well known novel—*A Voyage to the Centre of the Earth*—first appeared in 1863 and was based on the assumption that the interior of the planet was riddled with passages. Today investigations are more prosaic. Geophysical studies of materials in the laboratory supposedly analogous to those of the crust and mantle, and mathematical modelling have replaced imaginative prose, but a degree of uncertainty nevertheless remains. The Earth's interior can only be studied indirectly and this produces models that are more conjectural than those based on direct observation and measurement (see Bolt 1982, pp. 13–21). Because analogy and inference are involved, data may be interpreted in a number of ways and models derived from them are often in conflict; though apparently in accord with logic and the laws of physics. Clearly as more data are collected, the range over which speculation may range is reduced, but major shifts in the balance of scientific opinion over relatively short periods of time are still commonplace.

Some books (e.g. Middlemost 1985) place volcanology within the context of a wider discussion of igneous petrology. This is not the aim of the present volume, which focuses on the products of volcanoes, their effects on the physical environment and upon human activities. It is the case, however, that any appreciation of the issues involved in magma generation and eruption requires some knowledge of igneous petrology, and for this reason section 3.2 is devoted to this theme.

3.2 Igneous rocks

There are many excellent books which deal in detail with igneous processes and rocks (e.g. Hatch *et al.* 1972; Cox *et al.* 1979; Middlemost 1985; M. Wilson 1989). Moorhouse (1970) and MacKenzie *et al.* (1982) are high-quality atlases of rock textures and reference should be made to them if expansion and illustration of the material in this section is required. Those readers who have a working knowledge of igneous petrology can safely skip what follows.

Igneous petrology is a science that contains a profusion of technical terms, some of which are misleading, but are retained because they are long established and widely understood by the *cognoscenti.* The terms 'acid' and 'basic' are used, for instance, to describe rocks with respectively high and low percentages of silica. They were coined in the nineteenth century when silica was thought to form from silicic acid. Table 3.1 is a brief glossary.

Volcanic rocks are cooled and solidified magmas, chemically they contain ten or more elements—including oxygen, silicon, aluminium, iron and magnesium—which combine to produce complex melts in which silicate and oxide compounds predominate (Macdonald 1972). In most igneous rocks the chemical constituents usually occur as distinctive minerals, or phases, but in some rocks glass may be the major—in a few the only—component (Table 3.1). Igneous rocks have textures which are defined by the size, shape and relationships between crystals. Textures are determined by the nature of the magma from which crystals are derived and its cooling history. Sometimes a rock has coarser crystals, set in a finer matrix. Coarse crystals, known as *phenocrysts*, may be important in interpreting the cooling history of a melt and its possible origin.

Three criteria of primary, or first order, classification are used by petrologists and are based on differences in texture, chemical composition and mineralogy (Fig. 3.1). There is a distinction between extrusive (or volcanic) rocks produced by lava flows (Chapter 4) and/or pyroclastic materials (Chapter 5), and those intruded below the ground surface. The latter, also known as *plutonic* from Pluto the Roman god of the netherworld, normally cool slowly so allowing large crystals to form. Such coarse-textured rocks are termed *phaneritic.* In contrast volcanic rocks exposed at the Earth's surface, cool rapidly and have fine textures, which may be invisible to the naked eye. They are termed *aphanitic.* Since the nineteenth century the amount of silica (SiO_2) has been known to be an important chemical variable and rocks are commonly described as being *acidic*, *intermediate*, *basic* and *ultrabasic.* Today the terms silica-rich (*acidic* or *silicic*) and silica-poor (*basic*) are sometimes used, but more commonly rocks are described as *felsic* or *mafic*; a distinction based on the occurrence of light and dark minerals. Felsic rocks are rich in quartz and feldspars giving a light appearance, whereas mafics are dark, dominated by pyroxenes, amphiboles and olivines (all of which are rich in magnesium and iron). Although felsic rocks are usually acidic and mafic rocks basic, the terms are not synonymous (see Cas and Wright 1987, p. 17). The final criterion of classification is mineralogical and again this relates to

Table 3.1: A glossary of terms commonly used in igneous petrology (based on the references cited)

Term	Definition
Mineral (phase)	'A natural occurring, solid, inorganic element or compound, with a definite composition or range of compositions, usually possessing a regular internal crystalline structure' (Press and Siever 1986, p. 638).
Volcanic glass	A rock formed from mixtures of silicates that have had insufficient time for crystals to form. This may occur through either rapid cooling (quenching), and/or because the melt was very viscous.
Texture	'The size, shape and fabric [pattern of arrangement] of particles that form a rock' (Fisher and Schmincke 1984, p. 116).
Alkalic (alkaline) rocks	Igneous rocks that are rich in sodium and potassium.
Phaneritic	A textural term applied to igneous rocks. Crystals are visible without magnification. Plutonic rocks are commonly phaneritic.
Aphanitic	A textural term applied to igneous rocks. Crystals are visible only with the aid of magnification. Volcanic rocks are often aphanitic.
Silica-saturation	A magma is silica-saturated if it crystallizes to form mixtures of olivine, pyroxene and feldspars. If a magma is oversaturated it has excess silica and silica minerals, normally quartz, are present. If a magma is undersaturated, then a different suite of minerals is formed. These may include feldspathoids such as nepheline and leucite (based on Cox *et al.* 1979, p. 69).
Incompatible element	An incompatible element is one which, for a given composition of melt, is excluded from the minerals being crystallized and becomes concentrated in the liquid (Cox *et al.* 1979, p. 31).

chemical composition, different melts being associated with distinctive suites of minerals. In the past some 1500 names have been used to describe igneous rocks (Le Bas and Streckeisen 1991), but applying the three criteria of texture, chemical composition and mineralogy a small number of major groups emerge (Fig. 3.1).

Looking at volcanic rocks in more detail it is apparent that more varieties are found in nature than can be accommodated on an axis of silica percentage. In particular it is known that, in addition to varying amounts of silica, potassium, sodium and other elements may influence greatly the mineralogy of volcanic rocks. There are no natural boundaries between rocks derived from silicate melts; they represent a continuous series, and it is not surprising that many systems of classification have been suggested over the years with varying degrees of success (see Le Bas and Streckeisen 1991). The question of classification has been considered for more than 20 years by a subcommission of the International Union of Geological Sciences (IUGS) and a full record of its proceedings is to be found in Le Maitre *et al.* (1989) and a summary in Le Bas and Streckeisen (1991). The classification proposed by IUGS (Fig. 3.2) is one of simplicity and convenience. It adopts a procedure which has been in use for many years of defining major

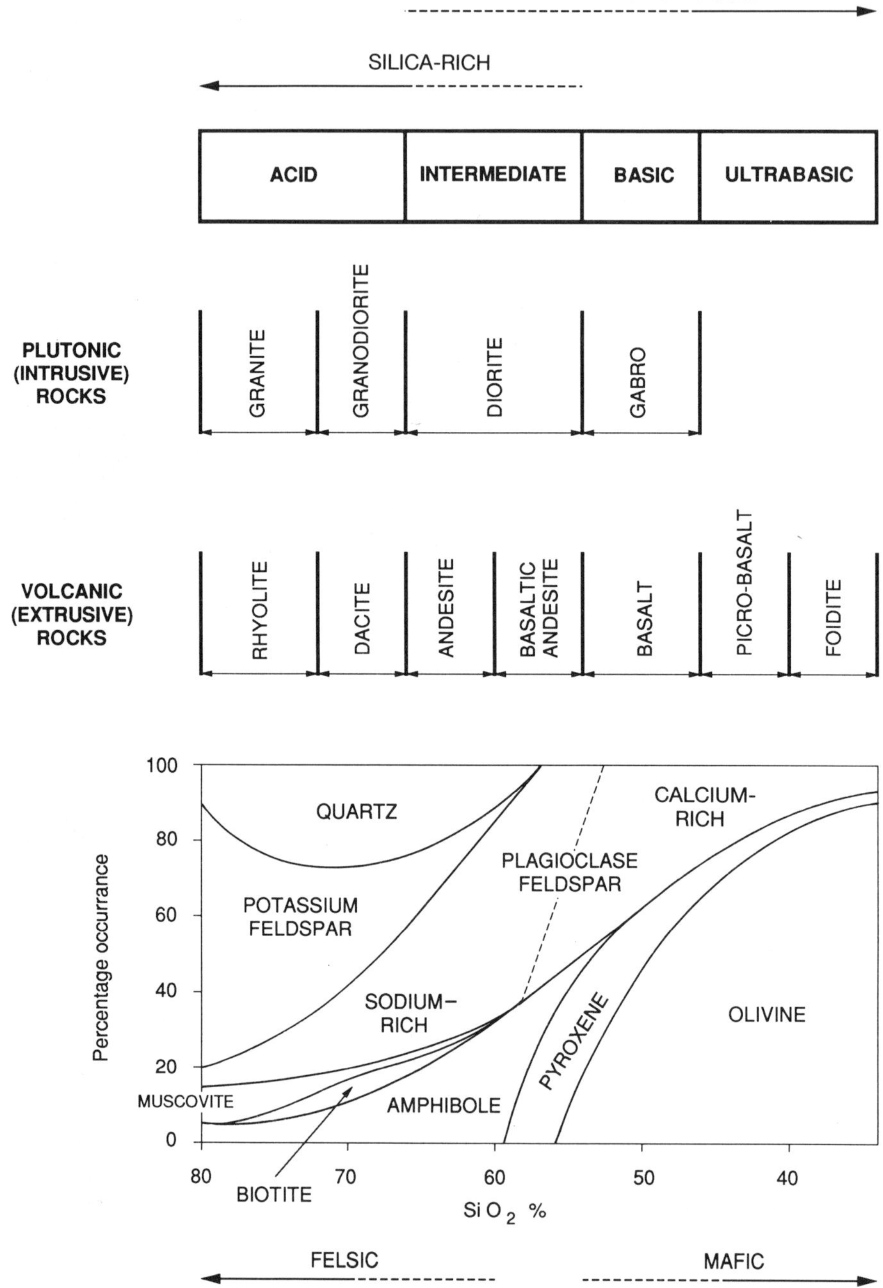

Fig. 3.1 A simple classification of igneous rocks based on the criteria of texture, chemical composition and mineralogy. It should be noted that the correspondence between plutonic and volcanic rocks is approximate (based on a number of sources including: Macdonald 1972; Press and Siever 1986; Le Bas and Streckeisen 1991; and Summerfield 1991).

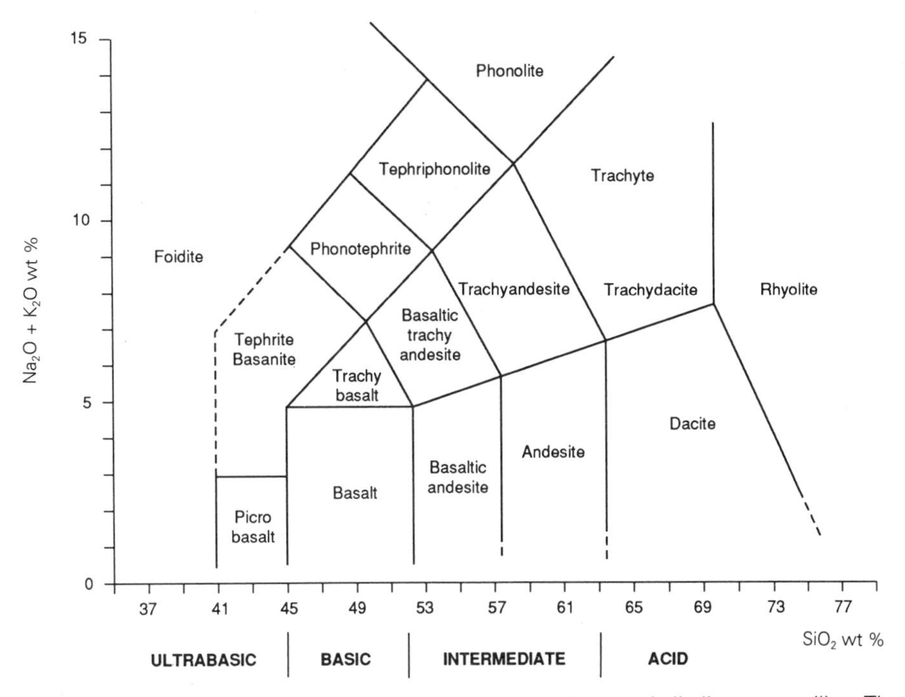

Fig. 3.2 The IUGS classification of volcanic rocks based on total alkalis versus silica. The line between foidite and basanite is dashed because other criteria are required to distinguish between these rock types (from Le Bas and Streckeisen 1991, Fig. 5, p. 830).

volcanic rocks on plots of total alkalis ($Na_2O + K_2O$ wt wt%) against silica (SiO_2). Fig. 3.2 is used throughout the rest of the chapter and a summary of the minerals found in the more common volcanic rocks is provided in Table 3.2.

Relating different types of magma to tectonic setting (Chapter 2, sections 2.3 to 2.5; and Table 3.3) is an alternative way of classifying igneous rocks (Middlemost 1985, p. 73). One example is Ringwood's (1974) term *orogenic volcanic series*, to describe rocks generated above subduction zones and whose members range from basalt to rhyolite, but with andesite, basaltic andesite and dacite predominating. *Calc-alkaline* is another example of a 'family name' and is often used interchangeably with the terms *orogenic volcanic series* and *orogenic suite* (Macdonald 1972 pp. 364–5).

Basalts are also commonly subdivided. Although *tholeiites* are predominantly rocks of ocean ridges and floors (Chapter 2), they are also encountered at subduction zones and in oceanic/continental intraplate situations (Table 3.3). Tholeiite is a term which has changed its definition over the years (Hatch *et al.* 1972) and, indeed, some petrologists have doubted its usefulness (e.g. Chayes 1966). It continues to be used widely (e.g. Basaltic Volcanism Study Project 1981), nevertheless, to describe a range of rocks which contain significant amounts of pyroxene usually in the form of hypersthene and/or pigeonite. Virtually all tholeiites are basalts in terms of their chemical compositions, but some are intermediate (Fig. 3.1). Ocean-ridge and floor tholeiites are notable for their low concentrations of 'incompatible elements' (Cox *et al.* 1979, p. 31; and

Table 3.2: The principal minerals found in some major volcanic rocks as shown on Fig. 3.2 (based on information in Le Maitre 1976; Condie 1982; and other sources)

Rock type	SiO_2 (%)	Na_2O (%)	K_2O (%)	Principal minerals in decreasing order of abundance[1]
Basalt	50	3	1	Plagioclase feldspar, pyroxene ± olivine.
Andesite	57	3	2	Plagioclase feldspar, pyroxene, amphibole.
Phonolite	57	8	5	Potassium feldspar, nepheline, plagioclase feldspar, ±amphibole, ±pyroxene.
Trachyte	63	6	5	Potassium feldspar, plagioclase feldspar ± biotite ± amphibole.
Dacite	66	4	2	Plagioclase feldspar, pyroxene, quartz, amphibole, biotite.
Rhyolite	74	4	4	Potassium feldspar, quartz, plagioclase, biotite.

[1] These are common orders of abundances and vary with different forms of these rocks. For instance the information in the table is for tholeiitic basalt, alkali basalts have typical orders of abundance of: plagioclase feldspar, olivine, pyroxene ± feldspathoids such as nepheline and leucite.

Table 3.3: A 'tectonic' classification of igneous rocks

Plate margin		Intraplate	
Subduction (convergence)	Ocean ridge (divergent)	Oceanic	Continental (divergence)
Calc-alkaline Tholeiite	Tholeiite	Alkalic basalt Tholeiite	Alkalic basalt Tholeiite

Table 3.1). These include potassium (K) and phosphorus (P) among the major elements, and rubidium (Rb), zirconium (Zr) and uranium (U) among the trace elements (Fisher and Schmincke 1984, p. 19). Tholeiites generated at island-arcs are more silica-rich, whereas those produced in regions of continental flood volcanism are more alkalic (see below) and richer in titanium (Ti) and phosphorous (P). The tholeittes of Hawaii and other oceanic islands contrast with those found on ocean floors. They are enriched in incompatible elements and have dissimilar mineral assemblages.

Alkali basalts are common in both oceanic and continental intraplate volcanic provinces (Table 3.3). They are more silica-undersaturated (see Table 3.1) than tholeittes, are enriched in volatiles (gases) and in incompatible elements. In contrast to tholeiites, olivine is very common within the matrices of many of these rocks.

Before leaving the topic of igneous rocks brief reference should be made to the *komatiitic magmatic suite* (Viljoen and Viljoen 1982). Komaiites are notable for several distinctive features. Virtually all are of Archean age and have percentages of MgO which exceed 18 wt %, and may reach 33 wt %. The origin of komatiites is discussed in Chapter 4, section 4.5).

3.3 The generation and movement of magma

As mentioned at the start of the chapter, consensus views on the processes responsible for the generation and movement of magma have been subject to change over the past century. Nevertheless the period from around 1970, the time when plate tectonics became accepted by the majority of earth scientists, and the mid-1980s saw the emergence of a distinctive paradigm, consisting of an interlocking set of mutually reinforcing models. Models were developed from investigations of erupted products, laboratory experiments and geophysical studies and until the middle of the 1980s the paradigm, although added to and modified, remained largely unchallenged (section 3.3.1). From around 1984/85 a threat has been posed to some aspects of the post plate-tectonics consensus by the publication of a number of important research papers (see section 3.3.2). The lasting impact of these papers may, however, only be judged with the passage of time and the wisdom of hindsight.

3.3.1 The plate-tectonics consensus

The consensus was based on accumulated insights from hundreds of individual pieces of empirical research and synthesis. The account which follows draws upon a subset of the literature and includes sources which the author has found particularly useful. These are: Ringwood (1969, 1974, 1975); Wyllie (1971); Carmichael *et al.* (1974); Yoder (1976); Basaltic Volcanism Study Project (1981); Gill (1981); Best (1982); Marsh (1982, 1984); Watson (1982); McKenzie (1985) and Thompson (1984). Reference should be made to these works if a more detailed treatment is required.

Melting of parts of the Earth's interior to produce magma is complex and involves a number of variables. For over a century and beginning with studies carried out in deep mines, it has been known that temperatures within the Earth increase with depth. All terrestrial planets of the Solar System have been volcanically active in the past and it has been hypothesized (Lowman 1976) that differential cooling has been a major, according to some workers the principal, process responsible for the timing and style of activity (Chapter 1, section 1.3.5). Research has focused on attempts to pinpoint the source of planetary heat. Over the years a number of mechanisms of heat production have been posited (Clapperton 1977) for both the terrestrial planets in general and the Earth in particular. These have included:

1) gravitational energy converted to heat during the formation of planetary cores;
2) energy generated by tidal movement of the Earth's crust under the gravitational attraction of other planetary bodies (Shaw 1970);
3) frictional energy produced by differential movements between the mantle and outer core, and the outer and inner cores;
4) heat generated by meteoric impact on planetary surfaces; and
5) the decay of radioactive isotopes (of K, U and Th) found within minerals.

Although some processes may have been important in the early stages of planetary evolution (i.e. 1 and 4), others are thought to be of relatively minor significance (i.e. 2 and 3) and it is generally accepted that radioactivity is the most important contemporary mechanism on Earth.

If a graph of temperature versus depth (a geotherm) is constructed, based on surface and near-surface measurements, its projection implies a temperature of around 25,000°C at the centre of the Earth (Press and Siever 1986). This is clearly a gross overestimate as such a temperature would render most of the Earth molten. Radioactive elements are concentrated in the outer layers of the Earth and it is assumed that below this zone the rate of increase in temperature declines. As well as making estimates of temperature, geophysicists use seismology to obtain additional information about the Earth's interior. Such information includes the division of the Earth into a number of shells or layers based on the refraction and/or reflection of elastic waves (Chapter 2, Table 2.1), estimates of the rate of increase in pressure with depth and, within broad limits, changes in density (Press 1972). Melting is strongly influenced by pressure and density. Work in the laboratory and measurements in the field, show that the melting point of basalt at sea level is between 1000 and 1200°C (Cas and Wright 1987). Melting point is raised by pressure, and hence by depth.

Synthesizing allows a melting curve to be constructed showing the temperature required to melt rocks at increasing pressures and depths. At point 'X' (Fig. 3.3)

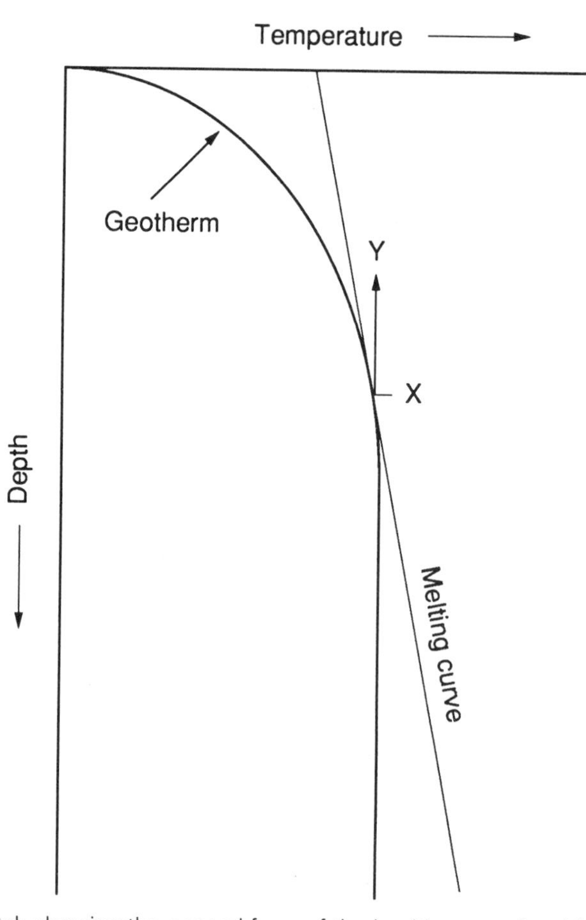

Fig. 3.3 Sketch showing the general form of the 'melting curve' and 'geotherm' (based upon Williams and McBirney 1979, Fig. 3.2, p. 40).

conditions are suitable for melting and there are three ways in which this can occur (Williams and McBirney 1979, pp. 40–42). First, heat may be added by external agencies such as decay of the radioactive isotopes of K, U and Th and frictional heating along subduction zones. Secondly, there can be a reduction in pressure, due to convection in the mantle and/or uplift and erosion of rocks at the surface. This will move the rocks from point 'X' to point 'Y' (Fig. 3.3)—i.e. up the adiabatic gradient—and magma will start to form through heat released from crystals. No additional heat source is required. Finally melting temperatures may be reduced by the presence of volatiles such as H_2O (Yoder 1976). Although debates about the relative importance of the ways in which melts form were common in the 1970s, ten years later few workers would disagree with L. Wilson *et al.* (1987, p. 74) that 'volumetrically, most melts are probably caused by adiabatic decompression of rocks moving to lower pressure sites. . . . Other . . . mechanisms include direct heating of rocks being carried . . . into subduction zones . . . ; volatile-induced melting . . . ; and local concentrations of radiogenic heat production'. Also by the mid-1980s it was held that melting originated along the edges of individual grain and their points of contact (Watson 1982). Melt infiltration occurs along grain-edges when the angle between crystal faces, the *dihedral angle*, is less than a critical value of ~60°. Laboratory experiments imply values of ~50° for partially molten basalts and 44–49° for more silica-rich melts (Watson 1982; Jurewicz and Watson 1984).

It is necessary briefly to consider trace elements and radio-isotopes, both of which have helped reveal the origin of magmas and their distribution with respect to tectonic province. Trace elements comprise less than 0.1% by weight of a melt. Their incorporation into minerals is known to vary not only with chemical composition and crystal structure, but also in relation to conditions of temperature and pressure in regions of melt production. It is because of the latter that they are of value to petrologists, since they can help unravel many problems of magma genesis. Barium, zirconium, rubidium, strontium and the rare earths[1] are very important, with rubidium and strontium often 'remain(ing) in a melt when the olivine and pyroxene of a peridotite are crystallised. . . . On the other hand, nickel fits easily into these mafic mineral structures and so tends to be incorporated in the peridotite' (Press and Siever 1986, p. 390). Trace-element concentrations in melts may be up to 1000 times greater than those in residual peridotites (Fig. 3.4).

An element may exist in several forms, each having the same configuration of electrons, and thus similar chemical properties, but diverse atomic weights because of a different number of protons (Middlemost 1985, p. 45). Some isotopes are radioactive and decay over time, others are stable, but for students of magma generation certain radiogenic isotopes have proved especially valuable in answering questions about how magmas form and why certain magmas are formed in particular tectonic situations. The ratio of strontium 87 (^{87}Sr), which is formed from the radioactive decay of ^{87}Rb, to the stable isotope ^{86}Sr is contingent upon the original Rb content and age of the sample. Analyses of samples from magmas erupted in different tectonic provinces give dissimilar $^{87}Sr/^{86}Sr$ ratios (see Condie 1982, p. 132) and, once corrected for age, they can be interpreted in a number of ways. The recognition that the mantle is inhomogeneous, that certain

[1] The 'rare earths' are metallic elements with atomic numbers 57 to 71.

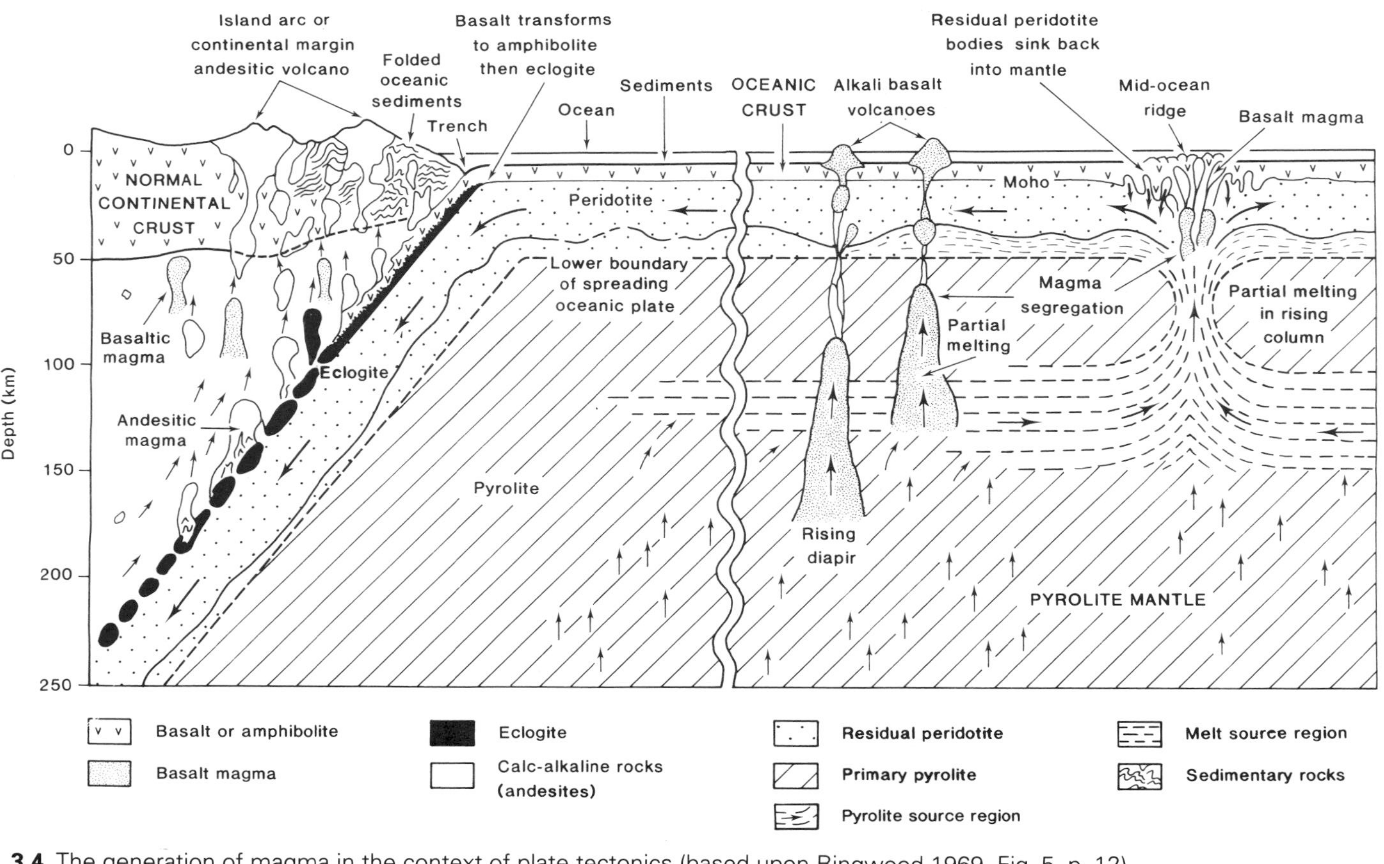

Fig. 3.4 The generation of magma in the context of plate tectonics (based upon Ringwood 1969, Fig. 5, p. 12).

magmas are derived from melts that have not been altered significantly during their movement to the surface, that some magmas are, conversely, the products of mixing and others reflective of interactions with crustal materials are insights which come from, or are supported by, isotope studies.

Figure 3.4, taken from the work of A. E. Ringwood (1969, 1974), provides an excellent summary of the processes of magma generation and movement, as they were understood following general acceptance of plate tectonics. Not surprisingly the diagram has been reproduced many times. Looking at the diagram in more detail, it is clear that major magma groups are the same as those which emerge on petrological grounds (section 3.2): tholeiites at ridges; alkalic basalts in intraplate situations and members of the calc-alkaline suite above subduction zones. To these may be added tholeiites associated with marginal basins.

Unfortunately things are not so simple as a cursory reading of Fig. 3.4 might suggest. Magmas appear to rise to the surface without undergoing any change. Such magmas are sometimes termed 'primary', but this is a misnomer since very few, if any, erupted magmas are unchanged. A more detailed examination of the figure shows that changes—'segregation' (*sic*)—during transit may occur and that this is due to cooling, chemical reactions, mixing with other magmas and interactions with crustal rocks during transit (Fisher and Schmincke 1984). *Parental* is a better term than primary and magmas produced by differentiation from, and the mixing of, parental magmas are referred to as 'derivative'. Magma generation, therefore, involves two aspects: the production of parental magmas and their evolution to more evolved compositions during transit.

3.3.1.1 Parental magmas

The consensus of the 1970s and early 1980s held that basalts were derived from partial melting in the upper mantle, through a combination of: adiabatic decompression; lowering of melting temperatures by volatiles; liberation of heat from radiogenic sources and the upwelling of mantle plumes (L. Wilson *et al.* 1987). An important link was established between the occurrence of volatiles and the amount of melting. The relative abundance of volatiles is controlled by the mantle source and the degree of partial melting. Volatiles lower melting temperatures, but large amounts of melt produced by other mechanisms will cause volatile dilution and this may reduce the explosiveness of subsequent eruptions.

When it is formed a melt must be in equilibrium with prevailing conditions of temperature and pressure. Profiles of temperature and pressure with depth imply that the rocks of the upper mantle are ultramafic (i.e. ultrabasic, see Fig. 3.2), rich in iron and magnesium and having mineral assemblages dominated by olivine, pyroxene and garnet.

Chemical data and experimental findings imply that ocean-ridge tholeiites are produced at shallow depths. These vary from 20–40 km (or even less) at one extreme, to 60–70 km at the other (Green *et al.* 1979); though relatively little melting is believed to occur below 30 km (Middlemost 1985, p. 44). Because depths are shallow, pressure is relatively low and melting may involve some 20–30% of the source material. Because of the amount of melting, volatiles will be diluted and low volatile content explains in part the lack of explosiveness of most tholeiitic eruptions. Melting takes place above ascending limbs of convection cells and removes, selectively, those elements required to form tholeiites. This leaves a residuum of depleted-source rock and in the 'classic model' (Fig. 3.4) a layered oceanic lithosphere is produced. This comprises ocean-floor tholeiite above 'depleted' peridotite. At greater depths *pyrolite* is inferred. *Pyrolite* is a term introduced into petrology by Ringwood (1962) to describe the hypothetical composition of material from which tholeiite melts are derived. Its composition is estimated by 'mixing' depleted peridotite and the magmatic material derived from it. Depletion was claimed to have occurred not only through the formation of tholeitte, but also by the extraction of crustal materials during the early history of the Earth.

Alkalic basalt, prominent on ocean islands like Hawaii and at continental rifts, is associated with 'hot spots' at the surface and is generated under different conditions. Geophysical, geochemical and laboratory studies imply limited partial melting (around 5–10%), at greater depths (around 80–150 km; Fig. 3.4), than those where ocean-ridge tholeiites form. Higher temperatures and pressures impart distinct 'signatures' to the chemistry and mineralogy of melts (Table 3.2). Limited partial melting is one reason why eruptions involving alkalic basalt may be more explosive. Because mantle plumes are derived from a lower portion of the mantle than those associated with mid-ocean ridges, it has generally been assumed that ocean-island alkalic basalts are the more primitive, since they sample material 'which does not participate in the plate-tectonic cycle, and thus retains its primordial composition' (Garvin 1991, p. 699). This view was supported by isotope studies. The isotope ^{3}He, for instance, is neither produced by radioactive decay nor recycled from the Earth's surface to its interior and is assumed to have survived since the time the Earth formed. It is common in certain ocean-island basalts.

Calc-alkaline magmas are produced by processes which contrast markedly with those forming tholeiite and alkalic basalt. Magmas are not produced in the upper mantle in the manner already discussed and there is general agreement that the primary process is one in which lithospheric material is carried down a subduction zone. When conditions of temperature and pressure are appropriate, a melt is produced (Clapperton 1977). Disagreements emerge when the processes of melt formation are examined in more detail. Some researchers hold that andesites are produced directly by partial melting of rocks such as peridotite (e.g. Kushiro 1974), but more petrologically complex models have proved the more popular. The model proposed by Ringwood (1969, 1974, 1975; see Fig. 3.4) deals with the issue of why there exists an evolutionary sequence of petrologically varied magmas at different stages in the development of subduction zones (see Chapter 2, section 2.5): the tholeiites of young arcs giving way to a predominance of more evolved melts in 'mature' arcs and along continental margins (Table 3.4).

In a review of calc-alkaline magma genesis, Fisher and Schmincke (1984, p. 20) argued that andesites may be produced by several processes, but the consensus of the time emphasized 'differentiation from basaltic magmas modified by fluids from the subducted slabs and locally by partial crustal melts'. The realization that water was important came from experimental studies such as those carried out by Eggler (1972) and analyses of volcanic glass (A. T. Anderson 1974). Fisher and Schmincke (*op. cit.*) concluded their review by raising the question of 'whether the explosive nature of many andesite eruptions is due mainly to high primary volatile content, high viscosity, or the interaction of magma with external water' (see Chapter 5, section 5.3.6.2). At the time of writing they could not give a definite answer.

Table 3.4: Ringwood's model for the production of parental magmas at subduction zones. The model emphasizes the role of water and its effects on melting. Water may be held in pore fluids, or be bound within hydrous minerals (based on Ringwood 1969, 1974, 1975)

Stages	Petrogenic processes
1	Amphibolite[1] in the oceanic crust becomes dehydrated along subduction zones at depths of 70–100 km.
2	The water liberated allows partial melting of pyrolite in the 'wedge' above the subduction zone.
3	Partial melting occurs under high water-vapour pressures and these conditions produce tholeiites, which are features of young arcs (see Chapter 2, section 2.5.2.1).
4	At greater depths, around 100–150 km, a high water-vapour pressure is maintained by the dehydration of serpentinite[2].
5	Partial melting of eclogite[3] oceanic crust (see Fig. 3.4) produces silica-rich rhyodacite melt.
6	The calc-alkaline suite is derivative (see section 3.3.1.2) and is formed through complex reactions during the upward transit of the rhyodacitic melt. These include reactions with overlying mantle pyrolite (Fig. 3.4), and further changes at depths of around 80–150 km and 30–100 km.

[1] Amphibolite is a metamorphic rock in which amphibole and plagioclase feldspar are dominant.
[2] Serpentinite, a rock composed almost exclusively of serpentine, and is produced by the alteration of minerals such as olivine and pyroxene.
[3] Eclogite is a metamorphic rock produced at very high pressures. Its mineralogy is dominated by garnet and pyroxene.

3.3.1.2 Derived magmas

Once formed a magma will evolve, it will cool, may mix with other rocks and change its characteristics. Many of these changes occur during transit to sites of eruption. One of the major breakthroughs of the 1970s and early 1980s was the development and testing of models of magma transport (e.g. Marsh and Kantha 1978; Marsh 1982, 1984). Once formed a magma will be buoyant in comparison with its surroundings and will rise towards the surface at a rate controlled by density contrasts and the deformation of the rock through which it passes (L. Wilson *et al.* 1987, p. 74). At high pressures (i.e. depths of several kilometres or more), surrounding rocks will act as plastic substances (see Chapter 4, section 4.2) and movement will be *diapiric*. A *diapir* is a discrete body of buoyant magma which moves upwards through the crust and mantle. Diapirs occur: beneath ocean-ridges; below intraplate volcanic belts—both oceanic and continental—and above subduction zones (Fig. 3.4). Another process, which may assist magma movement, is fracturing of the surrounding rock (Spera 1980). The progress of a diapir will be regulated by its rate of heat loss and this will determine whether or not magma reaches the ground surface or forms an intrusion. The earlier passage of diapirs may be crucial. These will heat surrounding rocks so reducing subsequent heat losses (Marsh 1982) and resulting in rates of magma production which are probably between 10 and 1000 times higher than rates of volcanic output (Crisp 1984). Within a diapir parental magma normally changes. Crystals form as the magma cools and, since they are usually heavier than the magma which remains, they will sink through the liquid. The effects are to: increase the buoyancy of the melt; alter its speed of ascent; and cause changes in its composition.

As a diapir approaches the surface, its characteristics change. Because diapiric movement requires plastic deformation of crustal rocks, it cannot be sustained at high levels. Near-surface rocks are under relatively low pressure and do not deform plastically. Magma is 'ponded' in magma chambers. The internal workings of magma chambers are amongst the most important processes in igneous petrology, for it is in these reservoirs that many derived magmas are formed.

Magma chambers may be inferred from a variety of geophysical techniques including: seismology (e.g. Sharp *et al.* 1980); geomagnetic techniques (e.g. Stuart and Johnston 1975); measures of gravity anomalies (e.g. Eggers *et al.* 1976) and precise levelling of the ground surface (e.g. Fiske and Kinoshita 1969). Geological study of extinct volcanoes which have been heavily eroded may also reveal evidence of former magma chambers. Information on the processes operating within magma chambers can be obtained by constructing models based on the analysis of erupted products, but these are clearly inferential. Many pyroclastic *fall* deposits and *ignimbrites* show, for instance, time-progressive changes in mineralogy and chemistry which may relate to the 'compositional zoning' of the magma chambers from which they are derived (e.g. R. L. Smith 1979; Hildreth 1981). Changes in lava geochemistry over the history of a volcano may also hold valuable clues to magma storage and evolution (Guest and Duncan 1981). When all the evidence is put together it emerges that magma chambers:

1) are found at depths of ~3–20 km below the surface ground;
2) have volumes ranging from more than 500 km^3 to less than 30 km^3;
3) are morphologically complicated, three-dimensional entities which, although frequently modelled as spherical, are normally more complex in form (Fig. 3.5);
4) are zones where involved chemical and physical processes take place; and
5) are open systems, into which magma may be added from zones of melting, or removed by eruption (O'Hara and Mathews 1981).

The evolution of derived magmas within magma chambers has been a major focus of petrological research since the early decades of the present century and pioneer work by Norman L. Bowen (1922, 1928) provided the first mechanism to explain how this could occur. Bowen's model, known as *fractional crystallization* (Fig. 3.6), takes as its starting-point a magma which is cooling and crystallizing. Bowen argued that the first crystals to form will use only some of the elements present in the melt and, once formed, will leave the magma depleted in these elements and, thus, changed (Martin 1989). Crystals will sink to the bottom of the chamber. Later a second set of minerals will form, again changing the melt which remains. In the case of many basaltic parental magmas the early minerals to form are mafic ones like olivine and this leaves the residual melt enriched in silicic components. Provided no new magma is added and no eruption takes place, the process will continue until the rock is solidified. Over the years many geologists have supported the model through studies of the geochemistry/mineralogy of igneous intrusions and volcanic products; especially pyroclastic falls and ignimbrites.

Though it was modified and refined over the years (Osborn 1979), it became apparent by the 1970s that fractional crystallization could not on its own account

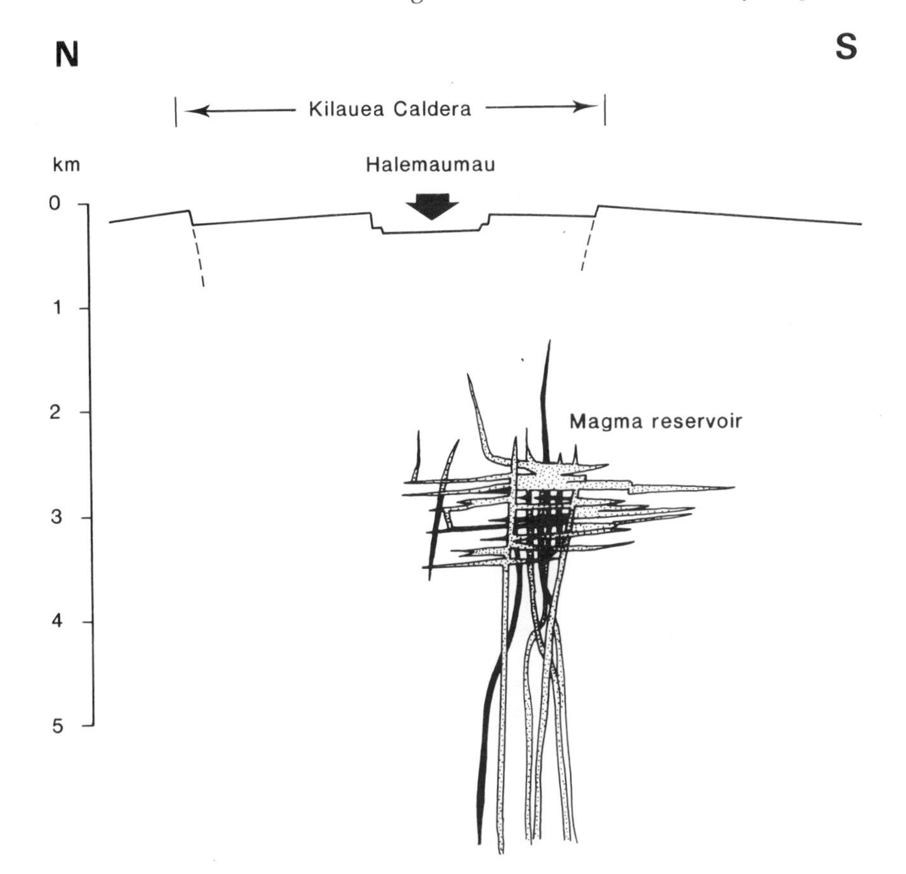

Fig. 3.5 A speculative cross-section of the magma chamber below Kilauea volcano, Hawaii (based on ground-deformation studies from Fiske and Kinoshita 1969, Fig. 8B, p. 347).

for all the changes which must take place within magma chambers. More advanced experimental research and more detailed study of volcanic products, implied that very complex processes were the norm, not the exception, within magma chambers and required additional explanatory models. These models are discussed in considerable detail by Middlemost (1985, pp. 51–70), but briefly they include the notions that: crystals may form, not only in the melt, but also in contact with the floor, walls and roof of the chamber; that magma is not always, or usually, a Newtonian substance, but may have a yield strength (see Chapter 4, section 4.2) so that crystals can float as well as sink; that hot magma may partially melt and assimilate the walls and roof of the chamber; that volatiles may be gained and/or lost to the surrounding rock; that gas bubbles rising through a melt may 'scavenge' other volatile components and so alter the chemical composition of the melt which remains and that major changes in the geochemistry of the chamber may be caused by mixing of old and new magma (McBirney 1980; Huppert and Sparks 1980; Sparks *et al.* 1980). Magma mixing, for instance, was invoked by Sigurdsson and Sparks (1978) to account for a rhyolitic eruption of Askja volcano in Iceland. Finally, it was argued that a special type of

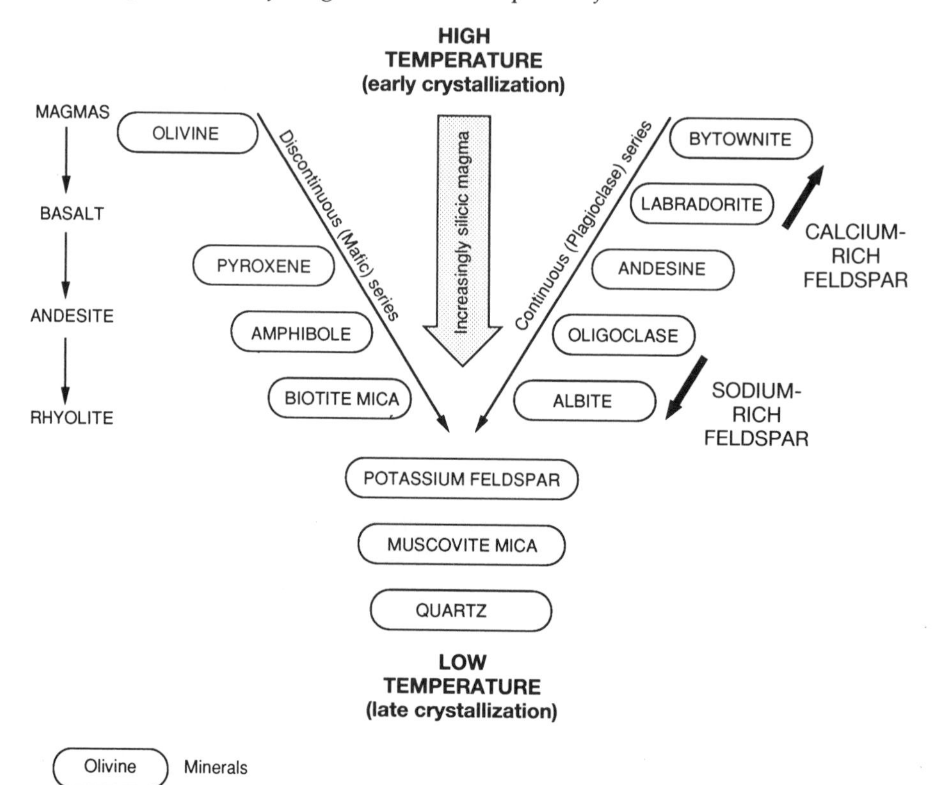

Fig. 3.6 Fractional crystallization of a basaltic parental magma. The diagram shows the order in which minerals crystallize during cooling of a magma. This regular, ordered sequence of crystallization is known as *Bowen's reaction series.* Two convergent series are involved: one *continuous* and involving different forms of a given mineral (e.g. plagioclase feldspar); and the other *discontinuous,* in which different mafic minerals are crystallized (based on a number of sources).

convection, known as *double-diffusive* convection, could cause magma to separate into compositionally distinct, subhorizontal layers (Huppert and Sparks 1984).

3.3.2 Challenges to the consensus

During the past few years scores of papers have been published on the generation and movement of magma. Some of these have served merely to clarify and expand upon what was already known in the mid-1980s. For instance, there has been further clarification of the mechanisms of diapiric movement. In a review Bruce Marsh (1987, p. 1044) notes that, 'hot diapirs can go faster, but they also cool faster, and their penetration distance is limited to a few body diameters, at most. Penetration of the entire lithosphere thus requires successive bodies'. Additionally there have been attempts to refine models explaining the operation of magma chambers (e.g. Spera *et al.* 1986; Sparks and Marshall 1986; Pyle *et al.* 1988) and the processes producing subduction zone melts (e.g. Brophy and Marsh 1986; Wheller *et al.* 1987; Hamilton 1988; Ringwood and Irifune 1988; M. Wilson

1989). There remain three closely related themes where recent publications have seriously challenged accepted wisdom and led to the modification of consensus views.

Two of these themes are dealt with elsewhere in the book. In Chapter 4 (section 4.3.1.1) melting by means of decompression is discussed and the new light this casts on the generation of flood basalts is highlighted (White and McKenzie 1989), while in Chapter 2 (section 2.4) the impact of the ideas of J. T. Wilson (1988) and others on 'supercontinents' and plate-boundary volcanism are reviewed. The third area where the consensus has been challenged is potentially more fundamental, though at the moment more speculative. It concerns the generation of parental magmas. Recently the question of whether ocean-island volcanoes sample a more primitive source region than those of ocean ridges has been brought into focus by the publication of a number of papers on isotopes. As well as being based on familiar isotopes like strontium, the interpretation of which have been the stock-in-trade of geochemists for many years, more recent research has extended the range so enabling new insights to emerge. Reporting on the Mantle Plume Symposium held in California in 1991, Laura Garwin (1991, p. 699) writes that 'the evidence for the existence of primitive mantle has been crumbling, and the idea had few advocates at the meeting'. If mantle plumes are not derived from the primitive mantle then this raises questions about their origin. On this issue the delegates at the symposium were more equivocal, many suggesting that the mantle convects as two distinct layers separated by the seismic discontinuity at around 650–670 km (see Table 2.1; and Ringwood and Irifune 1988). Plumes, it was argued, originated at the boundaries of either the upper/lower mantle—(i.e. at a depth of ~650–670 km)—or from the mantle/core—or possibly both. One provocative contribution to the symposium noted that the heat flux from plumes is about 6–10% of the Earth's total and that this is similar to the heat assumed to be lost by the core. If plumes originate at the core/mantle boundary then this may allow a simplification of established ideas about mantle convection and plate tectonics, with equilibrium being maintained by subducted slabs cooling the mantle and plumes cooling the core (see Garwin 1991).

Further research on plumes by Larson (1991), though focused on ancient volcanic successions, has wider implications. The author adduces evidence and identifies a 'superplume' some 6000 km across which affected a large area of the Pacific in the mid-Cretaceous, 120–180 million years ago. The author argues that convection on such a large scale must be related to events in the core and a plume origin at the core–mantle boundary is suggested. The Cretaceous is a long time ago and many of the assumptions the author makes about the crustal geology of the time are open to question as are some of his findings, but in reviewing the paper Keith Cox (1991, p. 564) points to its contemporary relevance by concluding that 'if the coupling of core, mantle and surface events . . . can be firmly established, a substantial unifying framework will be available for the study of most geological processes'. The research horizons are wide indeed.

3.4 The eruption of magma

Most of the processes operating within magma chambers lead to increases in volume (Wilson *et al.* 1987), and this in turn places greater stress on crustal rocks

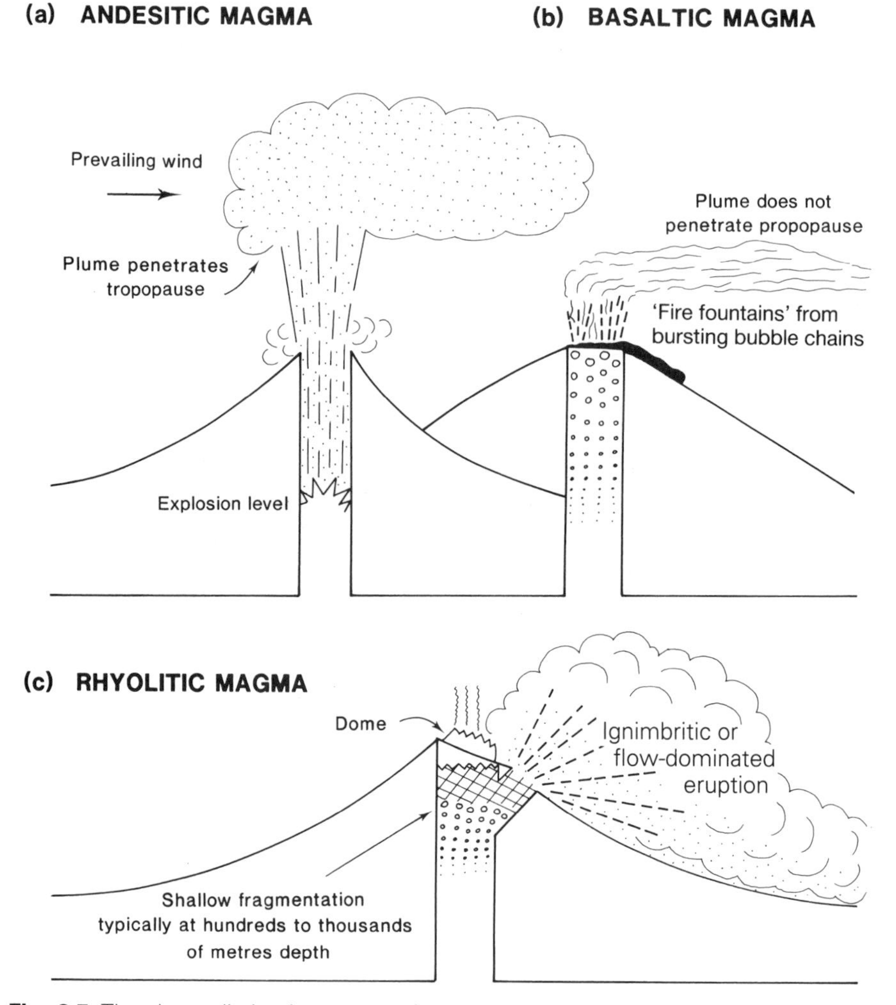

Fig. 3.7 The three distinctive types of eruption produced by basaltic, andesitic and rhyolitic magmas (modified from Clapperton 1977, Fig. 7, p. 394).

which eventually fail (Blake 1981b). Further movement of magma requires new pathways and these may trend either vertically and/or horizontally; being known as dykes (sometimes dikes) and sills respectively. Dykes and sills are preserved in the geological record as igneous intrusions, regardless of whether they were actually linked to volcanoes at the Earth's surface. As Fig. 3.5 shows magma chambers are not simple in configuration and are often associated with plexuses of dykes and sills. The development of and magma movement along dykes and sills is complex and its discussion beyond the scope of the present volume. Readers who are interested should refer to papers by: Weertman (1971); Pollard (1976); and Spence and Turcotte (1985).

The pathways linking subterranean magma chambers, dykes and sills to the surface are known as conduits and processes operating within them largely

Fig. 3.8 The Plinian column produced by the eruption of Mount St Helens on 22 July 1980. The column, which was over 14,000 m in height, introduced gases and dust into stratosphere. Although the plume was rich in dust, it had a relatively low sulphate content and this limited its global climatic impact. The 'convective' and 'umbrella' regions of the plume may be seen clearly (see Chapter 5, section 5.3.2 and Fig. 5.4) (photograph is reproduced by courtesy of the United States Geological Survey, no. 942).

determine the style of eruptions. As the surface is approached it is hypothesized that the expansion of volatiles, particularly H_2O, CO_2 and SO_2, is the major control on magma movement (L. Wilson and Head 1981a). At depths greater than a few kilometres CO_2 and SO_2 are important, though nearer to the surface

Fig. 3.9 A small pyroclastic flow moving down the slopes of Mount St Helens on 7 August 1980 (photograph reproduced by courtesy of the United States Geological Survey, no. 941).

H_2O, CO_2 and SO_2 in combination may be of greater significance (Macpherson 1984).

The solubility of volatiles in magma is strongly affected by pressure. As movement towards the surface occurs pressure is reduced and gases exsolve. Silica-rich melts are viscous (see Chapter 4, section 4.2), the higher the viscosity the more restricted will be the migration of bubbles and, if the volume occupied by bubbles is greater than ~75%, then explosive fragmentation of the magma into pyroclastic materials and gases will occur at shallow crustal depths.

Fragmentation depths may vary from hundreds to thousands of metres (see McBirney and Murase 1971; L. Wilson *et al.* 1987; and Fig. 3.7). Eruptions are typified by dome building and the development of pyroclastic flows, including ignimbrites (see Chapter 5, section 5.3). When eruptions of this type occur in populated regions the effects are often devastating (see Chapters 7, 8 and 9).

In silica-poor basaltic magmas, gases are free to exsolve as pressure is reduced and more restricted fragmentation takes place (Dobran *et al.* 1990). Eruptions are characterized by the quiet release of gases, plumes with restricted heights are formed and produce localized climatic effects (see Fig. 3.7; Chapter 6, section 6.4.1.1). Dispersal of the predominantly coarse pyroclastic falls is also spatially confined (see Chapter 5, section 5.2) and the principal volcanic products are lava flows (Fig. 3.7).

Between these two extreme are eruptions where magma is intermediate in composition; typically an andesite (Fig. 3.7). Here processes are more complex (see Chapter 5), with pyroclastic materials being 'dynamically locked (into) the gas(es)' (L. Wilson *et al.* 1987, p. 86). The main features of these eruptions are high eruption columns, which spread gases and pyroclastic fall deposits over vast distances and may have significant and widespread climatic impacts (see Chapter 6 and Fig. 3.8). As eruptions wane columns collapse and often generate pyroclastic flows and surges (see Chapter 5, section 5.3.2, and Fig. 3.9).

The rhyolitic, andesitic and basaltic eruptions depicted in Fig. 3.7, equate to distinct volcanic styles, these being known as *Peléan/ignimbritic*, *Plinian* and *Hawaiian/Strombolian*, respectively. Magma type is not the only variable controlling eruptions and different styles are found: when magma interacts with water stored in subterranean aquifers, crater lakes or ice; when activity occurs on the sea-bed and where eruptions take place in *vacuo*, as would have been the case on many of the planets of the Solar System. Sections 5.3 to 5.5 of Chapter 5 are devoted to the detailed discussion of eruptive styles, to which further reference should be made.

To conclude it is necessary to say something about the landforms produced by volcanoes. These range in size over several orders of magnitude from the surfaces of whole planets at one extreme, to the micro-morphology of individual lava flows at the other. Volcanic landforms form over time scales ranging from less than a day, to millions and even billions of years. In view of their importance, it is strange that volcanic landforms have not received more attention from geomorphologists and sobering to recall that the last major book on the topic was Charles A. Cotton's *Volcanoes as Landscape Forms* first published in 1944; though a more respectable number of research papers have been published. The reasons why geomorphologists, especially those from English-speaking traditions, have neglected volcanoes are many and varied, but the emphasis in recent decades on short-term process studies concerned amongst other things with rivers, slopes, coasts and glaciers has had much to do with it (Gregory 1985; Tinkler 1985). In the last few years there have been signs of change and geomorphology has become concerned again with structural landforms in general and with volcanoes in particular (e.g. Ollier 1984, 1988; Clapperton 1990). It remains the case, however, that most recent advances have been carried out by workers who are not geomophologists in any formal sense.

Volcanic geomorphology is an implicit, at times an explicit, theme running through much of the book and is discussed where appropriate in Chapters 1 to 5. The landforms produced at the sites where typical basaltic, andesitic and rhyolitic

Table 3.5: Major landforms produced at sites of basaltic, andesitic and rhyolitic volcanism (based on numerous sources)

Magma type[1]	Typical landform or construct
Basalt	*Lava shields* are produced by successive eruptions of lava flows. They are low-angle landforms, usually with slope angles of 6° or less. The summit of Mauna Loa in Hawaii is 4000 m above sea level and over 100 km across. Often lava is not erupted from the summit, but from *parasitic* cones on the flanks. Some more silica-rich basalts, such as those producing Strombolian activity, may build *strato-shield* cones like that of Mount Etna, Sicily. These are intermediate between shield and strato-volcanoes.
Andesite	*Strato-volcanoes* are landforms produced by calc-alkaline magmas and Plinian activity. Examples include Vesuvius (Italy), Mount St Helens (USA) and Fujiyama (Japan). They are built of mixtures of lava flows and pyroclastic materials. Strato-volcanoes are much steeper than lava shields and slope angles of ~30° are not uncommon. Because pyroclastic materials are more easily eroded than lavas and Plinian eruptions fairly infrequent, strato-volcanoes are often heavily dissected.
Rhyolite	Very silica-rich magmas usually produce domes (see Chapter 4, section 4.3.2.1).

[1] These are typical magmas. Other magmas with the same comparable silica contents produce similar landforms.

magmas are erupted are omitted from these chapters and are, therefore, summarized in Table 3.5.

4

Lava flows and lava landscapes

Whereout pure springs of unapproachable fire are vomited from the inmost depths: in the daytime the lava streams pour forth a lurid rush of smoke; but in the darkness a red rolling flame sweepeth rocks with uproar to the wide deep sea (the Greek poet Pindar (c. 522–c. 438 BC) writing of a lava flow on Mount Etna, Sicily).

4.1 Introduction

Lava occurs in many forms. On Earth a fundamental distinction may be drawn between lava produced on land and that erupted subaqueously. Further subdivision is possible depending upon whether lava is derived from basaltic or more silica-rich magmas (Table 4.1). Lava also covers large areas of the terrestrial, or inner, planets of the Solar System (see Chapter 1, section 1.3); and occurs as komatiite (see section 4.5), those 'unique' and not yet fully understood volcanic rocks of the Precambrian.

Until the 1970s research on lava flows was predominantly petrological and descriptive in character, with only limited consideration being given to what are now considered important areas of research; including the physical mechanisms of lava flowage. Gordon A. Macdonald's classic text—*Volcanology*—published in 1972 is a worthy summary of these petrological and descriptive traditions that go back to the birth of geology in the mid-nineteenth century and which have involved contributions by a distinguished cast including: George Poulett-Scrope (1825); Sir Charles Lyell (1830, 1858); Thomas Jaggar (e.g. 1904, 1917); A. I. Du Toit (1920); Howel Williams (1932); Chester K. Wentworth (e.g. Wentworth *et al.* 1945); Macdonald himself and many others. It is upon this foundation of meticulous description that the considerable progress made in the

Table 4.1: Classification of lava flows and landscapes on Earth. The numbers within the table refer to the sections in the text where the topics are discussed

	Lavas produced on land (section 4.3)	Lavas produced subaqueously (section 4.4)
Lavas derived from basaltic magmas	Section 4.3.1	Section 4.4.1
Lavas derived from silica-rich magmas	Section 4.3.2	Section 4.4.2

last two decades has been built. Not only have the petrological and descriptive traditions continued—often as responses to closely observed eruptions (e.g. Guest *et al.* 1980; see also Decker and Decker 1991)—but to these have been added important new insights which have come about through the exploration of the inner planets of the Solar System (e.g. Greeley 1987) and by relating the physical properties of flows to processes of magma generation and eruption (e.g. Peterson and Tilling 1980; Kilburn 1981; Dragoni *et al.* 1986; Fink 1990; Morgan 1991). Additionally, there has been the rediscovery after a long limbo (Cotton 1944) of lava landscapes as rewarding topics worthy of study by geomorphologists (e.g. Greeley 1977a; Ollier 1988).

In view of its importance to all that follows initial discussion of lavas will focus on their physical properties.

4.2 The physical properties of lava

Although the physical properties of lava are interrelated and, for instance, composition has a profound effect on viscosity, it is possible to consider these properties under two heads: *attributes* and *motion*. Under the former falls composition, volume, temperature, dimensions, rate of discharge, texture, velocity and cooling, whilst under the latter are to be found the mechanisms and physical principles controlling motion.

4.2.1 The principal physical attributes of lava

These are summarized in Table 4.2, which was compiled from the sources acknowledged and any further information should be obtained from these. Hopefully the table is self-explanatory, but it should be borne in mind that most of the information has been obtained from basalts and knowledge of more silica-rich flows is far more limited because of their less frequent eruption.

4.2.2 The factors controlling the motion of lava

The term *rheology* was first used in 1929 to describe the study of the deformation (i.e. change in shape or size) and flow (i.e. the rate of change in deformation) of materials under *stress* (i.e. force per unit area). *Strain* is defined as the deformation experienced by a material because of an *applied stress* and it produces a relative change in shape and/or volume. Three 'classic' models (Fig. 4.1) have been used to describe the rheological properties of a variety of materials, which range 'from foodstuff and printing ink to face creams to lava flows' (Chester *et al.* 1985, p. 189). The application of rheological concepts to lava flows is extremely complex and what appears is merely a summary. For a more detailed treatment reference should be made to the references cited.

As Fig. 4.1a shows viscous behaviour applies to fluids, like water and other liquids, and the rate of deformation (i.e. *strain rate*) is proportional to the stress applied. This is an ideal Newtonian fluid and deformation can occur under very small stresses (Summerfield 1991, p. 522). Elastic materials (Fig. 4.1b) recover their original form when the deforming stress is removed and Hooke's Law states that strain is proportional to applied stress, provided the *elastic limit* of the material is not exceeded. Once this is exceeded, then Hooke's Law ceases to apply

Table 4.2: The principal physical attributes of lava (based on information in Williams and McBirney 1979 and the sources cited)

Attribute	Comments
Composition	The basic relationship is that the volume and length of a flow is inversely correlated with the silica content of the parental magma. Hence, rhyolitic, andesitic and basaltic lavas are in a sequence of increasing volume and length. The factors, particularly the role of volatiles in controlling whether, or not, an eruption produces effusions of lava and/or pyroclastic materials, are discussed in Chapters 2, 3 and 5.
Volumes	Volumes are correlated directly with composition and it is estimated that some 90% of subaerial flows are of a basaltic composition. Whereas Holocene basaltic flows over 0.5 km^3 are common, those over 4 km^3 are extremely rare. Siliceous lavas more than 1 km^3 are very uncommon, but there are exceptions such as a flow in Chile (volume 24 km^3; Guest and Sanchez 1969). Over longer periods of geological time very large volumes of lava have been erupted such as the Roza flow of the Columbia River flood basalts—1200 km^3.
Temperatures	Measurements made at Hawaiian volcanoes suggest temperatures of around 1200°C and it is inferred that the komatiites of South Africa were emplaced at temperatures of around, or in excess of, 1400°C. Most lavas are much cooler than this and, although measurements are sparse, temperatures of more silica-rich lavas are probably much lower.
Dimensions	The largest lava effusions are continental flood and mid-ocean ridge basalts. In the East African Rift Valley there are also large flood basalts of more silica-rich composition. The largest flow of this type to be erupted in recent historical times is the 1783 Laki fissure flow (Iceland) which travelled 40 km. The Miocene Rosa flow (Columbia River, USA) is 300 km long (Swanson and Wright 1981). Basaltic lavas erupted from central volcanoes are much shorter than those from Hawaiian volcanoes, rarely exceeding 50 km (mean for Mauna Loa about 25 km, Kilauea less than 11 km). Silica-rich lavas are of more limited extent and typical lengths of rhyolitic flows are less than 4 km. Thicknesses vary greatly. Columbia River basalts average 25 m, Hawaiian lavas vary from a few centimetres to around 15 m—tending to thicken downslope—whereas silica-rich flows are short and thick. Andesitic flows may be around 30 m in thickness. Walker (1973a) introduced the concept of 'aspect ratio' to describe lava-flow dimensions; this being defined as the ratio of average thickness to horizontal extent. Hence, low-viscosity basaltic lavas have low-aspect ratios, while highly viscous trachytic, andesitic and rhyolitic lavas have high-aspect ratios (section based on Williams and McBirney 1979; and Cas and Wright 1987).
Rates of discharge	The rate of discharge, or effusion rate, depends in the main on the viscosity of the lava and the efficiency of eruptive conduit. In most eruptions, rates of discharge either decline and/or fluctuate and this is most pronounced in basaltic eruptions.

Table 4.2 – *cont.*

Attribute	Comments
	High initial rates of discharge are due to high initial gas contents, but the decline in rates during eruptions may be due to this factor and others, including widening of the conduit. Typical historic rates of discharge for basaltic volcanoes range from around 0.3 to over 45 m^3s^{-1} and those of more silica-rich effusions are normally less than 1 m^3s^{-1}, although exceptions do occur (e.g. Augustine volcano, Alaska estimated discharge 11.6 m^3s^{-1}; Newhall and Melsom 1983). Many prehistoric flood basalts have effusion rates orders of magnitude higher than these values (Williams and McBirney 1979).
Textures	These depend on a large number of factors: rate of cooling; composition; gas content; and temperature. Most lavas have large crystals (phenocrysts)—formed during the rise of magma or its residence in magma chambers (see Chapter 3)—set in a granular groundmass. In more silica-rich lavas (i.e. andesites and rhyolites) there may be alignment of elongated crystals, due to viscosity and velocity gradients. Many other textures are often found (section based on Hatch *et al.* 1972).
Other (cooling, velocity)	Lavas have a low thermal conductivity and high heat capacity. They cool slowly and may still be hot years after eruption. The principal heat loss is radiation from the surface, but cooling by rain may be important. Conduction is a minor factor. Exsolution and expansion of gases—in particular H_2O is endothermic and exerts a major cooling effect. Velocity depends on such factors as effusion rate, viscosity, volume and ground slope. These factors tend to cause basaltic flows to be the fastest moving. Rates may be as high as 64 km/h, but generally velocities are much lower. More silica-rich lavas move more slowly; typically tens of metres per hour. All lava flows show a reduction in velocity with distance from the source.

and the material shows plastic behaviour. A coil spring is a good example of an elastic substance. Finally there is plastic flow. In plastic substances no deformation takes place until the *yield strength* (i.e. a threshold of stress) of the material is exceeded. Once the yield strength is exceeded then, in a perfect plastic substance, strain occurs at a constant rate (Fig. 4.1c).

Many years ago the eminent rheologist Reiner (1960, p. 11) noted that, 'real substances possess all rheological properties, although in varying degrees' and the three models shown in Fig. 4.1 should be seen as representing idealized substances. As far as lava flows are concerned some form of plastic behaviour is clearly predominant (Hulme 1974), since it is a matter of observation that a yield strength has to be exceeded before flowage can take place and a lava flow does not appear to act as either a fluid or an elastic substance. 'A Newtonian lava flow would spread indefinitely both down and across a slope. The fact that lava flow lobes commonly have a well-defined cross-sectional shape is evidence for a finite yield strength in essentially all lavas' (L. Wilson *et al.* 1987, p. 81). Within the category of plastic flow, however, the 'perfect' plastic model described above is merely one of several and Robson (1967) and Hulme (1974) have favoured what is known as a 'Bingham' plastic model as being most appropriate for lavas. Once

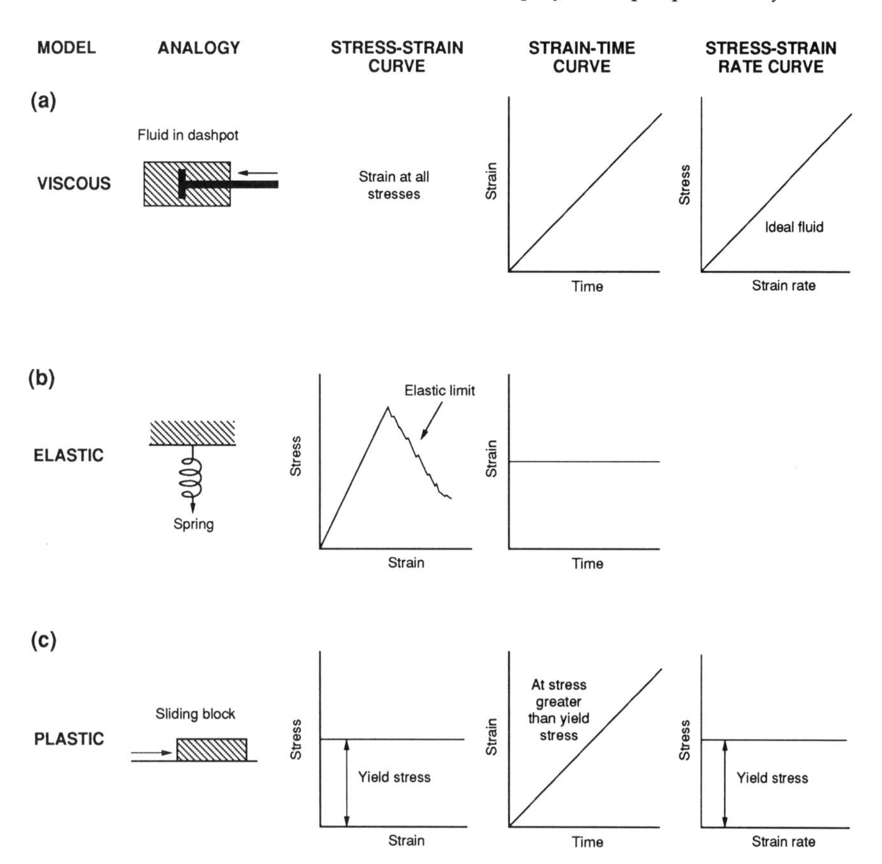

Fig. 4.1 'Classic' models of the responses of materials to stress: (a) viscous; (b) elastic; and (c) plastic (modified from Statham 1977, Fig. 2.13, p. 53).

the yield strength of Bingham substances is exceeded, then the relationship between shear stress and strain rate is linear. Using suspensions of kaolin, Hulme was able to model unconfined flows on an inclined plane and reproduced many features commonly found on lava flows; including *levées*. For Bingham materials to flow down a slope they must be thick enough for shear stresses at the base to exceed the yield strength of the material and, since near to their margins flows are generally thinner, then stationary zones occur in which levées form. Hulme also related the volumes, centre-line thicknesses, widths of flows and the thicknesses of levées to yield strength, with viscosity, density, the force of gravity and angles of slope of the ground over which flows move emerging as additional controlling variables.

The ability of Hulme's model to explain the morphologies and dimensions of lava flows on the basis of yield strength and plastic flow has been tested by several authors, both on Earth and on the other inner planets of the Solar System (e.g. Hulme and Fielder 1977; Moore *et al.* 1978; Zimbelman 1985; Head and Wilson 1986). These investigations have also highlighted some of the limitations of Hulme's model. H. J. Moore *et al.* (1978) and L. Wilson *et al.* (1987), for instance, point out that the model is applicable only to flows in which widths are much greater than thicknesses, that have high effusion rates (i.e. volumes per unit time) and in which cooling during emplacement is unimportant. For other types of flow different rheological models are required. In particular and as noted by L. Wilson *et al.* (1987, p. 82), special attention has to be given 'to the motions of very high-viscosity lavas, which may take a very long time to come to equilibrium with the stresses induced by gravity; Huppert *et al.* (1982a) successfully applied a Newtonian model to the growth of a lava dome' (see section 4.3.2.1). There is also evidence that some lavas may approximate to so called pseudo-plastic behaviour, which differs from both Newtonian and Bingham motions because when compared to the former the relationship between shear stress and strain rate is variable, while in contrast to the latter the substance does not possess a yield strength. Laboratory analysis by Shaw (1969) using samples of Hawaiian basaltic lava indicated that at high shear rates a Bingham model was the more appropriate, whereas at lower rates a pseudo-plastic model provided a better rheological explanation.

Viscosity is a second important consideration when it comes to understanding the physics of lava flows. To the layman a viscid substance is a thick, sticky material that does not flow easily; the more viscous the substance the more resistant it is to flowage. To the physicist viscosity means the internal resistance of a substance to flow and is 'caused by molecular or ionic cohesion' (Cas and Wright 1987, p. 21). For silicate melts (both magmas and lavas) it is not only vital in its influence upon mobility, but is also an extremely complicated topic; as is clear from discussions in standard texts (e.g. Fisher and Schmincke 1984, pp. 52–5; and Chester *et al.* 1985, pp. 196–8). The definition of the viscosity of a substance is the ratio of shear stress to the rate of strain[1]. For Newtonian substances the ratio of shear stress to strain rate is constant (Fig. 4.1a) and early attempts at measuring the viscosities of lava flows—summarized by Macdonald (1972)—assumed Newtonian flow properties. Viscosities were calculated using an equation (the 'Jeffrey's formula') first derived by the eminent physicist Sir Harold Jeffreys (Cas and Wright 1987, p. 22). The accuracy of measurements made using this formula was low, because of the problems involved in measuring both velocities and flow dimensions in the field; a particular difficulty being that the surface of a flow is known to move faster than its interior (Williams and McBirney 1979).

When lava is assumed more realistically to be non-Newtonian (i.e. Bingham or pseudo-plastic), measurements become even more complicated, because the ratios of shear stresses to strain rates are not constant (McBirney and Murase 1984). In the case of a pseudo-plastic substance, for example, apparent viscosity decreases with increasing strain rate. Two techniques have been used to measure viscosity. For very fluid lavas it was recognized many years ago by Einarsson (1949) that, once calibrated for temperature, the rate at which a steel rod (penetrometer) could be pushed into a lava flow would give a measure of viscosity, but more recently emphasis has been placed on using shear vane viscometers. In this apparatus a blade is rotated in liquid lava and the frequency of rotation under a constant force noted. Today much is known about the viscosity of basaltic lavas from two well-studied cases: the Makaopuhi lava lake, Hawaii (e.g. Shaw *et al.* 1968) and Etna (e.g. Pinkerton and Sparks 1978; Chester *et al.* 1985). On Etna Bingham viscosities were around 9.4×10^3 poises, whereas on Hawaii values of

[1] Viscosity is measured in Newton seconds per square metre (Ns/m^2) in SI units or, more commonly, by the older cgs unit the 'poise'. One poise is equal to 0.1 Ns/m^2.

Table 4.3: The factors thought to control the viscosity of lava flows and magmas (based on the sources noted)

Factor	Comments
Temperature	Work in the field and laboratory by Murase and McBirney (1973), show that the viscosities of both magmas and lavas increase as temperatures fall. This is due in part to crystallization, but graphs of the relationships of viscosity and temperature indicate that, at similar temperatures, magmas of different composition have different viscosities, with rhyolites being the most viscous, followed by andesites and basalts. This implies that composition is a very important control.
Chemical composition	The relationships between chemical composition and viscosity are very complex and are explained in detail by Cas and Wright (1987, p. 26). Briefly silica and to lesser extent aluminium are important, for when they form bonds with oxygen (Si–O and Al–O bonds), these are far stronger than other oxygen bonds. It is suggested by Mysen *et al.* (1982), that Si–O bonds form complex structural units within magmas and lava flows and that the complexity of these correlates with viscosity. Silica-rich magmas are, hence, generally more viscous than silica-poor ones.
Crystals	A general relationship is that the greater the crystal content the higher the viscosity. Estimates of the viscosities of both magmas and lavas may be made using a method suggested by McBirney and Murase (1984).
Volatiles	According to Murase (1962) and Williams and McBirney (1979), dissolved water has a profound effect on viscosity. At a given temperature, the viscosity of a magma will decrease with increasing water content. However, the situation is more complex than this because solubility depends on other factors including pressure, temperature and the presence of other volatiles. Dissolved water has a greater effect on decreasing the viscosities of silica-rich than silica-poor magmas. Relatively little is known about the effects on viscosity of other volatiles.
Pressure	In laboratory experiments (e.g. Kushiro 1976; and Kushiro *et al.* 1976), it was found that viscosity was reduced as pressure was increased.
Bubbles	This is complex and is more fully explained by Cas and Wright (1987, pp. 26–7). In basalts exsolution of water and other volatiles has little effect on producing low viscosity because this is already low and caused mainly by temperature and composition. The effect of bubbles is to lower the viscosity further. In more acidic melts the high viscosity produced by other factors may be influenced by the presence of bubbles.

$6.5–7.5 \times 10^2$ poises were recorded. It is significant that the Etna lava was at a lower temperature and had a higher crystal content. It is also well known and, indeed not surprising, that as lava cools along its length viscosity increases (Walker 1967). A note of caution, however, is sounded by Cas and Wright (1987, p. 23) who emphasize how much is still to be learnt about viscosity when they note that, 'there are no field measurements of the Bingham viscosities for more felsic or salic lavas'.

In later sections of this chapter it will become apparent that viscosity is a major influence on lava mobility, geometry and geomorphology. It is also critical in any discussion of the mobility of magma (Chapter 3). In the last paragraph some of the factors which control viscosity were mentioned briefly: crystal content, temperature and cooling rate, but the fact remains that these and others are still not fully understood and are the subject of much debate in the literature. A summary of what is currently understood about the influence of a variety of factors on the viscosity of lavas and magmas is tabulated in Table 4.3.

The factors controlling the nature of lava movement have been discussed for many years. More than a hundred years ago an English physicist carried out an experiment in which a dye was introduced into a glass tube through which water was flowing. At low velocities the dye was merely moved in the direction of flow as a smooth unbroken stream, but as velocity increased and exceeded a threshold value, then the dye was seen to produce eddies and vortices and became mixed

Table 4.4: The principal differences between pahoehoe and aa lava flows (based on Macdonald 1972; and Williams and McBirney 1979). Note: there are many minor morphological differences between pahoehoe and aa lava flows and these are fully discussed by Macdonald, McBirney and Williams and Kilburn (1990); tumuli formation is reviewed in detail by Walker (1991)

Type of lava flow	Surface morphology	Internal composition	Nature of flow
Pahoehoe	Smooth, and sometimes rope-like. The ropy morphology reflects dragging of the partially solidified plastic crust by the still-liquid lava beneath. When the lava flow is a narrow stream the 'ropes' are bent into curves that are convex; pointing towards the distal margins. Cracks in the surface are often sites where liquid lava is 'squeezed up' to form ridges. Sometimes mounds are formed (*tumuli*) by both squeezing up under hydrostatic pressure and differential sagging of the surface following the drainage of lava tubes.	Lava is fed to the flow front by means of 'tubes' and channels. At the close of an eruption lava often drains from the tube, to produce 'cave' systems.	Following eruption the upper tens of centimetres cool quickly and become highly viscous. The flow remains liquid beneath this crust and, while the eruption continues, frontal flow is maintained. Some less active parts of the interior solidify and movement is often confined to distinct 'tubes' within the flow, which feed the flow front.

Table 4.4 – *cont.*

Type of lava flow	Surface morphology	Internal composition	Nature of flow
Aa	Lava has a sharp, rubbly surface, made up of angular jagged fragments (i.e. clinker). Aa flows do not normally flow in the same way as pahoehoe, but by means of open 'streams' or channels of lava which build their own embankments or 'levées'. At one time it was often stated that aa flows did not contain lava tubes. It is now known that, although relatively uncommon, this is not the case.	Aa flows usually comprise: an upper fragmental layer; a massive interior; and a rubble-rich base.	The sharp, broken surface is produced by fragmentation of the upper viscous material by the movement of the flow beneath. Overall the flow advances 'like a caterpillar tread dumping talus over its front and then overriding the fragments' so accounting for three-fold internal composition (Williams and McBirney 1979, p. 110).

with the water. To these two states the physicist Reynolds gave, respectively, the names *laminar* and *turbulent* flow and defined a dimensionless number—the *Reynolds' number*—to describe the change from one state to the other. Defined in terms of the experiment (i.e. for Newtonian fluids—like water flowing in tubes) this is the mean diameter of the tube, multiplied by the mean velocity, multiplied by the density of the fluid, divided by the viscosity. Modifications were made subsequently for open-channel flows and in both cases transitions from laminar and turbulent flow were found to occur at Reynolds' numbers of between 500 and 2000. Much later the Reynolds' number was refined further to take account of non-Newtonian flows, and of relevance to lava flows, is a value number known as the *Hampton number*, which marks the transition from laminar to turbulent behaviour in Bingham substances (Hampton 1972). As Cas and Wright (1987, p. 28) note the important feature of the Reynolds' number 'is that it is inversely proportional to viscosity, and directly proportional to velocity. Because of this, lavas may move by laminar flow or turbulently, but generally and because of their relatively high viscosities, *most lavas flow in a laminar fashion*'. Turbulent flow may be of importance when komatiitic lavas are considered (section 4.5) and it is well known that some very fluid flows may show turbulent motion close to their eruptive vents, but become rapidly laminar as viscosity increases downslope (Williams and McBirney 1979).

Laminar flow imparts distinctive textures to lavas (Williams and McBirney 1979, p. 118). These include the orientation of crystals, vesicles or bubbles (i.e. cavities within the flow caused by the exsolution of gases) and inclusions parallel to the direction of flow. Sometimes the whole flow has these features within distinct bands. During final stages of motion a flow may show *auto-becciation.* This is the breaking of the flow into angular fragments as viscosity increases and is due to the 'stresses set up by flowage' (Macdonald 1972, p. 96). Laminar flow may become discontinuous and, if this occurs, then the flow may show shearing of one band over another. Because of high viscosity this is particularly characteristic of silica-rich flows and, if the shear zones are finely banded, the rock is described as a *laminite* (Williams and McBirney *op. cit.*).

4.3 Lavas produced on land

As Table 4.1 shows there is a distinction between those flows produced by basaltic and those erupted from more silica-rich magmas.

4.3.1 Lavas produced from basaltic magmas on land

Two situations are involved in the eruption of basaltic magmas. There is first the eruption of flood basalts, which at various times in geological history have produced large expanses of lava and, secondly, there are the more spatially limited effusions of individual flows that build central volcanoes.

4.3.1.1 Flood basalts

As Table 4.2 and Fig. 4.2 show flood basalts not only cover large areas of the Earth's surface, but have also been erupted over long periods of geological time. In the past flood basalts have received relatively little attention from writers of general texts on volcanology, but since the mid-1980s the publication of a number of research papers has transformed understanding of their eruptive processes and they are now viewed as crucial elements in the large-scale geomorphological evolution of continental land masses (Macdougall 1988).

The principal attributes of flood basalts are easy to describe. In common with ocean floors and intraplate volcanoes like those of the Hawaiian Islands, flood basalts are composed predominantly of tholeiitic basalts; though in Ethiopia and the Deccan (India) they are alkali basalts (see Chapter 3, section 3.2; and Cas and Wright 1987, p. 16). Flood basalts are predominantly *pahoehoe* flows (see section 4.3.1.2) and are typified by well-developed columnar joints. Columnar joints (Fig. 4.3a and b) are exceptionally common in areas of flood basalt, but are not exclusive to them because they are also features of thick flows from central volcanoes (Williams and McBirney 1979). Flood basalts were presumably erupted as vast lakes of lava taking many years to cool and solidify and it is in this situation that columnar jointing developed.

Traditionally columnar jointing has been explained in terms of the thermal stresses developed during cooling (Spray 1962; Holmes 1965; Williams and McBirney 1979) and it has become conventional to recognize a three-fold morphological subdivision in vertical sections: the upper and lower tiers being straight sided, with well-developed vertical joints, or *colonnades* and a complex of irregularly curved and fan-shaped columns, called *entablature*, separating them (Tomkeieff 1940; and Fig. 4.3a and Fig. 4.4). The distinctive columnar form is explained by Macdonald (1972, p. 99) and most other writers in terms of contraction, which produces fracturing. Fractures or joints tend to develop at 90° to the top and bottom of the flows, these being the principal surfaces of cooling.

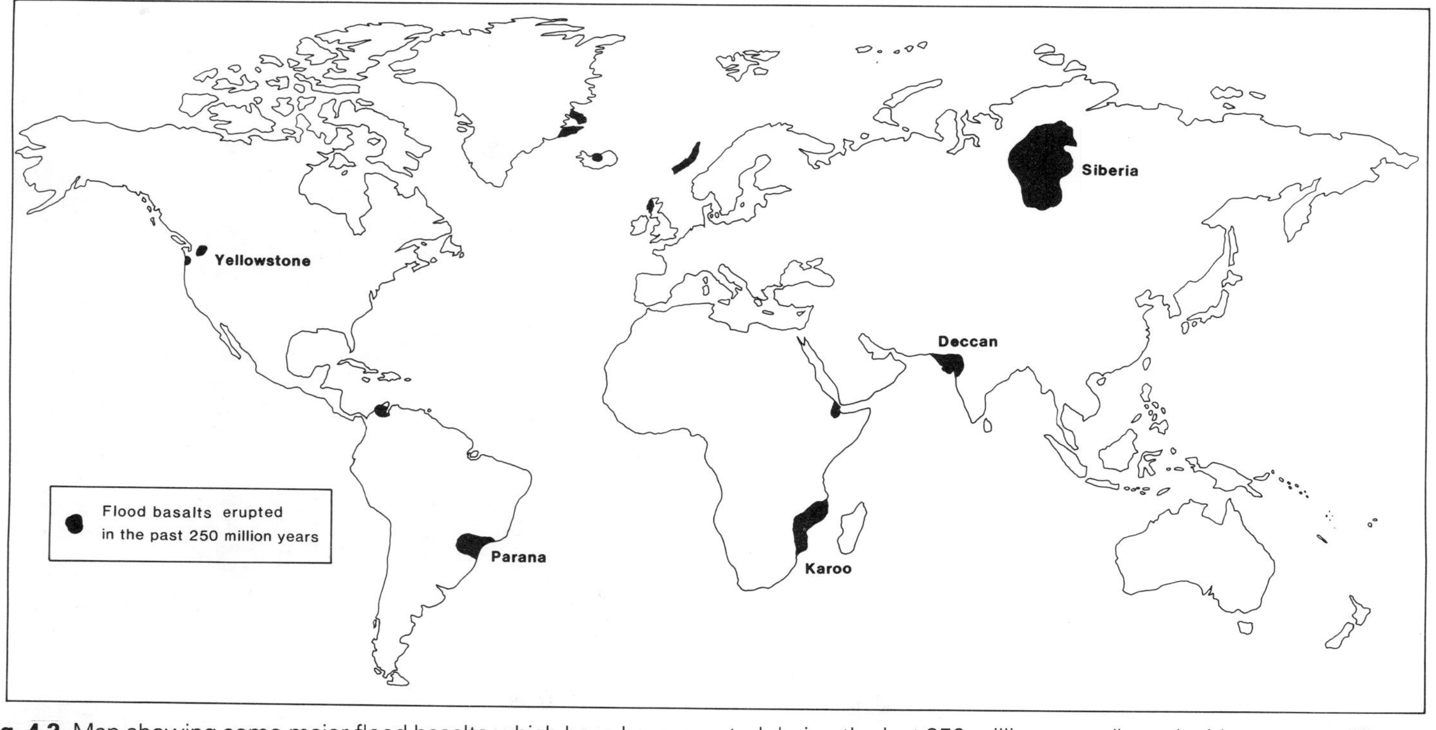

Fig. 4.2 Map showing some major flood basalts which have been erupted during the last 250 million years (based with some modifications on Morgan 1991, p. 44).

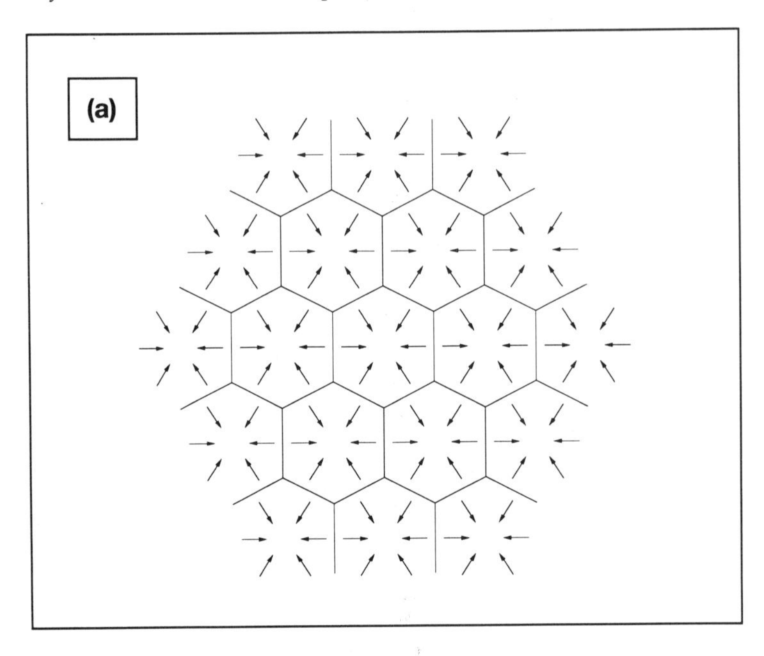

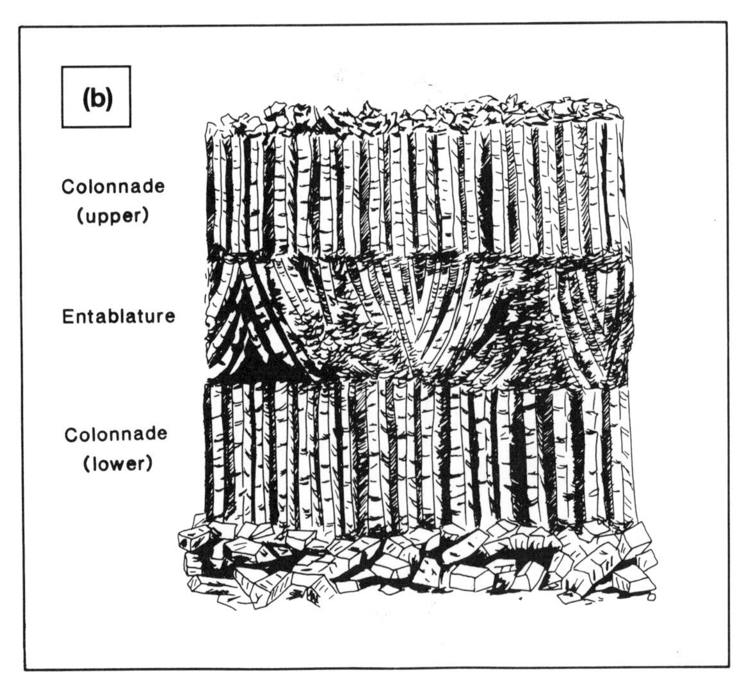

Fig. 4.3 Columnar jointing in lava flows.
(a) Plan view showing the formation of the 'ideal' hexagonal pattern of joints by uniform contraction towards evenly spaced centres (based on Holmes 1965, Fig. 63, p. 101).
(b) Sketch showing the three zones commonly found in vertical section (modified from Williams and McBirney 1979, Fig. 5.10, p. 115, and based on Tomkeieff 1940).

Fig. 4.4 Columnar jointing in a tholeiite plug at Edinburgh Hill, South Shetland Islands. The plug is 110 m in height (photograph reproduced with the permission of Dr J.L. Smellie, British Antarctic Survey).

Other thermal stresses, acting parallel to the cooling surfaces, produce intersecting joints and regular polygonal columns are formed; ideally hexagonal but ranging from four- to eight-sided. Colonnades reflect stresses developed during early-stage, rapid cooling, whereas the entablature form later when cooling is much slower. An alternative thermodynamic explanation has been proposed by Kantha (1981). According to this model convection is critical and it is suggested that dissimilarities in temperatures and chemistry between the top and bottom of a stagnating flow is sufficient to initiate columnar 'finger like' convective motions through the melt (Cas and Wright 1987, p. 71). As cooling proceeds, solidification occurs and contraction joints develop along boundaries between adjacent convective 'fingers'. Kantha tested his model in the laboratory using salt, but for lava flows further verification is required.

One notable feature of flood basalts is that the mainly *pahoehoe* flows do not normally possess lava tubes and channels. This is in marked contrast to those erupted from central volcanoes (section 4.3.1.2) and whether tubes failed to form or were destroyed later by convective movements within lava lakes is unknown (Cas and Wright 1987, pp. 72–3).

The generation of magma by mantle plumes is discussed in Chapters 3 (section 3.3), and from the time of the publication of a seminal paper by Kevin Burke and John Dewey, plumes have been linked to the generation of flood basalts (Burke and Dewey 1973). During the 1980s these ideas were taken further and in less than a decade notions of flood-basalt generation have been completely transformed. Particularly significant has been a paper by White and McKenzie (1989), which develops the Burke/Dewey model and proposes that plume-derived magmas rise towards the surface as relatively narrow columns. These spread laterally at high levels within the crust to form areas of melt up to 2000 km across. If this process occurs under a continent, then it will cause the uplift of a large dome which may be raised as much as 2 km towards its centre. This will in turn cause further melting through decompression (see Chapter 3, section 3.3) and vast amounts of basaltic magma will be produced. Although uplift through plume activity is temporary, the process causes what is known as *underplating* of continents and they remain elevated (McKenzie 1984). In an important paper Cox (1989) tested the persistence of 'topographic highs' over long periods of geological time, through an examination of the relationships between plume-induced uplift, basaltic flood volcanism and the long-term effects on geomorphology, particularly patterns of continental drainage. The areas examined included: Arabia and the Red Sea; the Indian subcontinent, southern Brazil and the Karoo District of southern Africa. In each he found that the effects of the plume could be traced in the structural geomorphology of these areas for up to 200 million years following initial uplift. Figure 4.5 illustrates Cox's research in India and his suggested links between the continental plume, the Deccan flood basalts and drainage patterns.

The Cox model represents a major new avenue of research. It is possible, however, to criticize some of its geomorphological aspects. As Nina Morgan (1991, p. 44) notes in a review article in the *New Scientist* in which she elicits the opinions of the well-known structural geomorphologist Michael Summerfield, 'the main features of drainage patterns are recent ones; the features of ancient drainage patterns are difficult to recognize with certainty'. She adds that Summerfield believes that much more rigorous testing of the model is required, especially through the study and dating of sediments eroded from the putative 'topographic highs'.

Before leaving flood basalts it is important to note a distinction drawn by Greeley (1977b) between the continental flood basalts already discussed and plains basalts such as those of the Snake River, Idaho, USA. The latter were erupted from central vents and fissures and are low-angle shields and sheets. They are transitional between true flood basalts and lava flows produced by volcanic edifices like those of Hawaii (Cas and Wright 1987).

4.3.1.2 *Basaltic lava flows from central volcanoes*

Far more has been written about the generation, movement, emplacement and morphology of basaltic flows than any other type of lava. The reason is not hard to find, for they are being currently produced in such well-studied regions as Iceland, Hawaii and Mount Etna (Sicily). Indeed a high proportion of total published research has been concerned with just one of these regions, Hawaii, and just one volcano, Kilauea.

Although some eruptions which build central volcanoes start along fissures, it is usual for these to become localized at a few points or vents. As Cas and Wright (1987, p. 64, my emphasis) note 'basaltic lavas can issue from vents as:

a) coherent flows from small *boccas* (openings), or from the overspill or breaching of a lava lake ponding in a crater; or

b) fire fountains that reconstitute around the vent and then flow away'.

Repeated eruptions of lava from the same or closely spaced vents build two principal types of volcanic edifice, or construct: the classic low-angle *shields* of Hawaii and other oceanic islands; and the much steeper and higher *composite* (or

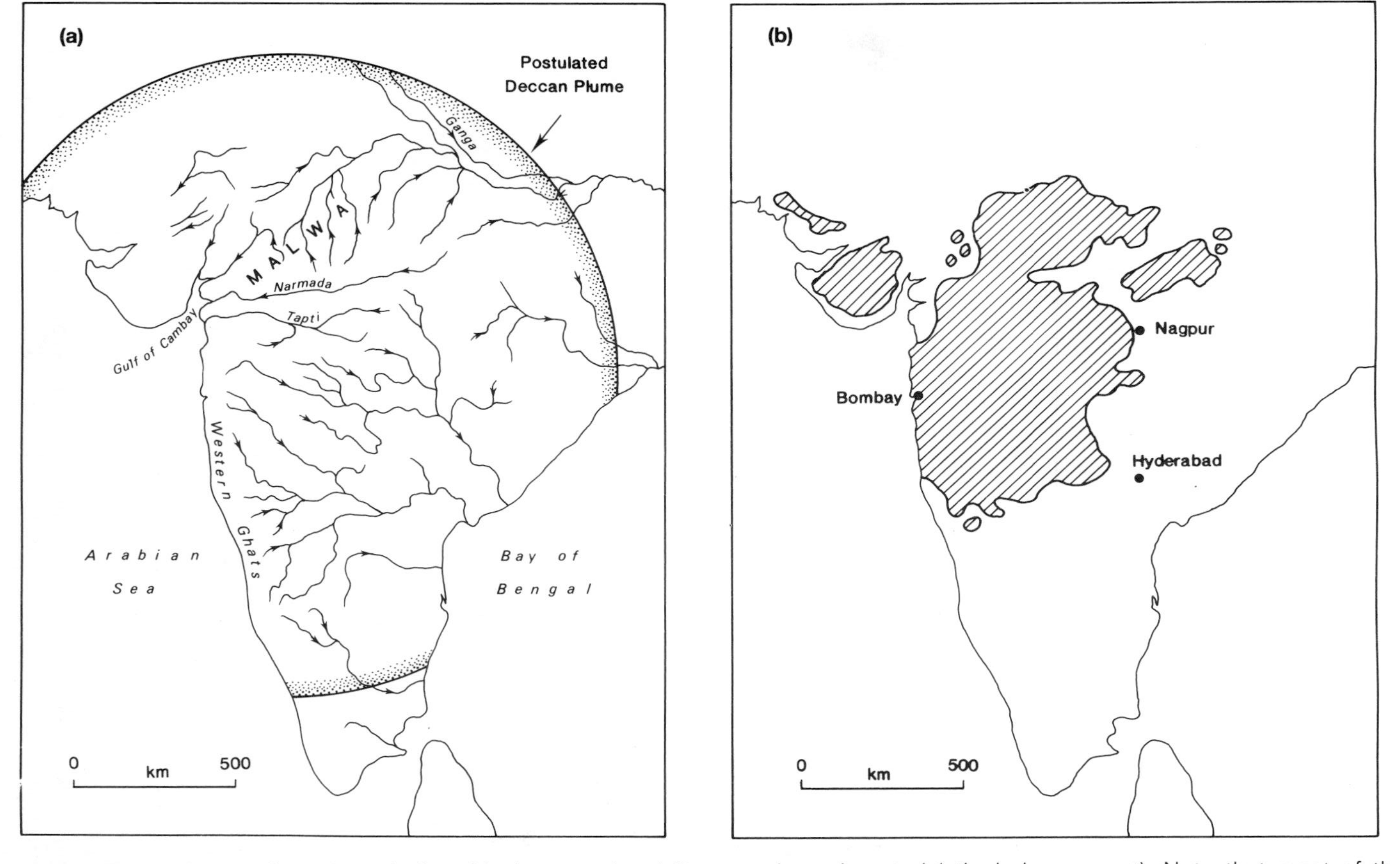

Fig. 4.5 (a) The drainage of southern India with the postulated Deccan plume inserted (stippled ornament). Note that most of the continent shows dome flank drainage, i.e. the Malwa system to the north. The Gulf of Cambray and the Narmada and Tapti drainage is related to rifting (from Cox 1989, Fig. 2, p. 874; see also Cox 1988; and Mahoney 1988).
(b) The Deccan Trap flood basalts (based on Williams and McBirney 1979, Fig. 11.7a, p. 268; see also Beane *et al.* 1986).

strato-shields) so common in continental situations and formed from admixtures of lava and pyroclastic materials (see Chapter 3, section 3.4).

On basaltic volcanoes most flows are described as being either *pahoehoe* or *aa*. These terms, first introduced into the volcanological literature over a century ago by the pioneer American geologist and volcanologist C. E. Dutton, were used at the time to describe surface form, but their definition is now much wider and includes differences in flow behaviour and internal composition (Table 4.4). A question of fundamental importance which was not answered until the early 1980s concerned the reasons why basaltic eruptions sometimes produce pahoehoe, whereas others erupt aa flows. On Hawaii it has been observed that most flows begin as pahoehoe and change to aa as they move down the sides of a volcano (Fig. 4.6), while on slopes of the same angle pahoehoe flows are the hotter, more gas-rich and faster moving. On other volcanoes these simple generalizations do not seem to hold (Williams and McBirney 1979, p. 111). On the basis of observations made in Hawaii, Peterson and Tilling (1980) argue that two variables have an affect upon whether flows are pahoehoe or aa: rate of shear and viscosity. As Fig. 4.7 shows, the change from pahoehoe to aa occurs as viscosity and/or rate of shear increase and a threshold is crossed. This may be caused by such factors as increased viscosity due to cooling and increased rates of shear as a flow encounters a steeper slope. If a flow is aa throughout then it is produced by high viscosities and/or high rates of shear—usually involving a high eruption rate—whereas the converse holds for flows that are formed solely of pahoehoe (Fig. 4.8).

Since its publication the Peterson/Tilling model has found widespread acceptance, but has been modified in detail. Kilburn (1981, p. 373), for instance, emphasizes that the evidence from several volcanoes implies that the transition between the two states is unidirectional (i.e. from pahoehoe to aa) and that the reverse 'rarely, if ever occurs' and he gives a rheological explanation. A later refinement by Rowland and Walker (1987) involves detailed study of a type of flow known as *toothpaste lava*. Toothpaste lava has been known for many years (Einarsson 1949; Macdonald 1967), either by this title or its synonyms: 'drawn-surface pahoehoe' (Foster and Mason 1955); 'spiny pahoehoe' (Peterson and Tilling 1980); 'semi-hoe' (Malin 1980) and 'fine aa' (Jones 1943). As its name implies, lava of this type is viscous and appears from an eruptive vent, like toothpaste from a tube (Fig. 4.9), and has 'surface grooves and drawn-out spines imparted at the orifice where the lava encounters the air' (Rowland and Walker *op. cit.*, p. 631). Rowland and Walker argue that these morphological features represent the transitional state in the Peterson/Tilling model from pahoehoe to aa (Fig. 4.7) and are produced with a viscosity higher than the former and a rate of flow (i.e. rate of shear) less than the latter. Toothpaste lava has been described on many basaltic volcanoes including those of Hawaii and Mount Etna.

As well as the distinction between pahoehoe and aa, other morphological features of basaltic lava flows are of interest to both the pure and applied volcanologist. Once erupted a lava flow may be described as being either *simple* (i.e. made up of a single flow unit) or *compound*, (i.e. made up of several flow units, or overlapping lobes). In 1970 George Walker asked the important question of why certain flows are simple, whereas others are compound, and found that two variables were important. The first was effusion rate (volume per unit time) and the second viscosity. He noted that basaltic lavas erupted rapidly tended to form simple flows, whereas those erupted at low effusion rates were

Table 4.5: Principal morphological features of domes of different types (based on Macdonald 1972; and Williams and McBirney 1979)

Type	Principal features	Selected examples
Upheaved plugs	Formed when lava forms a steep-sided column. Rarely more than 800 m in height. Many are circular, but others are elongated due to migration of the eruption along a fissure. The ratio of the height to diameter is commonly 0.3 and 0.5. Summit depressions are common, due to withdrawal of magma from the feeding conduit, or escape of lava from the flanks. Surfaces are fractured.	Lassen Peak (California), Santiaguito (Guatemala) and Tarawera (New Zealand).
Peléan domes	Diameters are much larger than their feeding pipes. The surfaces are fractured during growth 'the viscous lava develops a carapace that is repeatedly fractured . . . as new magma rises from below' (Williams and McBirney 1979, p. 192). Talus slopes often occur on the flanks of these domes.	Like spines these are common in the Caribbean Islands, where they often occur as steep-sided conical masses of lava known as 'piton' e.g. Petit Piton (St Lucia).
Spines	Columns of lava, emerging from the tops of domes. They grade to Peléan domes and are in many respects a subtype.	The 'classic' example was the spine of Mont Pelée that grew during 1902–03. Some of the larger piton of the Caribbean may be described as spines.
Exogenous domes	These are built at summit vents, by repeated outflows of lava. Sometimes a distinction is made between these and *endogenous* domes which form by expansion from within (Macdonald 1972).	Castello dome of Ischia (Italy).
Intrusive domes	Formed by shallow intrusions. They produce doming of the country rock and are not exposed at the surface.	Sutter Buttes of California.

Fig. 4.6 Flows of pahoehoe (left) and aa lava (right) on Kilauea volcano, Hawaii (photograph reproduced with the permission of Dr M.A. Summerfield, University of Edinburgh).

normally compound and produced many small flow units concentrated around the eruptive vents. Viscosity may also be a control over whether flows are simple or compound. On Etna (Chester *et al.* 1985) effusion rates are much lower than on oceanic basaltic volcanoes, such as Hawaii and Iceland, and compound lava flows are very common. If multiple compound flows issue from the same vent, both surface morphology and overall dimensions become very complex and produce so-called 'flow fields' (Wadge 1978).

Later in Chapters 7 and 9 the prediction of volcanic hazards and the effects of eruptions on people and their activities will be reviewed. Knowledge of factors governing the length of flows is of vital importance to the hazard analyst because

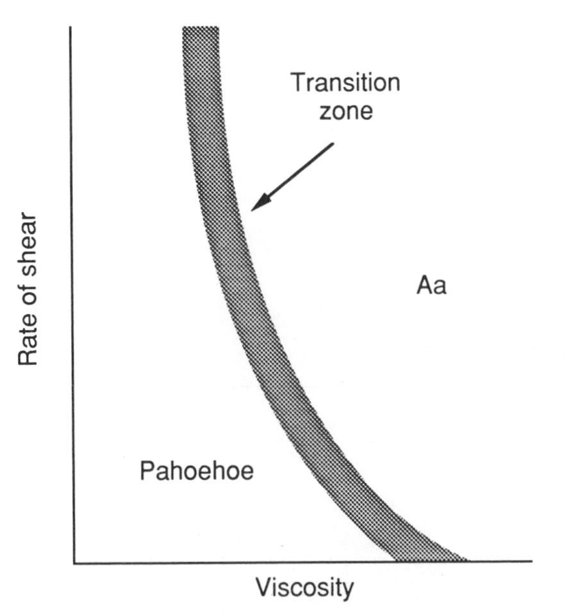

Fig. 4.7 The relationship between viscosity and rates of shear for pahoehoe and aa lava flows (based on Peterson and Tilling 1980, Fig. 9, p. 285).

this can make the difference between a settlement being unaffected or else being destroyed by lava. It would be easy to dissemble and claim that the inspiration behind the majority of published papers has been the requirements of hazard analysis, but in truth most authors have sought to fulfil the traditional 'pure' research goal of advancing knowledge for its own sake. The fact remains, though, that many publications are directly relevant to applied volcanology (see Chapter 7, section 7.3.1; and Duncan *et al.* 1981).

It may be stated with confidence that viscosity is important in controlling the dimensions of lava flows (Table 4.2), since it is a matter of observation that viscous silica-rich magmas produce short, thick flows, whereas less viscous basaltic magmas produce much longer, thinner ones. Over the last 30 years, however, attention has focused on basaltic flows and in a 'benchmark' paper, George Walker (Walker 1973a) considered 40 flows on 19 volcanoes and came to the conclusion that effusion rate (i.e. volume per unit time) was the single most important factor in controlling flow length, though others including rheology, volume, ground slope, channelling into pre-existing valleys and whether the flow was pahoehoe or aa were also of significance. Later Wadge (1978) and Lopes and Guest (1982) working on Mount Etna, confirmed Walker's conclusions, but developed the analysis further by showing that:

1) there had to be sufficient volume for the effusion rate/length relationship to apply and that it was not relevant for flows of less than about 1 km in length (Fig. 4.10); and

2) that the relationship between effusion rate and length depended upon cooling rate. As cooling takes place this will cause the crust of a flow to thicken and, once a critical thickness is reached, the flow will stop (Wadge 1978).

Fig. 4.8 A tumulus on a pahoehoe lava flow of Mount Etna (author's photograph).

Two studies from Hawaii have caused additional modifications to be made to Walker's model. Michael Malin (1980) produced findings which were at variance with those of Lopes, Guest, Wadge and Walker. No strong relationship between effusion rate and length was found. With similar effusion rates to those of Etna, Hawaiian flows tended to be longer and he found a good correlation between length and volume. One possible reason for these aberrant findings was that a greater number of the flows in Malin's data set were 'tube fed' (i.e. pahoehoe) than was the case for flows considered by the other writers. Since a tube feeding a flow front is contained within the body of the flow it will be relatively unaffected by atmospheric cooling and, hence, high effusion rates are not a prerequisite of

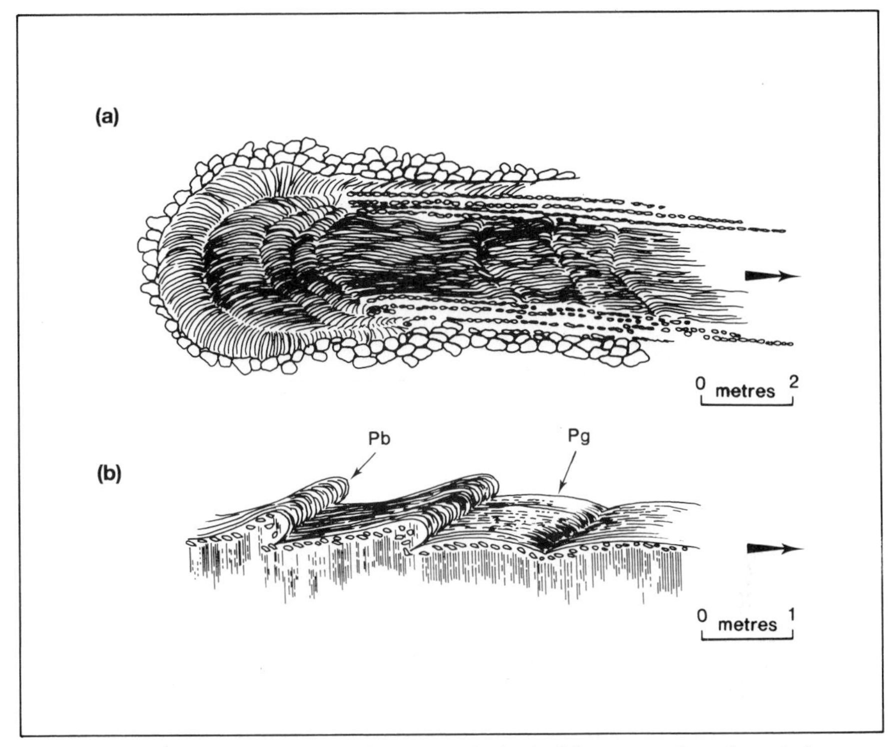

Fig. 4.9 Sketches showing some of the morphological features of toothpaste lava.
(a) Emergence of lava from an eruptive vent. Note the grooved and ridged outline imposed by the vent.
(b) Close-up showing morphology, in particular cross-ridges thought to be due to extrusion of viscous lava in a discontinuous manner. Pb are known as 'pulse buckles' and Pg 'pulse flaps' (from Rowland and Walker 1987, Fig. 3a and b, p. 633).

length. In these cases total volume of material is critical. Malin also found that additional factors influenced length, namely: cross-sectional area; viscosity and cooling rate. Pieri and Bologa (1986) took a new look at the issues involved and studied both theoretical models of heat loss from lava flows and empirical data. The authors' argue that the most significant relationship is one between eruption (i.e. effusion) rate and the plan area of flows and that this, unlike some other suggested associations, had a sound physical basis. They note that the reason why Walker's correlation between eruption rate and length is commonly found, is because length usually correlates with plan area.

4.3.2 Lavas produced from silica-rich magmas on land

In terms of composition lavas produced from silica-rich magmas fall into two categories. There are those produced by very silica-rich rhyolitic magmas (>70% SiO_2) and those more characteristic of magmas of an intermediate composition, i.e. andesites (~57–63% SiO_2), trachyte (~58–70% SiO_2) and dacite (~63–70% SiO_2) (Le Bas and Streckeisen 1991).

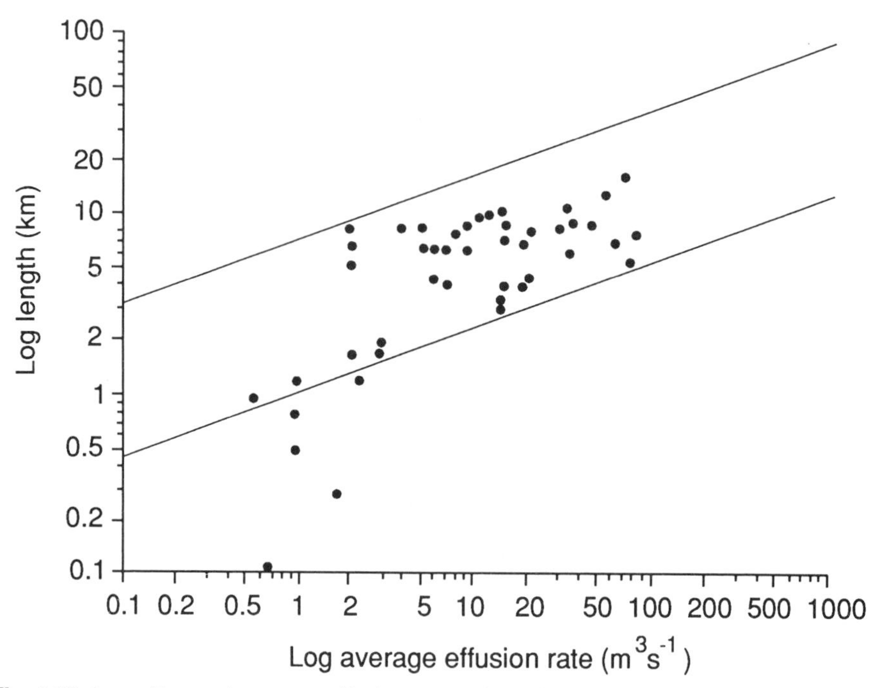

Fig. 4.10 Logarithms of average effusion rates plotted against logarithms of the lengths of historic flows on Mount Etna. It should be noted that for flows shorter than about 1 km, the relationship between effusion rate and length does not hold because the supply of lava has stopped before the full flow length could be developed (based on Lopes and Guest 1982; and from Chester, *et al.* 1985, Fig. 4.19, p. 162).

4.3.2.1 Lavas derived from rhyolitic magmas

Volcanoes formed from rhyolitic magmas erupt very violently, but extremely infrequently. Such eruptions are normally associated with the generation of pyroclastic materials (see Chapter 5) rather than lavas and, according to Cas and Wright (1987, p. 790), only one such event has been witnessed, this being the final subaerial phase of an eruption which formed two new islands between 1953 and 1957 in the St Andrew's Strait, northern Papua New Guinea (Reynolds *et al.* 1980). Many other examples of rhyolitic lava flows—both Holocene and older—have been studied and it is upon these that the information contained in this section is based. The St Andrew's Strait eruption began as a submarine event and it was only later that subaerial flows emerged. This seems to be a common sequence of events on land as well as at sea, for geological reconstructions imply that lava flows are also and commonly a late-stage product. For instance as discussed in Chapter 5 (section 5.3.4) and by Smith and Bailey (1968), the production of rhyolite often occurs after the eruption of large volume ignimbrites, in association with ring faulting and caldera collapse (see Fig. 5.8). The Mono craters in California show these associations, but at other volcanoes such as the rhyolitic domes of Lipari (Aeolian Islands) there are no simple correlations with caldera ring faults. Rhyolitic lava is, however, almost always associated with major phases of pyroclastic activity (Cas and Wright 1987, pp. 79–80).

It is estimated by Simkin *et al.* (1981) that during the last 10,000 years some

217 dome-forming events have occurred. They are common features on volcanoes erupting dacitic, trachytic and andesitic lavas and examples include the Castello dome (Island of Ischia, Italy) which is trachytic (Rittmann 1930) and many of the domes of the Qualibou caldera of St Lucia (Caribbean) which are dacitic (Robson and Tomblin 1966). Domes are notable features of silica-rich lavas, because high viscosity inhibits flowage far from eruptive vents. The terminology used to describe domes is confusing and no agreement has been reached amongst authors. The morphological classification of Williams and McBirney (1979) is clear and easy to use and is reproduced as Fig. 4.11a and Table 4.5. A more recent classification by Blake (1990) not only reduces the number of individual types of dome to just four, but also links these successfully to their processes of formation. As Fig. 4.11b shows the four types include Williams and McBirney's Peléan domes (taken to include spines) and upheaved plugs. Two additional types are defined: the *low lava dome* and the *coulée*, the former being similar to the endogenous domes described in Table 4.5 and the latter a recoining of a term first introduced by Cotton (1944) to describe stubby lava flows transitional between true flows and low lava domes. The intrusive dome (also known as a cryptodome) is not considered by Blake, which is understandable because it is not strictly a dome, but rather an igneous intrusion.

Using data from experiments with kaolin slurries and from field work, Blake has been able to make several statements about the morphology and modes of formation of domes, which expand upon the points made in Table 4.5. In the first place upheaved plugs can form only when emergent magma is very viscous and/or strong, as the lava does not deform when it emerges from the eruptive vent. A good example of an upheaved plug is the O'Usu dome on Usu volcano, Japan (Williams 1932), but other domes of this type are rare because they are unstable landforms and liable to collapse. Secondly, although it is an over-simplification (see section 4.2), a *Bingham* rheology may be used to model the growth of domes. When modelling is carried out it emerges that the principal reason why Peléan domes are often associated with the explosive generation of pyroclastic flows, whereas low lava domes like that produced by the 1979 eruption of Soufriere (St Vincent) are not, is because of differing yield strengths. Peléan domes are much stronger than low lava domes and 'such strengths are compatible with explosive release of pressurized magma which characterizes Peléan (*sic*) eruptions, but which is absent during low lava dome emplacement' (Blake 1990, p. 124). This distinction has major implications for the hazard analyst. The transition from low lava domes to coulée is a third point of discussion. As Fig. 4.11b shows, coulée form on sloping ground. Blake argues that as magma has a finite yield strength, a symmetrical low lava dome will form first, but will then start to flow down slope. For a given slope angle the higher the yield strength the less likely it is that a coulée will be produced. Finally, on Peléan domes there is an observed relationship between the talus apron and the overall dome geometry. Reference to Fig. 4.11b indicates an overall conformity of slope angle at the junction between the talus apron and the upper portions of the dome and measurements of the ratios of dome height to radius suggest values of 0.6–0.9, which are similar to the tangent of the angle of repose of the talus. According to Blake, Peléan domes have too high a yield strength to travel far from the vent under their own weight and the whole morphology, with the exception of any projecting spines, is controlled by repose angles of the talus, rather than the rheological properties of the lava.

Additional recent insights into the processes of dome growth and development

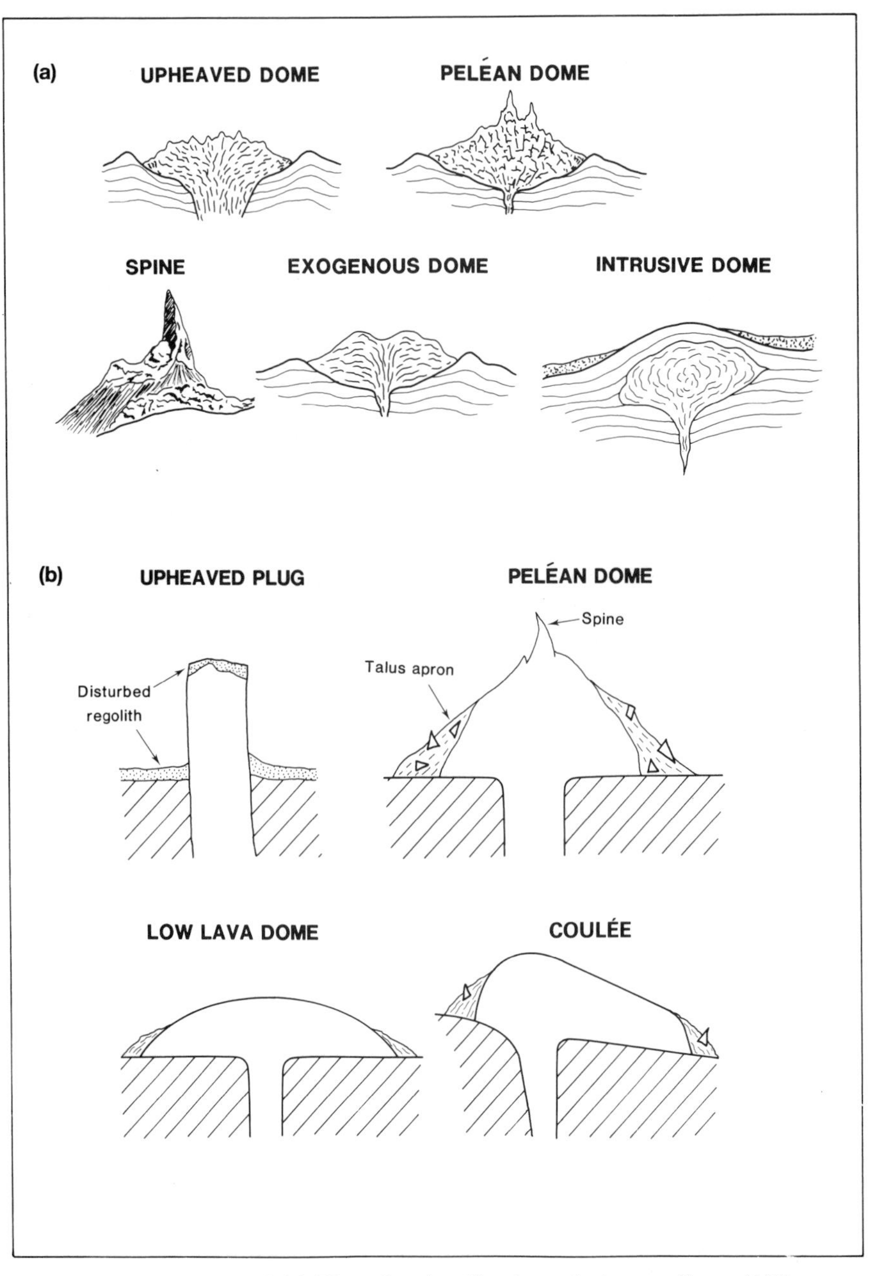

Fig. 4.11 (a) Williams and McBirney's classification of domes (from Williams and McBirney 1979, Fig. 8.10, p. 190).
(b) Blake's classification of lava domes (from Blake 1990, Fig. 1, p. 90).

have come through close study of the dacitic dome which formed on Mount St Helens between 1980 and 1986. Using Blake's classification (Fig. 4.11b) this was a Peléan dome and in a series of papers (Swanson and Holcomb 1990; Anderson and Fink 1990; Iverson 1990; Denlinger 1990), the conclusions of Blake (*op. cit.*) are both expanded and developed. A particularly valuable study was made by Steven Anderson and John Fink who studied the micro-morphology of the dome surface, which they classified into two categories: smooth and scoriaceous. They found that the relative occurrence of these micro-morphologies is determined in part by the underlying slope over which the material is erupted, but that water content is of crucial importance. Introducing a new concept—*critical water content*—they show that as this approaches 0.3–0.4 wt % the lava will vesiculate and increase the potential for explosive activity. Roger Denlinger makes the assumption that the major obstacle to lava extrusion on the Mount St Helens dome is its brittle exterior, or 'carapace', and extrusions only take place when the carapace is ruptured. He argues that rupturing may occur in response to two processes: an increase in magma pressure or a growth in cracks. Upon testing these ideas, he finds that the crack-growth hypothesis has the better support.

There can be little doubt that 1990 was a good year for lava dome studies and that knowledge was advanced significantly. It must not be forgotten, however, that Blake's research was based on just 20 domes and that the well-studied Mount St Helens dome cannot be taken as typical of even Peléan constructs. Much more testing of these recently proposed models is clearly required.

As well as domes, rhyolitic magmas are also known to produce true lava flows. For instance, the earliest rhyolites from the Long Valley caldera, California, USA, have flow units 50 m in thicknesses and flow lengths of around 6 km (Bailey *et al.* 1976), while Gibson and Walker (1963) described flows in Iceland with thicknesses up to 60 m. As Cas and Wright (1987, p. 81) note, both examples are exceptional, but for different reasons. In the case of the Long Valley rhyolites it is notable that the lavas are 'aphyric' (i.e. there is an absence of visible crystals, or phenocrysts, and the rock is fine-grained and/or glassy) with the implication that mobility was imparted by very high eruption temperatures. The lavas of Iceland, in contrast, come from a mixed series which includes basalts as well as rhyolites and it is probable that viscosity was reduced and, hence, mobility increased, by superheating following magma mixing (see Chapter 3, section 3.3.1.3).

4.3.2.2 Lavas derived from dacitic and andesitic magmas

Like the magmas from which they are derived, so lavas of this composition are in several respects intermediate and have characteristics in common with both basaltic flows and rhyolitic flows and domes. As mentioned in section 4.3.2.1 dacite and andesite commonly produce domes, but the 'classic' landform is the block lava flow (Fig. 4.12). In the past the term *block lava* was used as a synonym for aa flows, but in recent years it has been recognized as a distinctive category of flow in its own right (Macdonald 1972; Williams and McBirney 1979). Although block flows are produced by basaltic magmas and even rhyolites, typically they are of andesitic and dacitic composition (Macdonald 1972), of small volume, short in length, slow moving and have a surface of partially solidified, angular detached blocks riding above a molten interior (Decker and Decker 1991). After cooling the blocks pass down into massive flows; sometimes columnar jointed. Flow fronts are steep and falls of lava from high on the front are common. Mechanisms of movement are very similar to those of aa flows (see section

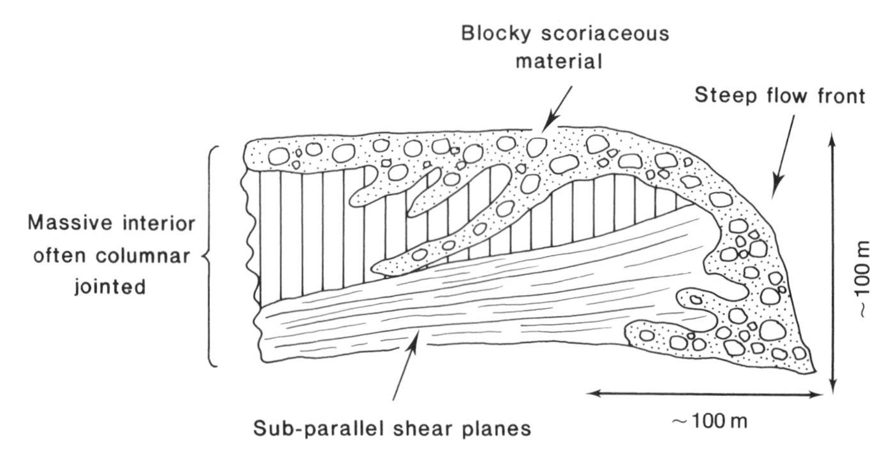

Fig. 4.12 Idealized section through a block lava flow (composite of several sources).

4.3.1.2), but because of higher viscosity 'angular blocks are formed by breaking . . . of the partly to wholly congealed upper part of the flow as the still-mobile part beneath it continues to move' (Macdonald 1972, p. 92). Higher viscosity also results in greater internal shearing along planes subparallel to the surface over which the flow passes (Fig. 4.12) and these may be preserved as 'plains of separation' once the flow cools.

Occasionally effusions of andesite and dacite can achieve impressive dimensions. One well-known example is the Chao lava of northern Chile (maximum age 100,000 years), a dacite flow some 14.5 km long and with flow fronts up to 350–400 m in height (Guest and Sanchez 1969). According to de Silva and Francis (1991) this is the largest coulée known and its surface is notable for the development of flow ridges—'ogives'—which give the flow an 'elephant skin' appearance in plan view, similar in fact to the texture of basaltic pahoehoe flows. The reasons why a dacite magma should produce such a large flow are not known, but it has been suggested (see de Silva and Francis 1991) that like all coulée, ground slope was probably of importance in addition to the large volume of material involved (~26 km^3).

4.4 Lavas produced subaqueously

In textbooks published up to 20 years ago relatively little attention was paid to lavas erupted below, or flowing into, water. Chapters 2 and 3 of the present volume consider the reasons why vast amounts of theoliitic basalt are found on ocean floors and the manner in which volcanic islands are built above mantle plumes, while in Chapter 5 (section 5.5) submarine pyroclastic activity is considered. Today the theme of subaqueous volcanic activity has been given renewed emphasis not only because of its importance in terms of the tectonic evolution of the planet, but also as a result of the wealth of data now available from both the direct and indirect (i.e. remotely sensed) exploration of both shallow waters and ocean floors. Although clearly information is neither as good nor as comprehensive as it is for volcanoes on land, it is possible to make useful statements about the distinctive types of flows and other landforms produced by both basaltic and more silica-rich magmas.

4.4.1 Subaqueous basaltic lavas

It has been known for over a century that the most characteristic form of basaltic lava to be found in subaqueous environments is *pillow flows* (see Snyder and Fraser 1963). Pillows are roughly elliptical bodies of lava separated from each other and ranging from less than 0.3 to 1 m or more across. Although pillows appear as discrete entities in both plan view and section, in three-dimensional reconstructions it is clear that most, if not all, are interconnected and represent cross-sections through lava tubes (Hargreaves and Ayres 1979). Study of oceanic flows and ancient successions has demonstrated that pillows are closely associated with submarine sheet flows, but it was only in 1971 that actual processes of formation were observed (J. G. Moore 1975); it being clear that, like pahoehoe flows on land, pillow flows are also fed by branching lava tubes. The principal features of and nomenclature used to describe pillow lavas are summarized in Fig. 4.13. As the figure shows, the thin *selvedges* (i.e. edges or 'skins') of individual pillows are often glassy. Window glass is rapidly cooled silicate melt and, when lava is quickly quenched by water, either opaque or clear glass (sideromelane) is formed (Macdonald 1972; Williams and McBirney 1979). Weathering and other changes known as palagonitization normally cause secondary minerals such as zeolite to be dominant (Heiken and Wohletz 1985).

Often the term *hyaloclastite* is used as a synonym for volcanic glass, especially when found between individual pillows (Fig. 4.13), but sometimes it is used to describe a distinctive type of flow produced by the 'drastic chilling of fluid lava discharged onto the sea floor and at shallow depths beneath it, or into rivers and lakes, or beneath glaciers' (Williams and McBirney 1979, p. 113). Although commonly associated with pillow lavas, hyaloclastite sometimes forms massive deposits which may grade laterally into terrestrial lavas and laterally and downwards into pillow flows (Fig. 4.14). It is thought that the formation of hyaloclastite may protect underlying pillow flows from very rapid quenching (Macdonald 1972, p. 105).

Whereas mid-ocean ridges produce effusive pillow and sheet flows and generally a restricted range of glassy materials, seamounts often have very extensive suites of hyaloclastite. These comprise not only *in situ* deposits, but also vast debris flows, or 'stone streams', on the sides of the seamount (Lonsdale and Batiza 1980).

4.4.2 Subaqueous silica-rich lavas

Silica-rich lavas erupted under ice are sometimes accessible to study and a small number of papers from Iceland (e.g. Furnes *et al.* 1980) and Antarctica (e.g. Hole 1990) have served to specify their principal attributes and mechanisms of emplacement. It must not be forgotten, however, that what is known is based on a very small number of recently published studies and that there are many areas of the world, such as the Central Andes (de Silva and Francis 1991), where interactions between glacial ice and silica-rich lavas have occurred, but about which little is known.

The most comprehensive study of silica-rich lavas erupted in the subglacial

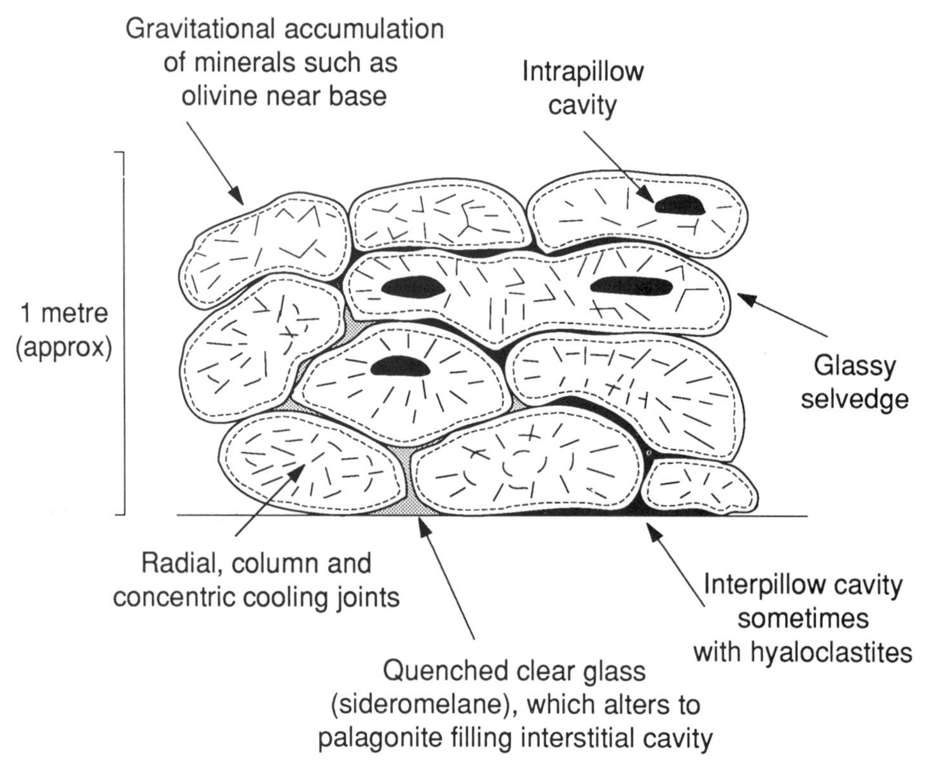

Fig. 4.13 Section through pillow lavas (based on a number of sources including Hargreaves and Ayres 1979; Williams and McBirney 1979; and Cas and Wright 1987).

environment is a study by Furnes *et al.* (1980) of rhyolitic flows in Iceland. Although the authors admit that the direct observation of subglacial eruptions is impossible, from the products generated they are able to demonstrate that two components are commonly encountered, these being rhyolitic hyaloclastites and lava lobes. They also show that the hyaloclastites may be subdivided into:
Type 1—consisting of fragments of pumice and glass; and
Type 2—comprising, angular fragments of obsidian (i.e. rhyolite containing more than 80% glass), together with flow banded and folded pumice and lithic-rich rhyolite.

Arguing by analogy from subaerial rhyolitic events, they suggest that the initial stages of subglacial eruptions are likely to have been explosive and to have produced pumice-rich—Type 1—hyaloclastites (Fig. 4.15a). Later as eruptions progress, the authors posit that the magmas become depleted of volatiles and rhyolitic lavas are able to flow along the surfaces formed between the hyaloclastites and meltwater (Fig. 4.15bi), as well as being able to intrude the hyaloclastite piles (Fig. 4.15bii). In both cases flows will cool extremely rapidly and obsidian 'skins' will form, effectively sealing the tops of the flows/intrusions. Still, later as more magma is supplied, bubbles will form beneath the 'skins' (Fig. 4.15bi and bii), the flows/intrusions will develop laterally, producing flow banding and folded lava lobes (Fig. 4.15ci). Eventually the pressure of the magma will exceed the strength of the 'skins', which will be broken (Fig. 4.15cii and ciii) and Type 2

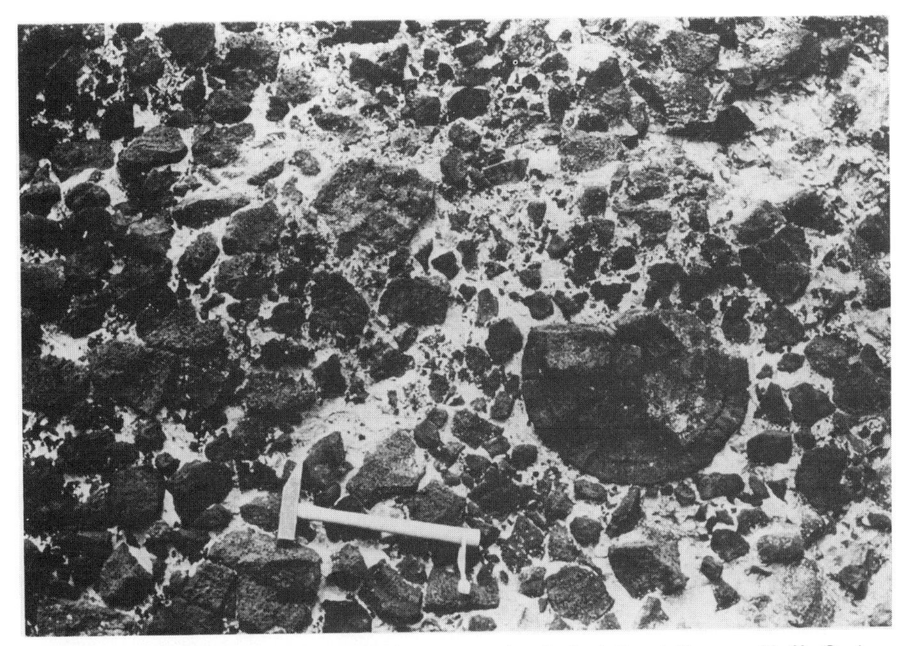

Fig. 4.14 Pillow lavas and hyaloclastite erupted subglacially at Brown Bluff, Graham Land, Antarctic Peninsula (photograph by Dr A.J. Milne and reproduced with the permission of the British Antarctic Survey).

hyaloclastites will be formed. The whole process, so the authors suggest, is capable of being repeated several times in a given eruption.

The model proposed by Furnes and his colleagues is both comprehensive and convincing, but it should not be forgotten that until it is tested and, if necessary modified, by other closely studied examples it must remain provisional. It is known that similar lobate lava flows are encountered in Alaska, but whether the actual processes proposed by Furnes *et al.* (1980) can be substantiated has yet to be resolved.

Submarine silica-rich lavas and hyaloclastites have not been described extensively in the literature. The principal reason is their rarity, especially in submarine environments, but another factor is also of importance. It is apparent that many early accounts failed to recognize these distinctive products and it is interesting that on the Island of Ponza (Tyrrhenian Sea, Italy), distinctive rhyolitic products were formerly classified as conglomerates and tuffs (Abich 1841), conglomerates (Judd 1875) and subaerial breccias (Doelter 1876). Following detailed research by Hans Pichler (1965), it is now accepted that the deposits on and around Ponza were generated by shallow-water activity. Pichler found that rhyolites were extruded into shallow water as dykes and formed domes up to 200 m in height (Fig. 4.16), which were chilled rapidly to produce outer glassy margins and extensive breaking (i.e. becciation) and jointing. The domes are covered by a thick layer of hyaloclastite, some of it massive but, especially on dome flanks, showing bedding and the incorporation of fragments (i.e. *hyaloclastite breccia*). As with the products of subglacial eruptions, so Pichler's account of Ponza is one of the very few detailed studies of shallow submarine activity and there is no way of

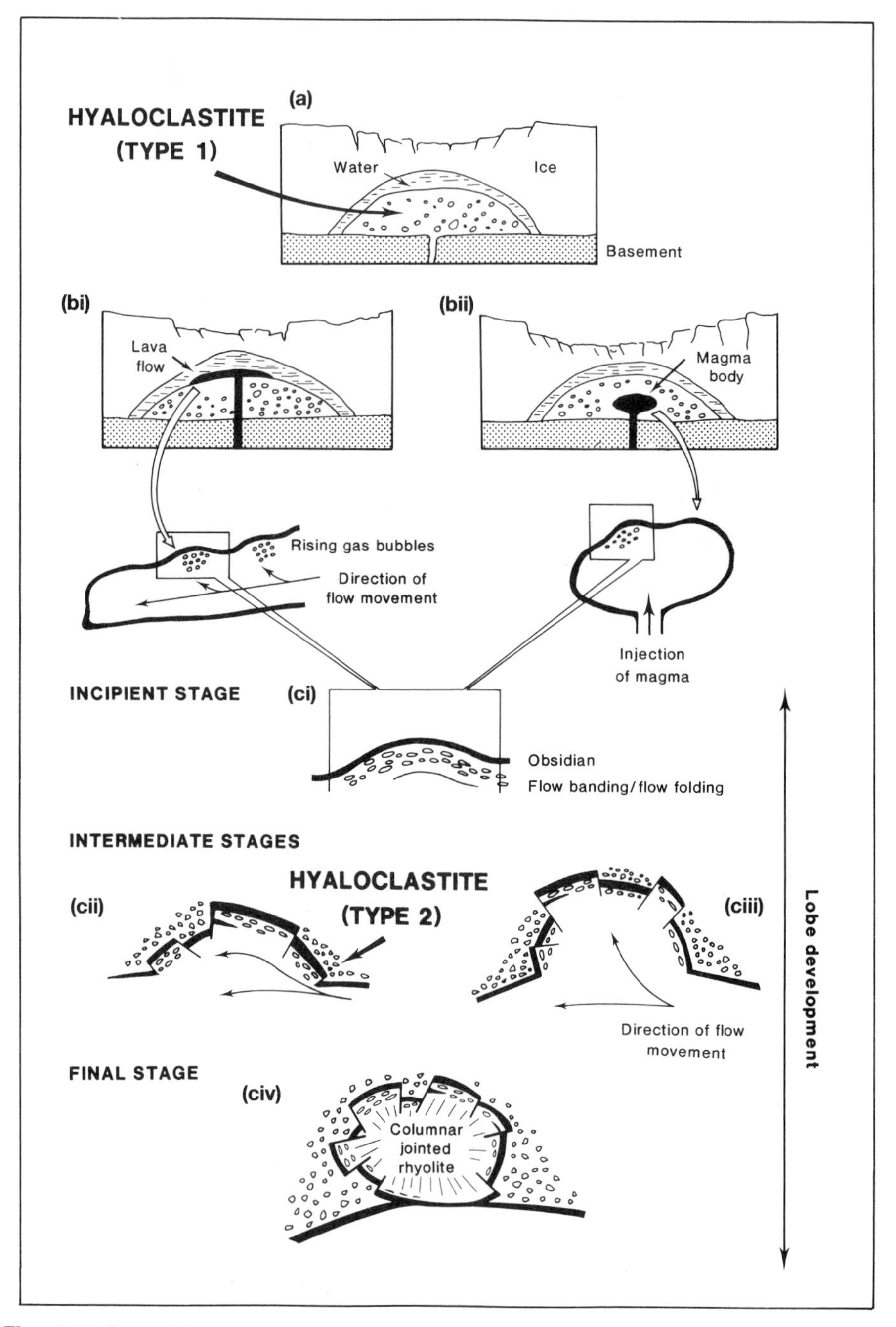

Fig. 4.15 A model showing the formation of two types of silica-rich hyaloclastite and lava lobes in subglacial environments. For explanation see text (from Furnes, Friedleifsson and Atkins 1980, pp. 95–110).

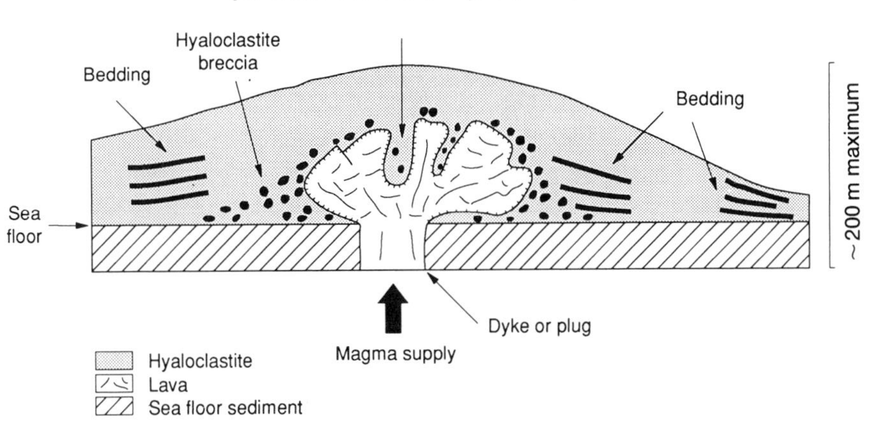

Fig. 4.16 Silica-rich lava dome extruded into shallow water at Ponza, Italy (based on Pichler 1965).

knowing whether the landforms described are typical.

Because of the volcano/tectonic situation (Chapter 2), it comes as no surprise to find that studies of ocean basins have yet to find any major silica-rich lava flows, domes and hyaloclastites, although small volumes of such materials are known from oceanic islands (Cas and Wright 1987).

4.5 Komatiites

According to Arndt and Nisbet (1982, p. 1) 'komatiites are fascinating rocks, spectacular and varied in outcrop and in hand specimen, attractive in thin section, and of great geochemical, tectonic and economic significance'. Until recently komatiites were little understood and, indeed, the term itself was not used in a consistent fashion (Condie 1982). Although a few later examples are known, komatiites are almost exclusive to Archaean era of the Precambrian, were erupted between $2.5–4.6 \times 10^9$ years ago and are ultrabasic having normally 43.5% or less SiO_2, with unusually high amounts—often over 18%—of MgO. There is also an intermediate category of rocks termed *komatiitic basalts* with MgO contents of between 10–18% (Middlemost 1985). They are by no means rare rocks and occur in a number of Precambrian provinces including: the Barberton Mountains in South Africa; Munro Township, Canada; Belingwe, Zimbabwe; and several areas of Western Australia. They are not only unique compositionally, but also in terms of their mineralogy, physical properties, behaviour and field characteristics. Komatiite magmas gave rise to a range of volcanic products, with lava being very common, but hyaloclastite and lithified fine ash (tuff) is also encountered. Mineralogically komatiites are ultramafic, dominated by olivine, pyroxene and chromite (see Chapter 3, section 3.2) and possess very distinctive textures, with needle-like crystals criss-crossing, radiating and clustering (Cas and Wright 1987). This texture, often called *spinifex*, is commonly accepted by petrologists as representing rapid cooling. Komatiitic lavas frequently display internal layering and, when examined in field section, appear to have caused erosion of the substrates over which they flowed.

In the 1980s, understanding the processes responsible for producing komatiites was advanced greatly by the publication of a number of books and papers (e.g. Arndt and Nisbet 1982; Huppert *et al.* 1984; Huppert and Sparks 1985) and from these works it is evident that eruption temperatures of between 1400 and 1700°C are implied, compared with values of only around 1200°C for recent basalts. Because of high temperatures Huppert *et al.* (1984) argue that flows would have melted the ground over which they flowed, so accounting for the observation that komatiites are responsible for erosion. High temperatures are associated with low viscosities and, because of this conjunction of circumstances, Huppert and his colleagues argue that lavas would not have moved as laminar Bingham, but rather as turbulent pseudoplastic or even Newtonian substances. Reynolds numbers would have been well in excess of the critical value (see section 4.2.2) required for turbulent flow.

Huppert *et al.* (1984) was a 'benchmark' paper, not only because it integrated geological observations with physical principles, but it also modelled the effects of cooling by means of crystallized, saturated, aqueous solutions of Na_2CO_3 which were cooled from above. These experiments indicated that rapid rates of cooling were probably responsible for the development of spinifex textures.

4.6 Concluding remarks

Recent advances in the study of lava flows and lava landscapes have come about through the integration of 'traditional' field-based geological and geomorphological approaches, with new insights from physical modelling and laboratory experiments. It may be contended that the momentum that has built up over the last decade and a half has yet to run its course and that additional advances will be made in the near future.

5

The pyroclastic suite

In an incredibly short span of time a red-hot avalanche swept down to the sea. . . . It was dull red, with a billowy surface, reminding one of a snow avalanche. In it there were large stones, which stood out as streaks of bright red, tumbling down and emitting showers of sparks. In a few minutes it was over (The first good description of a pyroclastic flow—Mont Pelée, Martinique—Anderson and Flett 1903)

5.1 Introduction

Since 1970 considerable progress has been made in understanding the pyroclastic suite, with hundreds of important research papers and several seminal texts being published (e.g. Fisher and Schmincke 1984; Heiken and Wohletz 1985; Cas and Wright 1987). Traditionally research on the pyroclastic suite has been dominated by igneous petrologists (e.g. Williams *et al.* 1982) and rapid progress has coincided with new contributions to old problems from a wide range of sciences which have included physics (e.g. Wilson 1980), sedimentology (e.g. Cas 1978) and geomorphology (e.g. Clapperton 1990). Since 1970 the opportunities for first-hand on-site investigation of eruptions have been considerable and important new clues to the processes responsible for producing pyroclastic materials have come from volcanoes as diverse in location and style as: Heimaey (Iceland); Mount St Helens (USA); Nevado del Ruiz (Colombia); El Chichón (Mexico); Unzen (Japan); and Pinatubo (Philippines). Additional insights have emerged from the theoretical and mathematical modelling of eruptions (e.g. Wilson 1976; Carey and Sparks 1986), laboratory experiments (e.g. Wohletz and McQueen 1984; Whitham and Sparks 1986), computer simulations of eruptive behaviour and the transport of particles (e.g. Cornell *et al.* 1983; Wohletz and Valentine 1990) and the reinterpretation of the products of historic eruptions (e.g. Barberi *et al.* 1978; Sigurdsson *et al.* 1985). What has been achieved is considerable, but many problems remain unsolved and many issues are still subjects of discussion and dispute.

5.2 The classification of pyroclastic materials

One of the major difficulties involved in reviewing recent work on pyroclastic materials is nomenclature and classification. Several attempts have been made to produce simple classifications, but confusion in the use of terms is still evident in the literature. Pyroclastic fragments are derived from explosive volcanism and are produced by the rapid expansion of volatiles during eruption. There are two situations in which explosive volcanism can occur, though they may and often do operate together (Wohletz and McQueen 1984). The first is the *magmatic*, 'in which volatiles in the melt (dominantly H_2O) exsolve and explosively fragment the magma by rapid decompression' (Wohletz and McQueen *op. cit.*, p. 158) and the second is *hydrovolcanic* or *hydromagmatic*, in which hot magma interacts with external water at and/or near the ground surface.

In 1980 an important paper was published which simplified the confusing array of terms which were then current in the literature (Wright *et al.* 1980). In their paper, *A working terminology of pyroclastic deposits*, John Wright and his colleagues recognized that the principal problem of classification is that two systems are required: one based on sediment origin (or genesis) and the other on sediment lithology (e.g. particle-size distribution, types of constituent fragment and amount of welding). Wright proposed that the simple scheme first proposed by Sparks and Walker (1973), which placed pyroclastic materials into *fall*, *flow* and *surge* categories, was a reasonable classification by origin, was in accord with accepted theory and should be adopted.

Fall deposits are defined as particles which are 'abruptly transported through the air or water and fall back to the surface' of the Earth (Fisher and Schmincke 1984, p. 8). Building on research by George Walker (1973b), Wright *et al.* (1980) used two indices to link different types of fall to Hawaiian, Strombolian, Sub-Plinian and Plinian eruptive styles (see Chapter 3, section 3.4); these indices, dispersal (D) and fragmentation (F), being defined in Fig. 5.1. Additional styles were introduced:

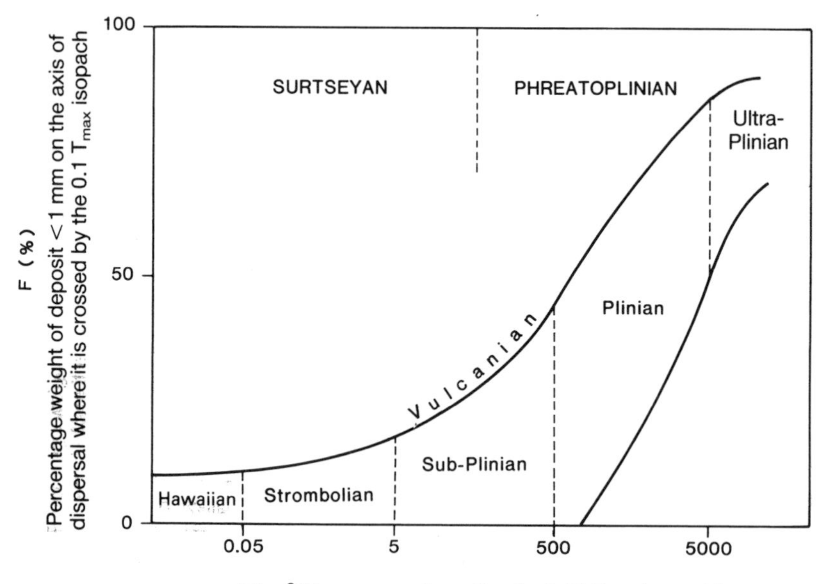

Fig. 5.1 A genetic classification of pyroclastic fall materials. Fragmentation (F) is the weight % of sub-millimetre-sized material found at a fixed point on the axis of dispersal (D) and dispersal is the area around the volcano enclosed by the 0.01 T_{max} isopach (line of equal thickness), where T_{max} is equal to the maximum thickness of the deposit (based on Walker 1973, Fig. 5, p. 439; with additional information from Walker 1980 and Wright *et al.* 1980).

1) Ultra-Plinian (Walker 1980) for the most widely dispersed magmatic falls; and 2) Surtseyan (Walker 1973b) and Phreatoplinian (Self and Sparks 1978) for two types of hydrovolcanic eruption.

Whereas falls are relatively easy to classify, *flows* presented more difficulties to John Wright and his co-workers. This was due to the vast array of terms and usages that were at the time and remain current in the literature. Flows are derived from 'an eruption cloud consisting of hot pyroclasts and gas, driven by gravity and moving across the ground as a density current. Many are generated by collapse of a particle-laden eruption column. Most pyroclastic flows move at high velocity down slope and along drainage systems' (Heiken and Wohletz 1985, p. 6). The classification proposed by Wright *et al.* (1980, 1981a) makes a distinction based on size between *ignimbrites* (Marshall 1935)—large volume pumice flows often referred to as ash-flow tuffs (e.g. Smith 1960)—and much smaller flows. Sometimes ignimbrites are considered a separate class of flow (Walker 1981a, 1982), but it is sensible to view them as end members of the flow continuum. As far as smaller flows are concerned, Cas and Wright (1987) recognize several types including: scoria flows and three subtypes of block and ash flow. Descriptions of these flows, their eruptive mechanisms, principal sedimentological characteristics and compositions are considered later in the chapter. One term, however, which continues to cause confusion is *nuée ardente*. Another is its synonym, 'glowing avalanche'. These terms have been used collectively to refer to all pyroclastic flows, have been restricted to all flows, except ignimbrites (Walker 1982), and have been restricted further to limited types of flow such as small volume block and ash flows produced by the collapse of an actively growing lava flow or dome (Wright *et al.* 1980, 1981a; Cas and Wright 1987). Both Wright *et al.* (1980) and Cas and Wright (1987) argue that it is best to avoid both terms and this convention is followed in the rest of the chapter.

Surges are the most recent major form of pyroclastic deposit to be identified, the term only becoming common in the volcanological literature from 1976 (Sparks 1976). Three subtypes of surge are recognized: *base*; *ground* and *ash-cloud*. *Base surges*, also known as *cold surges*, are generated by hydrovolcanic activity and are turbulent, low density, highly inflated flows of material. Such deposits were first recognized following nuclear test explosions (Brinkley *et al.* 1950) and only later were similar sediments described in connection with hydrovolcanic explosions such as the 1965 Taal eruption in the Philippines (Moore *et al.* 1966). A base surge consists of a ground-hugging blast of fragmental material travelling at hurricane velocities radially away from an explosive eruption. *Ground surges* and *ash-cloud surges* are considered separate entities by Wright *et al.* (1981a), but may be grouped together as *hot surges* and this convention is followed by many subsequent writers (e.g. Crandell *et al.* 1984a). *Hot surges* are highly dangerous to people and their activities (see Chapters 7 and 9), move at great speed and kill by heat and asphyxiation. They may be produced by a number of processes including explosive disruption of volcanic domes, gravitational collapse of domes (see Chapter 4, section 4.3.2.1), collapse of vertical eruption columns and lateral blasts. Often, but not always, they are associated with pyroclastic flows and may form basal units.

In addition to falls, flows and surges several other terms need to be defined. *Lahar* is word of Indonesian origin and has been used for many years to describe volcanic fragments transported by water (van Bemmelen 1949). This definition, which has found favour with several recent authors (e.g. Fisher and Schmincke 1984), encompasses all water-transported debris-flows regardless of grain size and water content and includes all flows regardless of whether they were formed during an eruption or later by processes of slope instability. This usage will be followed in section 5.6.4.

Following the 1980 eruptions of Mount St Helens a new term, *lateral blast*, entered the literature. According to Crandell *et al.* (1984a) this is an ill-defined collective term for the deposits associated with cataclysmic blasts, such as those which occurred on Mount St Helens. More recently it has been realized that the deposits produced by such events may be very varied and that the term lateral blast should be confined to the explosive mechanism and not the sediments resulting from it.

A *debris avalanche* (Ui *et al.* 1986) is defined 'as a large mass of material ... falling or sliding rapidly under the force of gravity' (Foxworthy and Hill 1982, p. 122) and is used to describe the sediments resulting from the formation of large amphitheatre-shaped hollows on the sides of volcanoes. Slope failure on a large scale is seen as the principal process of formation. Debris avalanches and their formation will be discussed in section 5.6.5.

If the distinction between *magmatic* and *hydrovolcanic* eruptions and the grouping of deposits into fall, flow, surge and laharic categories is accepted, then it is possible to define four principal situations in which pyroclastic materials are produced (Fisher and Schmincke 1984; Fig. 5.2a, b, c and d). As Fig. 5.2 shows, two are subaerial, one is concerned with eruptions under water and one relates to pyroclastic materials erupted from subaerial volcanoes and being deposited in water.

Before leaving the topic of classification by origin it is necessary to add a note of caution. Several recent papers have questioned the distinctions that are made between different types of pyroclastic materials; particularly between flows and (hot) surges (e.g. Fisher 1986; Druitt 1991). The argument, which will be more fully developed later, is that processes of formation and the deposits which result from them are members of a continuous series. In a large eruption such as the blast from Mount St Helens rapid transitions are evident from typical flow facies to more surge-like deposits.

Lithological classification of pyroclastic materials is more straightforward than a genetic one and Wright *et al.* (1980) based their categorization on three criteria: grain-size distribution and limits; fragmental composition; and degree of welding. As far as grain size is concerned, Wright *et al.* (1980, 1981a) make extensive use of a grading scheme first proposed by Fisher (1961) and this has been adopted widely by subsequent writers, either with or without modification (Fisher and Schmincke 1984; Cas and Wright 1987). It is reproduced as Table 5.1 and is used in the rest of the chapter. Pyroclastic materials may be formed from either molten magma (magmatic ejecta) or solidified rock (non-magmatic ejecta) and, despite the fact that fragments were first described and classified more than a hundred years ago by the pioneer volcanologist H. J. Johnston-Lavis, terms have not been used consistently.

Essential (sometimes called juvenile) ejecta are derived from magma, are particles of chilled melt and can be glassy (vitric) because of their speed of cooling. Crystals present in a melt before an eruption may also be present. *Lithic* is a term used to describe rocky, dense fragments within pyroclastic deposits and may be divided into juvenile fragments solidified from the magma being erupted

Table 5.1: Grain-size limits for pyroclastic materials. The distinction between blocks and bombs relates to their origin. Blocks are coarse angular fragments ejected from the volcano in a solid condition, whereas bombs are of the same calibre, but ejected whilst still molten. Because of their fluidity they are commonly shaped during flight. Agglomerate refers to masses of tephra containing bombs and breccia is a general sedimentological term for a rock composed of large angular fragments (adapted from Fisher 1961; Wright *et al.* 1981a; and Fisher and Schmincke 1984). Traditionally volcanologists have measured grain size in millimetres, but recently many have adopted the conventions of sedimentology and used ϕ (phi) units. These are defined as the $-\log_2$ of the grain size in millimetres: 64 mm = -6ϕ, 2 mm = -1ϕ, and 0.0625 mm = 4ϕ (Folk 1974)

Clast size	Pyroclast	Pyroclast deposit	
		Mainly unconsolidated: tephra	Mainly consolidated: pyroclastic rock
	Block/bomb	Agglomerate, bed of blocks or bombs, or block or bomb tephra	Agglomerate, pyroclastic breccia
64 mm	Lapillus	Layer, bed of lapilli or lapilli tephra	Lapillistone
2 mm	Coarse ash	Coarse ash	Coarse (ash) tuff
0.0625 mm	Fine ash (dust)	Fine ash (dust)	Fine (ash) tuff (dust tuff)

(cognate lithics), country rock ejected by the eruption (accessory lithics) and clasts entrained by flows and surges (accidental lithics). Together *accessory* and *accidental lithics* are sometimes called *xenoliths*. They may be of any petrological composition and age. *Accretionary lapilli*, although not strictly speaking fragments but rather sedimentary features, are spherical masses of cemented ash and are thought to represent 'accretion of particles around a wet nucleus' (Macdonald 1972, p. 133). They are common in hydrovolcanic deposits, but may also be found in airfall ashes which have been 'washed' by rain.

In considering the fragments which comprise essential pyroclastic ejecta, the degree of vesiculation is of importance. Vesicles are cavities within volcanic rocks 'formed at lowered pressures near or at the ground surface by volatile phases (*gases*) diffusing from solution in a magma' (Heiken and Wohletz 1985, p. 7—my emphasis). Pumice is white to grey to light brown, is highly expanded, of low density (normally it will float on water) and has glassy cavities (Whitham and Sparks 1986). It is normally associated with silica-rich magmas. In contrast, scoria and cinders considered here to be synonymous (Fisher and Schmincke 1984), are less vesiculated, are normally formed from silica-poor (mafic) magmas and are of higher density. Fragments will sink in water. Summarizing allows a lithological classification to be made (Table 5.2) which gives, additionally, some information on origin.

The third criterion of classification is the degree of welding. This is a post-depositional process and involves the welding, or sintering, of fragments under conditions of hot compaction (R. L. Smith 1960a, 1960b; Ross and Smith

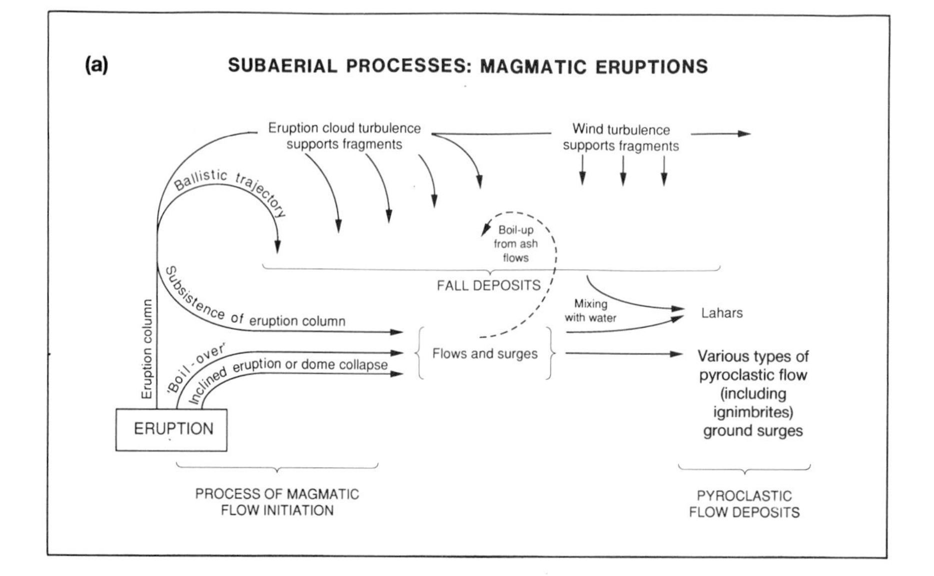

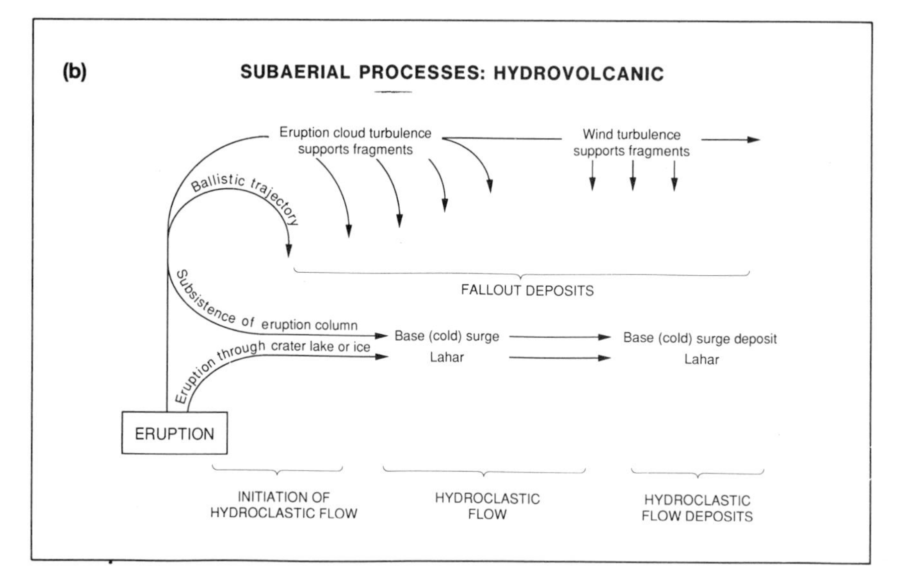

Fig. 5.2 Four model situations in which pyroclastic materials are produced and deposited: (a) subaerial magmatic eruptions; (b) subaerial hydrovolcanic eruptions;

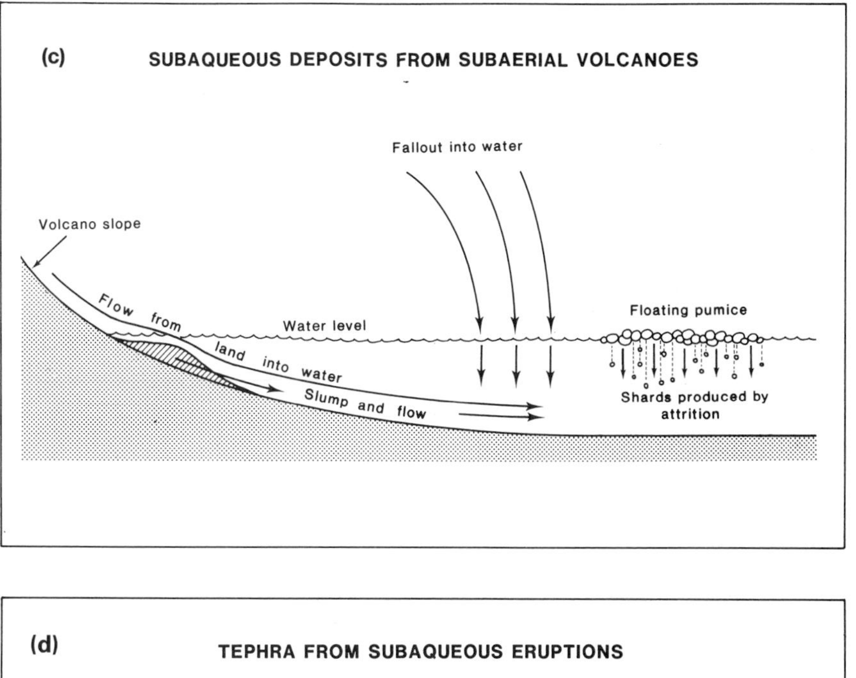

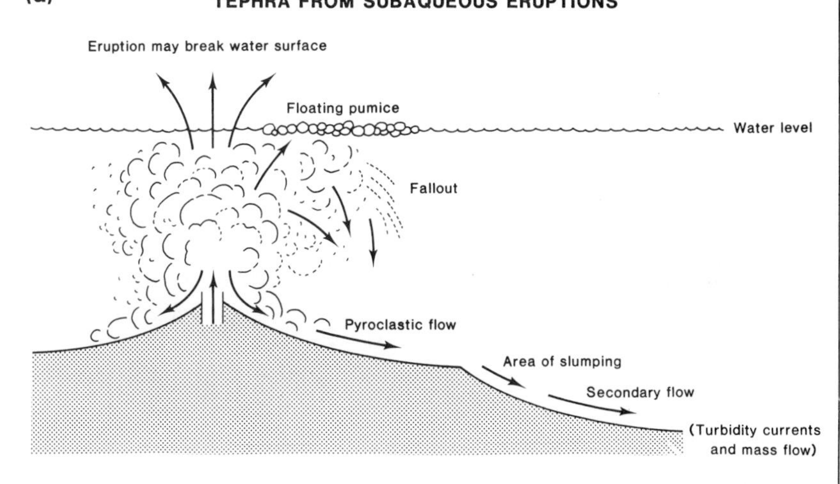

(c) subaqueous deposition from subaerial volcanoes; and (d) deposition from submarine eruptions (from Fisher and Schmincke 1984, Figs 1.3, 1.5 and 1.6, pp. 8, 9 and 10, with some amendments).

Table 5.2: Classification of pyroclastic materials based on fragmental components; (a) and (b) are based (with amendments) upon Wright *et al.* (1981a); (c) based on numerous sources

(a) Pyroclastic flows and hot surges

	Essential components		Other components
	Vesicular	Non-vesicular	
Ignimbrite/hot surge	Pumice	Crystals	Accessory and accidental lithics
Scoria flow/hot surge	Scoria	Crystals	Cognate, accessory and accidental lithics
Block and ash flow/hot surge	Poor/moderately vesicular juvenile clasts	Cognate lithics and crystals	Accidental lithics

(b) Pyroclastic falls

		Essential components		Other components
Predominant grain size	Type of fall	Vesicular	Non-vesicular	
>64 mm	Agglomerate breccia	Pumice or scoria	Cognate or accessory lithics, or both	Cognate and accessory lithics
>2 mm	Lapilli deposit	Pumice or scoria	Cognate or accessory lithics, or both	Crystals
<2 mm	Ash deposit	Pumice or scoria	Cognate or accessory lithics, or both	

(c) Wet surges

Eruptive mechanism	Temperature	Components	
		Essential	Other
Hydrovolcanic	Usually below 100°C	Mostly vitric, often with cracks Rapid cooling, may be vesicular or non-vesicular	Accretionary lapilli very common Accidental and accessory lithics are also found

1961). It is most common in ignimbrites, but examples may occur in certain fall deposits (Wright *et al.* 1980).

Before leaving the topic of classification two additional sets of terms which are commonly encountered in the literature require clarification. The first is terms used to describe materials that are neither magmatic nor hydromagmatic, and include:

1) *epiclastic* deposits formed by the weathering and erosion of volcanic rocks;
2) *autoclastic*, fragments broken by friction or gaseous explosion during movement of a lava flow or the collapse under gravity of spines or domes; and
3) *alloclastic*, 'disruption of pre-existing volcanic rocks by igneous processes beneath the earth's surface, with or without intrusion of fresh magma' (Fisher and Schmincke 1984, p. 89).

The second set is collective, the most common being:

1) *volcaniclastic*, all volcanic materials produced by any process of fragmentation even when mixed with non-volcanic sediments; and
2) *tephra* (reintroduced by Thorarinsson 1954 but originally coined by Aristotle), to describe all pyroclastic materials.

5.3 Subaerial pyroclastic eruptions

The processes involved in different types of eruption are dealt with in Chapter 3 (section 3.4) and what follows is focused on the generation of pyroclastic materials.

5.3.1 Hawaiian and Strombolian eruptions

Hawaiian eruptions produce relatively little tephra and what is produced is almost exclusively fall, which is confined to within a few kilometres of the vent (Fig. 5.1). Hawaiian magmas are basaltic and produce highly fluid lavas with low volatile contents. On Hawaii most eruptions start at fissures and produce a line of low-viscosity jets (fire or lava fountains), which are soon concentrated into one or more central vents. 'Most of the vesiculating lava falls back in a still molten condition, coalesces and moves away as lava flows' (Fisher and Schmincke 1984, p. 69). The majority of pyroclastic materials are produced early in an eruption and comprise coarse bombs, which range down in size to lapilli-sized fragments, and small volumes of glassy material. Fine-grained falls are rare. Because of the low viscosity of the lava, drops may be shaped during their flight through the air into distinctive forms as they solidify and on Hawaii tear-shaped pendants (Pele's tears—named after the Hawaiian goddess of fire) and threads of naturally spun glass (Pele's hair) are common. Lapilli and coarser fragments (known as spatter) often form accumulations around vents (spatter cones) and frequently display welding (Heiken and Wohletz 1985).

Strombolian eruptions involve more silica-rich basalts than Hawaiian activity and sometimes basaltic andesites are erupted (see Chapter 3, section 3.2). Strombolian eruptions differ from Hawaiian in that explosions are discrete events, punctuated by intervals of relative quiescence ranging from less than a second to several hours. The pyroclastic suite is again dominated by fall deposits, but the more varied conditions of eruption and their more violent character mean that the range of grain sizes is greater, fragmentation into fine grains more significant and dispersal over a wider area more commonplace (see Figs 5.1 and 5.3). Column heights are commonly 1 km or less and do not penetrate the tropopause (Walker 1981a). Except towards the more silica-rich end of the spectrum, flows and surges are rare, but examples do occur such as the Biancavilla and Montalto pyroclastic flows of Etna (see Duncan 1976; Chester *et al.* 1987), which were generated around 15,000 years ago when magmas were of a basaltic andesitic and andesitic composition (Le Bas and Streckeisen 1991). Whether, or not, activity at this time was truly Strombolian is a mute point.

Fig. 5.3 The Strombolian eruption column produced during the 1979 eruption of Mount Etna, Sicily. Typically Strombolian plumes do not penetrate the tropopause and fall deposits are not distributed far from eruptive vents. The dark areas of the plume are rich in fall materials (author's photograph).

Understanding the physics of eruption columns is vital in any consideration of eruptions producing pyroclastic materials. Important papers published in the 1970s and 1980s greatly increased knowledge in this area (e.g. Blackburn *et al.* 1976; L. Wilson 1976, 1980; Sparks and Wilson 1976; L. Wilson *et al.* 1978; Walker 1981c; Steinberg and Lorenz 1983; Carey and Sparks 1986; Carey and Sigurdsson 1989) and it is now common to separate eruption columns into three vertical zones:

1) a lower *gas thrust region*;
2) a middle *convective thrust region*; and
3) an upper *umbrella region* (Fig. 5.4 and Table 5.3).

In Hawaiian and Strombolian eruptions the restricted height of the plume means that the umbrella region is absent and falls are generated from the gas thrust region with ballistic trajectories being predominant. Falls are unaffected by wind drifting at the time of eruption (Cas and Wright 1987). When a particle falls through the air, it increases in speed until it reaches a constant, or terminal velocity, at which point forces of gravity and aerodynamic drag are in equilibrium. For a given height of removal from the gas thrust region, smaller particles with lower terminal velocities will travel further than larger ones with higher terminal velocities. This property not only explains many of the sedimentological characteristics of Strombolian and Hawaiian fall deposits (see section 5.6.1), but also allowed Wilson (1972) to construct tables linking the distances ballistic particles travel and initial 'muzzle' velocities at the vent. In Strombolian, in contrast to Hawaiian eruptions, some finer material is erupted, being derived from the convective thrust region of the plume. It may travel moderate distances and is affected by wind drifting. Many of the eruptions of Mount Etna are typified by this type of activity (Chester *et al.* 1985).

5.3.2 Sub-Plinian, Plinian and Ultra-Plinian eruptions

Magnitude of eruption, plus the distance of dispersal and fragmentation of airfall materials are the principal distinctions between the three subtypes of Plinian

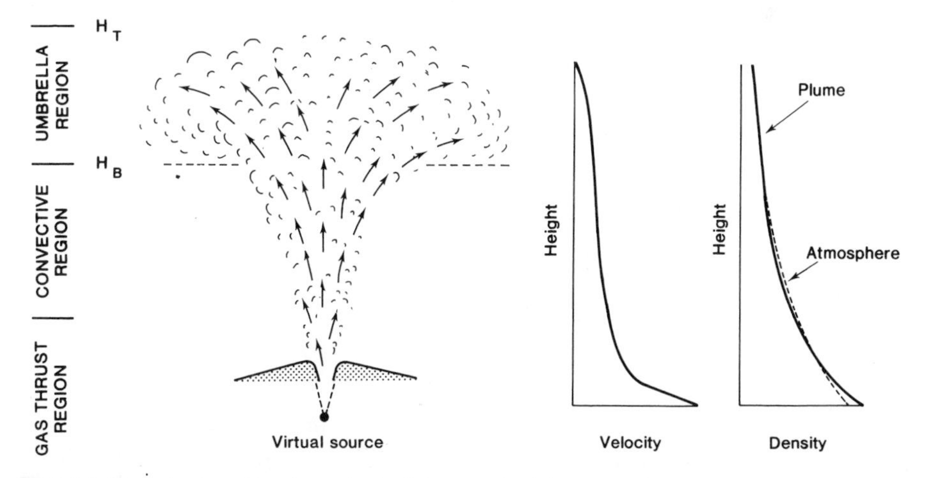

Fig. 5.4 Sketch showing the principal features of a convecting eruption column (from Carey and Sparks 1986, Fig. 1, p. 111).

Table 5.3: Summary of the processes operating within eruption columns (based on Fisher and Schmincke 1984; Carey and Sparks 1986; and numerous other sources. Additional information in notes after Embleton and Thornes 1979; Fisher and Schmincke 1984; and Summerfield 1991)

Region of column	Processes involved
Umbrella region	This is the uppermost part of the column and is notable for the lateral intrusion of material into the surrounding air. Its base is at point H_B (Fig. 5.4), where the density of the ascending plume, or 'jet' has the same density as its surroundings. The top (H_T) is defined by the momentum of the column at point H_B. Clasts entering this region will be transported laterally, often by strong winds such as 'jet streams'. Temperature inversion (i.e. an increase in temperature with height) is a feature of the troposphere/stratosphere boundary and often coincides with point H_B. Above this the plume will adopt an umbrella, or mushroom, shape.
Convective region	This region constitutes, typically, the greatest vertical extent of the plume and is dominated by forces of buoyancy. Velocities are lower than in the gas thrust region, but may still range from tens to over 200 ms^{-1}. Velocities decrease with height, apparently in a linear fashion (Sparks 1986), although this does not apply to the top and bottom of the region. Motions are highly *turbulent*[1] and velocities at any given sampling point will fluctuate irregularly and over time. Velocities are generally high enough to ensure that most ejecta are carried to the top of the region, but some will fall out of the edges of the plume where velocities are lower. Most clasts are *entrained*[2], and few but the largest (tens of centimetres) follow *ballistic trajectories*[3].
Gas thrust region	This is the region adjacent to the eruptive vent and is dominated by momentum derived from the force of the eruption. Velocities are high (hundreds of ms^{-1}) and deceleration is large. In this region the larger fragments are accelerated by gas thrust, follow ballistic trajectories[3] and may leave the column if their angles of ejection are high with respect to the vertical column axis. As larger fragments fall out of the column, density is reduced and the top of this region is defined as the point where the density of the column is less than that of the atmosphere. The grain-size range of particles is determined by initial velocity, ejection angle, and density (Wilson 1972).

[1] *Turbulence* is a property of fluid flows such as water, air and eruption plumes.

A fluid crossing a flat surface may act as a series of discrete 'layers' moving over one another; the resistance to movement being the result of *dynamic (or molecular) viscosity*, which is related to molecular composition. This is called *laminar flow* and is common in lava flows (see Chapter 4).

In certain other systems (including plumes) flow is always turbulent, i.e. velocity fluctuates in all directions within the fluid medium. Transition from laminar to turbulent flow is defined by a dimensionless—or Reynolds—number which is the ratio between velocity and *kinetic viscosity*. Kinetic viscosity is itself a ratio between *dynamic viscosity* and density. In volcanic plumes Reynolds numbers are high.

[2] *Entrainment* The initial setting into motion of a solid particle occurs when the stresses acting on it exceeds the resisting forces. With some exceptions, the higher the

activity (Fig. 5.1). They all involve the injection of large volumes of pyroclastic materials into eruption columns (Heiken and Wohletz 1985, p. 5). The term 'Ultra-Plinian' is of recent origin and was introduced by Walker (1980) to describe the most energetic Plinian eruptions with the most widely dispersed tephra, such as the AD 186 eruption of Lake Taupo, New Zealand. Sub-Plinian eruptions have lower degrees of fragmentation and less widespread dispersal of fall fragments (Fig. 5.1). They may be considered 'scaled down Plinian eruptions, and their mechanisms and dynamics can be treated as essentially the same' (Cas and Wright 1987, p. 151; see also L. Wilson 1976). In recent years much has been written about Plinian eruptions not only from the standpoint of eruptive mechanisms, but also in terms of the types of fragmental material they produce (e.g. Walker *et al.* 1971; Walker 1973b, 1981d; L. Wilson 1976, 1980; Sparks and Wilson 1976; Sparks *et al.* 1978; L. Wilson *et al.* 1978, 1980, 1987; Kieffer 1984; Carey and Sparks 1986; Carey and Sigurdsson 1987, 1989; Wohletz and Valentine 1990). Plinian eruptions not only produce large volumes of widely distributed fall materials, but also flows and surges. The factors controlling Plinian eruptions are many and varied, including the gas contents of magmas, their rheology, vent shape, vent radius and volume of magma (Fisher and Schmincke 1984), but most involve silica-rich rhyolitic, phonolitic and trachytic magmas (Chapters 3, section 3.2), being erupted at high velocities. The principal processes involved in the initiation, eruption and collapse of Plinian columns are summarized in Table 5.4 and Fig. 5.5.

During Plinian eruptions large fall fragments cannot be transported to great heights in the column and are released at low levels, to be deposited in the vicinity of the vent. Smaller fragments are transported to a height inversely proportional to particle size and are eventually deposited over a very wide area. Plinian columns penetrate the tropopause and, because particles are highly fragmented (Fig. 5.1), fallout of the finest dust may take months, or even years, and have significant climatic effects (Chapter 6). Spatially the pattern of deposition is one showing a decrease in grain size in all directions away from the vent, but modified by prevailing winds. In vertical section Plinian falls are observed to be both normal and reverse graded[1]. Normal grading is the expected outcome because particles falling to earth will form size and/or density layers, with the densest and heaviest fragments at the base. Self (1976) suggests several mechanisms working either singly or in combination which may promote reverse grading:

1) an increase in gas velocity taking coarse materials to greater heights in the later phases of eruption;
2) a change in vent morphology during eruption enabling particles to be ejected at lower angles and to travel further;
3) a widening of the vent and conduit as the eruption develops so lessening frictional drag, increasing exit velocities and causing coarse particles to be transported to a greater height; and

[1] Normal grading refers to deposits which become finer upwards while, conversely, reverse grading refers to deposits which become coarser upwards.

---

velocity the greater the calibre of material that can be entrained and transported.
[3] *Ballistic trajectory* Coarse fragments (commonly blocks and bombs) have motions determined by the power of the eruptive blast and act like shells from a cannon. They are little affected by wind.

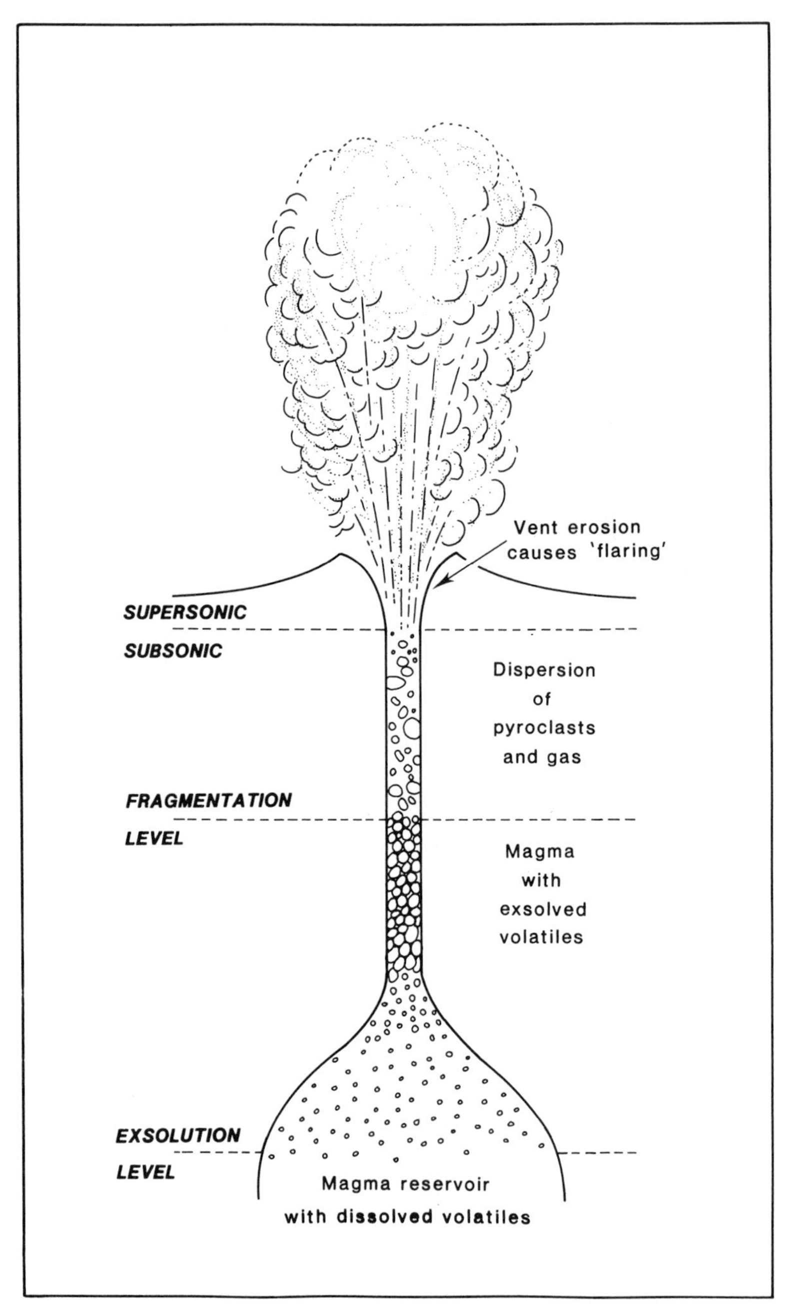

Fig. 5.5 Diagram showing different regions within a magma reservoir and conduit. Critical vertical levels are indicated (modified from Fisher and Schmincke 1984, Fig. 4.5, p. 65, and based upon ideas of L. Wilson *et al.* 1980).

Table 5.4: The initiation, eruption and collapse of Plinian eruption columns (after references in text, those acknowledged in the table, Wilson *et al.* 1987, and many other sources)

Stages in the life of a Plinian column	Comments
Initiation	Because of the magmas involved temperatures of around 850°C are common, viscosities are high and volatile contents around 5% (water and gases) are often to be found. Magmas tend to be saturated with volatiles. As magma moves towards the surface decompression takes place and, because of the high viscosity of the melt, migration of bubbles is inhibited and they form a relatively uniform 'foam'. When the volume occupied by bubbles exceeds ~75% then the melt is disrupted into highly vesiculated pumice and scoria. In this state the velocity of the pyroclast/gas mixture is very high. The roof is eventually fractured by the buoyancy of the melt and an eruption occurs. Within the magma column two critical levels of pressure occur (Fig. 5.5). The first—the 'exsolution level'—(L. Wilson 1980), separates that part of the magma reservoir where all volatiles are still dissolved from an upper portion where volatiles are being exsolved. The second—'fragmentation level'—is the level at which exsolved bubbles exceed ~75% of the total volume. Above this rapid acceleration occurs and there is dispersion of gases and pyroclastic fragments.
Eruption	Because volatile contents of magmas are high, so too are eruption velocities (typically 200–600 ms^{-1}). Surface rocks are of relatively low strength and exit velocities are initially so high then erosion of the vent occurs and the 'flared' shape may further enhance velocity and may lead to a transition from sub to supersonic flow at the point where the conduit is at its narrowest. Eruption columns commonly achieve a height of 30 km—possibly 60 km—for Ultra-Plinian events (Walker 1981a). Because velocities are so high, most fragments are 'locked' into the column and air is entrained making its bulk density much lower than the surrounding air. The 'convective region' dominates the eruption column (Fig. 5.4). Research modelling the dynamics of columns, makes allowances for air-entrainment and the progressive loss of fragments as velocity decreases, and indicates that cloud heights are proportional to the fourth root of mass eruption rates (L. Wilson *et al.* 1978). Column heights also depend critically on vent radius (L. Wilson *et al.* 1980).
Collapse	As the eruption wanes, gas content reduces and/or erosion causes the eruptive vent to increase in radius, an eruption column cannot be sustained and collapse occurs. Both flows and surges may be generated.

Note: Although pyroclastic materials are dominant in Plinian eruptions, domes and lava flows may also be features (Chapter 4).

4) an increase in column density—due to more fine material being present—causing large-calibre material to be released from a greater height.

Column collapse is a process of fundamental importance in Plinian eruptions, leading to the production of pyroclastic flows and surges. Instances such as the eruptions of Santa Maria volcano (Guatemala, 1902; see Williams and Self 1983) and Askja (Iceland; see Sparks *et al.* 1981), show that Plinian eruptions may occur without column collapse but these are exceptions. Whether or not a column collapses depends critically on a number of variables including gas velocity, vent radius and magmatic gas content (Sparks and Wilson 1976; L. Wilson 1976; Sparks *et al.* 1978; Wilson *et al.* 1980). When volatile content is high (e.g. water more than about 5%) collapse will not occur and the eruption will be dominated by convection, but when lower than about 1% collapsing columns will be produced. Between these two extremes vent radius becomes crucial. When vent radius exceeds a critical factor for a given gas velocity, then the column cannot be supported and collapse will occur (L. Wilson 1976) and flows and hot surges will move down the sides of a volcano at high speed. Further consideration of processes of movement and emplacement of flows and hot surges will follow in section 5.6.

It is clear from papers that have been published in the last two decades, that the relationships between flow and surge generation on the one hand and eruption column dynamics on the other are complex. In what has become known as the 'classic' model (Sparks *et al.* 1973), the transition between fall activity on the one hand and flow and surge on the other is attributed to a change from a predominance of convective conditions to ones favouring column collapse. There are, however, at least two problems with this model. First, it does not accord fully with the observational evidence. Carey and Sigurdsson (1989), for example, compiled a large set of data on Plinian eruptions and were able to show that for large eruptions when intensities exceeded a volume per unit time of $\sim 2 \times 10^8$ kg/s^{-1}, then eruptions tended to end with the generation of large volume flows. Many of these large eruptions were also associated with the collapse of the upper portions of volcanic edifices to produce calderas (Walker 1984). The argument of Carey and Sigurdsson (*op. cit.*) is that in these eruptions so much material is evacuated in such a short period of time that it cannot be accommodated within a simple eruptive column. Their model is in many ways similar to the one proposed by Druitt and Sparks (1984).

A second problem is that observational evidence of column collapse is not so unequivocal as is often thought. There is no doubt that collapse does occur, since it has been noted in the eruptions of Komagatake (Japan) in 1929, Mayon (Philippines) in 1968 and El Chichón (Mexico) in 1982 (Kozu 1934; Moore and Melson 1969; Tilling 1982), but the fact remains that actual observations remain limited and open to varying interpretation. Furthermore, in the Mount St Helens eruption of 1980, one of the most well observed of Plinian events, it was clear that not only did pyroclastic flows appear to precede the generation of a large column, but during the 'column phase' some small flows and surges were also generated by partial collapses (Rowley *et al.* 1981; Hoblitt 1986). The generation of flows before a column was established has led some writers to consider the possibility that another process may have been operating. This process known as 'boiling over' has been used to explain flows generated by certain Plinian and Peléan eruptions. Boiling over implies the steady outflow of material from a vent over which a stable column has not formed (L. Wilson *et al.* 1987) and involves flowage from a crater lip (Fisher and Schmincke 1984; see Fig. 5.6).

Partial collapse seems to be caused by short-term changes, or oscillations, in the properties of a column, with the result that not all material can be supported by convection (Hoblitt 1986). An alternative explanation for some of the Mount St Helens flows is that they were formed by a directed blast which occurred at the start of the eruption. The process of partial collapse was first invoked by Perret (1937) to account for flows generated by a directed (or lateral) blast from the base of the emerging dome or spine produced during the 1929 to 1932 eruptions on Mont Pelée and this model has found some favour in accounting for the early stages of the Mount St Helens eruption. The actual mechanisms remain unclear. What is certain is that flows were generated before, or at least during the early stages of, development of the convective column (Rowley *et al.* 1981).

Recently the debate about collapse has been given new impetus following the publication of several papers applying fluid and thermodynamics to eruption columns (e.g. L. Wilson and Walker 1987; Woods 1988; Woods and Bursik 1991). In the paper by Woods and Bursik the notion of thermal disequilibrium is introduced as a factor able to effect major changes in column behaviour. The authors argue that when particles are smaller than -2ϕ (4 mm—for definition of ϕ see Table 5.1), they are in thermal equilibrium with the gas and heat exchange is instantaneous. When they are larger, however, this does not apply: they are in thermal disequilibrium and this plays an important role in influencing column height and its change through time. 'Increasing the mean grain size . . . increases the degree of disequilibrium [and] for the same mass eruption rate, column height decreases steadily as mean grain size and hence thermal disequilibrium increase, until [a point] is reached at which a convective column can no longer develop' (Woods and Bursik 1991, p. 567). The column is replaced by a low fountain which is more dense than the surrounding air. An additional finding has emerged from this research. It is claimed that all columns must pass through a phase of what is termed 'superbuoyancy' before they collapse. The term introduced by Bursik and Woods, describes a column where the convective region (see Fig. 5.4) displays increasing velocity with height. Superbuoyancy in modelled eruption columns arises from either low initial velocity, or a large vent radius. A situation in which this is often encountered is when eruption rate (volume per unit time) increases but, because of vent erosion, initial (or vent) velocity decreases. The column may then evolve from being buoyant to superbuoyant. In section 5.6 the implications of this largely theoretical research in understanding the mechanisms of transport and emplacement of fall materials will be discussed.

5.3.3 Peléan eruptions

Peléan is a term used to describe eruptions involving viscous andesitic, dacitic and rhyolitic magmas, that are 'characterised by explosions of moderate to extreme violence in which solid or viscous hot fragments of new lava are ejected, commonly as . . . pyroclastic flows. These eruptions are usually associated with silicic lava dome growth' (Heiken and Wohletz 1985, p. 5). In 'traditional' classifications of volcanic styles, such as those of Macdonald (1972) and Williams and McBirney (1979), some prominence is given to Peléan activity, which is named after the cataclysmic eruption of Mont Pelée (Martinique) in 1902. In terms of the mechanisms producing pyroclastic materials, however, little can be added to the account already presented for Plinian eruptions and in fact more

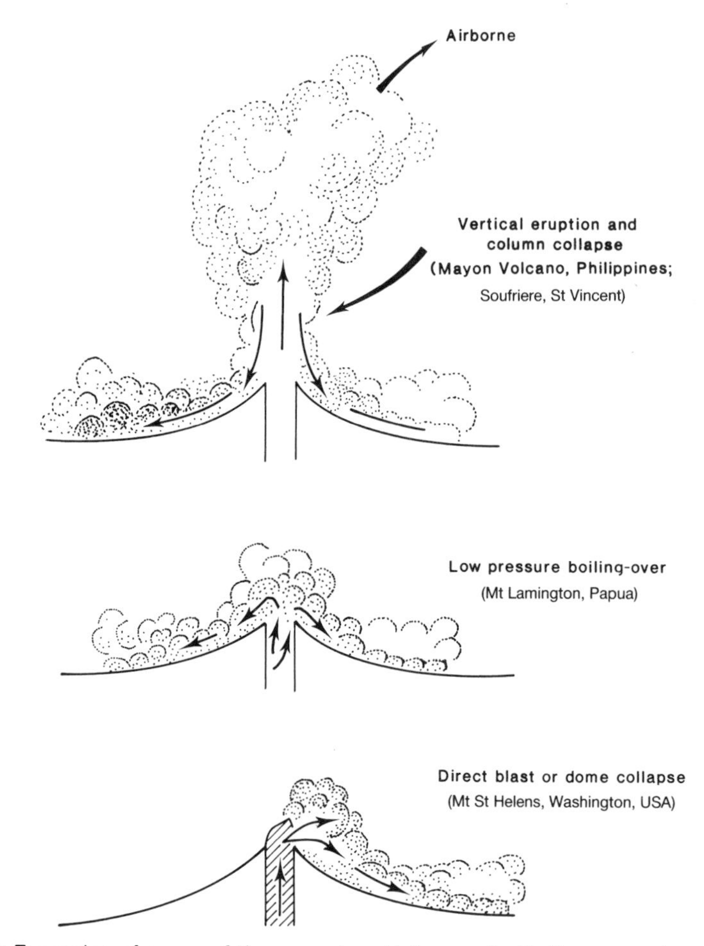

Fig. 5.6 Examples of some of the ways in which pyroclastic flows may be produced in Plinian and other eruptions (based on Macdonald 1972, Fig. 8.2, p. 149; and Fisher and Schmincke 1984).

recent, specialized works on the pyroclastic suite do not recognize any distinction (see Fisher and Schmincke 1984; Cas and Wright 1987).

5.3.4 Ignimbritic eruptions

According to George Walker (1981a, 1982) ignimbritic eruptions should be considered a separate class of explosive volcanic event. Large ignimbrites have been little understood until recently and as late as the 1930s large welded examples were commonly regarded as lava flows (Cas and Wright 1987, p. 224). Comprehending processes of formation has proved difficult because eruptions giving rise to the largest ignimbrites have never been witnessed, since they occur so infrequently. Ignimbrites range in size over several orders of magnitude from less than 1 to more than 2000 km^3 (e.g. the Fish Canyon Tuff, USA and Toba Tuff, Indonesia), occur in all volcano-tectonic settings and have compositions ranging from basaltic to rhyolitic, although those of more silicic composition are both more common and voluminous.

For small and medium-sized ignimbrites (i.e. <1 to a few hundred cubic kilometres)—like those of Vulsini (Italy) and Minoan Crete—models of column collapse like those outlined in section 5.3.2 have been often invoked to explain their generation (e.g. Sparks 1975; Sparks and Wilson 1976), but for the largest ignimbrites there has been fierce debate about whether such models explain all the features observed. In particular not all ignimbrites are derived from central vents, an essential requirement of the column-collapse model (Fisher and Schmincke 1984).

Many large ignimbrites are related to calderas (Williams 1941; Cas and Wright 1987) and there is in fact a strong positive correlation between caldera size and ignimbrite volume (Fig. 5.7). In Fig. 5.7 the most extreme case is that of the La Garita caldera (Colorado, USA) which gave rise to the largest ignimbrite known: the Fish Canyon Tuff with an estimated volume of around 3000 km^3.

In the past a model which enjoyed considerable popularity was based on observations made by Fenner (1920) of the relatively small (around 12 km^3) eruption of Katmai (Alaska) in 1912. It was thought at the time that the ignimbrite was derived from fissures in the floor of the valley now known—because of its fumarolic activity—as the 'Ten Thousand Smokes', but more recently a central vent has been located (Curtis 1968) and Fenner's hypothesis has lost ground; although it was still invoked by Nairn (1981) to explain certain New Zealand ignimbrites. In the 1960s the eminent American volcanologist R. L. Smith, suggested that large ignimbrites were not only related to large calderas, but also to fractures (Smith and Bailey 1968; Smith 1979). A diagram

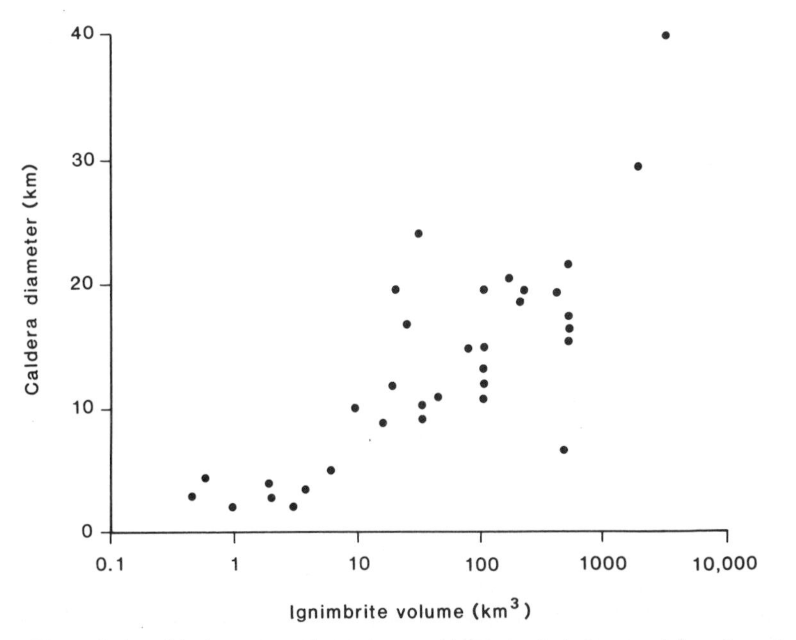

Fig. 5.7 The relationship between the volume of large ignimbrites and the diameters of associated calderas (from Cas and Wright 1987, Fig. 8.13, p. 233).

showing Smith's 'ring fracture model' is reproduced as Fig. 5.8 and this has been and remains a popular model. A later version of Smith's model posited that if fractures were outward, rather than inward dipping as shown on Fig. 5.8, then very large eruption rates would be produced by the subsidence of the caldera floor into the magma reservoir (Francis *et al.* 1983).

Close study of the Fish Canyon Tuff by Self and Wright (1983) led to yet another model being proposed, which appears to be more generally applicable to large ignimbrites. When studied in detail the authors observed that beneath the ignimbrite there was a thin Plinian fall and above it a thick surge with large wave-like bed forms on its upper margin. These bed forms were interpreted as representing a sudden release of energy during the eruption and the authors hypothesized that this could have been caused by foundering of a thin crust over a high-level magma chamber. With the magma chamber uncovered, a cataclysmic eruption was triggered and vast amounts of energy were released in a very short period of time, so producing the ignimbrite.

The principal problem with all models is that they have not been tested during eruptions. Unlike the model of column collapse, some alternatives have not been tested for their physical feasibility and in some cases even the detailed sequencing of events is still a matter of debate. For instance, in the ring fracture model, it is not known for certain whether fracturing occurs before or after caldera collapse (Fisher and Schmincke 1984). What is becoming clear and generally accepted is that column collapse on its own cannot account for all ignimbrites, especially those of large size and/or significant sedimentological complexity.

In passing it is worth noting that calderas are associated with a number of eruptive styles. They may form in Plinian, Peléan, as well as ignimbritic eruptions

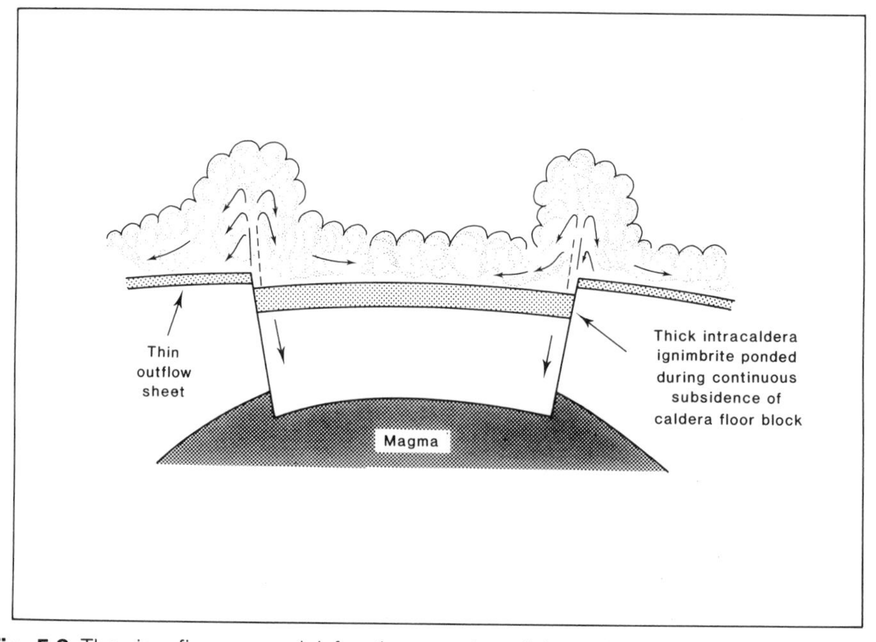

Fig. 5.8 The ring fissure model for the eruption of large ignimbrites (from Cas and Wright 1987, Fig. 8.15, p. 234).

and a flavour of current research on these major volcanic landforms may be obtained from the following sources: Francis (1983); Lipman *et al.* (1984); Wohletz *et al.* (1984); Francis and Self (1987); Gudmundsson (1988); and Scandone (1990).

5.3.5 Lateral blasts

The events at Mount St Helens in 1980 focused attention anew on the topic of lateral blasts, though examples of similar eruptions in the past are well known and include: Bezymianny, (former USSR, 1956); Lassen Peak (USA, 1915) and possibly Bandai san (Japan, 1988). Like all explosive volcanic blasts, those with a lateral component may be caused by either a sudden decompression of magmatic gases, or else the explosion of a high-pressure hydrovolcanic system (see section 5.3.6.2). They are triggered by gas pressure exceeding the load of the overlying rocks, or else by the sudden removal of these rocks by some outside agency, such as a landslide (Crandell *et al.* 1984a). In the case of Mount St Helens a tectonic earthquake was responsible for the failure of an area of significant swelling, or bulging, on the north side of the volcano which led to the 'unroofing' of the magma chamber (Foxworthy and Hill 1982), but other agencies of slope failure—such as gravity-controlled mass movement—or, alternatively, the structural weakness of one sector of a volcanic edifice could trigger a lateral blast if magma was in a pre-eruptive state.

The most important feature of lateral blasts is that pyroclastic materials move at speeds greatly in excess of those that occur under gravitational acceleration, since the force of eruption is focused into a restricted sector of the volcano and large volumes of material are ejected in a short period of time. According to Moore and Simon (1981) the initial velocity of the Mount St Helens blast was over 600 km/h^{-1}. Depending upon the circumstances of eruption, the temperature of deposits may range from almost cold to very hot and, indeed, weeks after the Mount St Helens eruption some debris still had a temperature of over 250°C (Banks and Hoblitt 1981), although 150°C after a few days was more common (Crandell *et al.* 1984a). Because of the violence of eruption and the number of factors which may have been involved (e.g. external water, simultaneous eruption of a vertical column and the possibility of channelling of the blast into pre-existing depressions—like radial valleys—on the sides of the volcano), conditions within the blast were complicated and variable (Kieffer 1981a, 1981b). A great variety of processes would have been operating simultaneously. When combined with differing conditions of transport and emplacement, it means that the suites of deposits produced by lateral blasts are complex both spatially around the volcano and in vertical section (Fig. 5.9) and for these reasons the term *lateral blast* should be restricted to eruptive processes alone.

Lateral blasts give rise to falls with ballistic trajectories, flows and surges (Figs 5.9 and 5.10). It is also common for the blast to be so violent that much of the pre-existing volcanic edifice is destroyed.

5.3.6 Open-system eruptions

In sections 5.3.1 to 5.3.5, emphasis has been placed on the role of magmas and volatiles as causal agents in the production of explosive eruptions and different types of pyroclastic material. In other words the magma reservoir and the conduit

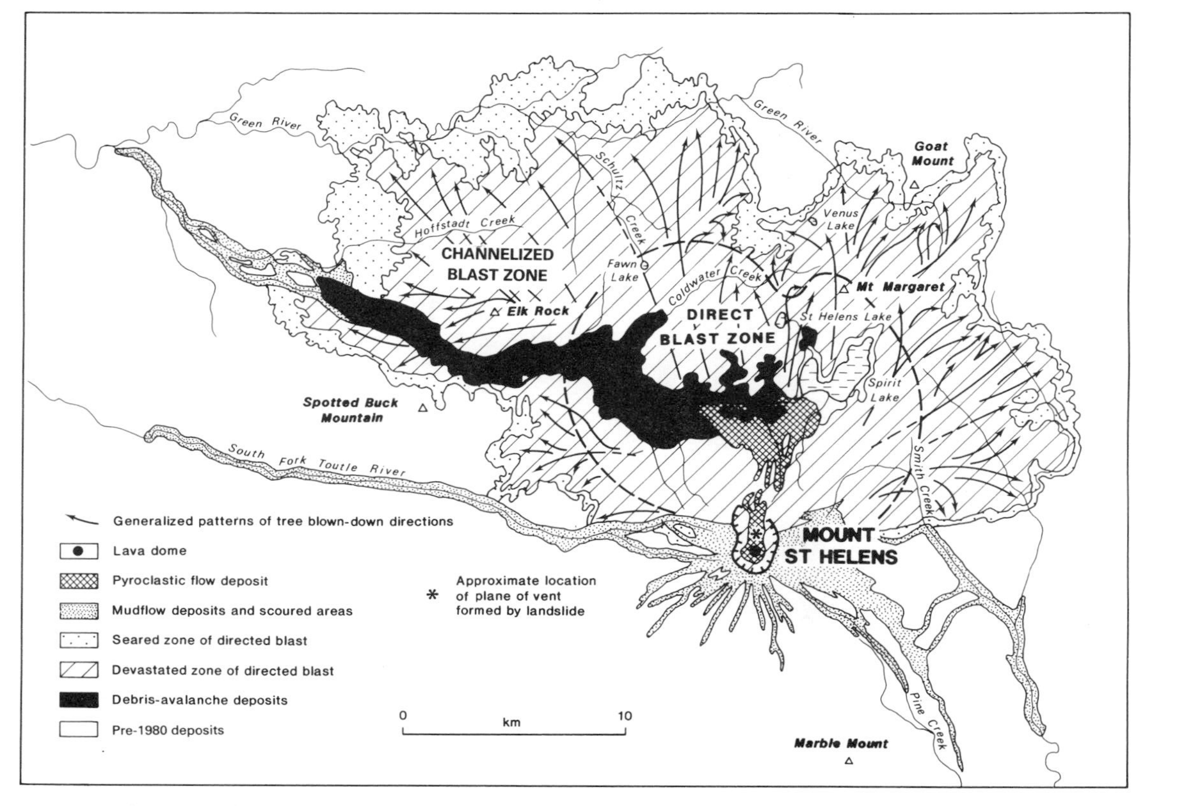

Fig. 5.9 The damage caused and the deposits produced by the lateral blast of Mount St Helens, 18 May 1980. In the 'blast zone' all life was destroyed and trees were pushed over or uprooted. The spatial complexity of depositional zones should be noted (after Kieffer 1981, Fig. 219, p. 381).

Fig. 5.10 Damage caused to forested land by the lateral blast of Mount St Helens in 1980. The upper photograph shows a group of trees which were pushed over and stripped of their leaves and bark. The lower photograph is a detailed view of a single log (photographs reproduced by courtesy of Dr J.P. Dickenson, University of Liverpool).

between it and the ground surface have been conceived as essentially closed systems (Cas and Wright 1987, p. 40)[2].

Since the mid-1960s, however, a number of studies have demonstrated that in certain cases this is an over-simplication and open-system behaviour can and does occur. Two situations in which it is more appropriate to conceive of explosive eruptions in terms of open systems are ones of magma mixing and hydrovolcanism.

5.3.6.1 Magma mixing

One of the principal advances made in understanding volcanoes during the decade to 1985 was the realization that the mixing of magmas of different compositions may trigger explosive eruptions (e.g. Sparks *et al.* 1977; Blake 1981a, 1981b; O'Hara and Matthews 1981; Huppert *et al.* 1982b; Turner *et al.* 1983; Blake *et al.* 1992). Observations of pumice within falls and ignimbrites by Sparks *et al.* (1977) showed that these are often petrologically varied and this, together with theoretical modelling, has led to the conclusion that magmas of differing composition may combine to produce an explosive eruption. Often silica-rich rhyolitic pumice is found to contain silica-poor basaltic phases and implies the movement of basic magma into a more silica-rich magma reservoir. The actual processes by which this results in eruption are complex and still disputed, but a brief summary of them is given in Table 5.5. A well-known example of magma mixing occurred during the 1875 eruption of Askja, Iceland (Sparks *et al.* 1977). This Plinian eruption is thought to have been caused by rhyolitic magma residing at the top of a magma chamber being 'injected' by basalt.

5.3.6.2 Hydrovolcanic eruptions

In magmatic eruptions rapid expansion of volatiles within the magma causes fragmentation of the melt and the production of hot pyroclastic materials of different types. They cool as they travel through the air or along the ground. In hydrovolcanic eruptions the sequence of events is different. Water may interact with magma in a number of ways and is commonly stored in the vicinity of volcanoes in lakes, as sea water, ice and ground water. If interaction takes place, then water cools the magma and this is accompanied by fragmentation of the melt (Fisher and Schmincke 1984).

The possible effects of water on explosive volcanism have been hinted at since the last century, but were 'generally regarded as minor local modifications or exceptions compared to magmatic mechanisms' (Fisher and Schmincke 1984, p. 74). The publication in 1949 of an important and scholarly monograph by Thomas Jaggar on 'steam-blast eruptions' was largely ignored, and it was only following the eruption of Taal volcano (Philippines) in 1965 that hydrovolcanism began to be accepted (Jaggar 1949; J. G. Moore *et al.* 1966). At Taal explosive activity was shown to have resulted from the mixing of magma with water held in a crater lake. Later intense programmes of research were undertaken involving: the field study of many volcanoes and their eruptions; laboratory experiments; and theoretical modelling. In a series of papers published between the early 1970s and the late 1980s, not only was the veracity of hydrovolcanism firmly estab-

[2] 'Closed systems are assumed to have boundaries which prevent the import and export of mass, but not energy. . . . Open systems, on the other hand, are characterised by an exchange of both mass and energy with their surroundings' (Chorley and Kennedy 1971, p. 2).

Table 5.5: Some of the ways in which explosive volcanic eruptions may be triggered by magma mixing (based on ideas in Cas and Wright 1987, p. 41, with some amendments)

Circumstances of mixing	Eruptive processes
Addition of volume	This is a simple mechanism and will mean that total magma pressure will increase and, if this exceeds the strength of the capping rocks, an eruption will result. The magnitude of the explosive eruption will depend on total volatile content.
Injection of a basic (basaltic) magma into a silicic magma chamber	If this occurs from below, then the lower part of the magma chamber will be heated, convect and rise through the chamber. As magma rises, decompression will occur, volatiles will exsolve and gas pressure will increase. If gas pressure exceeds the confining pressure of the capping rocks, an explosive eruption will result.
Injection of a basic (basaltic) magma with a high volatile content into a silicic magma chamber, with an initially low volatile content	Similar to above, but exsolved gases may be also transferred (by convection, diffusion and mixing) into the low volatile silicic melt. The possibility of an explosive eruption will be increased.
Hot basic (basaltic) magma contacts a cooler silicic magma	Heat transfer will occur and the basic magma will cool. This will cause crystallization and lead to further exsolution of gases in the residual basaltic melt. Gas pressure within the magma chamber will increase the possibilities of an explosive eruption.

lished, but knowledge of the physical mechanisms involved was also advanced (e.g. Colgate and Sigurgeirsson 1973; Peckover *et al.* 1973; Self and Sparks 1978; L. Wilson *et al.* 1978; Kokelaar 1983; Sheridan and Wohletz 1981, 1983; Wohletz 1983, 1986; Wohletz and Sheridan 1983; Decker and Christiansen 1984; Geophysics Study Group 1984; Wohletz and McQueen 1984; Lorenz 1987).

According to two of the most influential writers in the field, Michael Sheridan and Kenneth Wohletz, hydrovolcanism may be compared to industrial explosions between fuels and coolants. These fuel/coolant reactions (sometimes referred to as *contact-surface steam explosivity*; Kokelaar 1986) involve two fluids; the fuel having a temperature above the boiling point of the coolant. Generally, when fuels and coolants come into contact no explosion occurs and all that happens is that the coolant is vaporized and the fuel is cooled or quenched, but sometimes vaporization happens so rapidly that an explosion results (Sheridan and Wohletz 1983). Although details of the physical processes involved remain the subject of debate, it is generally agreed that two models—*superheating* and *pressure-induced detonation* (Wohletz 1983, 1986)—can explain both the observations made in laboratory experiments, when thermite melt ($Fe + Al_2O_3$) is detonated explosively with water (Wohletz and McQueen 1984), and in the study of actual hydrovolcanic eruptions and their products. Both superheating and pressure-induced detonation are complex, but it is possible to conceive of hydrovolcanic fragmentation as a cyclic, five-stage system exhibiting positive feedback (Table 5.6).

A critical variable controlling the explosive efficiency of hydrovolcanic fragmentation is the ratio of water to magma. As Fig. 5.11 shows, when the water/magma ratio is less than about 0.2 and assuming a basaltic magma, then water has little influence on fragmentation and the eruption is controlled by the characteristics of the magma alone. Activity is typically Strombolian and any water is ejected as steam (*phreatic activity*), with fragments of country rock sometimes being included. When the ratio approaches the point of maximum explosive efficiency (~0.35), then the eruption is dominated by both hydrovolcanic and magmatic fragmentation and is said to be *phreatomagmatic.* Large vertical eruption columns are formed and much fine-grained pyroclastic material is produced. Because large quantities of energy are required to heat the water, thermal efficiencies of phreatomagmatic eruptions tend to be lower than those of magmatic (e.g. Plinian) explosions for a given eruption rate. Columns are also

Table 5.6: Processes of hydrovolcanic fragmentation viewed as a five-stage cycle exhibiting positive feedback (based on Wohletz 1983, pp. 36–7; with additional information from Buchanan 1974; and Corradini 1981)

Stages	Description
1	Initial contact of magma and water creates a film of vapour at the interface. This usually occurs as magma rises into a zone of surface water, or water-saturated sediments.
2	The film of vapour expands, but at its outer (i.e. water) contact, condensation occurs and the film may collapse in places. Collapse releases energy. Expansion and collapse continue and increase, so that eventually sufficient energy is available to allow the melt to be penetrated and for fragmentation to occur. Penetration may occur in a number of ways: 1) ingress of linear jets of water into the magma; 2) impact of a mass of collapsing water with energy sufficient to produce a wave of stress capable of 'caving in' the surface layers of the melt; 3) direct interaction of water and magma—producing quenching, vaporization and explosive expansion; and 4) instability of the water/magma interface due to the transmission of a shock wave.
3	Penetration of water into the melt, or mixing of the collapse film with the melt, increases the surface area of the magma/water interface.
4	Increasingly rapid transfer of heat as water encloses fragments of melt.
5	Formation of a new film of vapour as water is vaporized by superheating and the reversion of the system to Stage 2. This further disperses the magma and *positive feedback*[1] occurs. 'An exponential increase in surface area leads to an exponential increase in total heat transfer' and explosions (Fisher and Schmincke 1984, p. 81).

[1] *Positive feedback* occurs when a change in input (in the system above at Stage 2) is magnified by the operation of the system (the output from Stage 5), so that its effect is continued and magnified. In this way the interaction of water and magma 'snowballs' and high-energy explosions occur.

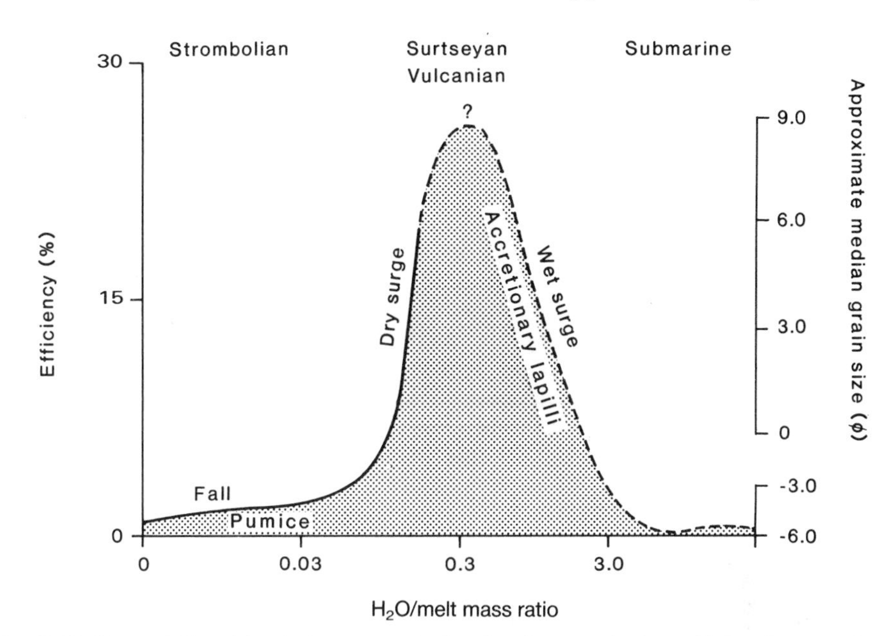

Fig. 5.11 The relationships between water/melt ratios, explosive efficiencies, median grain sizes of materials erupted (ϕ) and styles of eruption. Efficiency is expressed as a percentage of thermal energy converted to mechanical energy (based on Wohletz 1983, Fig. 1, p. 38).

lower (Wilson *et al.* 1978), but the effects may still be devastating (see Chapter 9). When ratios of water and magma exceed about 1, then water serves to 'damp down' the eruption and explosive efficiency is reduced. On land this is one of many environments in which *lahars* are generated (Sheridan and Wohletz 1983).

There has been a long-established tradition of using the term *Vulcanian* to describe 'highly explosive, short-lived eruptions that produce black, ash- and steam-laden eruption columns as witnessed during the 1888–90 eruptions of Vulcano, a small volcano in the Eolian (*sic*) Islands, Italy' (Fisher and Schmincke 1984, p. 82). In short a primarily magmatic type of eruption. Recently there has been a tendency to redefine Vulcanian and to consider it a type of hydrovolcanism (see Wohletz 1983; Fisher and Schmincke 1984). The principal reason for this change is that the 1888–90 eruption of Vulcano produced similar pyroclastic materials and had a similar style to the 1963–64 eruption of Surtsey (Iceland); an undoubted phreatomagmatic event[3]. The two terms are not synonyms. Vulcanian activity is distinctive because discrete explosions occur at intervals varying from hours to minutes and seem to be due to a repeated buildup of pressure underneath a capping plug (McBirney 1963). The relative contributions of exsolving magmatic gases and water in fragmenting the melt is still hotly debated (see Self *et al.*

[3] When originally defined phreatomagmatic applied only to magma reacting with ground water. It was not surprising, therefore, that a new term 'Surtseyan' was introduced by Walker and Croasdale (1971) to describe the interaction of basaltic magma with sea water. Later the term 'phreatoplinian' was used to describe the interaction of silica-rich magma and water (Self and Sparks 1978). In the present work these terms are avoided as much as possible, except where necessary—especially on figures—to convey the original meaning of authors.

1979). Because water is involved, the convention of Fisher and Schmincke (1984) will be followed and Vulcanian eruptions considered within the range of the hydrovolcanic activity.

As Figs 5.2b and 5.11 show the principal deposits produced by hydrovolcanic activity on land are: falls; surges; and lahars. As in magmatic eruptions, so hydrovolcanic fall deposits are produced by fragmentation of magma and are distributed as a result of processes operating within eruption columns. Surges have, in contrast, more complex modes of formation. It is clear from Fig. 5.11 that the amount of water has affects on the characteristics of surges, since on the ascending limb of the graph 'dry' surges are produced, but as the ratio of water to magma increases they become wetter. On the descending limb they are associated with accretionary lapilli and other features, such as bedding sags, indicative of transport in a saturated state. However, whether wet or dry, modes of formation are similar and surges usually form because of collapsing hydrovolcanic columns; a mechanism which has not only been observed at sites of eruption, but has also been modelled successfully in the laboratory using thermite explosive (Wohletz and McQueen 1984). In early papers on surge generation, vertical eruption columns and single collapses were assumed (e.g. eruptions of Capelinhos, Azores; Waters and Fisher 1971), but it is now accepted that eruptions are often more complex and Fisher and Schmincke (1984, pp. 247–8) quote the example of an eruption within Lake Ruapehu (New Zealand), in which a primary surge was generated at the start of the 1975 eruption by water jets and steam and only later did collapse produce a second surge. Phreatomagmatic surges are usually cold (section 5.2)—'dry or wet'—but in an eruption in Alaska in 1977, it is reported that surges were both 'dry and hot'; similar in fact to 'hot' surges produced in association with magmatic eruptions and pyroclastic flows (Self *et al.* 1980). Once again the perspicacious thoughts of Tim Druitt (1991) are of relevance: that all surges and flows are part of a common gradational series.

Phreatomagmatic surges are low particle concentration (dilute) mixtures of particles and steam, that expand rapidly in all directions away from a site of eruption and move at velocities of up to 100 ms^{-1}. The principal condition required for their initiation is the availability of steam in quantities sufficient to support the particles, but the physical processes of generation are in reality more complex; being often conceived as a chain linking one state of energy to another (see Fig. 5.12; and Cas and Wright 1987, p. 204). All energy is derived from the rising magma, but once this comes into contact with water it is transformed into mechanical (i.e. explosive) energy and then into momentum (i.e. the product of the mass and velocity of the particles in motion) plus kinetic energy (i.e. energy due to motion); as gases and fragments move away from the site of eruption. Potential or elevation energy, which is proportional to the mass of material and its position above the ground, is also of relevance because it can be translated into kinetic energy as the eruption column collapses. Finally, additional thermal energy is imparted to the surge from both the hot particles and the latent heat released by condensation of steam to water droplets (Cas and Wright 1987). The additional thermal energy affects the surge in its early stages and is responsible for high initial velocities and turbulent flow. As mentioned in Table 5.3, the change from laminar to turbulent flow is conditioned by the 'Reynolds' number'. Since surges travel at high speeds and have a low viscosity they are turbulent.

A distinctive form of hydrovolcanism occurs in environments in which the water/magma ratio is 'excessive' (Fig. 5.11; and Sheridan and Wohletz 1983,

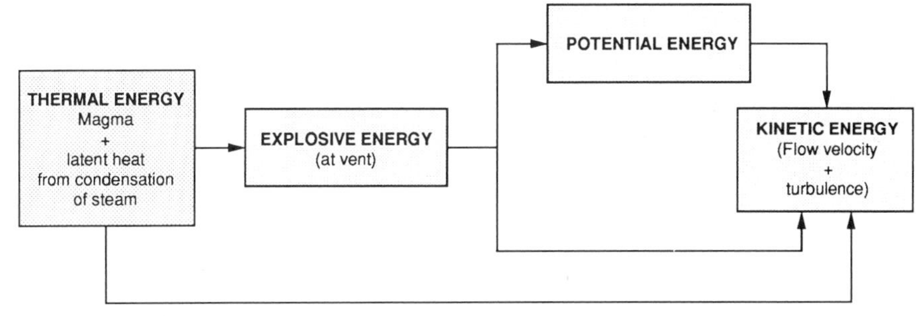

Fig. 5.12 The 'chain' of energy involved in the generation of surges by hydrovolcanism (from Cas and Wright 1987, Fig. 7.33, p. 204).

p. 22). Subglacial situations are ones in which this form of activity is most common and the best recent example is the generation of lahars during the 1985 eruption of Nevado del Ruiz, Colombia (Chapter 9, section 9.3.3).

5.4 Pyroclastic eruptions *in vacuo*

With increased interest in planetary geology (Chapter 1), it is not surprising that attention has been paid to pyroclastic eruptions on the terrestrial planets and other bodies in the Solar System. Atmospheres have been features of many of these bodies in the past and are features of some at present, but it is believed that many eruptions occurred when atmospheres were negligible. There are several important differences which occur when eruptions occur *in vacuo*.

1) Expansion of gases away from sites of eruption is essentially indefinite and high gas speeds are attained. Speeds are limited only by initial gas temperatures and molecular weight. As the density of the gas decreases with expansion, drag between the gas and the particles it contains decreases to a negligible value. Eventually clasts follow ballistic trajectories.

2) Most volatiles in magmatic or hydrovolcanic eruptions do not remain gaseous at low pressures and, not only condense to form liquids and solids, but also in so doing release latent heat. It is argued that the distinctive form of volcanism on Io (Chapter 1, section 1.2) may be due to silicate-melt reacting with sulphur and sulphur dioxide in the crust.

3) The heat-retaining properties of certain eruption clouds are probably a significant factor in planetary volcanism and may have been responsible for several features. On the Moon lava 'ponds' up to 5 km across were filled with clasts at close to magmatic temperatures and these probably caused the thermal erosion of the substrate and formed depressions which were possible source regions for sinuous rilles (see Chapter 1, section 1.3.1).

The list above, based on works by L. Wilson and Head (1981a, 1981b); Kieffer (1982); McEwen and Soderblom (1983); and L. Wilson *et al.* (1987), is by no means exhaustive and merely gives a flavour of the distinctiveness of planetary pyroclastic activity. For more detailed accounts reference should be made to: Guest *et al.* (1979); Murray *et al.* (1981); Glass (1982); Greeley (1985, 1987); and Chapter 1.

5.5 Submarine volcanic eruptions

Volcanoes occur in vast numbers beneath the oceans and seas of the world, being associated with intraplate volcanoes, mid-ocean ridges and back-arc (subduction) basins. Because of their location far less is known about submarine volcanoes than those which occur on land and, indeed, no deep-water eruption has ever been observed (Fisher and Schmincke 1984). What is known is derived from exploration by submersibles, studies of uplifted ocean floor and by drilling such as that carried out as part of the international Deep Sea Drilling Project between the 1960s and 1980s.

A crucial concept in submarine volcanology is *volatile fragmentation depth* (VFD). With increasing depth pressure becomes so high that exsolving volatiles cannot cause fragmentation of the melt. The VFD is rarely less than 500 m and, in addition to water depth, depends critically on volatile content and ultimately magma composition (Fisher and Schmincke 1984; Kokelaar 1986). Below the VFD the amount of pyroclastic activity is limited, especially when compared with the vast quantities of massive, sheet and pillow lavas which comprise most of the ocean floors. Hydrovolcanic fragmentation is also inhibited by high water pressures (Cas and Wright 1987, p. 45). Volcaniclastic material is, however, not insignificant and estimates vary around a median value of ~10% of total extruded products (Schmincke *et al.* 1983). The principal materials found in these deep-sea, predominantly mid-ocean ridge localities, are various types of hyaloclastite and pillow breccias. Hyaloclastites are volcaniclastic materials produced by 'the explosive granulation of volcanic glass when basaltic magmas are quenched on contact with water' (Heiken and Wohletz 1985, p. 4) and are an extremely common product of deep-sea volcanism (Honnorez and Kirst 1975; and Table 5.7). Pillow breccias form in a variety of situations including: collapse of pillow lavas on steep slopes; implosions of lava tubes due to high water pressures; and as talus slopes around seamounts and fault scarps. Often gaps between fragments are filled with a variety of secondary carbonate and clay minerals (Schmincke *et al.* 1983; Fisher and Schmincke 1984).

As water depth decreases the range of processes increases and both magmatic and hydrovolcanic fragmentation become more important. Two types of situations are found. The first is the shoaling volcano which is gradually building towards a few hundred metres, or less, of the surface. There is great uncertainty about the critical depth at which significant explosive activity can occur. According to Jones (1970), there is convincing geological evidence in the offshore regions of Iceland that explosive activity is common at depths of 100 m or less. Theoretical estimates by Allen (1980) produce figures of 100–200 m, but Fisher and Schmincke (1984, p. 274) believe that silica-rich submarine volcanoes as deep as ~500 m are capable of producing pumice floating on the sea surface. Whatever the critical depth the principal products are thought to be fall materials which may, or may not, break the surface and, if the volcano is silica-rich, floating pumice and pyroclastic flows (Fig. 5.2d). Processes of generation are self-explanatory except in the case of flows. Clearly an eruption column of the type formed on land cannot be produced, because of the weight and pressure exerted by sea water, and it seems probable that 'columns' lose energy very quickly, increase their density and undergo immediate collapse, thereby producing flows (Fiske 1963; Fiske and Matsuda 1964). Flows may pass laterally into density (turbidity) currents as they move down continental slopes and canyons

Table 5.7: Some additional processes of hydrovolcanic fragmentation relevant to the submarine environment (based on Kokelaar 1986, pp. 279–81)

Type of fragmentation	Description	Water depth
Bulk-interaction steam explosivity	This is like *contact-surface explosivity* and involves water that is trapped either close to the magma, or engulfed by it. The process of fragmentation involves the tearing apart of magma by steam expanding explosively and its shattering by pressure waves. It does not depend on a critical water/magma ratio.	The maximum depth at which this process operates is uncertain, but depends critically on the ease with which water can be transformed into vapour. It involves a large increase in volume. This volumetric increase may be inhibited by water pressure, at depths of around ~2 km in pure and ~3 km in sea water. Fragmentation may take place at greater depths if magma comes into contact with water held in porous rocks.
Cooling-contraction granulation	In deep water, magma will cool by conducting heat and this will cause chilling of the surface layers and solidification. The interior, when it cools, sets up stresses that cannot be accommodated by the solid surface and cracking occurs. Pillow lavas (see Chapter 4) are often fragmented on their margins, due in part to contraction, but also because of solidification before the lava stopped moving.	Can occur at any depth.

(Fig. 5.2d). Recently Cas and Wright (1991) have reviewed the whole question of subaqueous pyroclastic flows. Their studies have been focused not only on extant examples, but also many flows preserved in the geological record—such as the Ordovician flows of Ireland and Wales. They conclude that many of these are not flows, but have been transformed into water-supported debris-flows, avalanches and turbidity currents, due to their explosive breakup and/or ingestion of water. The number of true flows is, they argue, very limited.

The second situation is transitional and occurs when a submarine volcano is gradually building and emerging from the sea to become subaerial. Here it may be difficult to distinguish between products formed by processes operating on land

and/or at sea, especially since subaqueous materials may be introduced into the submarine environment by subaerial volcanoes (Fig. 5.2c). Important research on eruptions such as Surtsey (Iceland 1963–67) and Capelinhos (Azores 1957–58) has been carried out by Kokelaar (1983, 1986) and Kokelaar and Durant (1983), providing valuable insights into the processes operating at emergent volcanoes. Kokelaar (1986) shows that in the submarine environment several processes of fragmentation are possible. Two of these have been discussed elsewhere: *magmatic fragmentation* and *fluid/coolant interaction* (termed contact-surface steam explosivity by the author). In addition he introduces two new mechanisms (Table 5.7).

All processes of fragmentation increase in effectiveness as water becomes shallow and volcanic plumes increase in height and fall materials are distributed over greater distances. Kokelaar and Durant (1983) examined deposits at a depth of 45 m in the vicinity of Surtla—a subsidiary vent of Surtsey—and found a 'cupola', or concave dome, of steam during eruptions (Fig. 5.13). At even shallower depths magmatic and hydrovolcanic fragmentation cause materials to be ejected above the sea surface and such eruptions are notable for their black debris-rich clouds, which gradually turn first grey and then white as tephra-free bubbles break the surface (Kokelaar 1986, p. 283). This is typical early stage Surtseyan activity. At Capelinhos between 1957–58 events were similar to those at Surtsey, except that major surges were only generated after powerful columns had carried tephra to a considerable height.

5.6 Transport, emplacement and sedimentology of volcaniclastic materials

Just as considerable progress has been made in recent years in understanding the mechanisms responsible for producing volcaniclastic materials, so similar advances have occurred in linking sediment genesis to modes of transport and emplacement. In this section pyroclastic materials will be discussed first and then other volcaniclastic deposits such as lahars and debris avalanches will be considered.

5.6.1 Falls

Pyroclastic falls are produced by all subaerial eruptions and are transported by two mechanisms: ballistic action and turbulent suspension. Eruption column dynamics determine the height to which particles of different sizes are transported and, as discussed in section 5.3, this depends in large measure upon the type of eruption involved. Eruption columns can be deflected *en masse* by prevailing winds, but the particles within them act independently being suspended by turbulence. There are three important threshold-settling velocities which control the removal of particles from eruption columns (Fisher and Schmincke 1984, pp. 126–7). These are the settling velocities of:

1) large ballistic fragments (Table 5.3), which are not affected to any great extent by winds, or by processes within eruption columns;
2) particles which are suspended by turbulent eruption columns, but have a mass too great to be supported by winds; and

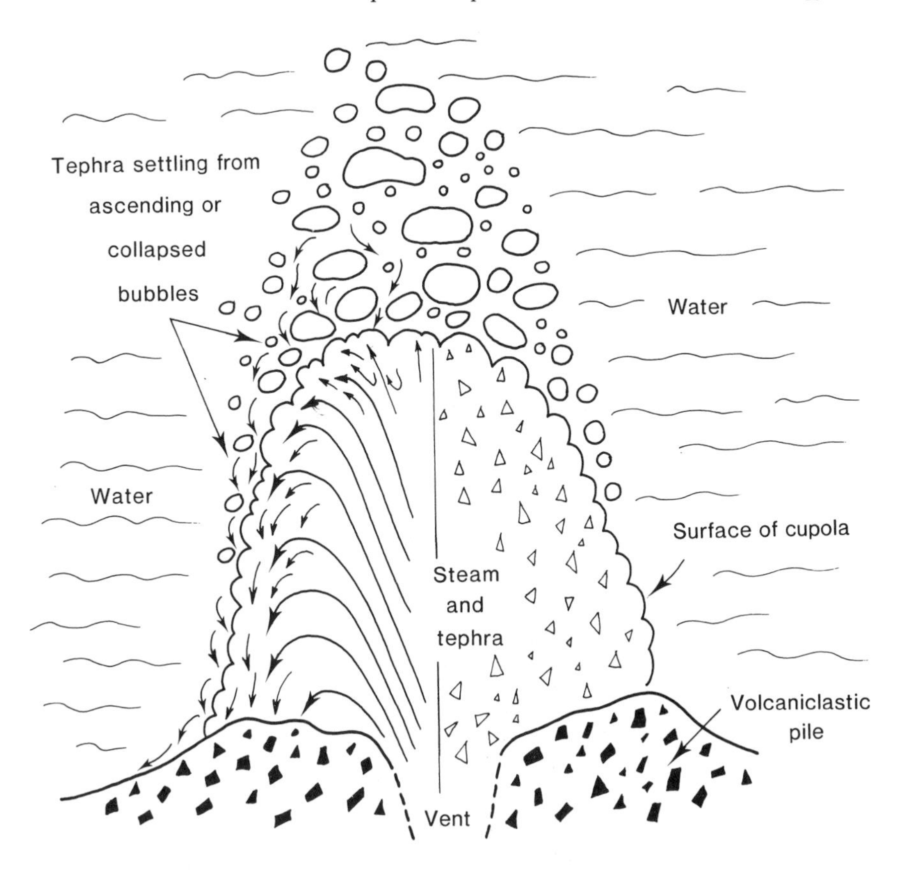

Fig. 5.13 Diagram showing a 'cupola' of steam enclosing a gas thrust jet at Surtla, Iceland (depth 45 m). Arrows show the movement of tephra and the fine fraction is commonly carried away by sea currents. The cupola is not a 'steam-filled void', but contains a high concentration of pyroclastic material (modified from Kokelaar 1986, Fig. 4, p. 282).

3) those fragments that are light and able to be suspended by both the column and by winds.

With the exception of ballistic fragments, particles are affected by winds once they leave a column and, in the case of the finest calibre, released from 'umbrella regions' (see Fig. 5.4) during Plinian and other energetic eruptions, they may spread around the Earth and have significant climatic effects (see Chapter 6).

Fall fragments return to Earth in an order determined by their settling velocities (Minakami 1942; Walker *et al.* 1971; L. Wilson 1972): coarsest in proximal locations (i.e. nearest to the volcano) and finest in distal ones (i.e. furthest from the volcano). Because the majority of eruptions are at their most energetic in their early stages, transporting power normally diminishes during an eruption and fall sequences tend to become finer upwards in vertical sections. Both generalizations do not always apply and writing of the 1980 Mount St Helens eruption, Carey and Sigurdsson (1982) noted that large numbers of small particles were removed from the eruption column because they were 'stuck together' by water. Also wide dispersal often depends on prevailing winds and, if these change in direction and/or strength during an eruption, then depositional sequences will be very complex indeed. Winds may also vary at different altitudes (Waitt and Dzurisin 1981). In the case of hydrovolcanic eruptions (section 5.3.6.2), the order of removal of particles from columns and their fall velocities through the air may be influenced by the occurrence of accretionary lapilli, which may be present if magmatic tephra are flushed by rain (see section 5.2). In short, near to the vent there may be a complete mixture of grain sizes and considerable sedimentological complexity. At distances beyond the immediate vicinity of vents all fall materials, regardless of origin, have certain common attributes (Table 5.8).

Table 5.8: Sedimentological attributes of fall deposits (based on an idea of Fisher and Schmincke 1984, Table 6.1, p. 126; the references cited and many other sources)

Attribute	Remarks
Composition	Falls may be of any composition, but those derived from intermediate and silica-rich magmas (e.g. andesites and rhyolites) are more widespread than those of more basic composition. There is a tendency for falls to become more silicic with increasing distances from the vent because of *aeolian fractionation*[1].
Texture[2]	Generally well sorted, regardless of grain size. Often better sorted at distance once large-calibre material has been deposited. Calibre (median grain size), varies with the type of eruption and distance from the vent. The greater the distance the finer the calibre.
Structure	Falls 'drape' existing topography, but against a steep slope or cliff they form 'wedge'-like accumulations. Grading is usually normal, reflecting decreased transporting power as an eruption wanes, but reverse grading is often encountered because of changes in the conditions of eruption, prevailing winds, and vent-radius. Contrasts often occur between beds and may reflect changes in texture and composition. These may, or may not, be marked by colour changes. A distinct *fabric*[3] is uncommon, because elongated particles are rare. Bedding planes are sharp at: 1) the base of falls when deposition occurs over weathered surfaces and/or a different rock type; and 2) when there are distinct changes in conditions of eruption, composition of magmas and wind strength/direction. Conversely, bedding is poorly developed and may be absent when deposition is: 1) fast with large and small particles being released together from an energetic column whose dynamics are varying; and 2) so gradual and slow that pedological processes disturb primary bedding structures. It may also be absent in old sections which have been weathered.
Distribution	Either circular around a vent or, more commonly, confined to distinct sectors because of the influence of winds. When confined to distinct sectors, distributions are normally 'fan'-shaped with apices either coinciding with vents or offset from them. Fans become wider with distance and materials are spread over greater areas.

Table 5.8 – *cont.*

Attribute	Remarks
Thickness	In general falls become thinner with distance. The maximum thickness may occur as a circular *isopach*[4] around the vent, but is commonly near to, or at, 'fan' apices. Because of the complexity of wind movements and the dynamics of eruption columns, it is common for 'secondary maxima' to occur at some distance downwind.
Associated facies	1) In proximal locations (within the vent and on the volcano itself), falls are associated with lavas (flows and domes), pyroclastic flows, deposits of debris avalanches, surges and lahars. 2) In intermediate locations, many of associations noted in (1) still occur, but an increasing proportion of reworked materials is evident. Coarse-grained materials (like debris avalanches) decrease in importance. 3) In distal locations fine-grained fall materials dominate and may be interbedded with lake, non-volcanic aeolian materials and soils. Coarse-grained pyroclastic materials are absent and so are lava flows.

[1] *Aeolian fractionation.* This refers to an observed increase in silica content in ashes with distance from source vents. It was first recognized by Larsson (1937) and is further discussed by Lirer *et al.* (1973). Within a gas/ash cloud, fractionation reflects the varying settling velocities of crystals, pumice, glassy fragments and rock particles. Fractionation is a complex process and involves interrelationships between magma chambers, eruption columns and plumes. It is discussed further in the text.
[2] *Texture.* Sorting is a measure of the standard deviation (i.e. the spread about the mean) of the grain-size distribution. The lower the value the better the sorting. The median size (or diameter) is the grain size at the 50 percentile point of the cumulative percentage distribution. Volcanologists generally calculate these measurements using cumulative percentile plots, the phi (ϕ) scale (see Table 5.1) and formulae developed by Inman (1952). Median diameter (Mdϕ) = ϕ_{50} and sorting = $(\phi_{84} - \phi_{16})/2$.
It should be noted that these are not the measures of median size and sorting which are used by most sedimentologists and, indeed, some volcanologists. These workers prefer formulae developed by Folk and Ward (1957). Sometimes values are also calculated for *skewness.* Skewness, the symmetry of the grain-size distribution about its centre, is described as negative if there is excess coarse material and positive if there is excess fine material.
[3] *Fabric* refers to the orientation, packing and boundaries between particles in a sediment.
[4] *Isopachs* (more correctly isopachytes) are lines of equal thickness.

There are two additional factors to be considered in the emplacement of fall deposits: aeolian fractionation and component composition. It is commonly observed that fall deposits become more silica-rich with distance from vents. In a pioneer study, Larsson (1937) established a correlation between the observed increase in silica content with distance and the decrease in the proportion of crystals and explained this by the early fallout of silica-poor crystals from plumes, leaving them dominated by the more silica-enriched pumice/glass fraction. More recently these arguments have been developed and show that chemical changes with distance are more complex than Larsson believed. As Fisher and Schmincke

(1984, p. 157) note three sets of variables have to be considered: those relating to magma chambers; those relating to eruption columns and those relating to ash clouds. With regards to magma chambers they note that most of those containing silica-rich melts are zoned compositionally, chemically, and in terms of their mineralogy. In many Plinian and ignimbritic eruptions silica-rich products lie high within magma chambers and are erupted first and in some cases are distributed furthest, whereas in other eruptions it is more silica-poor (mafic) materials that are distributed furthest (Lirer *et al.* 1973). Fisher and Schmincke (1984) also note that, although the actual processes within eruption columns are not well known, it has been frequently observed that flows are enriched with crystals because the lighter components are winnowed out by the eruption column, causing fall deposits to be an inaccurate reflection of original magma composition (see also Sparks and Walker 1977). Finally within the ash cloud itself fractionation can occur in the manner suggested by Larsson (1937) as a result of the differing settling velocities of lithics, crystals, pumice and glass.

The components making up fall deposits are: crystals; lithics; and pumice (including glass). As Walker and Croasdale (1971) found, these occur in different proportions depending on the type of eruption. In hydrovolcanic eruptions, for instance, there will tend to be a predominance of lithic materials with a high proportion of fine material, whereas in Plinian explosions the magmatic fragmentation of the melt will mean that the proportion of pumice will be much higher.

5.6.1.1 Hawaiian and Strombolian falls

As mentioned in section 5.3.1, the principal fall deposits associated with eruptions of these types are confined to proximal locations and comprise coarse bombs, air-shaped fragments like Pele's hair and Pele's tears and cones of *spatter* (i.e. coarse fragments which are partially molten when they came to rest and may be welded to each other). In Stombolian eruptions, the suite is more varied. At eruptive vents *scoria* (i.e. coarse basaltic fragments) and spatter often accumulate and build cones. The internal composition of such a cone on Etna is shown on Fig. 5.14.

In distal localities Strombolian falls are finer grained, composed predominantly of scoria, have a limited areal distribution and are of restricted volume; all these characteristics relating directly to low exit velocities from vents and restricted column heights. One distinctive feature is the occurrence of what Walker and Croasdale (1972) termed 'achneliths'. These are fragments with 'smooth, glassy surfaces formed from solidified lava spray' (Cas and Wright 1987, p. 134; Heiken and Wohletz 1985).

5.6.1.2 Sub-Plinian, Plinian and Ultra-Plinian falls

Deposits from Plinian eruptions are widely distributed because of high eruption columns and very energetic eruptions. In addition, the shear volume of material ejected—figures of 0.17 to 28 km^3 are quoted in the literature (Walker 1981d)—means that in proximal locations thicknesses of over 10 m are common and that, because of the silica-rich nature of magmas, pumice is a major constituent. Near to vents falls may be very coarse with pumice blocks of up to 30 cm or more and lithic blocks exceeding 1 m. With distance deposits tend to become thinner, finer grained, better sorted and the component population changes, with the proportions of dense lithics decreasing and of pumice increasing. Although in section many Plinian falls appear to be homogeneous especially near to vents, in detail

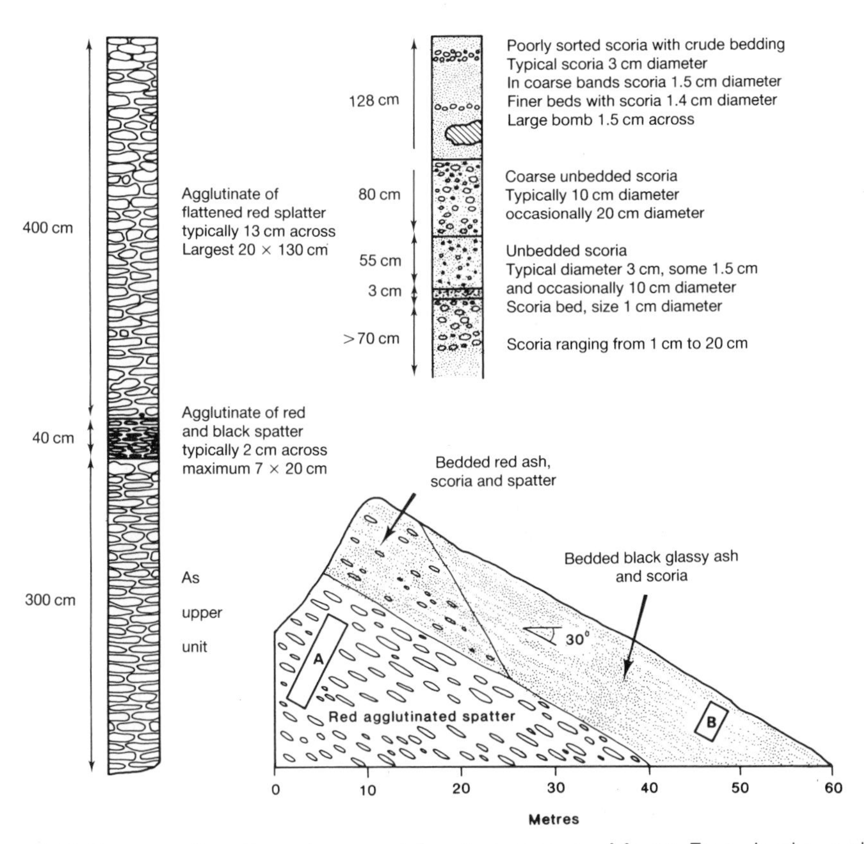

Fig. 5.14 Cross-section through a scoria/spatter cone on Mount Etna. In the early phases of the eruption large amounts of spatter were produced which welded themselves together to form agglutinate. The upper part of the sequence shows greater complexity with spatter close to the vent and the rest comprises well-sorted beds of scoria of different sizes. Although bombs are present they are relatively rare, since most of them emplaced during eruption will roll to the bottom of the cone. Typically a cone is black, but away from the vent fumarolic activity can cause changes to shades of red, orange and yellow (from Chester *et al.* 1985, Fig. 4.6, p. 133).

this may be far from the case. In addition to progressive changes in composition as lower levels of magma chambers are tapped during an eruption, it is common to find such features as reverse-grading of large pumice clasts and sometimes of lithics as well (Bond and Sparks 1976), internal stratification into distinct beds or layers particularly near to the vent and the interbedding of flows and surges (Walker 1981d; Self *et al.* 1984). Many of these features represent changes in the eruption column during the eruption. Some eruptions are not continuous but rather 'pulse like' (Cas and Wright 1987, p. 145), discharge rates fluctuate, activity may cease for a time due to blockage of the vent and rain can remove fine material from an eruptive plume. Also it should not be forgotten that just as there is a progression from a convecting to a collapsing column, so there is often a progression from Plinian fall to flow (see section 5.3.2). Some idea of the complexity of depositional processes and the difficulties faced in their interpretation can be appreciated from Fig. 5.15, which shows a complicated sequence of Plinian fall and other deposits from the 79 AD eruption of Vesuvius (Italy).

5.6.1.3 Hydrovolcanic (hydroclastic) falls

Basaltic hydrovolcanic, or Surtseyan, deposits are highly fragmented, but not so widely distributed as Plinian falls (Fig. 5.1). The reason for this is that eruptions are highly explosive so producing fine materials, but column heights are much lower than those associated with Plinian events (see section 5.3.6.2). Within regions of basaltic volcanism, such as the now no longer active Quaternary Eifel District in Germany, distinctive landforms known as Maar volcanoes mark the sites of former eruptions. These are low cones, with 'bowl-shaped craters that are wide relative to rim height' (Fisher and Schmincke 1984, p. 257). They are composed of a mixture of surges and thin fall beds and have been observed in the process of formation at eruptions such as Capelinhos, Azores in 1957–58 (Waters and Fisher 1971) and Taal, Philippines in 1965–66 (J. G. Moore 1967). Following (Lorenz 1973), it is possible to recognize three subtypes: maars, tuff-rings and tuff-cones and a summary of their geomorphology and processes of formation is given in Table 5.9 and Fig. 5.16.

One feature of proximal, basaltic hydrovolcanic falls is that they are often interbedded with surges and, particularly when the latter are planar bedded (section 5.6.3), it is often difficult to distinguish between them. On several of the volcanoes of Italy, for instance, the author and his co-workers have encountered these problems (see Chester *et al.* 1987; Guest *et al.* 1988) and some of the sedimentological criteria which may be used to distinguish between such falls and surges are discussed by Cas and Wright (1987, pp. 211–9) to which reference should be made.

With increasing distance from eruptive vents deposits become thinner and much finer grained. Often they are bedded, or laminated[4], because hydrovolcanic activity often occurs as discrete and short-lived explosions. Because of the water-rich nature of eruptions, accretionary lapilli are common and ashes are more poorly sorted than many other fall deposits, due to the early removal of wet materials from the eruptive plume (Brazier *et al.* 1982). In recent years research has been carried out to investigate the micro-morphological features of fine-calibre materials in different types of pyroclastic deposits (e.g. Wohletz 1983; Heiken and Wohletz 1985). This research involving scanning electron microscopy (SEM), shows distinctive forms of micro-morphology associated with basaltic hydrovolcanism with angular fine ashes broken by explosive fragmentation. An example of the use of this technique in confirming a hydrovolcanic origin for basaltic fall deposits on Etna is given by Chester *et al.* (1987) and further details are contained in a comprehensive photomicrographic 'atlas': *Volcanic Ash* (Heiken and Wohletz 1985).

When silica-rich magmas are erupted, fall deposits have certain attributes in common with their more basaltic cousins, showing relatively poor sorting due to rain flushing, laminations, associations with surges in proximal locations and accretionary lapilli. They also have distinctive features in the fine fraction which may be recognized by SEM (Heiken and Wohletz 1985). As implied by Fig. 5.1

[4] Laminations are fine-grained layers only millimetres thick.

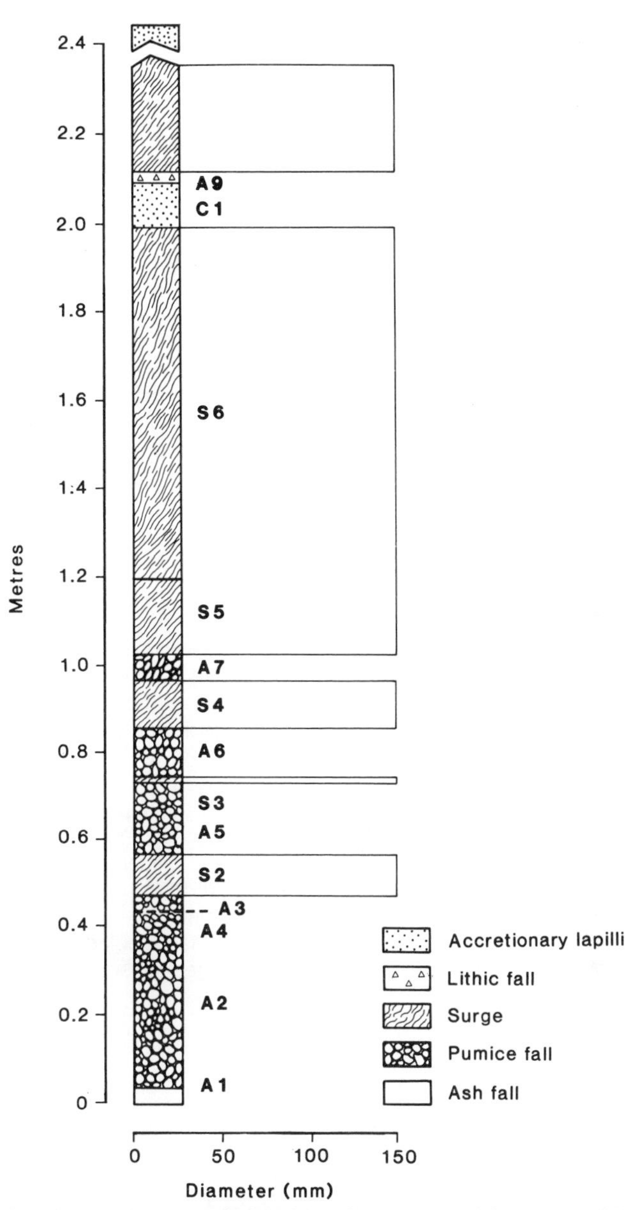

Fig. 5.15 Section through some of the deposits produced by the 79AD eruption of Vesuvius, Italy (simplified and modified after Sigurdsson, H. *et al.* 1985, Fig. 18, p. 349.) The figure illustrates some of the complexities within, and the relationships between, Plinian deposits of different types.

Table 5.9: Principal morphological features and processes involved in the formation of maars and tuff rings (based on Sheridan and Wohletz 1983, pp. 18–19)

Landform	Description	Process of formation
Maar volcano	A volcanic crater cut into country rock below the general level of the ground and with a low rim of coarse-grained surge and fall deposits. Size ranges: 1) width 100–3000 m; 2) depth 10–500 m; and 3) rim height less than 10 to nearly 100 m.	Although at one time it was thought that maars were produced by the explosive discharge of gases (especially CO_2), it is now generally accepted that they are of hydrovolcanic origin; the water being derived from below the ground. In some regions it is found that maars are restricted to river valleys, depressions and coasts where water is available, and that elsewhere cinder and scoria cones predominate. The distinctive morphology of maars is produced by a combination of explosive activity and subsidence.
Tuff rings (including cones)	A volcanic crater at or above general ground level. Normally much higher rims than maars and may be as high as 300 m.	These are thought to be transitional between cinder cones and pillow lavas and represent the central point in a continuum of landforms produced by hydrovolcanic activity. According to Sheridan and Wohletz (1983), tuff rings are formed in two stages: 1) initial deposition of explosive fall breccias; and 2) emplacement of thinly bedded surges as the eruption develops. Cones have a third stage of surge and fall emplacement. All three stages reflect differing water/melt ratios.

these ashes do differ from basaltic hydrovolcanic falls, being much finer even in proximal localities. In distal locations they often form very thin veneers over large areas. The reasons why fine grain size and wide dispersal is so characteristic of these falls is that they are deposited from much higher eruption columns, which may be of Plinian dimensions. Differentiating between silica-rich hydrovolcanic ashes and other fine-grained pyroclastic materials such as 'rain flushed' Plinian falls and co-ignimbrite tephra (see section 5.6.2) is again difficult, but not impossible and criteria are discussed by Cas and Wright (1987, pp. 159–60).

5.6.2 Flows, ignimbrites and associated 'hot' surges

Just as progress has been made in understanding the processes whereby flows are generated (see sections 5.3 and 5.5), so similar advances have been made in

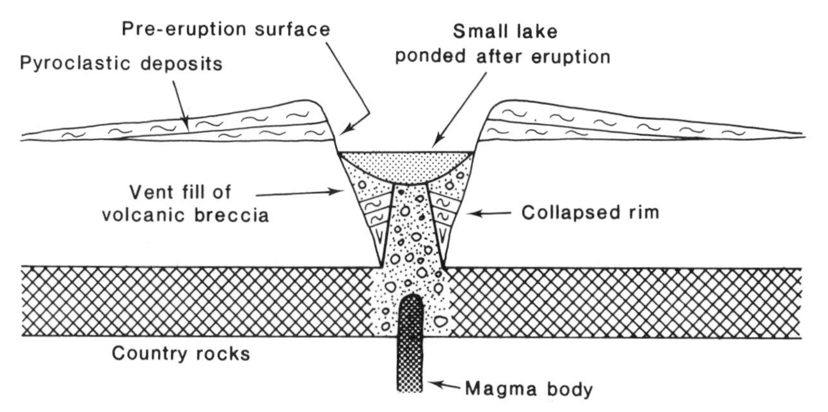

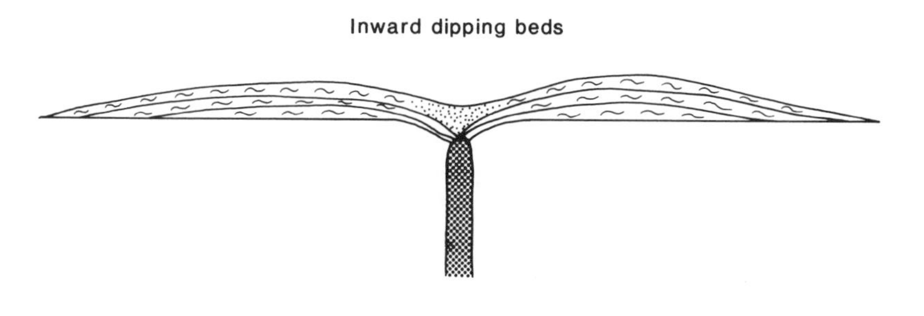

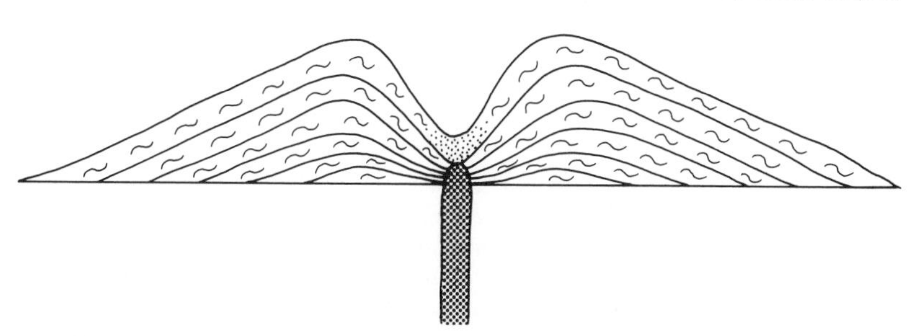

Fig. 5.16 Cross-sections showing the morphology of maars, tuff-rings and tuff-cones (from Cas and Wright 1987, Fig. 13.17, p. 377).

understanding the means by which these materials are transported and emplaced. However, as the following account indicates much is still uncertain and much is still to be learnt.

As far as emplacement mechanisms are concerned work by authors, who include Sparks *et al.* (1973); Sheridan (1979); Wright (1979); C. J. N. Wilson (1980, 1984, 1986); Wright *et al.* (1981b); C. J. N. Wilson and Walker (1982); Walker (1983); Bardintzeff (1985); Cas and Wright (1991) and many more, has done much to advance knowledge of flows produced by the collapse of Plinian and other columns. Flows not only vary in size, from small ones observed during closely monitored eruptions this century like Mount St Helens (Lipman and Mullineaux 1981) to massive ones such as the Pleistocene Campanian ignimbrite in Italy (Barberi *et al.* 1978), but also in the nature of their processes of movement (Fig. 5.17). Flows are gravity-controlled, mass-movement phenomena and follow topographic depressions—such as river valleys—radiating from volcanoes. For some small historic flows there it is probably only a question of a gradual transition before the resultant deposits become lahars. In these flows gas release only plays a small role in mobility and the processes involved are, as in lahars, a combination of laminar and plug flow (see section 5.6.4); allowing the transport of large blocks, the formation of flow fronts, levées and deposition on steep slopes. At the other end of the spectrum are large pumice-rich ignimbrites. They do not in general occur on steep slopes, have few if any surface features, are highly mobile (velocities of over 100 km/h^{-1} are common), are often observed to cross topographic barriers and are able to travel great distances (often more than 100 km). It should not be forgotten that high mobility and the ability to surmount barriers is not confined to pyroclastic flows and many cases are reported of large non-volcanic avalanches which have similar properties (Kent 1966). Given that potential energy depends upon source height and large eruption columns may be several kilometres high, whereas avalanches have a potential energy restricted by the height of the mountain from which they are generated, then pyroclastic flows are probably no more mobile than any other type of flow (see Sparks 1976). Sheridan (1979) and many subsequent workers have invoked the concept of an 'energy line' from the avalanche literature (Hsü 1975); this being defined as a line traced from the top of the gas thrust region of an eruptive column to the distal margin of the flow. Flows can surmount all topographic barriers below this hypothetical line.

Much debate has surrounded the mechanisms which impart mobility to flows, in addition to the simple conversion of potential to kinetic energy. Also and as argued before, many flows are not derived from column collapse and some other agency, or agencies, must be involved in order to account for their observed mobility. Until the mid-1970s it was thought that mobility could be explained by the turbulent flow of a highly expanded and, hence, dilute mixture of fragments and gas (see Smith 1960a; Murai 1961; Sheridan and Regan 1976), but the publication of an important paper by Sparks (1976) showed that this notion was untenable because the sedimentology of ignimbrites is suggestive of high particle concentrations and a dense mixture. More recently authors have emphasized the role of gases which may be derived from a number of sources (Fisher and Schmincke 1984, p. 226) and include:

'1. exsolution of gas from juvenile (glassy) particles which buoys up particles and reduces friction between them . . . ;

2. breakage of congealed, hot magma fragments (lithic) with the resultant release

of heated gases contained within . . . ; and
3. the heating and expansion of air engulfed at the front of the flow' (see also Sparks 1979; C. J. N. Wilson 1980; C. J. N. Wilson and Walker 1982). All these sources of gas assist in the fluidization of the flow.

Fluidization is now seen to be a critical process in explaining the mobility of flows and their emplacement far from the site of eruption and is well known in chemical engineering as well as in volcanology. It has been investigated in the field and in the laboratory (Sparks 1976, 1978; Sheridan 1979; C. J. N. Wilson 1980, 1984, 1986). Briefly, fluidization involves gas, or more rarely liquid, flowing through a bed of solid particles. As velocity increases a point is reached, the *minimum fluidization velocity*, when 'the drag force exterted across the bed by the fluid is equal to the buoyant weight of the bed' (Cas and Wright 1987, p. 180). The bed now acts as a fluid. Pyroclastic flows are not fully fluidized (Sparks 1976), because the size of grains are variable and the smaller particles reach their terminal velocities, before the larger ones are fluidized. The smaller fractions are removed—or elutriated—from the flow (e.g. C. J. N. Wilson 1980, 1984). Wilson's research showed that flows are very complex entities and that the mixture of grain sizes will have very significant effects on the nature of fluidization, the flow properties and the sedimentology of deposits. Although his arguments are too long to present in full, it is interesting to note that the interaction of the fluidization process with particles of different sizes may be responsible for both three different flow regimes and three types of deposit. These C. J. N. Wilson (1980) named:
Type 1, which are poorly fluidized and have features similar to non-volcanic debris flows and lahars—being ungraded, having a yield strength and the ability to support blocks on the surface (see section 5.6.4);
Type 2, which show some expansion and differential grading—normal and reverse respectively—of lithic and pumice clasts; and
Type 3, in which fluidization is at its most intense and there are, not only zones of major pumice and lithic concentration—usually near to the top and the base respectively—but also the elutriation of much of the fine-grained material. In the Type 3 flows, this process of fines depletion causes the deposition of tephra known as a co-ignimbrite ash in the vicinity and often on top of the flow.

Pyroclastic flow deposits show great vertical, longitudinal and lateral variations in their sedimentology, due to differing modes of formation and emplacement. A summary of the major sedimentological features of pyroclastic flows is provided in Table 5.10.

At the present time more is known about the relationships between eruption, emplacement and deposition of flows formed by the collapse of Plinian and other columns (see sections 5.3.2, 5.3.3, 5.3.4 and Fig. 5.6), than for any other type of eruption and in the 1970s and 1980s a model of emplacement was developed with considerable success (Sparks 1976; Fisher 1979; Sheridan 1979; Wright *et al.* 1981b; C. J. N. Wilson and Walker 1982; Walker 1983; C. J. N. Wilson 1986). Collapsing columns have been of major focus of research and have been observed during monitored eruptions. Figure 5.18 and caption summarizes the model and, although self-explanatory in the main, requires some further explanation.

First, it should not be forgotten that Fig. 5.18 shows a model. Even when flows are generated by collapse, deposits may and often do differ from the 'ideal sequence' (Fig. 5.18b). Secondly, in proximal situations a deposit known as

Table 5.10: Principal sedimentological factors of flow deposits (based on information in Fisher and Schmincke 1984; C.J.N. Wilson 1986; the specific references cited and many other sources)

Attributes	Remarks
Flow units, cooling units and cooling history	Flow units are sedimentological entities deposited as distinct lobes and may vary from a few centimetres to tens of metres in thickness. Boundaries are marked by differences in grain-size, composition, concentration zones, etc. If several very hot flow units pile up on top of one another they may cool as a single cooling unit. Cooling may take many years. The lower and upper margins of cooling units are commonly unwelded because of, respectively, rapid conduction of heat to country rocks and radiation/convection to the atmosphere.
Welding and the effects of high temperatures	During flowage, temperatures may be close to those of the parental magma and, because flows are good insulators, they may retain heat for long periods of time. Days later they may still measure more than 500°C. The area of most intense welding is in the lower half of the flow, because this remains hottest for the longest period of time.
Texture	Flows are more poorly sorted than falls, but some overlap is evident (Walker 1971; Sparks 1976). Sorting improves from proximal to distal exposures. Median size of particles also decreases in a similar manner. Several additional grain-size measurements are commonly made: 1) Crystals, a measure of the distribution of this fraction in the magma and its breakage during eruption and transport; 2) Lithics, produced by primary fragmentation, from the vent, the pre-existing volcanic edifice and/or picked up from the ground over which the flow has passed; and 3) Pumice. Pumice is easily crushed and broken during eruption and flowage and decrease rapidly in size with distance from the vent. Vertical grading of clastic materials is important in flows, but is not always encountered (see text and C.J.N. Wilson 1980), but commonly pumice is reverse graded and lithics normally graded. Sometimes the largest pumice clasts may be several times larger at the top of sections than at their base. Elutriation of particles due to fluidization and segregation within the eruptive column, means that flows are enriched with pumice and lithics but depleted of fine-grained vitric materials. The whole rock geochemistry of flows, therefore, does not normally reflect the composition of the original magma.
Components and chemical composition	It is recognized that flows consist of varying amounts of crystals, vitric (glassy) fragments, pumice and lithics. Usually vitric ash and fine pumice are the principal constituents of matrices (often 50% or more), followed by crystals—which may range from zero to over 40%—and fine lithic fragments. Many crystals are only small enough to form the matrix because they have been broken during eruption, transport and compaction. Elutriation of less dense, fine particles, means that crystals tend to be more abundant in pyroclastic

Table 5.10 – *cont.*

Attributes	Remarks
	flows than in other deposits produced by the same eruption. All components also occur in the coarser fraction. Because most flows are generated from silica-rich magmas, the principal crystals are: quartz; sanidine; and plagioclase in calc-alkaline melts. In *trachytic, phonolitic* and *rhyolitic* melts, anorthoclase normally replaces plagioclase. Commonly larger flows and ignimbrites show vertical compositional zoning, due to eruption from zoned magma chambers.
Associated materials	Three zones commonly comprise the moving flow: 1) the dense flow itself; 2) hot surges, which are dilute and occur at the head and on the lateral margins of the flow. In some slow-moving flows, surges may be absent (C.J.N. Wilson 1986). Although hot surges are very destructive, they may only leave a very thin deposit and are often absent from the geological record. Alternatively they may form a thin ground surge layer at the base of flow units (Sparks 1976). 3) A dilute ash cloud consisting of material winnowed and elutriated from the flow and lying above it. This can form a buoyant plume kilometres high, which may combine with the plume formed at the eruption site to distribute fall materials over a considerable distance, as well as forming a co-ignimbrite ash. In vertical section, both types of fall are commonly found to lie above flow (Sparks 1976). Because flows are generated from silica-rich magmas, lava flows associated with them tend to be very short and domes often mark the eruption site (see Chapter 4).

co-ignimbrite lag-fall deposit (*lag* or *lithic breccia*) is often found. These lithic-rich accumulations were first recognized by Wright and Walker (1977, 1981) in the Acatlan ignimbrite in Mexico and they concluded that breccias formed at the same time as the main body of the flow within a zone of continuous collapse, representing materials that were too heavy and/or dense to be supported by the column. Since 1977 similar deposits have been recognized in a large number of flows. Subsequent work by Druitt and Sparks (1982) on Santorini (Greece) has shown, however, that additional processes may be involved in producing breccia concentrations in proximal situations these being: 'segregation of lithics through the body of the flow' and separation of lithics by the more fluidized 'head' during its movement (Cas and Wright 1987, pp. 240–1). In the latter case the main body of the flow rides over the lag deposits and the term *ground breccia* is used for the sediments so produced (see also Walker 1985).

Thirdly, many flows and most medium and large ignimbrites do not originate because of column collapse, and hence the model (Fig. 5.18) is not applicable. In particular, deposits exposed in the field do not always resemble 'ideal sections' (Fig. 5.18b) and may in fact be quite different. Fisher and Schmincke (1984, p. 206) quote such examples from Mt Pelée (1902 eruption), sequences in the Laacher volcanic region in Germany and Mount St Helens. Fig. 5.19 is taken from the work of Guest *et al.* (1988) and indicates just how complex pyroclastic

Fig. 5.17 Coarse 'lithic' blocks at the base of a large ignimbrite, Roccamonfina Volcano, Italy (author's photograph).

flow sequences can be. The basal ignimbrites of Monte Vulture volcano in Italy are shown in a proximal exposure. Only at the base of the second ignimbrite from the top are ground surges to be found, there is no evidence of reverse grading of large pumice clasts, normal grading of lithics, surges derived from an ash cloud or lithic breccias (Fig. 5.17).

Finally, much research has still to be undertaken before two subtypes of pyroclastic flow: 'low-aspect ratio' and 'fines depleted' ignimbrites are fully understood. Both terms were introduced by George Walker and neither can be reconciled easily with a simple model of column collapse. Low-aspect ratio ignimbrites (Walker *et al.* 1980a) are found around many volcanoes, are thin and may cover large areas. They seem to have been emplaced quickly, to represent very high magma discharge and may pose a significant and underestimated hazard. As its name implies, a fines-depleted ignimbrite has relatively little fine ash and may represent a very turbulent, highly gas-charged, fluidized flow causing elutriation of most of the fines (Walker *et al.* 1980b).

5.6.3 Cold (base) surges

As mentioned previously (i.e. section 5.3.6.2) deposits of this type are produced by hydrovolcanic activity and, as with flows, so cold surges are far better understood today than was the case even a decade and a half ago.

Cold surges have distinctive sedimentological features both in plan view and outcrop and these are related to mechanisms of transport and emplacement. According to Wohletz and Sheridan (1979, p. 197) surges form 'an apron of

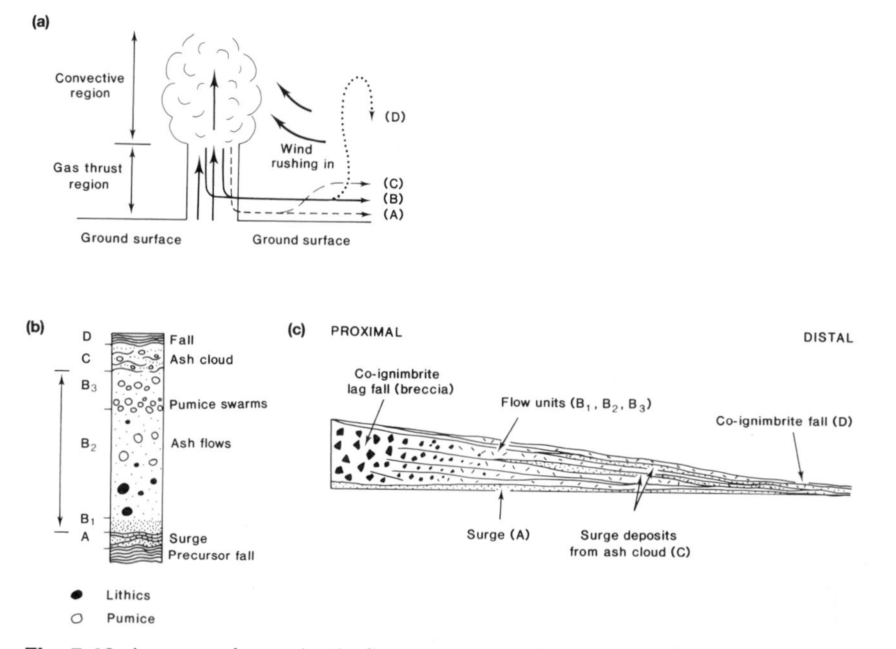

Fig. 5.18 Aspects of pyroclastic *flows* generated by column collapse.
(a) Collapse of a Plinian column showing: collapse of the outer margins to produce a 'dilute' surge (A); further collapse of the central region of the column to produce a *flow* (B); dilute ash cloud (*surge*) produced by elutriation of fines from the flow (C) and very fine material carried to a considerable height and deposited on top of the whole sequence (D). It should be noted that the ash cloud (C) is very mobile and was responsible during the 1991 eruption of Mount Unzen (Japan) for the deaths of three volcanologists who were observing the eruption from a valley spur.
(b) 'Ideal' sedimentary sequence; (A) to (D) inclusive is a single 'flow unit'.
(c) Changes in deposits with distance.
From: Fisher, R.V. and Schmincke, H.U. 1984: *Pyroclastic Rocks.* Berlin, Springer-Verlag, Fig. 8.23, p. 204 and based on an original idea of Fisher 1979; b. Sparks, R.S.J. 1976: Grain size variations in ignimbrites and implications for the transport of pyroclastic flows. *Sedimentology* 23, Fig. 1, p. 1 and c. Wright, J.V. *et al.* 1980: Towards a facies model of ignimbrite-forming eruptions. In Self, S. and Sparks, R.S.J. 1981: *Tephra Studies,* Amsterdam, D. Reidel, Fig. 3, p. 437.

poorly sorted tephra around a central vent.... The deposits have an overall wedge shape in cross-section with a nearly logarithmic decrease in thickness away from a vent'. If exposures at increasing distance from the vent are examined, different bedforms are commonly encountered. In proximal locations, exposures tend to be dominated by what are termed *sandwaves* with subsidiary massive beds, while in distal outcrops planar beds predominate over massive beds. Intermediate exposures show all three bed forms, but massive beds predominate

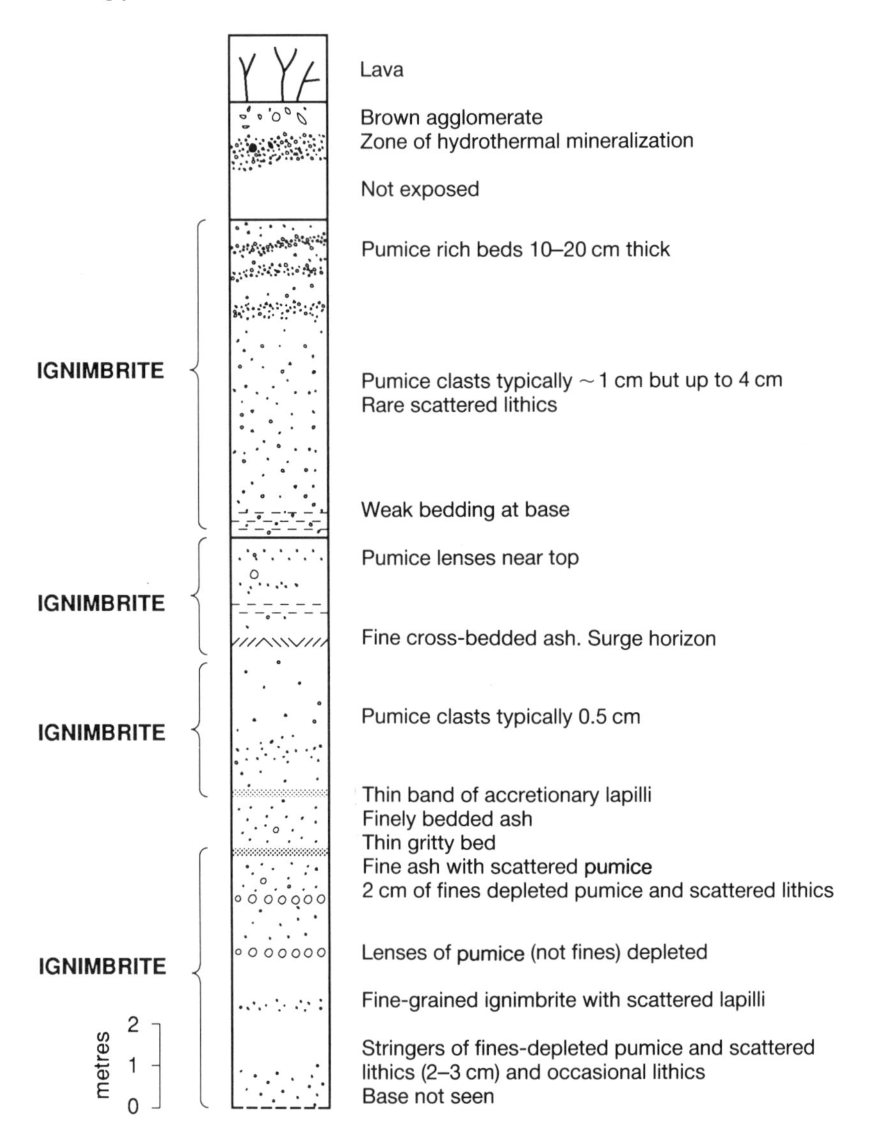

Fig. 5.19 Proximal section through the basal ignimbrites of Monte Vulture, Italy (from Guest *et al.* 1988, Fig. 4, p. 250).

(Sheridan and Updike 1975; Wohletz and Sheridan 1979). Sandwaves are generally fine-grained with median-size values of 0–4ϕ being typical. Massive and planar beds stand in an ascending sequence of increasing median size, although overlap between all the size 'envelopes' is evident (Wohletz 1983; and Fig. 5.20).

Accounting for the regularity of facies change has not proved to be an easy task. In what has become a classic paper, Wohletz and Sheridan (1979, pp. 189–93) studied surges at four volcanic centres in the USA and one in Mexico and argued that differences in bedform could be explained by differences in

Fig. 5.20 'Sandwave' surge beds on the Island of Vulcano, Aeolian Islands, Italy (photograph reproduced by courtesy of Dr A.M. Duncan, Luton College of Higher Education).

fluidization. They argued that when erupted surges are partially fluidized by steam, magmatic volatiles and the collapsing column, and that materials travel away from the eruptive vent under the control of gravity and/or the velocity imparted by the fluidized particles. Near to the vent surges are 'lean' (i.e. the particles are widely spaced), viscous and act as Newtonian fluids[5]. Because of high Reynolds numbers (see Table 5.3) surges are turbulent and able to build complex sequences of ripples, dunes, antidunes and cross-laminations, which together constitute the sandwave units. As in aeolian sequences, the sandwaves structures relate in the main to processes of saltation (Fig. 5.21), with detailed variations in bedding structures being caused by such factors as velocity and grain-size dissimilarities. In distal exposures, most gas has escaped and surges are non-fluidized, but zones of densely fluidized material may remain. In this situation, interactions between the grains are important and flow is laminar (see Table 5.3), producing planar beds, whilst the still fluidized portions create the subsidiary massive beds. In sections intermediate between proximal and distal exposures, both viscous and non-viscous forces are important, producing a transitional mixture of sandwaves, massive and planar beds; the dominance of massive beds being explained by the occurrence of densely fluidized parts of the surge which are formed by interaction between viscous and inertial flow (i.e. flow related to simple grain interaction, due to applied shear stress, particle density, and surface roughness).

[5] Viscous behaviour of a substance occurs when its rate of deformation (often called the strain rate) is proportional to the stress applied. The ideal case is known as a Newtonian fluid, e.g. water (Summerfield 1991).

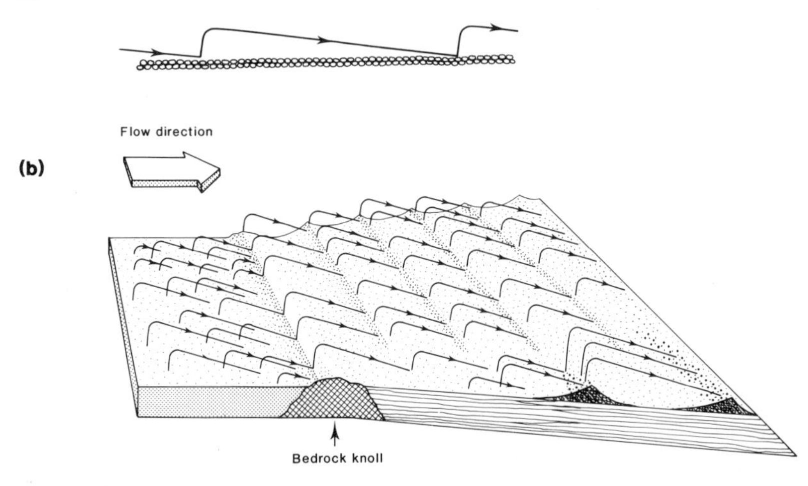

Fig. 5.21 (a) Saltation of a particle. When a particle rises into a flow, it encounters increasing speed with height and is carried forward. It returns to the surface along a trajectory defined by gravity and aerodynamic drag (from Summerfield 1991, Fig. 10.3, p. 238).
(b) Diagram showing the formation of ripples in arid regions by saltation and surface creep. The coarser grains travel by creep and collect at ripple crests. These asymmetric forms may build to form complex dunes and other structures (modified from Embleton and Thornes 1979, Fig. 10.6, p. 333).

The publication of the Wohletz/Sheridan model, has stimulated much debate about its general applicability and the veracity of the processes involved. For instance, Allen (1982) questioned whether the conventional interpretation of many of the sedimentary structures which constitute sandwaves may in fact be erroneous and argued that certain structures may be more typical of wet and cool and others of dry and hot surges. Specifically he argued that 'progressive' sandwaves bedforms (i.e. those that show migration structures in the direction of flow), were typical of dry and hot surges, whereas 'regressive' sandwave bedforms (i.e. those showing migration structures against the direction of flow), where typical of wet and cool surges. Fisher and Schmincke (1984) and Cas and Wright (1987) questioned the general applicability of the Wohletz/Sheridan model to all cold surges and concluded that 'given the complexities of surge mechanics . . . it may be yet premature to propose simple proximal to distal and up-sequence facies changes and models' (Cas and Wright 1987, p. 212). More recently Paul Cole has reopened the debate about progressive and regressive structures (Cole 1991). During detailed study of sandwave structures both at Roccamonfina volcano (Italy) and Sugarloaf Mountain (USA), he found evidence of progressive and regressive structures in the same suites of surges and concluded that migration of bedforms was determined by changes in the flow regime of the surge cloud and not by temperature and water content.

As well as the bed forms described above, wet surges also show additional sedimentary features, which are summarized in Table 5.11.

Table 5.11: Sedimentological attributes of cold (base) surge deposits (based on the references cited and many other sources)

Sedimentological attribute	Comments
Field relationships	Surges are thickest in valleys and are thin, or absent, on interfluves. In this respect they differ from flows and falls.
'U'-shaped channels	'U'-shaped channel forms are found in many deposits. They range from less than 1 m to 10 m or more across and may be up to 3 m, or even more in depth. A bed filling a channel reflects its shape. Channels are thought to represent lobes within the surge which act as semi-independent entities and are able to erode deposits already emplaced (Fisher 1977).
Bedding sags	Bedding, or 'bomb', sags are produced by the impact or large air fall blocks and/or bombs on wet—and still plastic—surge beds. This causes the beds to be penetrated, distorted and thinned. If three-dimensional exposure is available then sags are often observed to be asymmetric, indicating the direction of impact.
Accretionary lapilli	These are spherical or approximately spherical masses of concentrically layered, cemented ash; formed by accretion of fine material by moisture. They are not diagnostic of surge activity, since they can occur in rain-flushed fine falls.

5.6.4 Lahars

In section 5.2 the term lahar was defined as a collective noun to include all water-transported volcanic debris, regardless of origin and sedimentological properties. Further in section 5.3.6.2, it was noted that lahars can form during episodes of hydrovolcanic activity. This is, however, only one of a number of ways in which these distinctive deposits may be produced. Gordon Macdonald (1972) listed ten possible mechanisms and these are reproduced, with modification, in Table 5.12.

Lahars involve several different types of flow. At one extreme are simply particle-rich floods of water, many of which have been observed during eruptions and may be very destructive, as was evident during the 1980 Mount St Helens eruption (Foxworthy and Hill 1982). Debris flows are at the other end of the spectrum and are much more particle-rich, with 80% or more of their total mass being solids and the rest water. Between these two extremes are flows commonly referred to as hyperconcentrated.

The ratio of particles to water is a critical parameter influencing the nature of flow processes (Beverage and Culbertson 1964; Johnson 1970). Not surprisingly when water ratios exceed about 91% by volume, then the processes of flow are Newtonian (see section 5.6.3) and there are no differences between these and similar particle-rich floods in any subaerial environment. However, as Fisher and Schmincke (1984, p. 298) note, when solids exceed about 9% by volume then Newtonian properties begin to be altered and flows become progressively plastic in behaviour being similar to 'wet concrete', with high bulk densities and quite

Table 5.12: Some of the ways in which lahars are formed (mechanisms 1 to 10 are based on Macdonald 1972, p. 171)

Mechanism	Comments
Volcanic mechanisms	*Related directly to eruptions*
1	Mobilization of materials during eruption through a crater-lake.
2	Explosions cause the movement of old crater materials into rivers and streams draining a volcano.
3	Movement of pyroclastic flows into streams draining a volcano and the mobilization of debris.
4	The entrance and mass-movement of broken lava flows into streams draining a volcano.
5	Lava flowing over snow and/or ice and mobilizing loose materials.
6	Material ejected from the conduit of a volcano which has already been fragmented.
Associated mechanisms	*Related indirectly to eruptions*
7	Relase of water from a crater lake by the breaching of its walls.
8	Rapid melting of snow and/or ice on the slopes of a volcano.
9	Mobilization by heavy rain of fragmental material on the side of a volcano.
10	Mobilization of saturated material on the side of a volcano due to seismic activity.
11	Mobilization of materials due to other processes of slope instability.

different rheological attributes. Rheology is the study of the deformation and flowage of materials and for flows of high particle concentration it has long been accepted that quite different (i.e. non-Newtonian) flow models are necessary to explain their mobility and emplacement (Neall 1976; Innes 1983; Postma 1986).

In debris flows (particle concentrations over 80% by weight—about 60% by volume) movement can be modelled as being essentially plastic. It is a matter of observation that when water is split it will flow down slopes which are almost imperceptible because Newtonian substances move under the smallest of shear stresses. Plastic bodies only do so when applied shear stress exceeds a certain value; known as the *yield strength*. Initiation of a flow depends on yield strength being exceeded and this in turn is determined by factors which include: cohesion; mass; slope-angle; and moisture content. These and other variables may be combined in a mathematical relationship known as the *factor of safety* (Innes 1983) and many of the mechanisms listed in Table 5.12 initiate debris flows because they increase water contents significantly; so causing the yield strength of the particular material to be exceeded.

Debris flows consist of well-mixed masses of water and particles and, because of high viscosity, movement is laminar (see Table 5.3). As shear stress decreases due, perhaps, to a lessening of slope-angle or the draining away of water, then yield strength again becomes important and a flow may stop abruptly. In very thick flows a high shear stress may be maintained at the base even though the rest of the flow has congealed. This is known as *plug flow* (Johnson 1970) and involves the congealed mass being 'rafted' on a basal layer which is still capable of laminar movement. Debris flows have a considerable bearing strength and are able to support and carry large blocks. The initiation and movement of debris flows are related closely to the areal extent, morphology and sedimentology of the deposits produced. The nature of these relationships is summarized in Table 5.13.

Finally, one of the major problems of field research on pyroclastic materials is in distinguishing between lahars—particularly debris flows derived from pyroclastic flows—and true pyroclastic flows. One way in which this may be achieved is by using the technique of thermoremnant magnetism (Hoblitt and Kellogg 1979). An important consideration is the *Curie Point*, a temperature of ~400°C. As hot

Table 5.13: Some attributes of debris flows (based on the sources cited)

Attribute	Comments
Areal extent, thickness and distribution	Debris flows like pyroclastic flows, follow valleys and may be absent on interfluves. They may be absent, or thin, on steep slopes and when unconfined form lobate features. They vary in thickness from less than 1 m to over 200 m, but on moderate slopes may be relatively uniform with a steep front, reflecting abrupt coalescence due to shear stresses falling below the yield strength of the material.
Geomorphology	Surface morphology is generally flat, with gentle slopes towards the flow front. In places there may be a local relief of low amplitude ridges and valleys, often reflecting an irregular underlying surface topography (Crandell and Waldron 1956). In the past many descriptions of mounds and hummocks have appeared in the literature, but many of these are now thought to be debris avalanches (e.g. see Siebert 1984; Francis *et al.* 1985).
Sedimentology	Debris flows may be formed from a wide range of materials, reflective of the variety of processes by which they are generated (see Table 5.12). Debris flows derived from pumice-rich materials may be difficult to distinguish from pyroclastic flows (see text). Debris flows derived from pyroclastic flows often contain charcoal. Air-spaces (voids or vesicles) are common and are caused by trapped air (Crandell 1971). Generally debris flows have a very wide range of grain sizes and are poorly sorted. One characteristic feature is the presence of large blocks—many exceeding 1 m in size—which are supporting the flow. Sometimes these may be 'rafted' at and near to the surface. Except on steep slopes, lahars do not normally erode the ground over which they pass (Fisher and Schmincke 1984). A distinction is drawn between clasts (greater than 2 mm) and matrix. Clasts are predominantly matrix-supported, reflecting the strength of the material, and are typically weakly graded, although coarse materials are often absent from basal layers. Reverse grading with the coarse fraction concentrated at the top of the deposit is rarer, unless the clasts are pumiceous. Some deposits show no grading. Grading may reflect the ratio of solids to water.

particles in flows cool through this temperature they retain a uniform magnetic signature, reflecting the Earth's magnetic field at the time. The fragments in lahars, however, are derived from a variety of sources, are of differing ages and have been transported by processes well below the Curie Point. They will show random directions of remnant magnetism. Although this technique is useful, it is not without its problems since it is known that certain flows may be in part emplaced at below the Curie Point.

5.6.5 Debris avalanches

In recent years debris avalanches have excited considerable attention both in the volcanic and non-volcanic literature and important contributions have been made by a number of writers including: Hsü (1975, 1978); Voight *et al.* (1981); Ui (1983); Siebert (1984, 1992); Francis *et al.* (1985); Clapperton and Smyth (1986); Ui and Glicken (1986); Siebert *et al.* (1989); Clapperton (1990); and Smyth (1991). Much of this interest was stimulated by the closely observed eruption of Mount St Helens in 1980 (Voight *et al.* 1981) and the realization that distinctive avalanche deposits may be generated not only from rockfalls as occur on volcanic and non-volcanic mountains alike, but also from the collapse of whole sectors of volcanoes. Often such collapse is associated—as at Mount St Helens—with major eruptions, but there is evidence to suggest that it may in certain circumstances be independent of it. Following the Mount St Helens eruptions, volcanologists became increasingly aware of the likelihood of debris avalanches and research throughout the world (e.g. Siebert 1984), but especially in Japan (e.g. Ui *et al.* 1986) and South America (e.g. Francis *et al.* 1985; Clapperton and Smyth 1986; Clapperton 1990), has provided valuable insights into the nature of the processes involved and the types of deposit produced. One result has been that many deposits formerly classified as lahars are now thought to have been formed by debris avalanches.

In his comprehensive worldwide review of more than 80 possible events in the Quaternary, Lee Siebert (1984) showed that large debris avalanches have several features in common. First, they are associated with large amphitheatre-shaped scars within volcanic edifices. Secondly, the deposits in front of the scars are distinctive. Geomorphologically they consist of many hummocks, with small hills and closed depressions and sedimentologically two facies may be identified: large homogeneous blocks (megablocks) often showing close-fits one to another (jigsaw fits); and finer more heterogeneous matrix material. All features are commonly found in sediments produced by landslides (Coates 1977) and, what is more, the volume of deposits usually correlates closely with the volume of the amphitheatre-shaped scars; implying that the former were derived from the latter. Thirdly, the sizes of blocks and hummocks decrease with distance from source and, finally, the distance travelled by the avalanche (L) is largely a function of the vertical drop (H) between local topographic datum and the height of the scar. This is suggestive of high mobility, since H/L ratios are comparable with those of some pyroclastic flows. Generally distances travelled are greater than those of non-volcanic landslides and less than those of large lahars and flows.

The findings of Siebert (1984) have been confirmed by other authors and subsequent research, but detailed processes of formation, mobility and emplacement are still debated (see Smyth 1991). Debris avalanches are clearly of major importance on many volcanoes and present a significant additional hazard to many people living in their vicinity (see Chapter 7). Not unnaturally, because most research on debris avalanches has been undertaken by volcanologists with geological backgrounds, the emphasis in the literature has been on volcanic activity triggering off these slides. Such suggested triggers include: explosive eruptions; volcanic earthquakes; the intrusion of dykes; and hydrothermal alteration causing weakening of the cone. It is possible, however, to invoke many additional processes of which only one—tectonic earthquakes—is widely recognized. Volcanoes are intrinsically unstable landforms with high slope-angles, interbedding of materials on high-angle dip-planes, the accumulation of considerable masses of material at high altitudes and the ready availability of water. It may well be that the timing of many debris avalanches is not the result of any volcanic trigger, but rather snow-melt, rain storms, undercutting of unstable masses by streams and the operation of other subaerial processes. Debris avalanches remain phenomena about which much is still to be learnt.

5.7 Concluding remarks

In the introduction it was emphasized that research on the pyroclastic suite had made considerable progress on an interdisciplinary front over the last two decades. Such a rate of progress will be hard to sustain indefinitely, and many of the models which have been proposed require detailed testing during actual eruptions so that they can be refined further. In addition a major problem is that some of the highest magnitude pyroclastic events (e.g. ignimbrites) also have the lowest frequency of occurrence and some of the models used to explain their generation and products are not confirmed by first-hand observational evidence, but rather by the not always unambiguous interpretation of geological records.

Some would claim that study of the pyroclastic suite has been the volcanological 'bandwagon' of the last 20 years, with its researchers a sort of 'tuffiosi'—as the author has heard them so described. A fairer summary would be that major contributions have been made by a relatively small number of writers who have transformed a whole field of research from one that was almost moribund to one which is lively and exciting.

6

Volcanic gases and the effects of volcanoes on climate

There is no doubt that the most important changes which nature produces in the transparency of the atmosphere from time to time, over durations that directly concern us, are those due to variations in the amount of volcanic dust present (H. H. Lamb 1982).

6.1 Preface

Some of the information contained in this chapter first appeared in a review article published in *Progress in Physical Geography* 12 (1), 1988. Chapter 6 is an expanded and updated version of the review and contains much new material, reflecting the advances and new insights which have emerged over the past five years. The author wishes to thank the publishers of *Progress in Physical Geography* for agreeing to this republication.

6.2 Introduction

More than a hundred years ago the pioneer geologist and volcanologist George Poulett-Scrope (1862) drew attention to the importance of juvenile gases (i.e. the volatile elements within melts) in the movement of magma to the Earth's surface. In earlier chapters attention has focused on the effects of gases on styles of eruption (Chapter 3, section 3.4), the types of volcanic products erupted (Chapters 4 and 5) and in Chapter 7 the role of gas-monitoring in the prediction of eruptions will be considered. Volcanoes and climate is a theme which requires detailed consideration. Volcanoes affect climate through their emission of gases particularly sulphur dioxide, though it was thought for many years that the main factor causing change was in fact fine airfall ash (Newell 1981a, 1981b).

Several authors have highlighted the problems, both methodological and practical, of obtaining estimates of the amount and species of gases vented into the atmosphere by eruptions (Decker and Decker 1989; Fisher and Schmincke 1984; Stothers *et al.* 1988). Several approaches have been adopted, none of them entirely satisfactory.

The most obvious, often known as the *direct* method, is to sample gases emitted from eruption plumes, or at vents. Measurements taken at vents are beset with many dangers, both personal and scientific, in what are often hostile field environments. Not least are problems of sample contamination by atmospheric gases and bias. Techniques of remote sensing are often employed to sample plumes (Stoiber and Jepsen 1973; Rose *et al.* 1986; Malinconico 1987) and airborne correlation spectrometry can produce reasonable estimates of SO_2, but errors are possible (Stoiber *et al.* 1983; and Fig. 6.1). *Indirect* methods include: 1) estimates made from chemical analyses of volcanic rocks, a technique which can give only a rough prediction of volatile composition because magma is known to lose volatiles during eruption;

Fig. 6.1 A scientist from the United States Geological Survey using a correlation spectrometer (COSPEC) to measure the SO_2 content of the Mount St Helens plume. The photograph shows the late Dr David A. Johnstone, who was sadly killed by the eruption (photograph reproduced by courtesy of the United States Geological Survey—ref. Mount St Helens no. 32).

2) studies of volcanic glass and gases trapped within erupted products (e.g. H. G. Moore and Schilling 1973); and
3) the analysis of less volatile elements, whose concentrations are known to correlate with volatiles—especially the H_2O contents of magmas (J. G. Moore 1970; J. G. Moore *et al.* 1977; Aoki *et al.* 1981).

Despite the difficulties, Fisher and Schmincke (1984, p. 37) were able to conclude that 'most workers agree that H_2O (35 to 90 mol %), CO_2 (5–50 mol %) and SO_2 (2–30 mol %) are the most abundant volatiles in basaltic and andesitic magmas, whereas H_2, CO, COS, H_2S, S_2, O_2, HCL, N_2, HF, HB, HI, metal halogens and noble gases are minor constituents (<2 mol %)'. In terms of elemental abundance the most common element is hydrogen, followed by oxygen, sulphur, chlorine and nitrogen. With the exception of sulphur, this order of abundance is in close agreement with that found in the atmosphere, biosphere and surficial rocks of the Earth (Fig. 6.2, and see section 6.4.2.2).

In 1789 Benjamin Franklin read a paper to the Manchester Literary and Philosophical Society in which he discussed the severe winter of 1783/84. In 1783 Franklin had been United States Ambassador to France and observed the blue haze which drifted across Europe and into Asia, following the eruption at the Laki fissure, Iceland, in that year. In claiming a direct connection between this eruption, the haze and the severe winter that followed, he became the first scientist to propose a link between a volcanic eruption and the weather experienced in the months following the event (Franklin 1789; Mitchell 1982). Further interest in this possible relationship was stimulated by the violent eruption of Kratatau (Indonesia) in 1883, the general conclusions being contained in a comprehensive report prepared for The Royal Society of London (Symons 1888; see also Wexler 1951a; Simkin and Fiske 1983). The conclusions were that the worldwide optical effects, including spectacular sunsets which were noted in the months following the eruption, the cloud haze and lower temperatures recorded in many countries distant from the eruption were all due to this event. In this century several writers have suggested relationships between episodes of violent volcanism on the one hand and decreases in global temperatures in the years that follow on the other, these being considered for both the last few hundred years for which temperature and eruption records have been compiled and, more speculatively, over longer periods of geological time. With respect to the latter, particular attention has been focused on the possible influence of widespread volcanic activity upon the initiation and timing of Pleistocene glacial fluctuations (Humphreys 1940; Gentilli 1948; Wexler 1951b, 1952; Porter 1981).

The worker, however, who has had the greatest influence on subsequence research has been H. H. Lamb (1970, 1972, 1977, 1982a, 1982b). In the context of the wider study of climatic history, he proposed that it was possible over the period for which records were available to calculate an index (the Dust Veil Index), which when related to temperature records could be used to posit a causal link between these time series. Because of the importance of his research to both atmospheric scientists and volcanologists, Lamb's contribution will be taken as the starting-point for discussion.

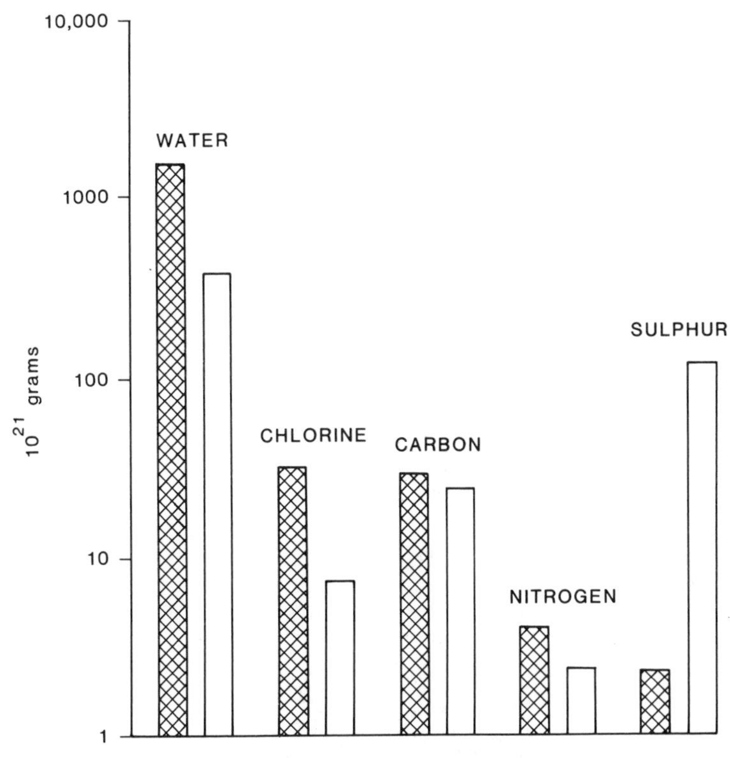

Fig. 6.2 Comparison of the volatile molecules and elements in the Earth's oceans and atmosphere and their volcanic production. For each substance, the bar on the left (hachured) represents the mass of that substance in the air, oceans and sedimentary rocks, whilst that on the right (unshaded) represents the total mass of the substance being currently produced by volcanoes, multiplied by the age of the Earth (from Decker and Decker 1989, Fig. 101, p. 179).

6.3 Dust produced by volcanoes and Lamb's Dust Veil Index

Lamb's Dust Veil Index was first defined and discussed in a comprehensive paper published in the *Philosophical Transactions of the Royal Society of London* (Lamb 1970) and was later used and refined in this author's two-volume work *Climate Present, Past and Future* (Lamb 1972, 1977). His argument is briefly as follows: first, explosive volcanism is able to produce worldwide veils of fine dust, which remain in the stratosphere for a number of years and serve to diminish the brightness of the solar disc and brighten the background sky. Dust from explosive volcanic eruptions can reach varying heights, but has to penetrate the lower stratosphere (20–27 km) before it can produce persistent veils; the normal size of fragments in these veils being 0.5–2 m^{-3}, with a residence time being determined by Stokes Law. The second part of his discourse concerns the dispersal of dust, which Lamb argues is carried around the world by the upper circulation; the dust taking from 2–6 weeks to circle the Earth in mid to low latitudes and from 1–4 months to become uniform over the whole latitudinal zone affected by the wind system. Because of the nature of this circulation, dust injected into the low stratosphere at low latitudes will gradually spread over the whole Earth, whilst that introduced at high latitudes remains within the hemisphere of origin. The dust has effects on the appearance of the sun (dimming and reddening), the radiation received and upon surface temperatures, for periods varying from months to several years.

Lamb compiled a Dust Veil Index (DVI) for volcanic eruptions of the historic past in order to examine and assess in detail the possible influence of volcanic dust. Recognizing with notable perspicacity that many historical eruptions are not well documented, Lamb proposed three alternative methods whereby the DVI could be calculated. In each case these formulae were normalized so that a DVI of 1000 was produced for the Kratatau eruption of 1883 and Lamb noted that for the most reliable results as many of these formulae as possible should be employed. These are:

1) $DVI = 0.97\ R_{DMAX}E_{MAX}t_{MO}$
2) $DVI = 52.5\ T_{DMAX}E_{MAX}t_{MO}$
3) $DVI = 4.4\ qE_{MAX}t_{MO}$

where R_{DMAX} = the greatest percentage depletion of the direct solar radiation as shown by any monthly average for the mid-latitudes of the hemisphere concerned after the eruption;
E_{MAX} = the maximum extent (greatest proportion) of the Earth at some time affected by the dust veil (taken as 1 for eruptions between 20° north and south, 0.7 for eruptions between 20 and 35° north and south, 0.5 for latitudes 35–42° north and south and 0.3 for latitudes greater than 42° north and south of the equator);
T_{DMAX} = the estimated lowering of average temperatures for the most affected year over the middle-latitude zone of the hemisphere affected (in °C);
q = the estimated volume in km^3 of solid matter dispersed in the atmosphere; and
t_{MO} = the total time in months between the eruption and the last observation of the dust veil, or its effect upon monthly radiation (or temperature) values in middle latitudes.

By means of these formulae it proved possible to rank eruptions in order of magnitude and Lamb considered that only those with a DVI of over 100 were likely to have had any climatological significance but it should be noted that several smaller eruptions, closely grouped in time, may have a similar effect to one large one. These data were used finally to draw relationships between dust veils and climatic change for areas of the world which have long records of temperature. With respect to the British Isles and the North Atlantic, Lamb concluded that there is 'evident logic in the apparent expectation that the following two or three winters (after an eruption) have a tendency to enhanced atmospheric circulation, more prevalent westerlies and hence mildness, and that this tendency is set up immediately after high-latitude eruptions: for these are the situations in which the chilling effect of volcanic dust is likely to be concentrated over the higher latitudes, with an enhanced thermal gradient between the region so affected and the lower latitudes that are either unaffected or becoming free of the volcanic dust veil' (Lamb 1972, p. 424).

Lamb's DVI should not only be viewed in terms of his research on past, present and future climates, but also with reference to the major impact his ideas have had on the research of others. Broadly this research falls into two categories. First, there have been many attempts to investigate the relationships between volcanic activity and climatic fluctuations at a number of scales and from varying perspectives. Secondly, many workers have used the DVI and other more recent indices, as forcing factors in models of climatic change. The present account is concerned with the first category. The principal reason for concentrating on volcanological aspects, rather than on climatic modelling, is because many models have used volcanic inputs in a rather uncritical fashion. This point of view is supported to some extent by two leading workers in the field. Bryson and Goodman (1982), in a stimulating account of the role of volcanic activity on climate during historical times, argue that because past temperatures can be simulated very well in models with inputs to the atmosphere from volcanic ejecta and CO_2 from natural and anthropogenic sources, then climatic forecasting over this time scale may ultimately depend, in part at least, upon the prediction of large explosive eruptions that can cause climatic perturbations. They conclude that 'it is not enough to have every refined general circulation model with all feedback loops identified and correctly interlinked. The extrinsic inputs that modulate the behaviour of the climate system must also be specified' (Bryson and Goodman 1982, p. 194). A further reason is that the role of volcanic ejecta as a forcing mechanism has been discussed extensively elsewhere and that any further statement would be merely repetitive. For those interested in this research, the following should be consulted: Schneider and Mass (1975a, 1975b); Pollack *et al.* (1976); Mass and Schneider (1977); Robock (1978, 1979, 1981); Miles and Gildersleeves (1978); Gilliland (1982); Lamb (1982a, 1982b); Pantic and Stefanovic (1982).

6.4 Volcanological perspectives on volcanoes and climate

Volcanological and wider earth-science perspectives on volcanoes and climate published since Lamb's articulation of the DVI may be split into groups. Some workers have accepted the thesis that volcanic eruptions, especially those that are explosive, can cause climatic change and have been concerned to improve the ways in which injections into the stratosphere can be measured and monitored, whilst others have been engaged actively with broader issues of causal inference over differing temporal and spatial scales. For convenience these will be considered separately, although they are closely related.

6.4.1 Measuring and monitoring volcanic injections

This has taken three forms. Volcanic eruptions which have occurred over the last 15 years have been studied in an increasingly detailed fashion in order to understand more fully the nature of injection of volcanic debris into the stratosphere. This has been supplemented by more thorough analysis of the records of past eruptions. A second research thrust has been to trace volcanic particles in terrestrial sediments, while a third has endeavoured to improve Lamb's DVI.

6.4.1.1 *The detailed study of past and present eruptions*

Information on aspects of contemporary volcanic activity has improved greatly over the last two decades, but is still somewhat scattered through the literature. In addition many of these sources of information are essentially volcanological and contain relatively little on the possible impact of eruptions on climate. This applies in varying degrees to well-known catalogues such as those published by: *Bulletin Volcanologique* (1941 to 1986); the Volcanological Society of Japan (*Bulletin of Volcanic Eruptions* 1961–1978); IAVCEI (*Catalogue of Active Volcanoes and Solfatara Areas of the World* 1950–1975); the Smithsonian Institution (*Reports of Volcanic Eruptions* 1968–1975 and monthly reports of the *Smithsonian Scientific Event Alert Network* (SEAN) 1975-date), and the Bulletin of Volcanology's *Global Volcanism Bulletin* (1986-date). In 1981 the Smithsonian Institution sponsored the publication of a comprehensive catalogue of volcanic eruptions over the last 10,000 years (Simkin *et al.* 1981). Although this contains a new index of volcanic explosivity (see section 6.4.1.3), it does not contain much on the nature of volcanic injections. This information tends to be published in two sources; as detailed post-eruption reports, such as those produced by the United States Geological Survey following the eruption of Mount St Helens (USA) in 1980 (Lipman and Mullineaux 1981; Foxworthy and Hill 1982) and as one-off studies of both contemporary and past eruptions (see Sigurdssen *et al.* 1985; Williams *et al.* 1986; Self *et al.* 1989; Vogel *et al.* 1990).

One result of the study of past and present eruptions has been to confirm and quantify the effects of large eruptions on climate. In two reports Newell (1970, 1981a; see also Newell and Weare 1976) has studied the effects on the atmosphere of the Mt Agung or Gunung Agung (Bali) eruption in 1963. In terms of ejecta introduced into the stratosphere, this was one of the largest eruptions this century and, unlike earlier events of similar or greater magnitude, occurred at a time when its effects could be closely monitored by a large number of meteorological stations. The principal conclusion of Newell's study was that the maximum effect was a decrease of some 1°K in the middle and upper tropospheric temperature of the tropics in late 1964 and early 1965. This finding was confirmed by Hansen *et al.* in 1978. Similar reductions in temperature have been posited following earlier eruptions. For instance, Hoyt (1978) argued that two previously little known eruptions may have been larger events than was supposed at the time, these being the eruptions of Paluweh (East Indies) in 1928 and Reventador (Ecuador) in 1929. These eruptions, Hoyt maintains, were responsible for a Northern Hemisphere temperature decline of about 0.3°C in the years which followed, while an eruption of Cerro Azul, Argentina, may have been responsible for a decrease in atmospheric transmission, presumably in the Southern Hemisphere, in 1932. In an intriguing retrospective study of the large eruption of Tambora (Indonesia) in 1815, Stommel and Stommel (1979), draw a direct link between this event and the cold summer of 1816, experienced in North America and Western Europe. Other authors who confirm the effect of volcanoes either directly on temperatures, or indirectly as atmospheric turbidity anomalies, include Volz (1975a, 1975b—1912 eruption of Katmai, Alaska) and Cronin (1971) and Deirmendjian (1973), who compared the effects of the eruptions of Krakatau (1883), Katmai (1912) and Agung (1963). In an interesting and scholarly paper, Stothers and Rampino (1983b) used written and archaeological sources to reconstruct major eruptions in the Mediterranean region which occurred before 630 AD. They then looked for correlations with episodes of atmospheric cooling. The authors support Lamb's contention, that for eruptions to have maximum climatic influence requires high eruption columns that penetrate the tropopause, large stratospheric injection of materials (not necessarily dust alone) and, for worldwide effects, volcanoes located in tropical latitudes.

Detailed study has not always confirmed the climatic impact of past eruptions. For many years the 1835 eruption of Cosegüina in Nicaragua was thought to have been extremely violent and to have had a major impact on world climate (e.g. H. Williams 1952; Lamb 1970; Kondo 1988). However, detailed study of the eruption and its products by Self *et al.* (1989) shows that, although violent, its magnitude was modest and had little effect on climatic change in the 1830s.

In the early 1970s research emphasized ashes rather than gases as being the principal agents of volcano-induced climatic change and this is explicit in Lamb's DVI (Lamb 1970). As Newell (1981b, p. 1) has noted, however, during the decade 'the view has evolved that dust or ash itself contributes to the early heating with sulphate contributing later, as oxidation occurs, to both stratospheric heating and tropospheric cooling'. Much of this new understanding has come through studying the effects of eruptions which have occurred in recent years. There can be little doubt that some of the most significant new insights have emerged from monitoring the plume produced by the Mount St Helens eruption in 1980. This was the first eruption whose atmospheric effects were measured on the ground, by aircraft and from space, with the latter involving the collection of data by NASA satellite systems such as SAGE (Stratospheric Aerosol and Gas Experiment) (Newell 1983). Results of the research are discussed by a number of writers and Deepak (1982) has compiled a comprehensive review, but the most relevant accounts are those by Kerr (1981) and Toon (1982).

Kerr (1981) argues that the Mount St Helens eruption introduced roughly half the amount of material into the stratosphere as Agung and some 10% as much as Krakatau yet, despite its relatively high magnitude, it had little appreciable effect on climate. According to Kerr (1981), this was because only a small amount of material stayed in the stratosphere for a sufficient time, while in the case of Agung the increase in material was some 20 times greater over time scales of several years. This author concludes that the magnitude of an explosive eruption is not the sole (or even the principal) consideration in terms of its climatic impact, but that the ratio of silicates (ash) to sulphates is a crucial parameter. It is aerosols formed of sub-micrometre droplets of concentrated sulphuric acid which stay in the stratosphere for years and cause depression of tropospheric temperatures. More sulphur-rich magmas rarely produce the most powerful eruptions (see Chapter 3, section 3.4), whereas high silica content is normally strongly correlated with explosivity and Kerr (1981) claims that the Agung eruption struck a balance between the sulphates present in the magma and the explosive capacity to transport material into the stratosphere. By direct measurement of the plume Vossler *et al.* (1981) and Toon (1982) have confirmed that a low ratio of sulphate to silicate limited the climatic impact of Mount St Helens and physical modelling of the eruption column has further added to knowledge by providing information on plume dynamics (Sparks *et al.* 1986).

The question of the sulphate/silicate ratio has been more thoroughly investigated in a retrospective study by Rampino and Self (1982) of three of the largest eruptions that have occurred in the last two centuries: Tambora (1815); Krakatau (1883); Agung (1963). As Table 6.1 shows, whereas the climatic response to these three eruptions was similar, the amounts of fine ash injected into the stratosphere

Table 6.1: Summary of the principal characterstics of the magma, fine-ash production, sulphate aerosol ratios and climatic impact of three large historical eruptions (after Hammer *et al.* 1981; and Rampino and Self 1982)

	Tambora (1815)	Krakatau (1883)	Agung (1963)
Climatic effect (max)	−1°C, 1814–16 mid-latitude	−1°C, 1883–85 60–90°N	−1.3°C, 1963–65 60–90°N
Magma type	Phonolite	Dacite	Basaltic andesite
Silica content of magma (SiO_2 %)	56	65–68	52
Ratio of fine-ash volumes	150	20	1
Ratio of masses of sulphate aerosols	7.5	3	1

differed greatly. According to Rampino and Self this is due to two circumstances. The first is that relatively sulphur-rich, yet relatively low-magnitude, events like the Agung eruption can cause atmospheric perturbations as great as those produced by much larger events such as Tambora and Krakatau. Secondly, there may be a limiting mechanism on stratospheric aerosol loading and its effects on climate. This is due to the residence time of aerosols in the stratosphere being controlled by a dynamic equilibrium between particle growth and removal, which itself depends on the mass density of the particles. They argue that the residence time of aerosols decreases by a factor two or three for a factor of ten mass density increase. Thus, dust introduced into the stratosphere ranging over several orders of magnitude cannot increase substantially the amount of sulphate that remains for longer than one or two years.

On 28 March 1982 there was a large eruption of El Chichón in Mexico. Unlike the Mount St Helens eruption two years before this was not a media event, since the eruption took place in a remote area of a poor country, but within minutes the presence of an eruptive plume had been registered by satellite imagery. At first it seemed that the eruption would have little climatic significance because its eruption column did not penetrate the tropopause, but later a high-level cloud was recognized over Hawaii and over the western United States by June the following year. The cloud was monitored from the ground, by aircraft, by meteorological balloons and from satellites; confirming the importance of sulphate droplets in long-residing stratospheric eruption clouds (Mitchell 1982; Anon. 1983). Moreover, certain additional features were noted these being: the great height attained by aerosols (~30 km); the fact that concentrations of sulphuric acid (originally sulphur dioxide) were about 40 times greater than those produced by the Mount St Helens eruption; and the observation that the stratospheric cloud was slowly encircling the globe at a rate of 22 ms^{-1}. As early as December 1982, Mitchell (1982) claimed that a slight chilling effect could be recognized over the United States and Galindo *et al.* (1984) noted a similar trend at Mexican observatories from May/June 1982 to March/April 1983 (see Robock 1984).

Surprisingly when the effects of this eruption were reviewed some years later, it became clear that there had been relatively little depression in Northern Hemisphere temperatures. According to Angell and Korshover (1985) any cooling that might have been caused by this eruption, was more than compensated by the warming that resulted from an El Niño/Southern Oscillation (ENSO) which also occurred at the time. The El Niño is the appearance of warmer than normal (1–3°C) sea-surface temperatures in the eastern tropical Pacific Ocean. The maximum temperature deviation usually occurs near Christmas and is often associated with the arrival of above-average precipitation in the dry coastal plans of Peru. El Niño (the little one) refers to the Christ Child's arrival at Christmas (Handler 1989, p. 234; see also Rasmusson and Wallace 1983; Ramage 1986). More tentatively, it was suggested that a moderate El Niño in 1816 may also have lessened the effects of the Tambora eruption of 1815. This putative link between ENSO events and volcanic aerosols has been more fully examined by Handler (1986a), who classified large historic eruptions into two groups: low latitude (tropical); and high latitude (extra-tropical). He went on to argue that there is a strong association between the injection of low-latitude stratospheric aerosols and ENSO years and between high-latitude stratospheric aerosols and non-ENSO years.

In a later paper, Handler (1989) presented a comprehensive statistical analysis of climatic data and eruptions, taking his findings much further and modifying his conclusions significantly. Not only is a general model put forward (Table 6.2) to link volcanic aerosol injection to a number of global climatic changes, but pertinent comments are also made about the effects of high-latitude eruptions. With respect to these eruptions, Handler argues that volcanic aerosols produce colder than normal land masses, which act as 'sinks', so enhancing the transfer of air mass from southern Eurasia. He is also able to link what at first sight appear to be isolated cases of the effects of volcanic aerosols to his general global model. Such cases include aspects of the Indian and Sri Lankan monsoons (Handler 1986b; Mukherjee *et al.* 1987) and crop-yield variations in the USA (Handler 1985). The relationships between volcanic injections, particularly from El Chichón, and El Niño events are still debated (e.g. Kerr 1989), but Handler's model remains impressive and promises much for the future.

On 15 June 1991 a large eruption occurred at Mt Pinatubo in the Philippines and a column of ~30 km in height penetrated the tropopause. Within ten days aerosols stretched for some 11,000 km and formed a continuous band from Indonesia to Central Africa (*Global Volcanism Bulletin* 1991a, 1991b). Satellite sensing indicated that the plume was rich in sulphur and its total mass appeared to be some three times greater than that of El Chichón (Bernard *et al.* 1991). At the time of writing it is not possible fully to assess the climatic impact of Pinatubo but, in view of the size of the plume and its composition, a significant perturbation of both world and regional climates might be expected in the next few years (see Luhr 1991; and section 6.5). Recently Woodcock (1992) has argued that Pinatubo's climatic impact was more marked than that caused by burning Kuwaiti oil wells following the Gulf War. In terms of total particulates ejected Pinatubo was equalled by the burning of tropical rain forests in just six months, the peak—but not the long-term—emission of SO_2 was only ~15% of that from power stations in the USA and all volcanoes on Earth only emit ~1% of the CO_2 produced by human activities (see also Hobbs 1991).

The relationship proposed by Lamb (1970) and others, that a distinct volcanic signature may be discerned in records of temperature in the years following major eruptions has recently been questioned. During the 1980s more complete records of temperature became available and this led to the suggestion that cooling following major eruptions may be more rapid than had been assumed previously; occurring within months of major eruptions. Kelly and Sear (1984) based their arguments on just nine eruptions, but subsequent research by Sear *et al.* (1987) established them with more certainty and also claimed differences in response between the Northern and Southern Hemispheres.

The difficulties in compiling and using historical records of temperature,

Fig. 6.3 The plume produced by hydrovolcanic maar-forming eruptions on Deception Island, Antarctica, in 1967 (see Table 5.9 and Fig. 5.16). Such events, although spectacular, produce plumes of limited height and climatic effects are localized. Plumes which penetrate the tropopause are required if more widespread climatic effects are to be produced (see Table 6.3). Photograph by L.E. Willey and reproduced by permission of the British Antarctic Survey.

together with the problems of searching for causal mechanisms of climatic change, are dealt with by Ellsaesser (1983, 1986; see also Ellsaesser *et al.* 1986) in a most challenging fashion. His argument is well summarized in his own words, 'there is no question but that the available data suggest an unmistakable coincidence in time between the occurrence of major volcanic eruptions and an accompanying or subsequent minimum in curves of surface temperature. The important question remains whether this represents a causal relationship or mere coincidence in time' (Ellsaessar 1986, p. 1184). He raises two 'paradoxes' which he claims have been ignored by many reviewers. The first is that cooling following the critical eruption of Agung (1963) is claimed only for the Northern Hemisphere, even though the dust cloud was ten times more dense over the Southern Hemisphere. Hence, effects are not proportional to presumed causes. The second paradox is that the data often imply surface cooling *before* the eruption in question actually occurred. It could well be that the first paradox is explained by Rampino and Self's (1982) notion of a limiting mechanism on atmospheric loading, but the second paradox is more serious and fully justifies Ellsaessar's (1983) note of caution that at the present time historical climatic data are not reliable enough to enable the effects of volcanic aerosols to be evaluated with confidence. This opinion has relevance for workers using volcanic inputs within climatic models.

Although much has been learnt through the empirical study of past and present eruptions, other workers have taken an alternative line and have studied different types of eruption and have attempted to specify those which are likely to perturb climate (Fig. 6.4). This has involved classifying eruptions in terms of the quantity of aerosol likely to be injected into the stratosphere. The leading worker in this field is George Walker (1973b, 1981a) and he has successfully linked the conventional qualitative classification of volcanic eruptions (Chapter 3, section 3.4) to their dust-producing characteristics. Two parameters derived from measurements made on erupted pyroclastic products are used:

1) the *dispersal index* (D), the area covered by pyroclastic materials and which depends principally upon the height of the eruptive column; and
2) the *fragmentation index* (F), this being the degree to which erupted material has been broken into fine grains. 'F' depends on the eruptive environment and the rheology of magma.

Using these parameters, Walker was able to classify eruptions and point to their differing rates of dust production (Table 6.3 and Fig. 6.4). Only certain types of eruption have both the capacity to produce large quantities of fine material and the eruptive power (as measured by column height) to inject these fragments into the stratosphere. As Walker (1981a) concludes, although ignimbritic followed by Plinian eruptions inject the greatest individual amounts of ash into the stratosphere, these are infrequent events and over the long term the more frequent Vulcanian[1] eruptions may be equally important. These three eruptions types are consequently, the crucial ones as far as studies of the climatic effects of volcanoes are concerned.

---

[1]In Chapter 5 (section 5.3.6.2) the Vulcanian activity is discussed and it is argued that it should properly be considered a form of hydrovolcanism. Although this usage is adopted throughout the rest of the book, in Table 6.3 it is used in its traditional sense in order that Walker's argument may be followed. Vulcanian eruptions produce steam and ash-laden eruption columns which may penetrate the tropopause.

Table 6.2: Summary of Handler's model linking aerosol production by volcanoes in different parts of the world, to variations in global climate (based on Handler 1989)

Assumptions	
1)	Surface temperature decreases have now been established following major eruptions.
2)	Stratospheric aerosols are restricted initially to a narrow range of latitudes and induced temperature decreases are non-uniform over the surface of the Earth.
3)	Aerosols not only produce gradients of surface temperature, but also induce pressure differences between land masses and oceans.
Argument	
Stage 1	Following the injection of aerosols into the stratosphere, cooling of the land will occur before cooling of the sea. There will be an initial increase in sea-level pressure on land. Due to the conservation of air mass at the global scale, this means a reduction in sea-level pressure over the oceans.
Stage 2	Because most of the land on Earth is in the Northern Hemisphere, the cooling induced by volcanic activity will transfer air mass from oceanic anticyclones to the continents. This will be particularly apparent in the Northern Hemisphere and especially over southern Europe.
Stage 3	Opacity, the month of eruption, its latitude, subsequent transport and lifespan of aerosols are important factors in controlling the direction of redistribution, and the magnitude of anomalous air-mass transfer.
The model	
El Chichón is an example of an eruption which produced a large aerosol injection at the right time of the year and in the right latitude (low latitude—Northern Hemisphere), to produce a strong ENSO in 1982/83. The Agung and Krakatau eruptions occurred in the Southern Hemisphere and are examples of eruptions where aerosols were concentrated over the southern oceans, so that they produced minor ENSO events. High-latitude eruptions, produce cooling of the regions poleward of the subtropics and this provides an additional sink for air mass leaving southern Eurasia in summer. One result is stronger monsoons. If aerosol are injected by a volcano which lies outside these regions (i.e. neither high or low latitude), then the effects on weather will be mixed.	

The research carried out by Walker (1981a) was concerned with fine ash and dust and it is well known that the explosive ignimbritic, Plinian and Vulcanian eruptions are normally associated with relatively silica-rich magmas of andesitic (57–63% SiO_2), phonolitic (~50–61% SiO_2), trachytic (58–70% SiO_2), dacitic (63–70% SiO_2) and rhyolitic (>70% SiO_2) composition (Le Bas and Streckeisen 1991; see Chapter 3, sections 3.2 and 3.4), however no reference is made by Walker to the important issue of the amount of sulphur contained in these magmas. The reason for this is that the volatile content of magmas is poorly known at the present time (Self *et al.* 1981) because, unless rocks are extremely fresh and glassy, volatiles may have been either gained or lost after cooling. The problems of obtaining estimates of volatiles within magmas has already been discussed (section 6.2), but Haughton *et al.* (1974) argue that the volatile sulphur content may be estimated from the amount of ferrous iron in mafic magmas,

Table 6.3: Characteristics of different types of volcanic eruption and their probable climatic effects (based on Walker (1981a), with additional information from Heiken and Wohletz (1985))

Eruption style	Generation of fine material and fragmentation[1]	Typical column height and dispersal	Possible climatic effect
Hawaiian and Strombolian	Negligible	~0.01 Hawaiian, ~1 km Strombolian. Limited dispersal	Little or no effect
Plinian and Ultra-Plinian	10^{13}–10^{16} g Fragmentation 50–85%	~30 km Plinian, ~60 km Ultra-Plinian. Dispersal ~4–15 000 km or greater (Plinian)	Very significant
Ignimbrite	10^{16}–10^{17} g (not all fine) Fragmentation is high and up to 85% may be fine dust	Similar to Plinian	Very significant
Hydrovolcanic-phreatomagmatic (Surtseyan)	10^{13}–10^{15} g Fragmentation ~90%	Rarely penetrate tropopause. Limited dispersal	Not significant
Hydrovolcanic-phreatoplinian	10^{14}–10^{17} g ~34–47% may be fine dust	Typical columns penetrate the tropopause. Dispersal limited by flushing of ash by rain	Limited because of rain flushing
Vulcanian	10^{11}–10^{15} g Highly fragmented ~40% fine ash and dust	Sometimes columns penetrate the tropopause. Wide dispersal	Significant. Smaller than ignimbrite and Plinian eruptions

[1] Fragmentation is the proportion of sub-millimetre material at a fixed point on the dispersal axis. Fine dust is smaller than 4ϕ.

provided they are saturated with respect to SO_2. Using a mixture of techniques, and remembering the earlier notes of caution, a list of sulphur concentrations in different magmas was compiled by Rampino and Self (1982). Summarizing shows sulphur compositions of ~1000–2000 ppm for basalts, ~800–2800 ppm for basaltic andesites, two values of 380 and 1250 ppm for phonolitic/tephrites, ~100–500 ppm for dacites and <10 ppm for rhyolites. In other words the amounts of sulphur and silica are roughly inversely correlated, so supporting the conclusion that silica-rich explosive eruptions have had less effect on climate than

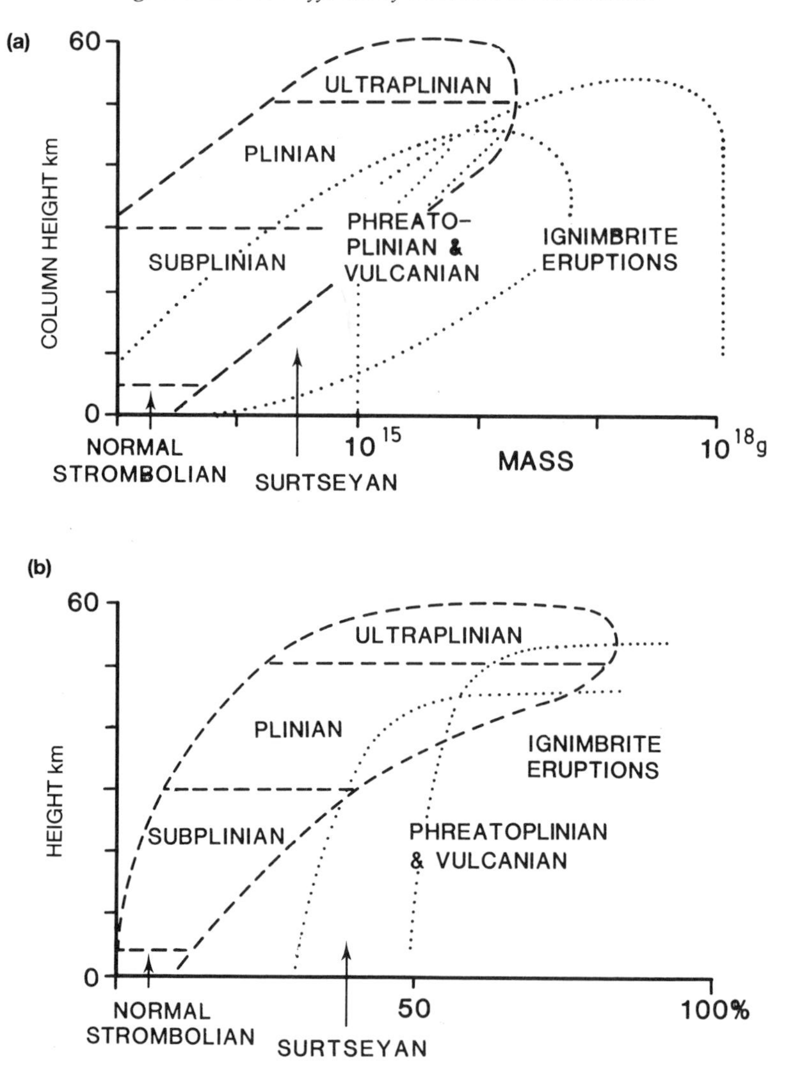

Fig. 6.4 Plots of fine ash and dust released from different types of volcanic eruption plotted against column height: (a) mass of fine ash and dust released by individual eruptions; (b) fine ash and dust released as a percentage of the total mass erupted in individual eruptions (from Walker 1981a, Fig. 3, p. 89).

was supposed formerly. Although it has been known for many years that basaltic fissure eruptions such as that at Laki in Iceland in 1783 produce large amounts of sulphur, the effects on global—as opposed to local and hemispheric—climate have been assumed to be minimal, due to the fact that eruptions columns are of limited height (Lamb 1970; Walker 1981a). More recently, however, it has been suggested that convective plumes can develop over these intense localized sources of heat and effectively carry sulphates to great heights (Sigurdsson 1982; Wood 1984; Stothers *et al.* 1986, 1988). It is clear that eruptions of this type must now be considered as potential influences on world climate. An additional point is that silica-rich volcanic dust is known to be capable of scrubbing sulphur and chlorine from eruption plumes and, during the 1974 eruption of Fuego Volcano (Guatemala), Rose (1977) noted that some 33% of the sulphur and 17% of the chlorine was removed by this process. The Fuego eruption, moreover, was of a basaltic andesite, implying that more explosive eruptions and more silica-rich magmas may make this process even more effective.

6.4.1.2 *Tracing volcanic particles in terrestrial sediments*

In addition to what might be described as 'direct' methods of tracing volcanic aerosols, concentrations of both silicates and sulphates are also found interbedded with both terrestrial and deep-sea sediments. Volcanic ash (tephra) has been recovered for many years from both land and deep-sea sites, but most research has been stratigraphical—concerned with tephrochronology—(Self and Sparks 1981; Watts 1984), or with the possible long-term relationships between volcanic activity and climate over time scales of 10^3–10^6 years (see section 6.4.2.2). Over the past two decades, however, important evidence about sulphate emissions from active volcanoes during historical times has been derived from cores extracted from the Arctic and Antarctic ice sheets. Advocates of this approach, such as Hammer (1977), argue that high-latitude ice sheets are the most favourable sediment traps when it comes to the study of past volcanism, because background concentrations of non-volcanic aerosols are low and, in the case of Greenland, the ice sheet is fed by heavy falls of snow so accumulation rates are rapid. Against this must be set a major difficulty that, whilst large eruptions at great distance inject material into the stratosphere and so contribute to the polar fallout, the record will also include the products of much smaller eruptions that have occurred near to the poles and in which materials have been transported to the site by circulation patterns in the troposphere. As Stothers and Rampino (1983a) make clear, the ice-core record is at its best and least ambiguous when local and global eruptions can be separated on the basis of historical evidence. This is not always possible.

The technique as developed by Hammer and others (Hammer 1977, 1980; Hammer *et al.* 1981) normally involves measuring changes in the electrical conductivity of melted samples taken from ice-cores; the argument being that the sulphate released from volcanoes is eventually deposited within the ice—about three years later—as H_2SO_4 and/or $(NH_4)_2$ SO_4 and that this may be measured indirectly as changes in electrical conductivity. However, these and other writers have also measured the liquid conductivity of meltwater, acidity and sulphate content directly. These methods, as Legrand and Delmas (1987) note, should be used with caution since they can give ambiguous results. The main findings of the research are two-fold. First, a broad relationship has been found between accepted features of Northern Hemisphere temperature history and the ice-core record and, for example, the peaks of volcanic activity since 553 AD are found between 1250–1500 and 1550–1700; the initial and culminating phases of the Neoglacial. Secondly, they are able to find statistical correlations between their volcanic impurity concentration index (vici) (derived from Lamb's Dust Veil Index but corrected for latitudinal effects and the residence time of aerosols in the stratosphere) and the occurrence of impurities in the ice-cores. The figures they quote are $R = 0.46$ (1500–1972 AD) and $R = 0.65$ (1770–1972), both being significant at high levels of confidence (Hammer *et al.* 1981). This is a surprising finding, since the DVI is not concerned with sulphate but with silicate particles and, as Lamb (1982b) himself notes, many large eruptions occurring between

1770 and 1972 could not have had any effect on Greenland. Furthermore, as pointed out later in section 6.4.1.3, the DVI is not without serious methodological and practical difficulties. It could well be that over the period studied large eruptions have produced both sufficient dust to rank highly on Lamb's Index and also sufficient sulphate to leave a trace in the Greenland ice.

In the period 1500 BC to 1500 AD, Stothers and Rampino (1983a) report a rather more complex picture with the ice-core record being dominated by the products from eruptions in Europe, a finding completely at variance with the relative unimportance of these volcanoes in more recent times. Their findings are, moreover, fully supported by historical research confirming the fact that these eruptions actually took place. It is their contention that these European eruptions (i.e. of Etna, Laki and Lanzarote volcanoes) had a high sulphur content, but were low in silica and that they had a disproportionate influence upon the ice-core record. Conversely, during the same period they argue that Northern Hemisphere volcanism outside Europe was less well represented, because eruptions involved sulphur-poor magmas, but produced much dust. Clearly if this section of the ice-core record were to be correlated with Lamb's Index, then much weaker relationships would emerge. A further complexity highlighted but not discussed in detail by Stothers and Rampino (1983a), concerns the nature of these European eruptions. If Walker's (1981a) thesis is accepted, that only explosive ignimbritic, Plinian and Vulcanian eruptions have the capacity to penetrate the tropopause and cause global climatic changes, then these rather 'mild' European eruptions cannot have had much effect; even though they are supposedly well represented in the ice-core record. The research already quoted concerning the role of convective plumes in introducing sulphur from fissure eruptions into the stratosphere may go some way to explain this apparent discrepancy. Also, although between 152 and 43 BC it is claimed that some 15% of the ice-core acidity peaks relate to Etna, there is no firm support for the view (despite sources quoted by the authors) that Etna was more active at this time than it is today (Chester *et al.* 1985). Hence, why is Etna so poorly represented in the ice-core record of more recent times? These are questions that remain to be answered and more detailed research is called for.

6.4.1.3 Refining Lamb's DVI

It has already been seen that one recurrent problem is that Lamb's DVI measures dust, but not sulphates. Over the last 15 years many authors have used the index and further problems have come to light.

At the outset it is only fair to note that many of these problems do not result from Lamb's published work, but rather its uncritical application. Kelly and Sear (1982) in a well-argued defence of Lamb's Index emphasize that his research was based on knowledge available in the 1960s, that it was intended to represent order of magnitude estimates alone and that for high degrees of reliability as many estimation formulae as possible should be used and the results averaged. They conclude that the index should only be used to suggest order of magnitude estimates of dust production, that any individual value of the DVI should be adopted only if its method of estimation is stated so that reliability may be estimated and that particular care must be employed when the DVI is based on climatic data, so that circularity of reasoning is avoided.

Against this view there still exists a powerful critique. First as Robock (1981) notes, Lamb used a number of simple assumptions about the spread and decay of volcanic dust to produce a Northern Hemisphere specific DVI (NHDVI); values of this index being presented as Appendix 2 in Lamb (1970). These figures have been used by many students of climatic change, including Robock himself (Robock 1978, 1979). Table 2a of Lamb's Appendix 2 eliminates those estimates of the NHDVI based *solely* on falls in temperatures, but this does not exclude the possibility of circular reasoning as might be assumed from a cursory reading of the table, since it still includes estimates based *partially* on temperature depressions. A further problem highlighted by Robock (1981) is that Lamb excluded all eruptions yielding a DVI of less than 100, which meant that many major eruptions of the last 70 years were excluded, such as: Hekla, Iceland (1947); Mt Spurr, Alaska (1954); and Bezymianny, Kamchatka (1956). The exclusion of these eruptions cannot be justified, especially since dust production is not the sole, or even the most important, variable producing climatic change. Further points of criticism are made by Newhall and Self (1982), who argue that only those values of the DVI that are based on opacity of the atmosphere (i.e. DVI/E_{MAX} or maximum extent; see Lamb 1970, p. 473 and Appendix 1) are realistic direct measures of dust injection. Unfortunately DVI/E_{MAX} cannot be calculated for eruptions that are closely spaced in time, that occurred before atmospheric records were available and are small. In addition they argue that non-volcanic phenomena leading to atmospheric opacity (like duststorms) cannot be eliminated. Of Lamb's 250 individual estimates of the DVI, they calculate that 5% are based on changes in radiation, 5% on temperature records, 12% on quantitative estimates of tephra volumes, 30% on a now-dated semi-quantitative estimate of tephra volumes and 48% on non-quantitative descriptions within the historical record. Hence, even with the elimination of those DVI values based on temperature changes, there still remain severe problems with the rest of the data set.

In order to overcome some of these problems other indices of dust injection were proposed, but these have been neither as widely used as Lamb's DVI nor have they fully come to terms with the aforementioned criticisms. Mitchell (1970) computed his own NHDVI using data supplied by Lamb for the period since 1850. In contrast to Lamb, he included many twentieth-century eruptions and many authors have argued strongly in support of this index in studies of climatic change over the last hundred years (Oliver 1976; Robock 1981), while Hirschboeck (1980) used volcanological criteria to classify about 5000 eruptions into three simple categories: small, moderate and great. Of these two schemes there seems little doubt that Hirschboeck's is better from a volcanological perspective because it excludes any possibility of circular reasoning, but is limited by the inclusion of at times incorrect information, with the category of great eruptions in particular including events of potentially great importance alongside those with little likely climatic impact.

In order to overcome these problems a completely new index has been proposed as part of a major study of historic eruptions sponsored by the Smithsonian Institution. This started in 1971 and culminated in 1981 with the publication of a substantial chronological list, directory and gazetteer of all known eruptions entitled *Volcanoes of the World* (Simkin *et al.* 1981). As part of the programme of research a Volcanic Explosivity Index (VEI), based solely on volcanological criteria, was derived for over 5000 eruptions and the results are published in full in Simkin *et al.* (1981). The derivation of the index is fully explained by Newhall and Self (1982), who argue that the size of an explosive eruption depends critically on five parameters, which were first identified by

Walker (1980, 1981a). These are:

1) magnitude (as determined by the volume of ejecta);
2) intensity volume of ejecta per unit time, as determined from column height and exit velocities;
3) dispersive power (as measured by column height);
4) violence of release of kinetic energy (similar to intensity, but meant for instantaneous rather than sustained eruptions); and
5) destructive potential.

Given that the historical record is poor and generally of variable quality both over time and spatially, the VEI is a composite index based on Walker's magnitude and/or intensity and/or destructiveness and/or (less frequently) dispersive power, violence and rate of energy release. The data used depends on what is available. The VEI has an eight-point scale (the maximum number that Newhall and Self believed could be justified by the historical record) and a given eruption is assigned to a point on the scale using as many as possible of the critieria listed in Table 6.4. Several conventions apply when the index is calculated and further details are given by Newhall and Self (1982).

More recently Legrand and Delmas (1987) has proposed a Glaciological Volcanic Index (GVI) based on ice-core measurements in Greenland and the Antarctic, but Bradley (1988) has put forward three arguments against its adoption at the present time. These are the bias towards high-latitude eruptions, the fact that it emphasizes sulphur-rich explosive events, and hence is not a complete catalogue, and that the distribution of 'acid' snowfall varies so that correlation between cores is by no means perfect. Nonetheless, Bradley does accept that none of these problems are insurmountable and the approach promises much for the future.

An ideal solution would be the direct measurement of radiation receipts at a set of evenly distributed, high-altitude sites around the world. Lamb (1970) did try and use the data available at the time to construct a record of direct radiation for the nineteenth century, while a similar approach has been attempted more recently by Pollack *et al.* (1976). The fact remains that neither attempt has been very successful (Bradley 1988), although again this approach may hold much for the future.

At present the VEI is the best available index of explosive volcanism. It is, nevertheless, far from ideal for studies of climatic change. As its originators admit, it says nothing about the sulphate/silicate ratio and corrections have not been made for either the latitude or altitude of a volcano, although both could be calculated and used to modify the index. As far as the sulphate/silicate ratio is concerned, these data are only just becoming available for well-studied eruptions of the recent past, while for earlier eruptions there exists the possibility of incorporating information from the study of ice-cores. Better knowledge of past eruptions can be achieved only through the detailed stratigraphical and chronological study of active volcanoes and the sediment traps in their vicinities. Despite these problems, the VEI is a better index of explosive volcanism that the DVI, since the latter with its seemingly high degree of precision has encouraged the uncritical acceptance of dust-injection estimates which are not justified by the historical data base. Although the VEI has found favour with some students of climatic change, the work of Stommel and Swallow (1983) on the possible effects of volcanism on late European grape harvests being a good example, the VEI has not been adopted by all workers. Because it includes eruptions that do not

Table 6.4: Criteria used in the estimation of the Volcanic Explosivity Index (VEI) (from Newhall and Self 1982, Table 1, p. 1232)

Criteria VEI*:	0	1	2	3	4	5	6	7	8
Description	Non-explosive	Small	Moderate	Mod-large	Large	Very large	—	—	—
Volume of ejecta (m^3)	$<10^4$	10^4–10^6	10^6–10^7	10^7–10^8	10^8–10^9	10^9–10^{10}	10^{10}–10^{11}	10^{11}–10^{12}	$>10^{12}$
(Tsuya classification)★	(I)	(II–III)	(IV)	(V)	(VI)	(VII)	(VIII)	(IX)	—
Column height (km)	<0.1	0.1–1	1–5	3–15	10–25	>25	—	—	—
Qualitative description	'gentle, effusive'	—	'explosive'	—	'cataclysmic, paroxysmal, colossal'	—	—	—	—
				'severe, violent, terrific'	—	—	—	—	—
Classification		'Strombolian'	—		'Plinian'	—			
	'Hawaiian'	—	'Vulcanian'	—	—	'Ultra-Plinian'	—	—	—
Duration (hours) of continuous blast	<1	—	—		>12	—	—	—	—
			1–6	—	—				
			6–12	—	—	—			
CAVW max explosivity●	Lava flows	—	Explosion or nuée ardente	—	—	—	—	—	—
		Phreatic	—	....	....	....			
	Dome or mudflow	—							
Tropospheric injection	Negligible	Minor	Moderate	Substantial	—	—	—	—	—
Stratospheric injection	None	None	None	Possible	Definite	Significant	—	—	—

★ If all eruptive products were pyroclastic ejecta.
* For VIE's 0–2, uses km above crater; VEI's 3–8, uses km above sea level.
● The most explosive activity indicated for the eruption in the Catalogue of Active Volcanoes.
Criteria are listed in decreasing order of reliability.

necessarily reach the stratosphere, Jakosky (1986) largely ignores the VEI in his influential study of volcanoes and climate, confining his attention to eruptions which are known to have introduced material into the stratosphere during the past hundred years. There are two problems with Jakosky's approach. The first is the short time span of accurate historical data on column heights (i.e. ~100 years at the most) and the second is that some of the eruptions he includes have quoted column heights which are highly questionable. The 1883 eruption of Etna, for instance, has a quoted height of 14 km, which is probably too high for a Strombolian event of this kind (Walker 1981a).

6.4.2 Causal inference over different temporal and spatial scales

The importance of scale in determining the nature and complexity of cause and effect has been recognized for many years by earth scientists (see Schumm and Lichty 1965; Tricart 1965; Thornes and Brunsden 1977). In the study of volcanoes and climate it assumes critical importance. So far the chapter has been concerned with hemispherical and worldwide changes in climate (particularly temperatures) over the last few hundred years and relationships with parallel records of volcanic activity. At these scales there is unanimity in the view that volcanic activity is the independent variable, whereas climatic parameters, like temperatures, are dependent. However, research at other scales has shown that the nature of cause and effect is far more complex. This complexity may be summarized by examining two specific spatial scales and then reviewing research with reference to differing time scales.

6.4.2.1 *The individual volcano*

In contrast to the large number of studies that have been published on the effects of volcanic activity on climate at the world scale, there have been few detailed studies of these two phenomena on individual volcanoes. Indeed, the majority of studies have been of a limited number of volcanoes and, hence, any conclusions must be considered as tentative at the present time.

Although rainfall enhancement is extremely difficult to demonstrate, even in controlled cloud-seeding experiments, it is nonetheless commonly believed that heavy rainfall usually accompanies explosive volcanism because of enhanced convection, the injection of vast quantities of aerosols and, if water is involved in the eruption, the great increase in water vapour. However, a review by Williams and McBirney (1979) indicates that this opinion is based on insecure foundations. Quoting detailed studies of past eruptions by Finch (1930) and Neumann van Padang (1934, 1935), they conclude that rainfall enhancement is an extremely rare event. Even the eruption of Krakatau in 1883 produced no increase in rainfall, even though its eruption cloud was discharging during a monsoon.

Studies on the effects of persistent volcanic plumes on climate have been carried out on Mount Etna (Sicily) and have produced interesting results. Etna is a continually active volcano, displaying mildly Strombolian activity and, because prevailing winds are from the west and northwest, the persistent plume is an important control over both rainfall and temperature on the lee side of the volcano (Affronti 1967, 1969a, 1969b; Durbin 1981; Durbin and Henderson-Sellers 1981). For all seasons temperatures at stations on the east and southeast are depressed in comparison with stations at a similar height on the western flank, due to enhanced cloud cover on the lee side. As with temperature, so rainfall also varies between the western and eastern flanks (Fig. 6.5). According to Durbin and Henderson-Sellers (1981), there are two reasons for enhanced precipitation on the lee side. The first is that air masses bringing winter rains have increased instability due to their forced accent over the northern mountains of Sicily and

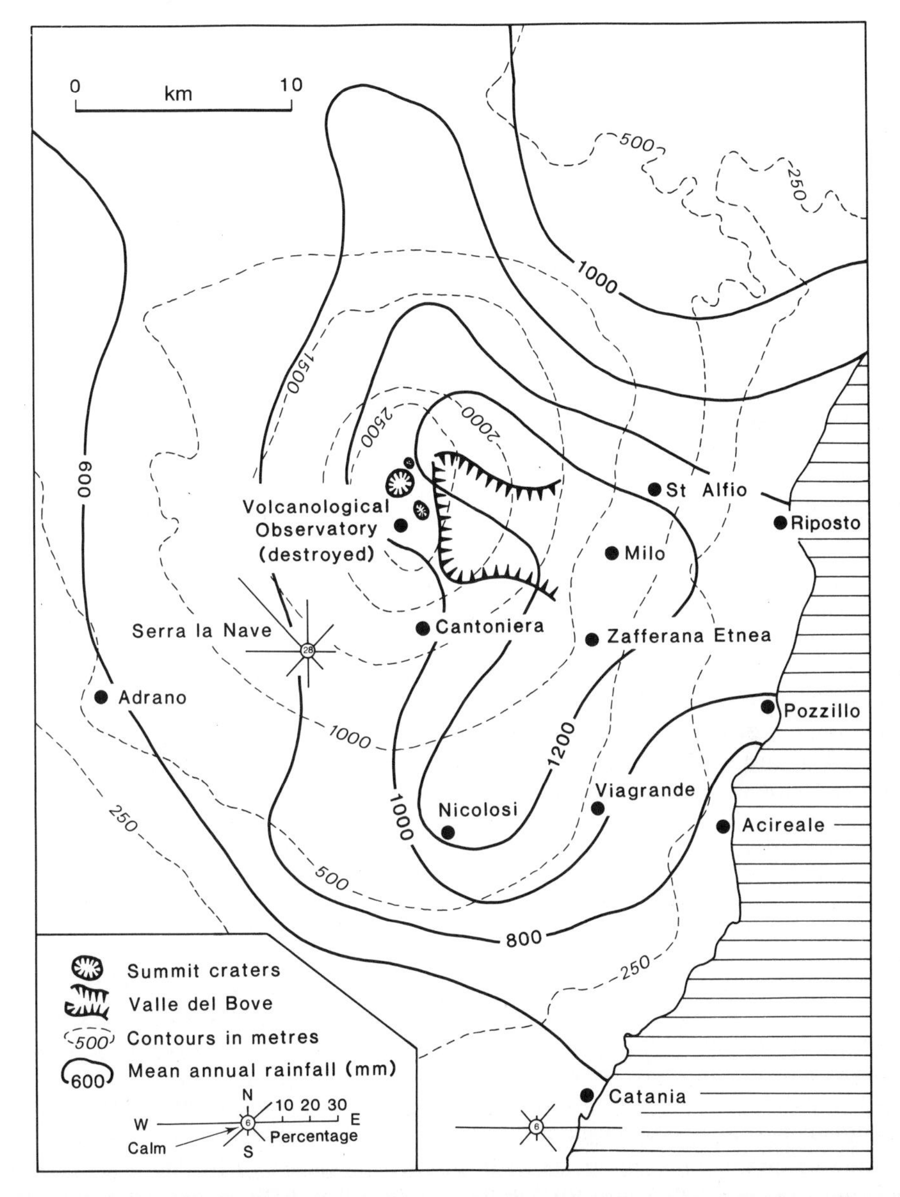

Fig. 6.5 Mount Etna. Mean annual precipitation and wind directions for selected stations. Prevailing winds are from west to east. Rainfall enhancement on the eastern flank is due to the persistent plume (modified from Durbin and Henderson-Sellers 1981, Fig. 1, p. 285).

that the north and northwest flanks of Etna lie in a rain shadow, while the second is that the eruptive plume and greater cloud cover on the lee side produce further enhancement. An additional microclimatological effect of eruptions that has become apparent through study of Etna, is that when activity increases and lava is generated through flank eruptions, the varying albedos of surface materials can have an influence on convectional rainfall. At high slopes on Etna in winter, fresh snow may lie next to fresh black lava and the heat emitted from the lava may increase microclimatological contrasts. Whitford-Stark and Wilson (1976), have noted that in these circumstances turbulence may reach the stage where small whirlwinds are developed and clouds evaporated when passing over a hot flow.

On individual volcanoes climate may be an important trigger mechanism for certain types of eruption. It has been known for many years that water held as snow and ice can generate lahars, and the eruption of Nevado del Ruiz volcano in Colombia (1985) is a good example of this activity (S. N. Williams *et al.* 1986). If populations are at risk then these are extremely dangerous events. However, in recent years it has become clear that interaction of water and magma can generate a particular type of explosive activity. As discussed in Chapter 5 (section 5.3.6.2) this is hydrovolcanism and involves not only normally explosive volcanoes, but also those that are generally effusive, producing much lava but little pyroclastic material from a basaltic magma.

Climatic induced hydrovolcanism may be important at specific stages in a volcano's history. Duncan *et al.* (1984) and Chester *et al.* (1987) have argued that on volcanoes such as Etna which are also high mountains (~3350 m), it is possible that over time scales of 10^3–10^5 years external factors may trigger hydrovolcanism. These external factors include water-table configuration, the height of the volcano at different times and climatic change. In particular it is suggested that the generation of lahars and hydrovolcanic fall deposits on Etna may be related to episodes of enhanced precipitation and ground-water storage at specific stages of the Pleistocene and Holocene. The authors emphasize the tentative nature of their findings and it should be noted that these relationships have not yet been established for other volcanoes that are high mountains and formed against a background of significant climatic change. Nevertheless the link between climatic parameters, especially rainfall, and hydrovolcanism is worth further research. The effects of hydrovolcanism are likely to be confined to the volcano and will have little knock-on effect on global climate, since eruption columns rarely penetrate the tropopause (Walker 1981a; and Table 6.3).

6.4.2.2 *The world scale*

At the world scale considerable debate about cause and effect has focused on certain time scales. In the late 1960s, Budyko (1968) continued a debate which had been current for many years; namely that volcanism could influence climate over much longer time scales than that represented by the historical record. Budyko's thesis is simple, based on probability theory and assumes that eruptions are independent events and subject to purely random variations in frequency. His argument is that if four large eruptions are taken as representative of the past hundred years and assuming this frequency coincides with the overall frequency throughout the Quaternary, then 40 large eruptions might be expected to occur every century, once every 10^4 years. Moreover, 130 large eruptions are to be expected every 10^5 years and 100 large eruptions within five years once every 10^6 years. It is Budyko's contention, that even 50–100 eruptions in a single century would reduce radiation received by the Earth by 10–20% and lower temperatures worldwide by 1–3°C. These views were not accepted by Lamb (1970, 1972), who argued that Budyko's assumptions were unrealistic, while one of his conclusions implies 'a more random distribution of ice-ages in time than most dated evidence suggests' (Lamb 1970, p. 495). Despite Lamb's comments on these and other ideas about long-term relationships between volcanoes and climate (see Auer 1956, 1958, 1959, 1965), this has not prevented subsequent workers dealing with them. One example is recent work by Caldeira and Rampino (1990), who have investigated the claims made by a number of writers that the large volume of gas released by the eruption of the Deccan basalts in India would have been sufficient to cause, or at least contribute to, mass extinctions at the close of the Cretaceous (Officer and Drake 1985; Officer *et al.* 1987; McLean 1985, 1988). In particular it has been argued that the amount of CO_2 generated would have caused significant greenhouse warming. Caldeira and Rampino tested these ideas using a numerical model and concluded that the effects of CO_2 emissions from the Deccan eruptions would have been too weak to have had any significant influence on late Cretaceous mass extinctions.

Over the longest time scale (i.e. 10^9 years) research has focused on two issues: the formation of atmospheres on the terrestrial planets; and climatic change in the later Precambrian. The former is much debated and views divergent, due to data being scanty and capable of varied interpretation. There is little disagreement about the crucial importance of volcanic activity, or that it was the independent variable upon which atmospheric characteristics depended. Today study of early atmospheres is a major field of research, but one about which a number of general statements may be made.

1) The terrestrial planets have varied atmospheres, with Mercury having virtually none due to its low gravity and high temperature, while Earth, Mars and Venus have ones composed of the heavier gases because of their higher gravities.

2) Atmospheres of the terrestrial planets are derived mainly from volcanic gases and, in the case of the Earth, this is regardless of whether its atmosphere is *primary*, developing slowly over time, or whether this was lost and a *secondary* atmosphere was formed by later degassing.

3) The distinctiveness of the Earth's atmosphere is a function of many factors. It reflects planetary size, position within the Solar System, the operation of biological processes including photosynthesis, storage of certain elements and compounds (e.g. carbon) in sediments—which if present in the atmosphere would give rise to serious 'greenhouse effects'—and the role of plate tectonics in removing substances harmful to life. The problem of excess sulphur (Fig. 6.2), for example, may be explained by its reaction with iron to form iron pyrites and its subsequent scavenging from the Earth's surface by processes of plate movement (see Stoiber and Jepsen 1973). Further information on planetary atmospheres may be obtained from Stoiber and Jepsen (1973); Holland (1976); Murray *et al.* (1981); Glass (1982); Cogley and Henderson-Sellers (1984); Kasting *et al.* (1988); Condie (1989); and Lunine (1989).

Later in the Precambrian there is much discussion of the contrasts between climate in the Archaean ($2.5–4.6 \times 10^9$ years ago) and the later Proterozoic eras ($5.7 \times 10^8–2.5 \times 10^9$ years ago). Over the past few hundred years the main gaseous emissions from volcanoes which are believed to have had climatic effects are sulphates and the role of carbon dioxide is considered to be an extremely limited and insignificant forcing factor, especially when compared to inputs from the consumption of fossil fuels (Lamb 1982a; Leavitt 1982; Hobbs 1991). This is in spite of the fact that recent studies of Etna have indicated that even between eruptions considerable amounts of CO_2 diffuse slowly from the flanks of this, and presumably other, active volcanoes (Allard *et al.* 1991; Gerlach 1991). In passing it is interesting to note that lead pollution from volcanoes is also thought to be of little importance in comparison with that released by internal combustion engines and by industry (Patterson and Settle 1987). However, in the Archaean, Wyrwoll and McConchie (1986) argue that, whereas the amount of heat lost by the Earth through conduction was little different from that occurring today, the rate of oceanic plate formation and cooling must have operated more efficiently; being achieved through faster spreading rates, greater length of oceanic ridges and subduction zones, smaller and more numerous plates or a combination of all of these (Bickle 1978; Nisbet 1984). Whatever the explanation this would have given rise to far more volcanic activity than occurs today, and hence a greater potential for climatic impact. Moreover, calculated spreading rates of some 4–25 times the present-day figure would increase atmospheric CO_2 emissions by more than 100-fold. The greenhouse effect created, so Wyroll and McConchie argue, was the principal reason why glaciations were absent at this time, yet were common later in the geological record (see also Christie-Blick 1982).

It is at much shorter time scales of 10^3–10^6 years and particularly with reference to climatic change in the Pleistocene, that there is the greatest debate about whether eruptions are the independent variables in relationships between volcanoes and climate, especially in the putative links between glacial advances and episodes of enhanced volcanic activity. Although the notion that volcanic activity may be one of several causes of glaciation has been around for many years, during the 1970s important new evidence was presented which was capable of being interpreted in a number of ways. In 1971 McBirney posited that episodes of enhanced igneous activity in the Mesozoic and Cenozoic occurred with an approximate unison in most of the world, with apparent disregard for local setting and contradicted the traditional view that igneous activity varied between provinces. Later Kennett and Thunell (1975) adduced evidence in support of this view, in what they claimed was the first attempt to examine long-term volcanic activity using ocean cores recovered during the Deep Sea Drilling Project. Their principal finding was that when corrected for recovery rate and other complicating factors, the cores indicated that worldwide Quaternary volcanism (as indicated by tephra layers) was some four times higher than the Neogene average and two times higher than the mid-Pliocene peak, low levels occurring in the early and late Miocene and the late Pliocene. From this they argued that increased volcanic activity coincided with glaciation. Though their results indicated a greater increase in volcanic dust than Chappel (1975) claimed was necessary to initiate serious and significant global cooling, they felt that the resolution of their data was insufficient to claim a definite cause/effect relationship. In reply to this paper, Ninkovich and Down (1976) argued that the increase in Quaternary volcanic activity was an artifact of the data and a consequence of sea-floor motion relative to the tephra horizons. Where sites had remained motionless, they found little sign of any increase.

Further controversy surrounds the interpretation of other evidence. Bray (1974, 1977) demonstrated a strong correlation between volcanic activity on land, especially in the Southern Hemisphere, and ice advances, claiming that massive eruptions were responsible. Again this evidence is capable of varying interpretation, since Rampino *et al.* (1979) argue that reanalysing Bray's data leads to the alternative conclusion that some pulses of glaciation began some 3–700 years before the volcanic events that supposedly triggered them, while in the past 2 million years major ice expansion often lagged behind the alleged triggering event by up to 10,000 years. The interpretation of records from polar ice-cores is, likewise, far from incontrovertible, with Gow and Williamson (1971) arguing that eruptions may have triggered worldwide cooling between 30 and 16,000 BP and Kyle *et al.* (1981) arguing strongly against it (see also Thompson and Mosley-Thompson 1981).

Over time scales of 10^3–10^6 years an important debate has been waged over whether the occurrence of large eruptions closely spaced in time could produce climatic conditions analogous to a 'nuclear winter'. It is argued (Turco *et al.* 1983, 1984; Thompson and Schneider 1986), that a thermo-nuclear exchange would pollute the atmosphere with vast amounts of soot, due to the fires which would be started, and it has been further suggested that the atmospheric effects of volcanoes might be used as a basis for modelling the likely effects and severity of such an apocalyptic event (Maddox 1984). Clearly, the relatively small amounts of aerosol produced by historic eruptions and the fact that there are significant differences between volcanic aerosols (sulphates and ash) and sooty smoke make such a link a tenuous proposition (Burke and Francis 1985; Stothers *et al.* 1988), but research in this area has focused attention afresh on the nature of the atmospheric loading which would have been produced by some of the large eruptions, which are known to have occurred in the Pleistocene (see also Schneider 1988). Attention has been focused on two major events: the eruption of Toba (Indonesia) ~75,000 BP—probably the greatest late Quaternary volcanic event—and the even older Roza flow of the Columbia River (USA) basalt group ($\sim 14 \times 10^6$ BP). According to Rampino and Stothers (1985) and Stothers *et al.* (1988) both eruptions would have had profound effects. 'Unless self-limiting mechanisms of stratospheric aerosol formation and removal are important, very large eruptions may lead to widespread darkness, cold weather, and acid precipitation. Even the minimum estimated effects would represent significant perturbations of the global atmosphere' (Stothers *et al.* 1988, p. 3). The link between this very interesting research and the wider issue of major climatic changes in the Tertiary and Pleistocene has yet to be established.

Although correlations of time series do not necessarily imply causality and despite the fact that data may be interpreted in a number of ways, it might be thought that there is little reason for supposing that the relationship between climatic and volcanic eruptions is any different at time scales of 10^3–10^6 years than it is at the others discussed so far (see Pollack *et al.* 1976). Several authors, however, have asked the fundamental question of whether at time scales of 10^3–10^6 years climate may, indeed, influence volcanic activity. As early as 1972, Lamb with characteristic insight noted that 'gradual change of world sea level brought about by . . . progressive melting of glaciers may build up strains in the Earth's crust which are ultimately released in volcanic outbreaks' (Lamb 1972, p. 432), but the issues are more fully discussed in an excellent and provocative article by Michael Rampino *et al.* (1979), in which it is suggested that several mechanisms exist whereby climate could increase volcanic activity. These mechanisms and others are listed in Table 6.5. As the table shows, there are both primary and secondary (trigger) mechanisms which may cause climate to exert an influence on volcanic activity. The authors admit that these mechanisms are

Table 6.5: Possible mechanisms by which climatic change may initiate, or trigger, explosive volcanic activity (based on Rampino *et al.* 1979)

Possible cause of increased volcanism	Processes involved
A) *Primary cause*	
1) Realignment of the geoid due to redistribution of water.	Redistribution of water during successive glaciations and deglaciations causes realignment of the geoid and leads to worldwide changes in crustal stress. This will be most acutely felt at plate margins and major fault intersections. The majority of Pleistocene volcanism was related to these zones (see Anderson 1974). Chappel (1975), argued that the stress gradients beneath continental margins in response to this process are some 10^5 times greater than the stress gradients associated with earth tides, which are known to affect the timing of some earthquakes and volcanic eruptions.
2) Hydrostatic unloading of ocean basins and its effect on magma movement.	This has been proposed by Matthews (1969) and suggests that the unloading of ocean basins during glacial episodes favours the upward movement of basaltic magmas.
B) *Trigger mechanism*	
3) Change in Earth orbit.	Some authors propose that the timing of glaciations is dependent upon variations in Earth orbit (see Hays *et al.* 1976). If this is the case then Earth tides will be created and may trigger eruptions. Hence, it is possible that both climatic change and volcanic activity in the Pleistocene is linked to extraterrestrial causes.
4) Magma mixing.	The isostatic processes noted above may not initiate volcanism, but may trigger it through magma mixing and greater activity at faults. Injection of basaltic magma into more silica-rich magma chambers would favour explosive volcanism. Mixed magma origin implied for many Quaternary tephra sheets (see Sparks *et al.* 1977).

somewhat speculative and that eruptions may be multi-causal, yet the fact remains that the question of cause is not as clear cut as is often assumed. At present the issue is impossible to resolve, because of the resolution of the data available, but in the future it is possible that the position will improve.

6.5 Conclusion and prospect

It is clear that since the publication of Lamb's 1970 paper, considerable progress has been made in understanding the climatic impact of volcanoes. Not only are certain historic eruptions crucial to an understanding of short-term climatic changes, but many of these changes have a direct bearing also on such crucial socio-economic issues as the nature and reliability of monsoons in the Indian subcontinent (Handler 1986b; Mukherjee *et al.* 1987), crop yields in the USA

(Handler 1985) and European grape harvests (Stommel and Swallow 1983). At the local scale of the individual volcano local climatic contrasts can also alter significantly the potential agricultural productivity of different sectors of a mountain. Mount Etna, Sicily, is a good example of the effects of a persistent volcanic plume on the potential and actual productivity of one of the most important agricultural regions of the Italian south (Chester *et al.* 1985). Over longer time scales it is also evident that volcanic gas and dust emissions may have played some role in the major climatic changes which occurred in the Pleistocene, while venting of gasses is crucial to a comprehension of the early histories of atmospheres, both on Earth and on the other terrestrial plants. Because of its relevance to other fields, study of the impact of volcanoes on climate is being carried out in many research centres and is interdisciplinary. Also because of contemporary concerns with the effects on people of global climatic change research is also normally well funded.

To conclude: a note of caution. Late in 1991 two important papers were published in the same issue of *Nature*, which reopened some of debates of the last 20 years. In reviewing the eruption of Pinatubo (see section 6.4.1.1), a joint team from France and the Philippines (Bernard *et al.* 1991) reported the presence of anhydrite ($CaSO_4$) phenocrysts within dacitic pumice. These authors together with the American James Luhr (1991) discussed the implications of this strange finding. Anhydrite, a mineral usually associated with sedimentary evaporite sequences, was important in the 1982 El Chichón eruption and this too had a major climatic impact. Anhydrite is very rare in the geological record of eruptions from subduction zones, since its high solubility makes its preservation highly unlikely. Even when it is present and has been measured it is known to be highly variable in its occurrence. Luhr (1991, p. 105) argues that 'at present . . . it is not possible to predict which volcanoes have the potential to erupt high-sulphur, anhydrite-bearing magmas and . . . produce large, climate-influencing . . . clouds. Nor can we address the issue for ancient eruptions'. Still more detailed monitoring of plumes during episodes of eruption will be called for if the role of anhydrite is to be fully understood. Pinatubo implies that in spite of the real advances reported in this chapter, less is perhaps known about the true impact of volcanoes on climate than is commonly supposed.

7

Predicting volcanic eruptions

Whenever I look at a mountain I always expect it to turn into a volcano (Italo Svevo *Confessions of Zeno* 1923).

7.1 Introduction

Historically volcanoes have not caused as much destruction as is commonly supposed. Indeed writing in 1984 and before the 1985 eruption of Nevado del Ruiz (Colombia) generated a vast lahar which claimed over 20,000 lives, Russell Blong estimated that from 1600 AD to 1982 just under 240,000 people had perished due to the direct and indirect effects of eruptions (Blong 1984; Fishlock and Matthews 1985; SEAN 1985). This compares with a mortality figure in excess of 250,000 for the single Tangshan earthquake (China) of July 1976 (Anon. 1978; Bolt 1988) and the estimated 24.4 million people affected in some way by drought each year in the 1970s (Wijkman and Timberlake 1984). In addition the Tangshan earthquake gave rise to breathtaking economic losses and in excess of 500,000 non-fatal casualties (Associated Press 1977). In Chapter 8 it is argued that volcanic hazards are but one side of the coin, the other being the benefits that accrue to both individuals and societies through the operation of volcanic processes. These include, not only the oft-quoted direct benefits of fertile and potentially fertile soils, but also the fact that in many cases volcanic activity provides reservoirs for the storage of ground water, areas of high scenic value—so allowing the development of tourism—and underground sources of heat for geothermal power. Over longer time scales 'volcanoes are nature's forges and stills where elements of the Earth, both rare and common, are moved and sorted' (Decker and Decker 1981a, p. 168) and many of the world's resources of fluorine, sulphur, zinc, copper, lead, arsenic, tin, molybdenum, uranium, tungsten, silver, mercury and gold, were formed by magmatic processes both direct and indirect. Additionally, the role of volcanoes in the formation of early atmospheres was crucial to the development and evolution of life on Earth, while not all climatic effects of today's volcanoes are negative (Chapter 6). Developing this theme of costs and benefits, Wijkman and Timberlake (1984, p. 100) go so far as to suggest that volcanic eruptions are not 'serious' disasters at all. When, however, the figures are examined in more detail and the hazard threat is evaluated, then it becomes clear that the considerable efforts currently being devoted to prediction are justified. Prediction is the key which enables the advantages of volcanoes to be maximized and the dangers minimized.

Many deaths are now avoidable often through the application of fairly simple methods of prediction, but more importantly the death toll falls very heavily on the inhabitants of just a few volcanic regions of the world. Looking at the death toll in more detail, Blong (1984) concludes that if the deaths due to disease and starvation are eliminated (on the basis that international aid is now normally available to prevent them), then there is only a limited number of volcanic styles which cause high rates of mortality. These are pyroclastic flows, some falls, lateral blasts and surges; together with such indirect yet significant effects of volcanic activity as lahars, flooding and tsunamis. Since many of these are associated with explosive volcanism it is not surprising that subduction-zone volcanoes in general, and Japan, Indonesia and the countries of the Caribbean, in particular, have in recent times proved to be the most hazardous volcanic regions. Loss of life is but one of the hazards associated with volcanic activity. Lava flows have only killed a relatively small number of people, yet in many areas they can have serious economic effects on agriculture, communications and settlements. On Mount Etna, Sicily, Chester *et al.* (1985) have estimated that it may take up to 900 years for land covered by lava to regain its former agricultural productivity (Fig. 7.1), while lava from the long eruption of Parícutin, Mexico (1943–52) eventually covered an area of 2400 ha, destroying the village of Parícutin and most of San Juan Parangaricutiro (Bolt *et al.* 1975; Nolan 1979). Recovery from this eruption was costly and slow. Perhaps the most important reason why volcanic hazards require detailed study and individual eruptions prediction is that over the past 300 years it is fortunate that casualties have not been greater. Several large eruptions have occurred in regions which were at the time and in some cases remain, relatively uninhabited. Examples include the pyroclastic flows produced by the 1912 eruption of Katmai volcano (Alaska) which generated the so-called 'Valley of the Ten Thousand Smokes' and the 1955/56 eruption at Bezymianny

Fig. 7.1 Lavas from the 1971 eruption of Mount Etna, Sicily. Lava flows do not generally cause loss of life, but as the photograph shows cause destruction of property and the 'sterilization' of land over which they pass (author's photograph).

(Kamchatka, Russia). If either had occurred in populated areas, like California or Japan, they would have caused major loss of life and had catastrophic effects on the economies of these regions and the world as a whole (Ollier 1988; and Fig. 7.2). Historical records even in areas like Greece, Italy and Japan, where knowledge of eruptions stretches back thousands of years, are of short duration in comparison with the estimated recurrence interval of certain types of volcanic eruption. The Yellowstone and Long Valley volcanic centres in the western USA are still capable of producing large ignimbrites (Cas and Wright 1987) and, if either did, then these would rank as amongst the greatest ever human disasters with up to 30,000 km^2 of land being devastated (Press and Siever 1986). Cas and Wright also make the valuable point that many silica-rich volcanoes that are capable of producing large ignimbrites have very long repose periods which may range up to 10^5, or even 10^6 years.

A point made strongly by many hazard analysts is also relevant (Burton *et al.* 1978). All hazards involve not only interactions between extreme events of nature and vulnerable populations, but changes in human geography both at the world scale and locally also increase the risks of disaster. These enhanced dangers are due to such factors as population growth, the migration of people into more hazardous regions and global shifts in the distribution of wealth and poverty. Such developments as the emergence of Japan as one of the world's foremost

Fig. 7.2 The total destruction of St Pierre, Martinique, during the eruption of Mt Pelée in 1902. Following several weeks of activity, pyroclastic flows were generated on 8 May and destroyed the town and its harbour. It is often stated that there was only one survivor, a prisoner in a dungeon, but in fact four avoided death in the town and several more on ships in the harbour (Macdonald 1972). The official death toll was estimated at 29,000. The photograph also shows the 'spine' produced by the eruption (see Fig. 4.11). Photograph reproduced with the permission of The National History Museum, London, ref. T03679/N.

economic powers and much denser settlement than hitherto in volcanically active regions, like the Bay of Naples (Italy), Central America and Indonesia, mean that the global toll from volcanic eruptions is likely to increase, both in terms of mortality and material losses. This point is crucial to an understanding of human responses and adjustments to volcanic hazards and is expanded upon in Chapter 8, but it also implies that the case for the prediction of eruptions is stronger than a cursory examination of historical losses might suggest.

7.2 Volcanic hazards which require prediction

In considering the volcanic hazards which require prediction, many writers have been concerned with the important question of when a volcano should be considered extinct and, therefore, of no further threat. Tazieff (1983a) estimates that only about 500 volcanoes have erupted in historical time, but admits that this figure is a gross underestimate of potential danger because in all parts of the world the historical record is very short in comparison with the repose period of certain volcanoes and types of activity. In some parts of the world written records only stretch back about a hundred years and there may be few if any cases of recorded eruptions. In 1951 Mount Lamington in Papua New Guinea erupted and caused some 5000 casualties, yet was considered from the evidence of both written and oral history to be dormant or extinct (Taylor 1958). This is not an isolated incident and many other damaging eruptions have occurred from volcanoes deemed to be inactive. Recent examples including Mount Arenal (Costa Rica, 1968) and Heimaey (Iceland, 1973). Clearly one way round the difficulty is to combine historical research with geological reconstruction of the 'life history' of a volcano and it is of significance that subsequent research at Mount Lamington has shown that previous activity occurred around 13,000 years ago (Baker 1979). The problem is, however, more complex as there are thousands of potential locations for renewed volcanic activity, yet only limited resources of trained earth scientists and funds available to mount the field work required. What is more, these human and material resources are strongly skewed in their distribution towards those parts of the world which are already at high levels of economic development. There is also the major difficulty of *monogenic* volcanoes (volcanoes which only erupt once) (Booth 1979). In the Auckland area of New Zealand some 50 eruptions have occurred over the last 40,000 years with an approximate 4000-year period of repose between events. Each event is associated with a different location (Fig. 7.3) and there is a remote possibility that the next could occur under the City of Auckland (Searle 1964; Booth 1979). Allied to this is the rarely mentioned but important question of when does prediction cease to be cost-effective. Devoting considerable effort to monitoring volcanoes which may not erupt is a poor use of scarce resources, when far more could be achieved by concentrating prediction on more likely prospects. This is not only a problem for policy-makers who have ultimately to accept the limited political risks of an eruption, but for scientists who have also to judge whether a volcano is still capable of threatening life and property over the normal time scales of human occupancy.

The ideal situation is outlined rather optimistically in a UNESCO publication, *Source Book for Volcanic Hazards Zonation* (Crandell *et al.* 1984a), in which it is suggested that volcanic hazards should be divided into two groups: *short term*

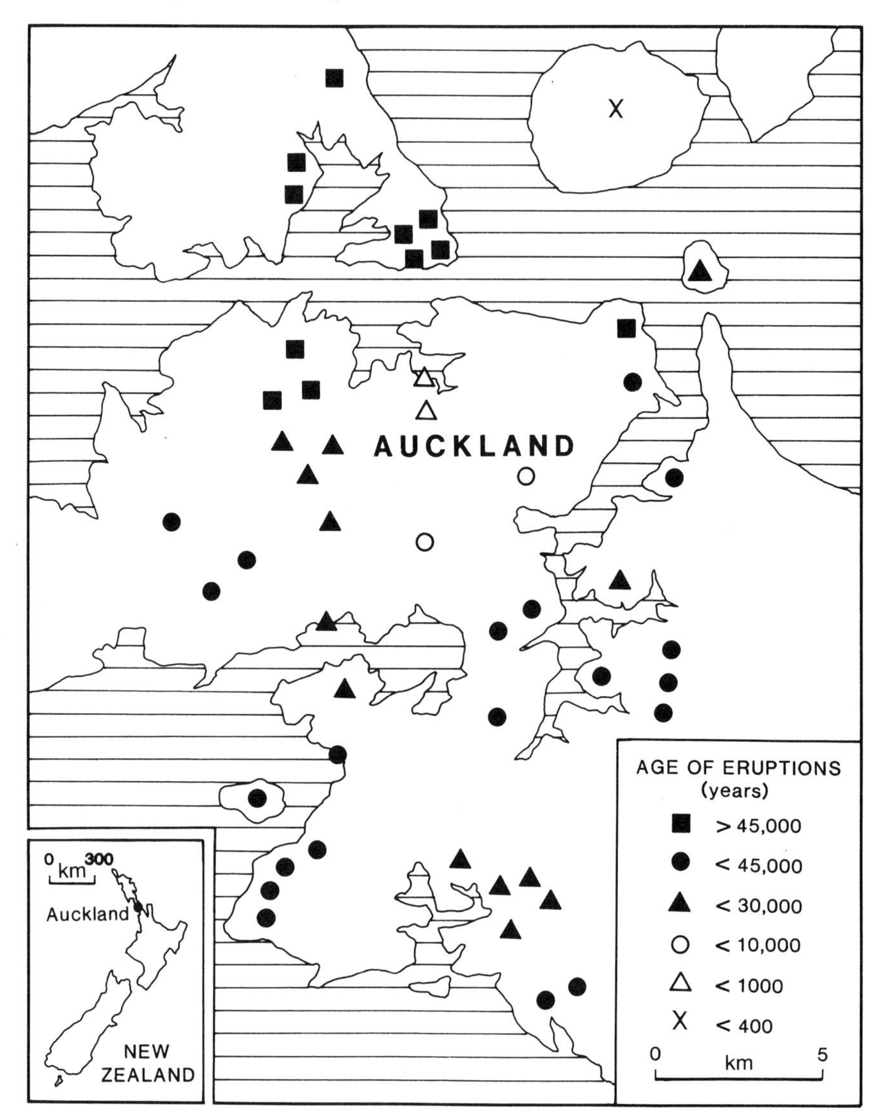

Fig. 7.3 The distribution and age of volcanoes near to the City of Auckland, New Zealand (after Booth 1979, Fig. 4, p. 335).

and based on events that occur more than once per century; and *long term*, related to events which are expected to occur less than once per century. For the former constant surveillance is suggested, whereas for the latter at least some instrumental surveillance is called for and high priority given to such tasks as comprehensive research into volcanic history and civil contingency planning. It is hard to argue against these views, but the fact remains that deciding between those volcanoes which will and those which will not erupt in the near future remains a difficult problem. There are excellent catalogues—*Volcanoes of the World*, produced by the Smithsonian Institution (Simkin *et al.* 1981), and an update, *Global Volcanism 1975–1985* (McClelland *et al.* 1989)—but even these show that there are many volcanoes for which little or no information is available. Furthermore, the amount of detail in *Volcanoes of the World* correlates strongly with those volcanoes that are located in regions which lie either within, or are accessible to, scientists from economically more developed countries of the world. Traditionally volcanoes have been classified as being *active*, *dormant* or *extinct* and, because of the problems already mentioned with the dormant and to some extent the extinct categories (see Tazieff 1967), the authors of the UNESCO source book prefer to use the terms *live* and *dead*. This seems sensible, but in the absence of other information fairly crude rules of thumb have to be employed in deciding whether a given volcano is capable of further activity. One such rule is the presence, or absence, of solfataric activity. This consists of thermal areas in the vicinity of a volcano, where temperatures are close to or above the boiling point of water and where water vapour and other gases are being discharged (Williams and McBirney 1979). Such activity can, however, be related to deep-seated processes as the roots of a volcano may be unrelated to future eruptions (Crandell *et al.* 1984a). Another not infallible guide is the 'freshness' of features and their lack of erosion which may be suggestive of fairly recent activity (Baker 1979), but again this may also reflect other processes, in particular the relative efficiency of subaerial denudation in different environments. Imperfect though these guides are they may be the only evidence of recent activity available to the volcanologist.

Some authors adopt a utopian stance when examining the issue of obtaining information on volcanoes which appear to be dead, but in fact may erupt in the remote future. J. V. Smith (1985) suggests a programme for future action. This would be of an international character and 'funded by perhaps 1% of expenditures on nuclear weapons systems' (Smith 1985, p. 676). Details of his suggested programme are summarized in Table 7.1, and many of the proposed actions are discussed in other parts of this book. The extent of international action and the amount of money required to put them into effect would, however, require degrees of cooperation and commitment that are not prominent features in the relationships between nations.

In Chapters 3, 4 and 5 the nature of different types of volcanic activity and the products which result from them have been discussed and it is clear that styles of activity are varied and, consequently, potential hazards are diverse. Volcanology is a young science and, as more eruptions have been observed in detail, the range of hazards which require prediction has increased considerably. Whilst the dangers posed by lava, certain types of pyroclastic flow and fall, lahars and tsunami have been known for many years, it is only with the observation of contemporary eruptions, the reconstruction of past events and theoretical modelling that the range of potentially hazardous eruption products and phenomena has become known. These include sector collapses, hydrovolcanism, directed blasts, newly recognized types of debris-flow, acid rain and volcanic earthquakes. It is salutary to recall that the greatest known loss of life due to the discharge of gas from a volcano occurred at Lake Nyos (Cameroons) in 1986. The potential hazard of gas discharges on this scale had not been recognized and this area was not catalogued as a volcanic region (Latter 1989a). There is doubt about whether the cause of this particular disaster was volcanic, or rather represented overturning and large-scale exsolution of carbon dioxide from water deep in the lake and it has been argued that rockfalls into the lake may have been responsible (SEAN

Table 7.1: Action that could be taken internationally to improve the monitoring of volcanoes (based on original ideas of J.V. Smith 1985)

Action	Further reference to these actions in other sections of this volume
1) Monitor the shape of all recognized volcanoes using altimetry from Earth-orbiting satellites.	Chapter 7, section 7.3.2.2
2) The mass production of seismographs, tiltmeters and gas detectors to allow the monitoring of seismic activity, ground deformation, and gas emission. These could be either placed in the field by scientists, or dropped by aircraft, with the data collected and relayed by satellite.	Chapter 7, section 7.3.2.1 (seismic) Chapter 7, section 7.3.2.2 (ground deformation) Chapter 7, section 7.3.2.3 (gas emission)
3) Increase the number of field teams for rapid deployment to imminent eruptions.	Chapters 7 and 8
4) Further development of seismic monitoring to determine volcanic plumbing systems.	Chapter 7, section 7.3.2.1
5) Continuous monitoring of volcanic aerosols, not just when an eruption has occurred, but by high-flying aircraft on a monthly basis.	Chapters 6 and 7
6) Further chemical analysis of ice-cores and tree-rings to reconstruct the emissions from past eruptions.	Chapter 6
7) Testing models of climatic change using data from (5) and (6) above.	—
8) Discussion of ways to alleviate the effects of volcanic eruptions.	Chapters 8 and 9

1986). More recent accounts of Lake Nyos consider both hypotheses, with the balance of argument currently favouring a volcanic cause and a deep-seated magmatic origin for the gas (Sigvaldason 1989; Tazieff 1989; A. B. Walker *et al.* 1992). Progress has been so fast over the past 20 years, that it is highly probable that additional hazards will be identified in the future.

Types of potentially hazardous volcanic event are listed in Table 7.2. Although the table is largely self-explanatory, certain additional features require discussion. The first is the frequency of recurrence of the various types of hazardous event. Using the historical record and being aware of the problems of interpreting these data sources, several authors have attempted to quantify hazards. Booth (1979) calculated that ~60 eruptions producing falls posed a hazard each century and that this is even greater if threats to aircraft are included (Kienle *et al.* 1990), lava flows have about the same frequency of recurrence, whilst pyroclastic flows, surges and lahars (~20 per century), structural collapse (~0.5–1 per century) and rare volcanic earthquakes are of decreasing importance. Although interesting, these figures are somewhat misleading, since the damage from one very infrequent tsunami may be many times greater than 20 lava eruptions—particularly in terms of loss of life. Figures compiled by Blong (1984) are instructive. They show that between 1600 and 1982 lava flows were only responsible for some 0.4% of

Table 7.2: Potentially hazardous volcanic events: characteristics and dangers (based mainly on Crandell *et al.* 1984a, with additions from numerous other sources)

Type of event	Notes
Lava flows	See Chapter 4
Domes and cryptodomes	See Chapter 4
Pyroclastic fall deposits	See Chapter 5
Lateral (directed) blasts	See Chapter 5
Volcanic gases	See Chapter 6
Pyroclastic flows	See Chapter 5
Pyroclastic surges	See Chapter 5
Hydrovolcanic (phreatic and phreatomagmatic) fall deposits	See Chapter 5
Lahars	See Chapter 5
Collapse	Structural collapse to form calderas is rare, but can be devastating (eg. Bandai San volcano, Japan in 1888). If the volcano is close to the sea tsunamis may be initiated. Collapse represents a high potential hazard, but one with a long recurrence interval.
Ground deformation and volcanic earthquakes	It is estimated that only 80 people have been killed by volcanic earthquakes during historical time and these constitute a minor hazard. Ground inflation before an eruption is also assumed to be a minimal hazard.
Tsunamis	Although a minor cause of death in the twentieth century, tsunamis or tidal waves have caused major losses of life in previous centuries. This is a particular hazard of explosive volcanoes located at and adjacent to coasts.
Other hazards	These include starvation, epidemic disease, contamination of water supplies and land, drowning, transport accidents, shock and exposure, cardiac arrests and breakdowns in civil authority. Many are preventable, given good civil defence.

Further information on the hazards shown in the table may be obtained from the following sources: Barberi and Gasparini 1976; Neall 1976; Booth 1977, 1979; United Nations 1977; Baker 1979; Baxter *et al.* 1982; Walker 1982; Tazieff and Sabroux 1983; Blong 1984; Crandell *et al.* 1984a; Druitt and Sparks 1984; Begoyavlenskaya *et al.* 1985; Crandell and Hoblitt 1986; Duncan *et al.* 1986; Francis and Slef 1987; Ollier 1988; Acharya 1989; Siebert *et al.* 1989; and Prata *et al.* 1991.

volcano-related deaths, whereas tsunamis accounted for nearly 19% of the deaths and pyroclastic flows, avalanches, lahars and flooding almost 30%. For the period 1900–82 the pattern changes with pyroclastic flows, avalanches, lahars and flooding accounting for more than 80%, tsumanis 0.8% and lava flows 0.2%. Part of the reason for these changes is improvements in data collection and the fact that relief aid and civil-defence measures have reduced the numbers suffering disease and starvation (nearly 40% 1600–1982; only just over 6% 1900–82), but the fact that exceptional events of low-recurrence interval may distort the figures in the short term is also of significance. The flooding and laharic activity associated with Nevado del Ruiz (Colombia 1985) and the putative gas emission in the Cameroons (1986) will 'distort' the figures for the 1980s.

A second feature of Table 7.2 is the fact that, although some hazards such as ground deformation and earthquakes may be considered minor even over long time scales, others which do not threaten life to any great extent are also of great human importance. Pyroclastic fall deposits cause major short-term agricultural and financial losses, while the sterilization of land by lava flows may be devastating to the long-term recovery of a region. In terms of long-term financial losses lava, pyroclastic flows, surges, lateral blasts, tsunamis and collapse are in one class and more serious than gas emissions and falls. Full recovery from the former group often takes centuries, for they often involve land sterilization, permanent land loss and major social dislocation. In contrast, although short-term losses from falls may be serious, in the long term the effects are normally minimal and, indeed, in some cases ashes may improve soil quality. Lavas of the 1669 eruption of Mount Etna covered large areas of the southern flanks of the volcano and today much land has not yet retained its full agricultural productivity, yet ashes from the same eruption have been cultivated with vines for a considerable time (Chester *et al.* 1985).

It is evident that a worldwide survey, whilst identifying potentially hazardous forms of volcanic activity, is only part of the information necessary to carry out prediction. For an individual volcano, general information has to be related to the geological and recorded history of activity and the ways in which people use a particular volcanic region. Linking the global and local scales of analysis is by no means easy, but in excellent reviews Decker (1973, 1978) shows some of the ways in which this may be done. For any volcano he identifies both positive and negative risk factors which should be considered in a predictive exercise and a modified and revised version of his table is reproduced (Table 7.3). It not only summarizes the risks that have been discussed in this section, but also points the way forward towards consideration of the methodologies and techniques that may be used for predicting eruptions.

7.3 The methodology and techniques used in prediction

The *raison d'être* of prediction is two-fold. In the first place, it is desirable to know the products which are going to be erupted over different time scales and what effects these will have on people living in an area. This approach is called *general prediction* and is defined by George Walker (1974) as, the study of 'the past behaviour record of a volcano so as to determine the frequency, magnitude and style of eruptions, and to delineate high risk areas' (Walker 1974, p. 23). The output of general prediction normally includes a hazards map, alternative titles

Table 7.3: Relative danger of different tectonic, geological, topographic, environmental, social and economic factors on individual volcanoes[1] (factors 1 to 8 are based on Decker 1973, Table 2, p. 377, and factors 9 and 10 are added by the author;[2] remarks compiled from a number of sources)

Danger[1] Negative (−)	Positive (+)	Remarks[2]
Tectonic and geological factors		
1) Rift zones	Subduction zones	Rift zones will tend to produce basaltic magma and effusive eruptions, subduction zones more evolved magma and more explosive activity.
2) Shield volcanoes	Stratovolcanoes	Shield volcanoes are formed of lava flows and stratovolcanoes pyroclastic materials. Hence, the former bear witness to a geological history of effusive activity, whilst the latter are more indicative of explosive activity.
3) Lavas	Pyroclastic materials	A volcano which has a history of producing pyroclastic flows, surges and large spreads of airfall is likely to be more dangerous than one which has only produced lava flows.
4) Low silica	High silica	If the volcanic products are silica-rich, then future activity is more likely to be explosive than if the products are silica-poor.
5) Little water	Much water	The occurrence of snow and ice, a crater lake, a coastal location and large quantities of ground water, makes a volcano more prone to hydrovolcanic activity and the generation of floods and lahars.
6) Short repose	Long repose	Many devastating eruptions have occurred at volcanoes, which have long repose periods. Many very dangerous forms of volcanic activity (e.g. pyroclastic flows, lateral blasts, surges and collapse) may have long repose periods.
Topographic and environmental factors		
7) Upwind	Downwind	Pyroclastic and hydrovolcanic fall deposits are distributed by winds prevailing at the time of an eruption and will be thicker downwind of the eruption vent.
8) Ridges	Valleys	Lava flows, pyroclastic flows, surges, lahars and floods are all, to some extent, controlled in their distribution by pre-existing topography. Valleys trending away from the eruption vent will be particularly dangerous locations.
Social and economic factors		
9) Sparse settlement	Dense settlement	Areas of dense settlement will, clearly, be at greater risk.
10) High economic status	Low economic status	In economically developed areas, absolute economic losses will be high, but because of good civil defence, administration, medical care and aid, loss of life is likely to be low. Recovery is also likely to be more rapid. In areas of more limited economic development loss of life is likely to be higher. Economic losses (in relation to the areas' wealth) will be high as well. Recovery may be very slow.

for this approach being *hazard mapping* and *assessment* (see Latter 1989a). General prediction may be used not only to highlight existing hazards, but may also be employed for planning purposes. The latter includes: the development of evacuation plans; civil-defence measures; zoning for insurance purposes; and steering new development into less hazardous locations (see Chapter 9).

A second purpose of prediction is to forecast the actual time and type of eruption. This is known as *specific prediction* and is based on '*surveillance* of the volcano, and the *monitoring* of changes, for example, in seismic activity or tilt, so as to forecast the time, place and magnitude of an eruption' (Walker 1974, p. 23, my emphases). This usually involves the identification of precursory signs of activity and necessitates careful observation of the volcano for months, and in some cases years, before an eruption occurs.

Although frequently described as separate approaches, which in one sense they are—general prediction being geological, historical, statistical and cartographic in character and specific prediction principally geophysical—the approaches are complementary and once an eruption has started both may be employed to determine how it will develop and what measures will be necessary to minimize its effects on the inhabitants of a region.

7.3.1 General prediction

Maps showing areas of a volcano most at risk in future eruptions is by no means new. In 1919 following an eruption of Kelut volcano (Indonesia) hazard maps were produced by The Volcanological Survey of the Netherlands Indies (Neumann van Padang 1960). Dangerous areas were defined on the basis of land devastated during historic eruptions and the influence of topography on the distribution of erupted products was shown to be critical. Following independence from the Netherlands further hazard maps were compiled by The Geological Survey of Indonesia and an example of this pioneer work is reproduced as Fig. 7.4. It shows the areas at risk from pyroclastic flows generated by Merapi volcano, Java (Neumann van Padang 1960; United Nations 1977). Over the last 50 years and particularly since the middle 1960s, considerable progress has been made. This has involved not only a greater number of hazard maps being compiled, but also considerable advances in improving methods of assessment.

All general predictions have at their heart a fundamental assumption. Uniformitarianism decrees that the 'present is the key to the past' and general prediction recasts this and assumes that on active volcanoes past activity largely determines what will happen in the future. Clearly this assumption accords with both common sense and observation, since volcanoes such as those on Hawaii and Mount Etna (Sicily) which have been historically associated with mainly quiet effusions of lava are unlikely suddenly to start erupting more evolved products such as pyroclastic flows. This is a reasonable assumption but one which should clearly be treated with caution because, as argued earlier, high-magnitude low-frequency events are features of many volcanoes and it is known that volcanoes can change their behaviour patterns fairly quickly. In the case of Etna, detailed geological reconstruction has indicated that around 15,000 BP the internal plumbing favoured high-level storage of magma and the eruption of more silica-rich products, including a pyroclastic flow. Around 5–6000 BP collapse of one sector of the volcano and hydrovolcanic activity is also implied by the geological record (Chester *et al.* 1985, 1987). General prediction has not

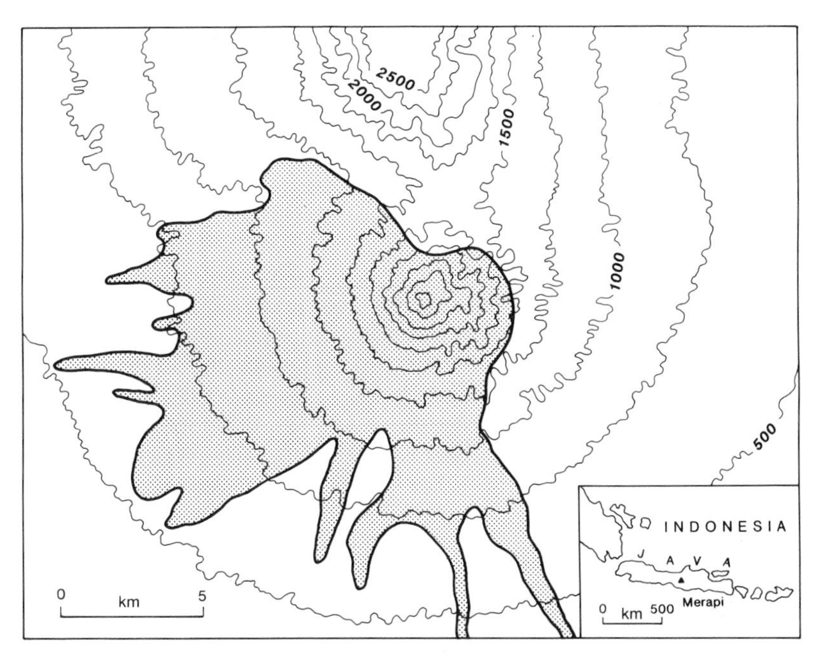

Fig. 7.4 Merapi volcano, Java (Indonesia). The shaded area shows the zone at particular risk from 'glowing avalanches' (*sic*) (i.e. pyroclastic flows). This map was compiled around 1934 by the Netherlands Indies Volcanological Survey (from United Nations 1977, Fig. 1, p. 2; and based upon ideas of Neumann van Padang 1960).

advanced in isolation and the identification of new potential hazards such as hydrovolcanism, certain types of collapse and lateral blasts means that the range of phenomena which have to be considered in any assessment is increasing and there are instances where hazard maps produced only a few years ago now require updating. Hydrovolcanism presents a particular problem for it 'involves many variables and presents volcanologists with a daunting task of eruption prediction' (Duncan *et al.* 1986, p. 379). Allied to this is the fact that once an eruption takes place, any existing general predictions will require modification. The new eruption will serve to test the predictive model and its products may well modify the topography of the construct so that new zoning is called for. For these reasons the past is not always the key to the future, only a guide to it.

The methodology for producing general predictions in the form of hazard maps has attracted much attention and is well summarized by Booth (1979, p. 335) 'as an exercise in the recognition and areal mapping of deposits known to have been formed during particular phases of volcanic eruptions and extrapolated to indicate the areas that could suffer identical hazards during a future, similar event. For volcanoes with a poor historical record, or for those where minimal information is available, only careful field work will supply the data necessary for significant zoning.' This statement, although a useful summary, requires expansion. There are relatively few volcanoes with long historical records—particularly of infrequent events—and the success of hazard mapping depends critically on the quality of geological studies and the reconstruction of the life history of the edifice. George Walker (1981b) points out two serious considerations. The first is that episodes in the history of the volcano require accurate dating. Normal techniques of geological mapping, supplemented with caution by evidence of weathering rates and erosion, can indicate the relative ages of different volcanic products, but absolute dating is also required. There are now a variety of techniques which may be used to date volcanic successions and their usefulness depends on the type of material and its age. Techniques include radiocarbon, tree-ring (dendrochronology), potassium/argon (K-Ar), uranium series, hydration rinds on natural glass, lichenometry, palaeomagnetic variations and thermo-luminescence. Detailed discussion of these techniques is beyond the scope of this volume and those interested should refer to Goudie (1981a, 1990) and Crandell *et al.* (1984a). Non-volcanic sediments are sometimes interbedded with volcanic successions which makes dating easier and the association of volcanic episodes with raised shorelines in the Aeolian Islands (Italy) is one example of this situation (Keller 1967), whilst the use of river terraces to date early phases of volcanic activity on the flanks of Mount Etna (Sicily) is another (Chester and Duncan 1979, 1982). A second consideration is that often powerful volcanic events leave very little trace in the geological record. The pyroclastic flow which laid waste to the town of St Pierre (Martinique) following the eruption of Mt Pelée in 1902 is now represented by a 20 cm ash layer, while by definition low-aspect ratio flows (see Chapter 5, section 5.6.2) are thin in relation to their volume (G. P. L. Walker *et al.* 1980a). Hence, geological reconstructions have to be carried out with diligence and at levels of detail commensurate with the precision required by the particular prediction.

There is also a human dimension. Volcanoes occurring in such places as the USA, Japan, Iceland, the former Soviet Union, the Azores, Italy, New Zealand and other countries that are highly developed economically, normally have one or more of the following characteristics which assist in the production of hazard maps. These characteristics are: the availability of high-quality topographic maps; carefully reconstructed historical records of activity and, more importantly, reserves of trained earth scientists, funded research programmes and enthusiastic civil authorities. In contrast, in many Andean, Central American, Caribbean, East Indies and oceanic island states, few if any of these characteristics are to be found and it is in these countries that some of the most potentially dangerous volcanoes are located. In recent years the position has improved both through aid programmes and by research carried out by university and research institute expatriate groups. To take one year—1986—by way of example, the United States Geological Survey (USGS) carried out volcanic hazard studies in Cameroon, Colombia, Indonesia and several South American countries, in addition to participating in collaborative programmes with host countries and through international bodies such as the United Nations Educational, Scientific and Cultural Organisation (USGS 1987). This is not an isolated case and teams from many economically more developed countries are involved in similar aid programmes, but the fact remains that it is in poorer countries that most still needs to be achieved. Basil Booth argues that reconnaissance-scale general prediction can be cost-effective and estimates that 'an average polygenic volcano with reasonable access can be zoned in approximately 100 man days' (Booth 1979, p. 335). In the author's experience this is an underestimate, but it does indicate what well-planned aid programmes can achieve.

The range of approaches to general prediction in societies at different levels of socio-economic development, may be compared by looking at three examples.

The first is the programme of hazard assessment being carried out in the USA. Here there are many active and potentially active volcanoes, with a number of different styles of activity in a country enjoying high levels of economic development, the largest reserve of skilled earth-science personnel in the world and strong government support. In the USA the historical record, though often of high quality, is relatively short. The second example is Mount Etna, Sicily. This is a volcano which has a long historical record of eruptions stretching back to before the time of Christ. It is a basaltic volcano—producing mainly lava flows—and has been studied by both Italian and other European research groups. It illustrates what can be achieved through the intensive study of one very active and economically important volcano. The third example is from an economically less developed part of the world. In the Lesser Antilles (Caribbean), the problems faced by governments of poor countries with dangerous volcanoes are well illustrated. In addition to these examples, there are many other countries in which mapping has been carried out and Table 7.4 shows a selection of these, together with references to allow further reading.

7.3.1.1 General prediction and hazard mapping in the USA
Since 1967, the USGS has been engaged in a programme designed to assess potential hazards at volcanoes (see Miller 1990; Wood and Kienle 1990). The output from this programme has been most impressive (Crandell 1984a) and has included not only hazard-zoning maps for the whole of the western United States (scale 1:5,000,000; Mullineaux 1975) and for the State of Washington (scale 1:1,000,000; Crandell 1976), but also in-depth studies of individual volcanoes and volcanic regions such as: Mount Rainier (Crandell 1973); Mount Baker (Hyde and Crandell 1978); Mount St Helens (Crandell and Mullineaux 1978); Mount Hood (Crandell 1980) and Mount Shasta (Miller 1980). Additionally attention has been paid to Hawaii (Mullineaux and Peterson 1974; Crandell

Table 7.4: A selection of examples of countries in which hazard maps have been produced. (The USA, Caribbean countries and Sicily are specifically excluded and are dealt with in the text)

Country and region	References
The Atlantic Ocean and Europe	
Iceland	Sigvaldason 1983; Imsland 1989.
Azores	Booth *et al.* 1983.
Tenerife (Canary Islands)	Booth 1984.
Italian peninsula	Barberi and Gasparini 1976; Barberi *et al.* 1983, 1984; Alexander 1987; Rosi and Sbrana 1987.
Aeolian Islands (Italy)	Sheridan and Malin 1983; Frazzetta *et al.* 1984.
The former USSR and Japan	
Kamchatka (the former USSR)	Melekestsev *et al.* 1989.
Japan	Shimozuru 1983b.
Indonesia	Zen 1983; Kasumadinata 1984.
New Zealand	Searle 1964; Dibble 1983; Dibble *et al.* 1985.
Central and South America	
Mexico (Popcatepetl volcano)	Boudal and Robin 1989.

1975), to Augustine volcano, Alaska (Kienle and Swanson 1983), and the Long Valley caldera, California (Miller *et al.* 1982; Bailey and Hill 1990). Not all research has been originated by the USGS and much valuable work has been carried out by university academics and state authorities (Tilling and Bailey 1985).

The USA has not been long settled with literate observers and, even with the addition of evidence from oral history, the record of eruptions only stretches back a couple of hundred years at the most. This record plus many geological reconstructions of the past activity, allows a broad classification of volcanoes to be made in terms of decreasing probability of eruption (Scott 1990). Following Bailey *et al.* (1983) these are:

Group 1 Volcanoes which have eruptions with return periods of 100–200 years and/or erupted within the past 200–300 years;

Group 2 Volcanoes which have eruptions with return periods of thousands of years and/or last erupted more than 1000 years ago; and

Group 3 Volcanoes which last erupted in pre-Holocene times, but still have shallow magma chambers.

The distribution of these three groups of volcanoes is shown on Fig. 7.5, and it is clear that the attention of scientists has been devoted largely to those with the highest eruption risk. This is a reasonable stratagem, since the probability of eruption from volcanoes in Group 3 is by definition very low, although some monitoring by methods of specific prediction is being carried out as a precaution. Some of the volcanoes in Group 2 and even in Group 1 if they were to erupt have the potential to cause major loss of life and severe economic disruption, since their magmas are relatively silica-rich. An event like the creation of Crater Lake (Mt Mazama, about 6600 BP) if it were to recur would be cataclysmic indeed (Tilling and Bailey 1985).

By selecting Mount St Helens, it is possible to see how the American approach operates. This is perhaps the most interesting of all volcanoes from the perspective of general prediction because only two years after a map showing potential hazards had been published (Crandell and Mullineaux 1978) the volcano erupted, so testing its reliability. Mount St Helens is a young volcano in the Cascade Range (Washington State), the products are of mainly dacitic composition (Chapter 3, section 3.2) and eruptions have been of an explosive character producing domes, pyroclastic flows, some lava flows and sheets of fall deposits hundreds of kilometres away (Crandell 1984a). Since the penultimate eruption occurred in 1857 (Crandell *et al.* 1979), Dwight Crandell and his colleagues had to reconstruct the 'life history' of the volcano. This was achieved through painstaking logging of deposits and dating by means of the radiocarbon (C^{14}) technique of organic materials interbedded with volcanic horizons (Crandell *et al.* 1975). Because the slopes of Mount St Helens were not so eroded as some of those of other volcanoes in the Cascade range, it had long been assumed that the volcano was relatively young and stratigraphic research and dating has confirmed this. Even though volcanic activity started ~37,000 years ago, virtually the whole volcano has been formed since ~2500 BP and most of its upper part within the last 500 years. Since ~2500 BC (~4500 BP), the volcano had never been inactive for more than five centuries and one to two centuries is more typical. On the basis of these findings, Crandell and his co-workers forecast that there was a strong probability of an eruption in the next thousand years and that 'an eruption is

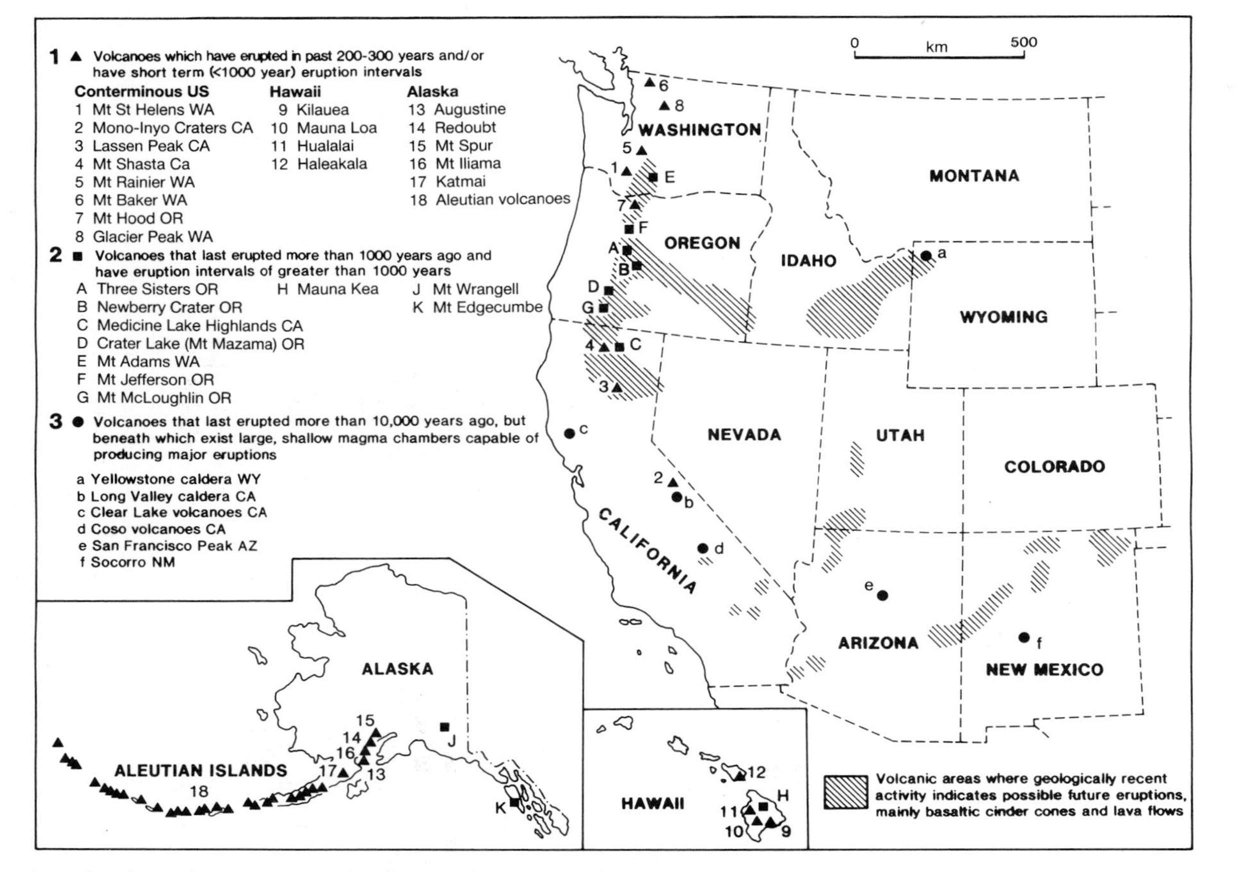

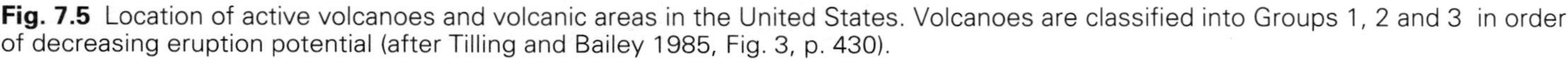

Fig. 7.5 Location of active volcanoes and volcanic areas in the United States. Volcanoes are classified into Groups 1, 2 and 3 in order of decreasing eruption potential (after Tilling and Bailey 1985, Fig. 3, p. 430).

likely within the next hundred years, possibly before the end of this century' (Crandell *et al.* 1975, p. 441). A hazard map followed in 1978 (Crandell and Mullineaux 1978) and it was assumed that any future eruption would produce lava flows, domes, pyroclastic flows and lahars, and that the distances travelled by these products would be similar to those achieved during previous eruptions. The topography of the volcano, the orientation of river valleys and prevailing wind directions were also incorporated and a simplified version of this hazard map is reproduced in Fig. 7.6. Later a second map showing areas that would probably be covered by airfall materials was published (Crandell *et al.* 1979; Fig. 7.6).

On 20 March 1980 a swarm of earthquakes began beneath Mount St Helens which increased and each day involved several of magnitude 4 or greater. Small eruptions at the summit began on 27 March and small ash and phreatic eruptions occurred until 21 April, and from 7–14 May. Throughout this period the volcano was under constant surveillance by techniques of specific prediction and from mid-April a swelling (or bulge) was observed high on the north side of the volcano. On the basis of monitoring it was assumed that the volcano was about to have a major eruption and the tourist and forested region on the volcano was evacuated. On the morning of 18 May at 08:32 hours (local time) an earthquake (magnitude 5.1) triggered slope-failure causing unroofing of the magma chamber, a gigantic explosion and a lateral blast which destroyed much of the northern flank and devastated more than 400 km^2. The lateral blast was accompanied by an eight-hour ash eruption, an avalanche of rock, lahars and pyroclastic flows (Decker and Decker 1981b; Foxworthy and Hill 1982). The eruption continued with decreasing violence for many months and caused the deaths of 65 people (including the volcanologist David Johnston), destroyed thousands of trees (many of which were literally blown over) and left the volcano with a large crater on its northern flank (Decker and Decker 1981b).

Reviewing the effectiveness of the hazard mapping following the eruption, Miller *et al.* (1981, p. 789) concluded that 'forecasts were generally accurate, although the magnitude of the catastrophic and unprecedented landslides and lateral blast of May 18 greatly exceeded our expectations'. One feature of the response to this eruption was its coordination, so that once there were signs that the volcano was about to erupt both general and specific techniques were used to assess its progress and likely impact. For example, as early as 23 April Dwight Crandell was warning of the possibility of a large landslide, while after the events of 18 May a modified hazards map was drawn, assimilating the lessons that had been learnt. As Fig. 7.7 shows, this map incorporated an area of danger to the north. Further explosive eruptions were considered to be a possibility and, accordingly, a new map was seen to be essential (Miller *et al.* 1981).

The approach of scientists and public officials to the Mount St Helens emergency illustrates many features of good practice. First, scientists and public officials were forewarned and could, therefore, be forearmed by hazard mapping

Fig. 7.6 Hazard maps showing the general prediction of future eruptions from Mount St Helens, USA. Upper map shows the areas at risk from lava flows, pyroclastic flows, lahars and floods (modified from Crandell *et al.* 1979, and based on Crandell and Mullineaux 1978). The lower map shows the zones at risk from airfall tephra. Potential thickness is greatest in Zone A and progressively less in Zones B and C. Tephra from most eruptions will fall into the shaded sector (modified from Crandell *et al.* 1979).

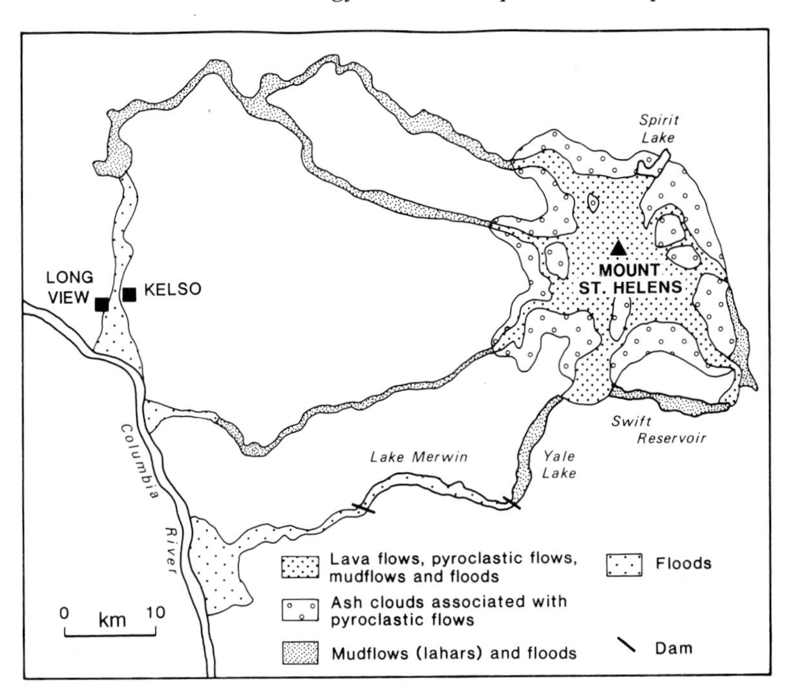
Spirit Lake
MOUNT ST. HELENS
LONG VIEW
KELSO
Columbia River
Swift Reservoir
Lake Merwin
Yale Lake
Lava flows, pyroclastic flows, mudflows and floods
Floods
Ash clouds associated with pyroclastic flows
Mudflows (lahars) and floods
Dam
0 km 10

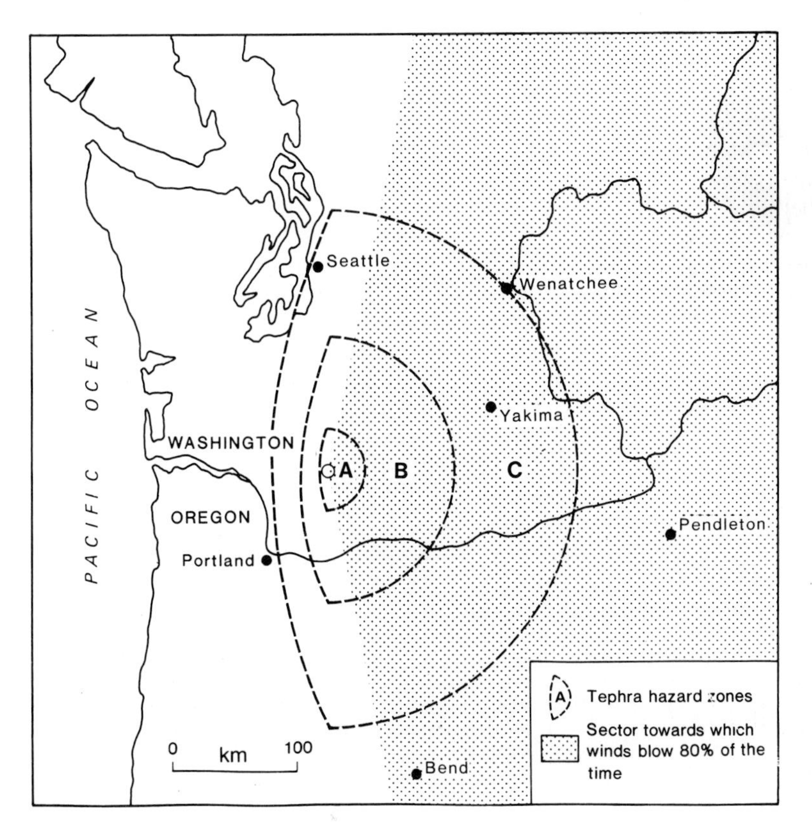
PACIFIC OCEAN
Seattle
Wenatchee
Yakima
WASHINGTON
OREGON
Portland
Pendleton
Bend
A
B
C
Tephra hazard zones
Sector towards which winds blow 80% of the time
0 km 100

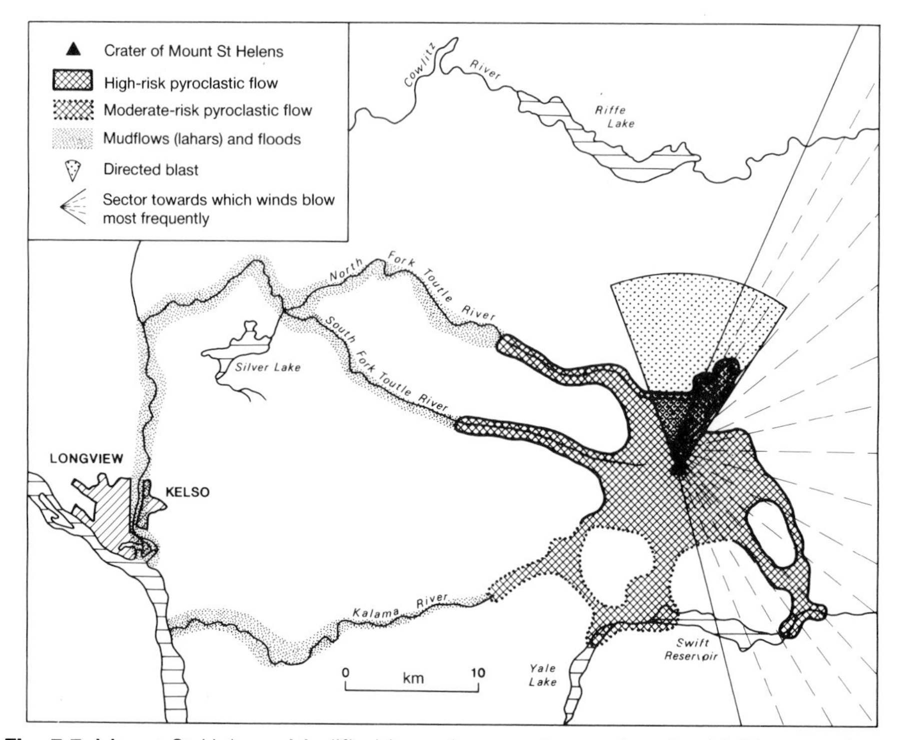

Fig. 7.7 Mount St Helens. Modified hazard zones drawn after the 18 May eruptions. This figure should be compared with Fig. 7.6, in particular note the area at risk from future directed (lateral) blasts (after Miller *et al.* 1981, Fig. 455, p. 796).

and could act successfully to reduce damages and minimize loss of life. Secondly, the close integration of general and specific prediction meant that emergency measures could be coordinated. Indeed as Swanson *et al.* (1985) noted in a review of the progress of prediction carried out on Mount St Helens between 1975 and 1984, the distinction between general prediction (termed forecasts) and specific prediction (termed prediction) was arbitrary, being a question of the time scale over and the precision with which estimates of future happenings could be made. When the volcano was thought to be in danger of eruption, specific techniques of geophysical monitoring could be instituted and once activity had begun both general and specific approaches could be used. Between June 1980 and the end of 1984, all but two extrusions of material from the volcano were predicted successfully.

The success of this integrated approach to prediction in the USA came from both a long history and development of volcanological expertise and, more recently, strong support from government. As early as 1912 Dr Thomas Jaggar of the Massachusetts Institute of Technology was instrumental in setting up an observatory to monitor the Hawaiian volcanoes (Decker and Decker 1988) and by the late 1960s this had become one of the leading, some would argue the principal, repository of volcanological expertise in the world. The motto of the Hawaiian Volcano Observatory—*Ne plus haustae aut obrutes urbus* (no more shall the cities be destroyed)—aptly sums up part of its role (quoted by Tilling and

Bailey 1985, p. 431). Under the stimulus of the Disaster Relief Act (1974), American applied volcanology was widened in scope and geographical coverage to embrace all the active and potentially active volcanoes within the USA. Under the Act the goal of policy and the chosen agent of government, the USGS, is 'to reduce the loss of life, property, and natural resources that can result from volcanic eruptions and related consequences' (Bailey *et al.* 1983, p. 14). It is this combination of strong government support, both legislative and financial (the USGS budget for the programme was over US$10 million in 1986; USGS 1987), together with a high level of scientific competence, which made the response to the eruption of Mount St Helens so successful.

7.3.1.2 General prediction at Mount Etna, Sicily

Like Mount St Helens, Mount Etna is a volcano which is located within an economically developed country (Italy) and one where central government and its agencies have devoted considerable resources, both financial and scientific, to understanding its activity. Here the similarity ends. Whilst Mount St Helens is 'normally' an 'inactive' volcano within a relatively remote region, Mount Etna is continually active—probably the largest continental volcano of its type in the world—and its flanks constitute the most economically important region within the whole of Sicily. Sicily is one of the poorest regions in western Europe with a legacy of organized crime, banditry and delinquency—including the heavy hand of the mafia (King 1973, 1975)—and it is an island where much of the population is dependent upon a fairly unproductive agricultural sector from which millions have emigrated. Emigration has been traditionally permanent and predominantly to the USA, but in recent years temporary migrations have been more common to find work in the cities of northern Italy, Switzerland, Germany, France and the United Kingdom. The coastal areas of the island, and particularly the east and the Etna region (Fig. 7.8), have been and remain distinctive. Inhabitants here are more outward-looking, the agricultural potential of the land is higher and organized crime has been less of a problem. Greater wealth and development have been caused by a combination of natural endowment and the skill and ingenuity of the people. The Etna region contains some 20% of Sicily's population of ~5 million on some 7% of its land area, most being concentrated in the agriculturally productive east, southeast and south sectors of the volcano, where population densities of 500 per km^2 are common and may reach up to 800 per km^2. The second city of the island, Catania (population over 400,000), is also located to the south of the volcano and is not only the focus of the rich agricultural region, but is also an industrial and administrative centre as well. The key to this relative prosperity is irrigated agriculture and volcanic soils of high potential fertility, which have been profitably exploited by the inhabitants for more than 2000 years (Chester *et al.* 1985).

Volcanologically Etna is a young volcano and, though activity began more than 300,000 years ago, it is estimated that some 98% of its volcanic products have been erupted over the last ~100,000 years, forming an edifice covering 1750 km^2 and rising to more than 3200 m (Chester *et al.* 1985). Historical records of activity are excellent and stretch back to before the time of Christ. A reasonably good catalogue is available from 1300 AD and an almost continuous record from around 1500 AD (von Wältershausen 1880; Romano and Sturiale 1982; Chester *et al.* 1985). On Etna a distinction is drawn between persistent activity at the summit and periodic flank eruptions and it is the latter which usually cause damage to people and property. Since 1500 AD there have been over 50 flank eruptions in addition to several tectonic earthquakes which have destroyed towns and villages, yet settlement of the region continues and, indeed, increases due to the actual and perceived economic benefits of the region outweighing the damage caused by volcanic eruptions and earthquakes.

In order to minimize the threat of hazards, both government-supported and independently funded Italian and foreign research groups have produced general predictions in the form of hazard maps (Frazzetta and Romano 1978; Guest and Murray 1979; Cristofolini and Romano 1980; Duncan *et al.* 1981; Chester *et al.* 1985; Forgione *et al.* 1989), and although details vary the approaches are very similar. The nature of volcanic activity on Etna is quite different from that of Mount St Helens and an alternative approach to hazard mapping has been adopted. Despite the number of eruptions Etna has killed very few people (less than 20 in the last 100 years) and the major losses from lava flows are to property, agricultural land, communications and the economic well-being of the region. During historic time activity has been of a mildly Strombolian character and lava flows have been of basaltic composition; rarely travelling more than about 15 km from their source vents (see Chapter 4, section 4.3.1.2).

All published hazard maps have common features. They recognize that on a volcano so active as Etna, no area may be classified as being absolutely safe and general prediction must seek to identify those zones where risks are particularly high. It is also acknowledged that over longer time spans more violent events may occur, including large collapses, lahars and pyroclastic flows. Hazard assessment assumes that the pattern of activity in the immediate future will follow that of the immediate past (Chester *et al.* 1985).

Taking the map produced by Duncan *et al.* (1981) as an example (Fig. 7.8), it is assumed that most flank eruptions will occur in a zone where vent densities exceeded more than 1 per km^2, since it is in this zone that virtually all flank eruptions have begun during the last 400 years. Guest and Murray (1979) calculated what they called *eruption catchment areas*. Remembering that lava flows are of restricted length, then this is the area in which an eruption would have to start in order to effect a given settlement. Coincidence of the eruption catchment area with the zone of high vent densities, suggests that the town or village in question is at some risk. The higher the coincidence the greater the risk. In addition, Guest and Murray recognized that lava flows follow ground contours and some areas are protected topographically from the effects of eruptions. Further analysis of these data and their focus on the settlement geography of the volcano allowed Duncan and his co-workers to produce their hazard map (Fig. 7.8). All authors who have produced hazard maps are aware that there are potential dangers of accepting general predictions at face value. These include the recognition that once an eruption starts it will alter ground contours and that over short periods of time (say less than 100 years) there may be changes in the preferred sector of eruption. Hence, the towns of Milo and Fornazzo (Fig. 7.8) have been threatened by lava several times since the Second World War. Also in the southern part of the zone of high vent densities around Nicolosi eruptions tend to be infrequent, yet historically these have been amongst the most voluminous and the large eruption in 1669 destroyed much of Catania.

Etna illustrates the use of general prediction on a volcano which is frequently active and produces lava flows. Such predictions have two purposes. The first is to identify where lava will invade once an eruption has started, and the second is to

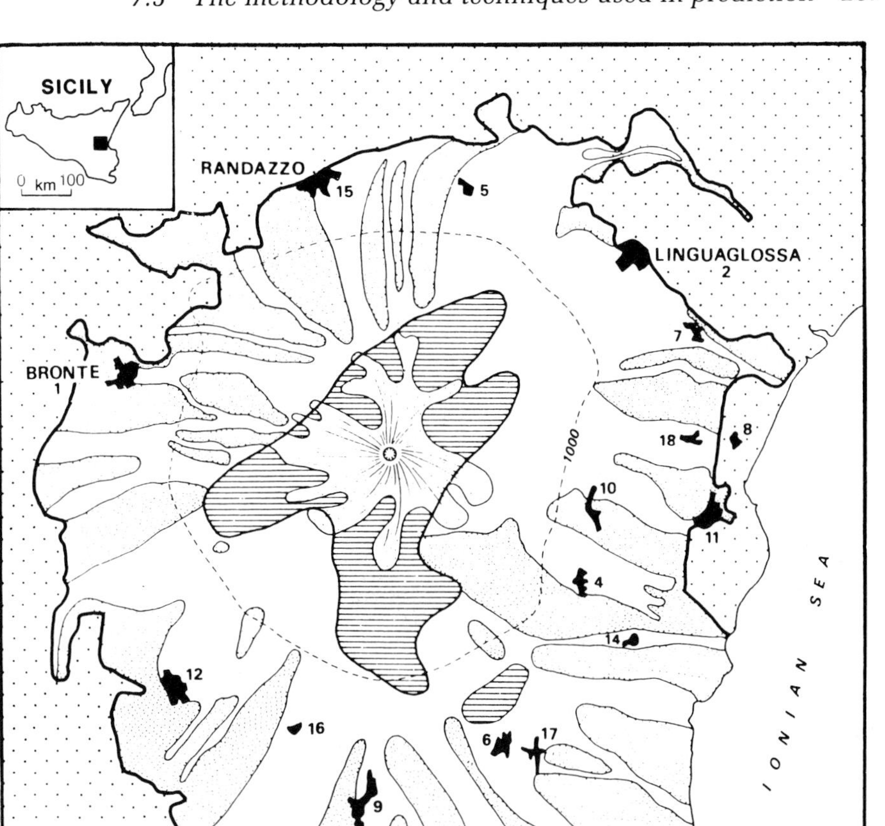

Fig. 7.8 Hazard map of Etna, showing areas topographically protected from lava flows, areas of high eruptive-vent density and settlements at risk (from Duncan *et al.* 1981, Fig. 6, p. 176).

enable new, often costly, developments to be steered to locations which are relatively safe, through a system of physical planning controls. Integration of general predication and physical planning requires far more detailed surveys and these have already been started by scientists from the University of Catania and the International Institute for Volcanology (Catania), both of which are funded by the Italian government (Chester *et al.* 1985).

7.3.1.3 General prediction in the Lesser Antilles (Caribbean)

Although the nature of the hazard threats on Mount St Helens and Mount Etna differ markedly both volcanoes have one thing in common; they are located in economically developed countries and have funded research programmes and indigenous scientific expertise. In the Lesser Antilles (Fig. 7.9), this does not apply and hazard assessments have been carried out by an uncoordinated group of resident and expatriate earth scientists. In general the islands are poor, wanting in trained earth scientists and have populations that are not educated about the effects of potential eruptions (Fiske 1981; Roobol and Smith 1989). The Lesser Antilles volcanic arc is socially very diverse. Some islands are administered by colonial powers (the UK, France and the Netherlands) others are independent, some are politically stable others racked by instability and unrest and some are still covered by forest vegetation whilst others are cleared and cultivated. In short,

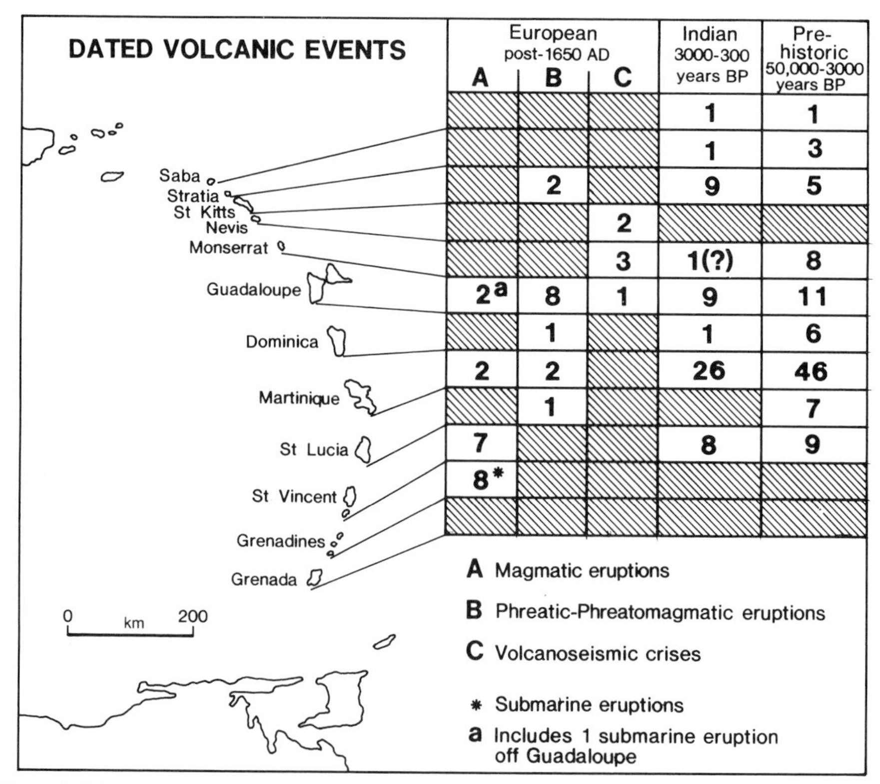

European post-1650 AD A	B	C	Indian 3000-300 years BP	Pre-historic 50,000-3000 years BP
			1	1
			1	3
	2		9	5
		2		
		3	1(?)	8
2[a]	8	1	9	11
	1		1	6
2	2		26	46
	1			7
7			8	9
8*				

Fig. 7.9 Map showing the active volcanic arc of the Lesser Antilles and dated activity classified according to European, pre-Columbian Indian, and prehistoric periods of settlement (from Roobol and Smith 1989, Fig. 1, p. 59).

these islands provide a microcosm of some of the problems involved in general prediction—indeed all hazard assessment—in economically less developed countries.

Like Mount St Helens the historical record of eruptions is of limited duration in the Lesser Antilles and, even though Columbus arrived at Dominica in 1493, accounts of eruptions are only available for the last 300 years. Using these records (Robson and Tomblin 1966), plus information from volcanic horizons associated with pre-Columbian archaeological sites, some of which are dated, and conventional geological and geochronological investigations, Roobol and Smith (1989) have been able to reconstruct a history of eruptions (Fig. 7.9) and an assessment of their characteristics (Table 7.5). Maps predicting future eruptions have been compiled by Roobol and Smith (1989) and others. Figure 7.10 is an example of one such map for part of the Island of Montserrat. It shows the threat posed by pyroclastic flows around the Soufriere Hills (Baker 1985). As its author admits, with the poor dating currently available it is not possible to quantify recurrence

Table 7.5: Volcanic hazards in the Lesser Antilles arc: categories, magnitudes and frequencies (after Roobol and Smith 1989)

Magnitude and frequency	Category of hazard	Examples
Increased magnitude ↓ Increased frequency ↑	I) Events occur every few decades and affect only the volcano's flanks. Magma is andesitic and activity Peléan.	All the Lesser Antilles arc volcanism of the last 300 years (e.g. six pyroclastic, 14 phreatic/phreatomagmatic and four lava/dome eruptions). Volumes erupted less than 2 km^2 (usually less 0.5 km^2). Simultaneous eruptions in 1902 at Mt Pelée (Martinique) and Soufriere (St Vincent) indicates that multiple hazards are possible. Volcanic/seismic crises also included in this category (six in last 300 years).
	II) Products extend beyond flanks of the volcano and on to smaller islands. The whole land area is affected.	Large Plinian eruptions which have produced extensive fall deposits and high- and low-aspect pyroclastic flows. Examples Mt Pelée (~2000 BP), Roseau tuff (Dominica ~30,000 BP).
	III) Effects not confined to one island, but events so violent that adjacent islands are affected. Events occur every few tens to hundreds of thousands of years.	Exemplified by the geology of St Lucia, Dominica and St Vincent. May involve sector and caldera collapse.

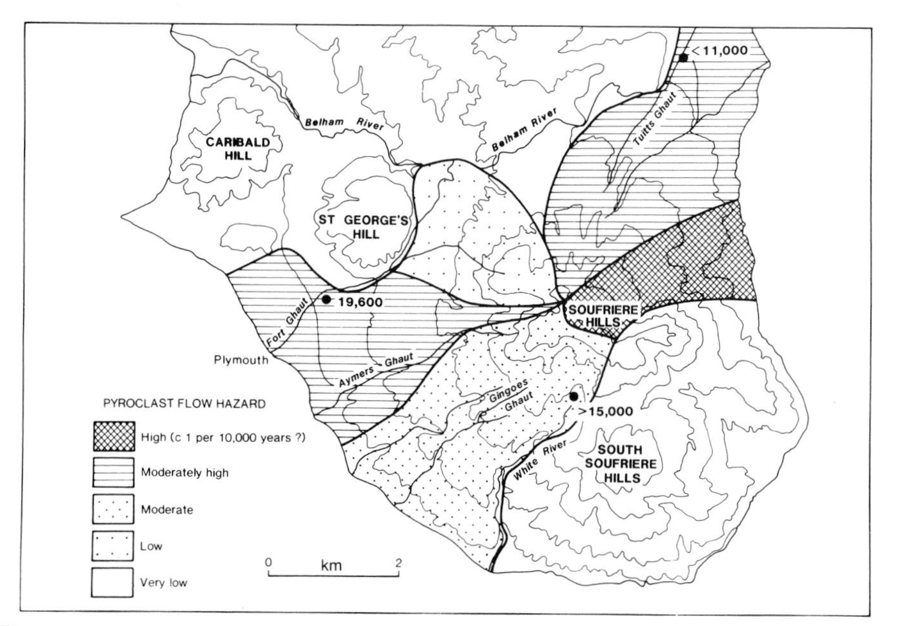

Fig. 7.10 Hazard map for southern Montserrat, showing the threat from pyroclastic flows. The numbers indicate the dated age of flows in years before present (from P.E. Baker 1985, Fig. 11, p. 293).

intervals with accuracy, but it is possible to identify those flanks of the hills which are most at risk and to conclude that the likely recurrence interval is of the order of once every 10,000 years.

The literature on hazard mapping in the Lesser Antilles, shows that much has been achieved through *ad hoc* studies by individuals and research groups with limited funds and research time. Other studies have been carried out under grants and consultancies awarded by outside agencies such as the Netherlands Geological Survey, National Science Foundation (USA) and the Natural Environment Research Council (UK) and this is typical of much general prediction in less economically developed countries. With respect to consultancy, Roobol and Smith (1989) highlight a difficulty. Experience, they argue 'suggests that ... reports have a tendency to be "lost" by island administrations or placed in "archival black holes". Thus the conscientious earth scientist may be placed in a moral dilemma if the government of the country concerned fails to follow up recommendations, and the report remains "lost", with the contents unavailable to the local administration and population' (Roobol and Smith 1989, p. 80; see also Wadge 1985). These authors go on to quote one case in which a consultancy report was covered by secrecy clauses and had not been acted upon some six years after its submission. It is the present author's experience that this is not a problem confined to economically less developed countries and that all contract research carries this danger. Poor countries do not have a monopoly on administrative secrecy. All governments have priorities and in poor countries remote, long-term volcanic hazards are often given a lower priority than attempts to improve living standards. Also if a population knew that a government had been advised of a threat and an eruption did occur, then this would involve much unpopularity—

even unrest—if priorities had ruled against the implementation of a contingency plan. Economically less developed countries are simply more extreme cases of a problem that applies with varying degrees in all states.

7.3.1.4 *Estimating eruption probabilities*

Although the research already reviewed in this chapter pays attention to the probability of certain types of eruption occurring with different frequencies, the focus has been on hazard mapping. Some writers, especially those concerned with general prediction on volcanoes which have a large number of individual eruptions and/or a long history of activity, have attempted to use statistical techniques to quantify the likelihood of an event occurring within a specified number of years. The quantification of hazards is by no means a new area of research and as early as the 1920s Imbò (1928), using earlier research by Riccò (1907), was seeking to find a pattern in the frequency of Mount Etna's flank eruptions. The greatest stimulus, however, to the application of statistical methods in estimating eruption probabilities has come from work by Wickman, who in a series of papers set out an approach which has been widely used by others (Wickman 1966a–e). Briefly, Wickman observed that on certain volcanoes eruptions are time-dependent, in other words the volcano is 'without memory' and the probability of eruptions will follow the *Poisson* distribution. This distribution is applied to data where the probability of an event is much less than the probability that it will not occur. Thus, the probability (P) of seeing (m) eruptions during an interval of time (t) is:

$$P(m,t)\ \frac{(t/a)^m \exp(-t/a)}{m!},\ (m = 0, 1, \ldots ; t>0)$$

Where (a) is the mean repose time and (P) may be considered as either a function of (t) or (m) (Klein 1982).

According to Wickman (1966a–c), volcanoes displaying this property of time-independent eruptions include: Etna (Italy); Kelut (Indonesia); Merapi (Indonesia); and Popocateptl (Mexico). Applying the model to Etna implies that there is a probability of 0.86 that a repose period will last one year, 0.21 that it will last for 10 years and only 0.04 that it will last for 20 years (Chester *et al.* 1985). The Poisson distribution has been applied to the historical records of eruptions at several volcanoes over the last 20 years (see Cruz-Reyna 1991 for a discussion of its applicability). Klein (1982) in a study of the Hawaiian volcanoes, first carried out careful checking to ensure that the data met the rigorous assumptions of the Poisson distribution (Davis 1973) and concluded that eruptions were random events because several causal processes were acting together. However, he also sounded a note of caution when he wrote that, 'large volume eruptions tend to be followed by longer reposes as shallow magma chambers refill. On Kilauea, both summit eruptions and rapid intrusions tend to cluster at times associated with other physical events on the volcano' (Klein 1982, p. 1).

In an important paper published in 1976, Wickman takes his analysis further and considers volcanoes that do have 'a memory' and do not conform to the Poisson distribution (Wickman 1976). Because of a lack of adequate historical records, his work is of a semi-quantitative nature and involves the application of *Markov chain models* (Davis 1973). He constructed simple models of different volcanoes where it was possible to consider eruptions as dissimilar activity states.

He then considered the probabilities of transition from one state to another (see Scandone 1983). Carta *et al.* (1981) applied this type of modelling to Vesuvius (Italy) and concluded that predictions of volcanic activity were possible if behaviour remained the same for long periods of time and transitions from one activity state to another were due to stochastic processes alone (see Scandone 1983).

Estimating eruption probabilities is an interesting research field and one in which progress is likely as volcanoes and their associated hazards are studied in more detail. Application of statistical modelling is difficult and some examples in the literature have excited controversy. One instance is an attempt by Mulargia *et al.* (1985) to calculate the probabilities of major flank eruptions on Etna, using extreme value (i.e. inferential) methods applied to a data set comprising the timings, durations and volumes of eruptions between 1605 and 1980. In the course of the analysis the authors had to check that data were random by examining the goodness of fit between them and the Poisson distribution. Once this had been achieved they were then able to proceed and make probability estimates of future events. In a subsequent comment their approach was criticized on two counts. First, the Poisson distribution assumes that there is no linear trend in the time series (i.e. it is stationary) and that events are independent of each other. This it was argued was not fully investigated by the authors. Secondly, the information used by Mulargia and his colleagues was incomplete, it contained data from more than one statistical population and was therefore biased (Chester 1986). A reply to these comments by the original authors was published (Mulargia *et al.* 1986). Regardless of the merits of this particular case, the substantive point remains that to be of value statistical modelling must be applied with care.

One interesting development has been to consider the risk, defined as 'the exposure of individuals to death and injury, and of structures to damage from volcanic hazards' (Newhall 1984, p. 1), over such shorter time periods as months or even weeks. In 1980 Mount St Helens started to erupt after a long period of inactivity and eruptions became frequent. Newhall (1982, 1984) calculated the probabilities of risk over short periods of time and found that for a given locality they were very variable. Quiet episodes after the initial eruption in 1980 had risks of a magnitude most would accept, while during active phases risk became unacceptable. Clearly this approach is of great value not only to the scientists who are required to monitor eruptions, but also to officials who have to decide when an area may be considered safe enough to allow public access.

7.3.1.5 *The impact of computer cartography and image processing*

The majority of general predictions are published maps and difficult to change; though as the Mount St Helens example shows they may be modified as an eruption progresses and more information becomes available. Procedures of computer data handling now allow the construction of models of future eruptions and their projection forward, so that their possible effects may be viewed in sequence. So far most research has focused on volcanoes which produce pyroclastic flows, surges and lahars. These three types of flow are controlled by gravity and in 1982 Michael Malin and Michael Sheridan introduced from the literature on landslide (see Hsu 1975), the concept of the 'energy line'. The 'energy line' is a device which allows energy losses away from an eruption site to be modelled. The acceleration (a) of the flow at any point (i) is given by the formula:

$$a_i = g[\sin(\beta) - \tan(\Theta)\cos(\beta)]$$

Where (g) is acceleration due to gravity, (β) is the slope of the land surface at point (i) and (Θ) is the angle between the horizontal and an energy line joining the top of the collapsing column and the end of the flow (Wadge and Isaacs 1988; and Chapter 5).

Clearly the formula may be modified to compute a three-dimensional view of energy losses (an 'energy cone') and the information required to do this includes: an estimate of possible column height; the topography around the volcano; the location of the eruption vent; and the angle the energy cone makes with the horizontal. Sheridan and Malin (1983), used this approach to generate computer-assisted hazard maps, which showed the effects of typical surge eruptions on three Italian volcanoes: Vulcano, Lipari and Vesuvius. More recently Wadge and Isaacs (1988) have applied it to the island of Montserrat and thereby demonstrated the potential application in economically less developed countries. The authors argue that the data required are simple, since only two parameters require estimation before the computer model may run, these being column height and energy-line angle. Topography around the vent can be retrieved from topographic maps and column heights and energy-line angles are easily estimated from previous eruptions in the Lesser Antilles and elsewhere. Alternative models with different values to show the effects of a variety of eruption scenarios may then be compared (Fig. 7.10). Using an image processor the authors not only showed the effects of some 32 eruption models, but were able also to project these forward in time so that the effects of eruptions could be assessed in advance and appropriate measures taken by the authorities.

Many of the problems of conventional hazard mapping are surmounted by this new approach, including the low spatial accuracy of boundaries and the sequencing of hazards over time, but caution is still required. It is a useful approach for those eruptions where the energy line is of relevance, but of less value—indeed may be totally irrelevant—for those volcanoes producing large volumes of, say, lava. Modelling is at its best when constrained by field evidence so that realistic values of column height may be used. Even without these data, probable values may be entered on the basis of similar eruptions elsewhere, but reliability is likely to be lower. Overall the value of these new approaches is high, since they are strongly orientated towards the day to day needs of the emergency planner. The fact that maps can be produced with even poor data and by using analogous eruptions from other areas, illustrates the future potential application to the requirements of poor countries. There is, however, a danger that computer displays may be over impressive and that disaster planners may use them uncritically, because they are unaware of the assumptions upon which they are based.

7.3.2 Specific prediction

As mentioned earlier in this chapter, specific prediction attempts to forecast eruptions over the short term (days to years) and is based on the surveillance and monitoring of changes in topographic, geophysical and geochemical parameters so that the time and place of an eruption may be determined. Before discussing specific prediction in detail, it is necessary to make two general points.

The first is to sound a note of caution and emphasize a problem of causal inference. This problem affects all forecasts of hazards, not just volcanic ones, and in the context of earthquakes the point has been well made by Bruce Bolt. Substituting the words in brackets for 'earthquake' and 'seismic' he writes, 'suppose that . . . measurements indicate that an [eruption] of a certain magnitude will occur in a certain area during a certain period of time. Now presumably this area is a [volcanic] one, or the study would not have been initiated in the first place. Therefore, it follows that by chance alone, the odds are not zero that an [eruption] will occur in the period suggested. Thus, if the [eruption] occurs it cannot be taken as decisive proof that the methods used to make the prediction are correct, and they may fail on future occasions. Of course, if a firm prediction is made and nothing happens, that must be taken as proof that the method is invalid' (Bolt 1988, p. 159). This is a critical point and one that should always be kept in mind when evaluating the claims of authors advocating a particular approach to specific prediction.

A second point is that specific prediction is even more costly than general prediction. Cost involves not only the capital charge of installing seismographic networks, carrying out tilt and ground-deformation studies, monitoring gases and measuring gravity changes, but also recurrent costs over many months and years. Additionally, highly trained and qualified personnel are required. Whereas in the case of general prediction much can be achieved through visits by expatriate research teams, many techniques of specific prediction require personnel to be in place for many months, or even years, before an eruption. It comes as no surprise to find that the worldwide distribution of volcanoes which are routinely monitored does not correspond to any realistic global rank order of risk. The majority of the world's most dangerous volcanoes are not monitored at all (Latter 1989a) and it is the economically more developed countries of the world that have the most comprehensive programmes of specific prediction (e.g. USA, Japan, the former Soviet Union, Italy, Iceland and New Zealand). Some techniques of specific prediction are capable of being carried out from aircraft and satellites and offer the prospect of improving the surveillance of volcanoes located in regions which are poor, remote or both.

7.3.2.1 Seismic monitoring

Seismographs record elastic waves which are produced by a number of processes related to volcanic activity. These include increased structural loading of the crust as the edifice grows, underground temperature changes, magma movements and gas flows (Tazieff 1983b). Although increased crustal loading is only linked indirectly to eruptions, the other processes subject rocks to mechanical stresses, vibrations and, ultimately, failure, producing elastic waves that can be recorded on seismographs (Decker and Decker 1981a). The fact that earthquake waves are generated by active volcanoes has been known for more than a century and as early as 1855 L. Palmieri recognized them at Vesuvius (Shimozuru *et al.* 1989), while the classic subdivision of earthquakes into three categories—collapse, dislocation (tectonic) and volcanic—dates from the 1890s (Hoernes 1893). Even earlier Aristotle speculated that winds within the Earth caused earthquakes and that volcanoes represented vents from which these were finally released, while in the *pre-uniformitarian* adolescence of geology as a science in the eighteenth and early nineteenth centuries, *Plutonists* believed that all earthquakes were of a volcanic origin, whilst *Neptunists* contended that earthquakes were caused by collapse (Schick 1981).

It is only in this century that earthquakes as precursory signs of impending volcanic activity have been studied in a systematic fashion and, according to Williams and McBirney (1979, p. 351), this 'outweighs all other methods in reliability and general utility'. Indeed, these authors quote from an unreferenced study (possibly Harlow 1971) to support their contention. They state that only 4% of a sample of documented volcanic eruptions occurred without any increase in earthquake activity, in 38% there was an increase without any eruption occurring and that in 58% there was a significant increase followed by an eruption.

In the 1950s and 1960s, three studies were published on seismic precursors which have become 'classics'. The first was the reporting by Professor T. Minakami and his colleagues of long-term investigations of Asama volcano (Japan). This study was based on over 2000 eruptions in the crater region between 1933 and 1961 and careful recording of the seismic events that preceded them (Minakami 1950, 1959a and b, 1960). Three types of volcanic earthquake were identified:

Type A with focal depths exceeding 2 km;
Type B which occurred at shallow depths beneath the central crater and were numerous both before and during eruptions; and
Explosion earthquakes located practically at the surface.

In the case of Type A both P and S seismic waves could be recognized, whereas in Type B and the explosion earthquakes, the shallow depths meant that P and S phases were not distinguishable. As Fig. 7.11 shows, this study also found a relationship between the number and amplitude of shallow earthquakes and the kinetic energy of the subsequent eruption.

The second study was a report of the seismic monitoring of Kilauea volcano (Hawaii) before the 1959 eruption (Eaton and Murata 1960). Over a period of several months earthquakes were noted first at depths of 50–60 km, but as time passed their depths of focus decreased, they increased in number and became localized beneath what was to become the site of eruption. Williams and McBirney (1979) in reviewing this study, make the point that the pattern is atypical of eruptions in Hawaii, since earthquakes normally begin at much shallower depths.

A third study was by Blot and Priam (1964) who suggested that in the New Hebrides deep earthquakes (focal depths of more than 350 km) were associated with eruptions several months later, and again confirmed the relationship between the depth of focus and the time remaining before an eruption, and between the magnitude of earthquakes and the eventual size of an eruption.

Since the early 1960s additional studies have been published which have in some ways improved knowledge, but in others have made the picture more confused. One improvement has been the recognition of a further type of characteristic seismic signature—the so-called harmonic or volcanic 'tremor'. This is a unique seismic phenomenon and comprises a more or less continuous oscillation of the ground with a frequency range of around 0.5–10 Hertz. It is probably generated by a number of processes including the formation of gas bubbles and their collapse, by turbulence in the magma column, by frictional forces between the ascending magma and the conduit, and by movement among layers in the magma (Shimozuru *et al.* 1969; Decker 1973) and, whatever its cause, may provide useful short-term predictions because it often precedes

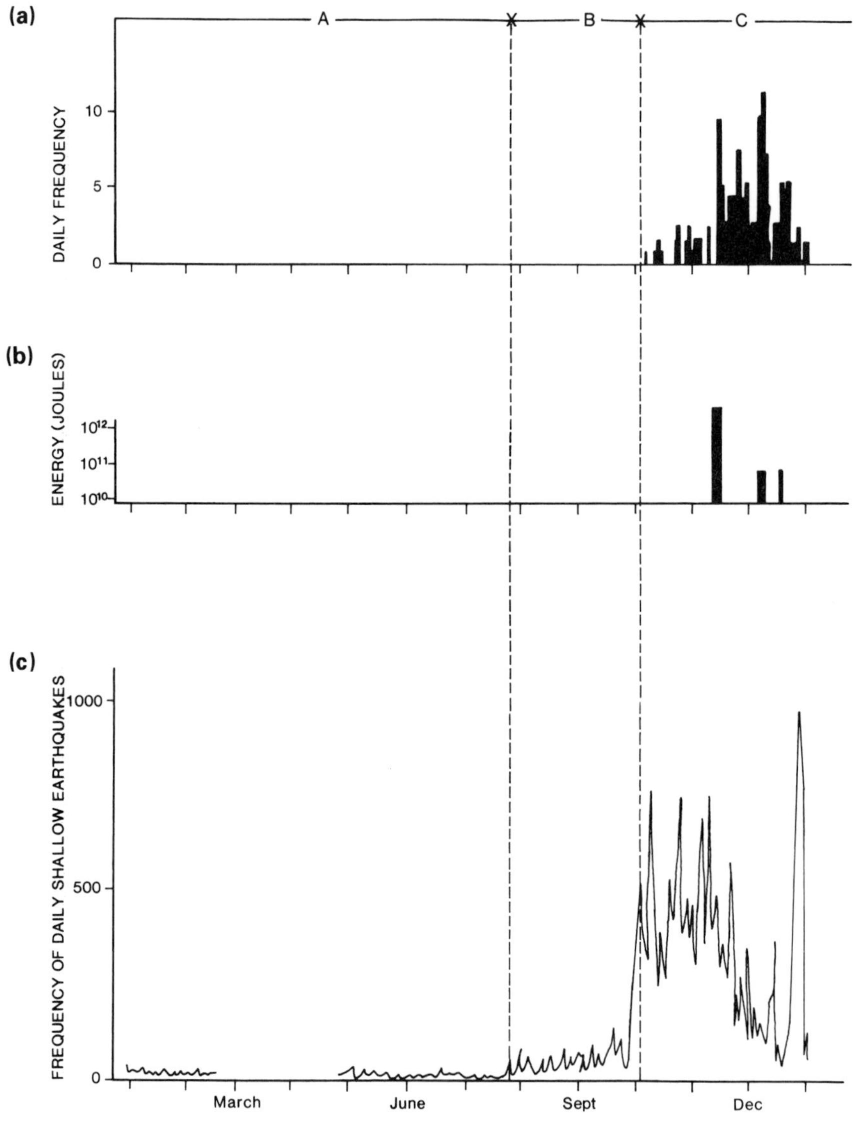

Fig. 7.11 Volcanic and seismic activity at Asama volcano, Japan, in 1958. Period A represents the inactive stage; B the pre-eruptive stage; and C the stage of active eruptions. (a) Shows the daily frequency of eruptions; (b) the kinetic energy (Joules); and (c) the daily frequency of shallow earthquakes (modified from Williams and McBirney 1979, Fig. 15.1, p. 353, after Minakami 1959a, 1959b, 1960).

eruptions by intervals ranging from hours to days. In Hawaii, high-amplitude volcanic tremor may be the best short-term indication that an eruption is about to begin and many houses occupied by the staff of the Hawaiian Volcano Observatory have tremor alarms fitted (Decker and Decker 1981a). As with all forms of volcanic earthquake, not all cases of volcanic tremor (even in Hawaii) are followed by eruptions (Decker and Decker 1988, 1989).

On the more negative side, many studies have shown that the relationships between seismic events and volcanic eruptions are more complex than was thought in the 1960s. Several examples will suffice to illustrate this point. First, there are many recorded instances in which major seismic 'crises' have not been followed by eruptions. Between 1982 and 1984 in the Campi Flegrei volcanic area northwest of Naples (Italy), intense seismic activity was accompanied by ground inflation in early 1983 of between 1 and 5 mm per day (Barberi *et al.* 1984; Duncan *et al.* 1986; Martini 1989), while similar enhanced seismicity was recorded adjacent to Rabaul volcano (Papua New Guinea) from 1983/85 (Mori *et al.* 1989). In neither case did an eruption follow, the conclusion being that seismic activity and ground inflation were related to deformation caused by localized magma intrusions, which did not reach the surface. Secondly, further work on volcanic tremor has shown that it is a far more complex phenomenon than thought previously. As mentioned above, in the case of basaltic volcanoes like Hawaii and Etna (Chester *et al.* 1985), the cause of tremor is thought to relate to the flowage of liquid magma, yet it was noted before the eruption of more viscous dacitic magma at Mount St Helens (Tazieff 1983b) and work at the Aso Volcanological Observatory (Japan) has allowed at least four classes of tremor to be recognized (Kubotera 1974). A third complexity is that increased knowledge has shown that individual volcanoes may have unique seismic signatures and that, although some general rules apply, it is difficult to make firm statements which are applicable across a range of volcanoes in different eruptive environments. Research on Pavlov volcano (Alaska) by McNutt (1989), shows that virtually no Type A earthquakes were recorded before eruptions, while more recent data from Asama volcano, Japan (Shimozuri and Kagiyama 1989), implies that some of the statements made by Minakami do not hold for more recent events. Following 1961, Asama entered an 11-year episode of quiescence and the next eruption was not preceded by any seismic activity. More recent eruptions have had seismic precursors, but these differ from those recorded by Minakami and cannot be applied, necessarily, to other volcanoes. A fourth point is that the relationships, between deep earthquakes on the one hand and volcanic eruptions (and volcanic earthquakes) on the other remain unclear. Following the pioneer work of Blot and Priam (1964), several studies of deep earthquakes along subduction zones and their relationships to both actual eruptions and more shallow focused volcanic earthquakes were carried out (e.g. Blot 1965; Latter 1971; Carr and Stoiber 1973). In a later study Blot (1981) summarizes his views in the context of a research project carried out on White Island volcano (New Zealand), in which he argues that in island arc and continental plate tectonic situations (see Chapter 2, section 2.5) two processes occur which give rise to seismic waves: a tectonic process—caused by the migration of strain released along the descending lithospheric plate—so producing deep focus earthquakes; and a magmatic process—in which there is a relatively fast ascent of magma—that causes volcanic earthquakes. Reviewing the question of deep earthquake precursors in the Lesser Antilles, Shephard (1989) concludes that, although there is a correlation between tectonic earthquakes and volcanic earthquakes (and hence eruptions), the reasons for this remain unclear and that there is no obvious physical mechanism to link them. He also notes that the correlation only holds for those volcanoes of the northern Lesser Antilles and not for those of the southern island group. Using an old idea of Robson and Barr (1963), he puts forward the tentative suggestion that this may be related to different regimes of stress.

The last paragraph gives both a flavour of the problems and a view of the research frontier in the field of volcanic seismology. In addition, research also includes investigations aimed at determining the size and characteristics of magma chambers (e.g. Sharp *et al.* 1980), and studies of earthquake-source mechanisms (e.g. Schick 1981). However, in the context of the present volume the words of Tazieff (1983b, p. 166) still remain true that, although 'seismic ... information is the most precious data for forecasting; it is absolutely insufficient to base a prediction on with safety'. It is also true that the conclusion of Williams and McBirney (1979, p. 352) retains its validity and force. When monitoring any potentially active volcano 'the most dependable evidence of a possible eruption is simply a marked increase, or in some cases a decrease, in the number of small shallow earthquakes in the vicinity of the vent. A change that is less than tenfold may not be significant, but a hundredfold change has a high probability of being a forerunner of an eruption'.

For volcanoes that are considered dormant and/or lie within economically less developed countries much of this discussion is theoretical, since the provision of even a rudimentary seismic network may be too costly both financially and in terms of skilled personnel. Nevertheless, there are seismographs available which are relatively cheap and may be left unattended for long periods. There is scope, as the experience of the United States Geological Survey shows, to link these instruments by satellite to a central monitoring facility (Williams and McBirney 1979; J. V. Smith 1985). Any unusual seismic signals can then alert scientists to the need for more detailed investigations into the possibilities of eruption. Recently at the Cascades Volcano Observatory in the USA a new computer program has been installed called, Raw Real-time Seismic Amplitude Measurement (RSAM) (Endo and Murray 1991). The system processes seismic signals, provides average amplitude information and is able to differentiate between changes in activity that are not precursory (e.g. weather, instrument problems) and those that are caused by dome building. Testing this system against actual events which occurred at Mount St Helens in 1985 and 1986, suggests that it may in future be a valuable procedure for seismic prediction.

Because tectonic earthquakes and volcanic eruptions have similar global distributions, it is not surprising that seismographs located on and near to active volcanoes record seismic waves originating from tectonic as well as volcanic causes. This is one of the principal practical difficulties involved in applying seismology to specific prediction. It should not be forgotten that a tectonic earthquake can trigger an eruption if one is imminent. Such a situation occurred in 1960 when Puhuehuen volcano erupted following the large Chilean earthquake, while in 1980 the main phase of the Mount St Helens eruption was triggered on 18 May by a magnitude 5.1 earthquake which shook the volcano, causing slope failure and bringing about the unroofing of the magma chamber (Foxworthy and Hill 1982). In many volcanic areas there is a persistent background of seismic activity, but it is change in the scale and character of the signature which is of importance.

Swanson and Casadevall (1983, p. 1423) have asked the intriguing question: 'Do eruptions trigger earthquakes, or *vice versa*?' They review several papers including Acharya (1981, 1982) and conclude that there is a direct relationship between the state of volcanic activity and the next large earthquake in a region of subduction. They also note that following the Mount St Helens eruption in 1980, seismic activity increased along a previously unknown fault to the north of the volcano. Only time will tell whether these relationships will be confirmed by further studies, so allowing the prediction of tectonic earthquakes from volcanic activity and eruptions from earthquakes.

7.3.2.2 Ground deformation

Although superficially a simple technique which should work well in predicting the timing of eruptions, and despite much optimism when the first long-term studies of deformation and eruptions were published, this approach to specific prediction is not without its problems. It appears to be a matter of common sense that as magma moves towards the surface in the period leading up to an eruption, this will be expressed by changes in surface elevation. As magma is fed into a reservoir, internal pressure will increase and layers of lava and pyroclastic materials will be displaced, causing a swelling or inflation at the surface. As Tilling (undated) notes, the net effects should be: slope steepening; increases in horizontal and vertical distances between points at the surface; and fracturing (producing pre-eruption earthquakes). Conversely following an eruption the reservoir should decrease in volume, causing a reduction in slope-angles, a shortening of distances at the surface and a fall in the number of volcanic earthquakes. Over the last 70 years studies of several volcanoes have confirmed changes of this type. The first author to recognize such changes was Omari (1911, 1920) and in the case of silicic volcanoes, there are many examples in the literature where large ground inflations have been noted some, but not all of which, have been followed by eruptions. A classic case is an inflation in 1910 of more than 150 m at Usu volcano (Japan) (Imamura 1930), while a more recent instance is the distinct bulge that developed on the north flank of Mount St Helens immediately before the eruption of 1980 (Christiansen and Peterson 1981). This bulge was first recognized visually and only later was it confirmed by measurement. The Hawaiian volcanoes, especially Kilauea, have been monitored since 1916 when Professor Thomas Jaggar first began measurements of ground tilt using a pendulum (Jaggar and Finch 1929) and over the years a general picture of deformation before, during and after eruptions has emerged (Fig. 7.12). This research is well reviewed by Fiske and Kinoshita (1969) and by Kinoshita *et al.* (1974) and, as Fig. 7.13 shows, has largely confirmed the common-sense impressions noted above.

From the discussion so far, it might be assumed that ground deformation has now achieved the status of a reliable prediction device. However, this impression is misleading because further studies of the Hawaiian and other volcanoes have shown that the relationships between magma movements and eruptions on the one hand and ground deformations on the other are far more complex. In the first place, if these correlations hold on Hawaii, then it should be possible to derive a simple model to link the size and shape of the magma reservoirs (intrusions) to the amount of vertical and horizontal displacement, by means of geometrical relationships. In the 1950s the Japanese geophysicist Mogi (1958) put forward such a model which related the vertical and horizontal displacement at the surface (at different radial distances from the 'epicentre' of uplift) to enhanced pressure of the magmatic intrusion at depth. His model, however, makes three assumptions all of which are open to question. The model assumes that the magma reservoir is small and spherical; that the Earth's crust is a semi-elastic body; and that the radius of the reservoir is small in relation to its depth. In the context of Kilauea, Fiske and Kinoshita (1969) found that surface layers do not always behave in a

Fig. 7.12 A scientist from the Hawaiian Volcano Observatory, makes measurements of horizontal distance with an electronic laser beam instrument known as a 'geodimeter'. On the Hawaiian volcanoes and several others, changes in ground elevation are being used successfully to predict eruptions (photograph reproduced by courtesy of the United States Geological Survey, ref. HVO 196).

perfect semi-elastic way and non-elastic deformation occurs before eruptions, while Dieterich and Decker (1975) showed that surface uplift may be the same for intrusions of differing shapes and depths, and that if horizontal displacements are also considered then different geometrical solutions are possible. For Kilauea they argue that the data imply a clustering of cylindrical bodies, which reach towards ~1 km of the summit; thus better explaining ground inflations and deflations. Nevertheless, the Mogi model has been reasonably successful in Hawaii in identifying the shallow storage of magma and the response at the surface to changes in reservoir size.

Not all volcanoes have shallow magma reservoirs and not all of them are of small radii in comparison with their depth below the surface. In these situations ground inflations and deflations do not describe such a simple pattern. On Etna, for instance, three research teams have monitored changes in elevation. The first study by Geoffrey Wadge (1976) and colleagues from Imperial College, London, used a series of benchmarks—one concentric around the summit, the other concentric and near to the base—and carried out electro-optical distance measurements between 1971 and 1974. In 1975 a team from University College, London, established an 11 km traverse across the summit which was levelled at regular intervals until the 1980s (Murray and Guest 1982). It is clear that Etna contrasts with the situation on Hawaii, the implication being that its 'plumbing' is different (Chester *et al.* 1985). As Table 7.6 shows, the evidence supported by geophysical studies implies no significant high-level storage of magma (unlike

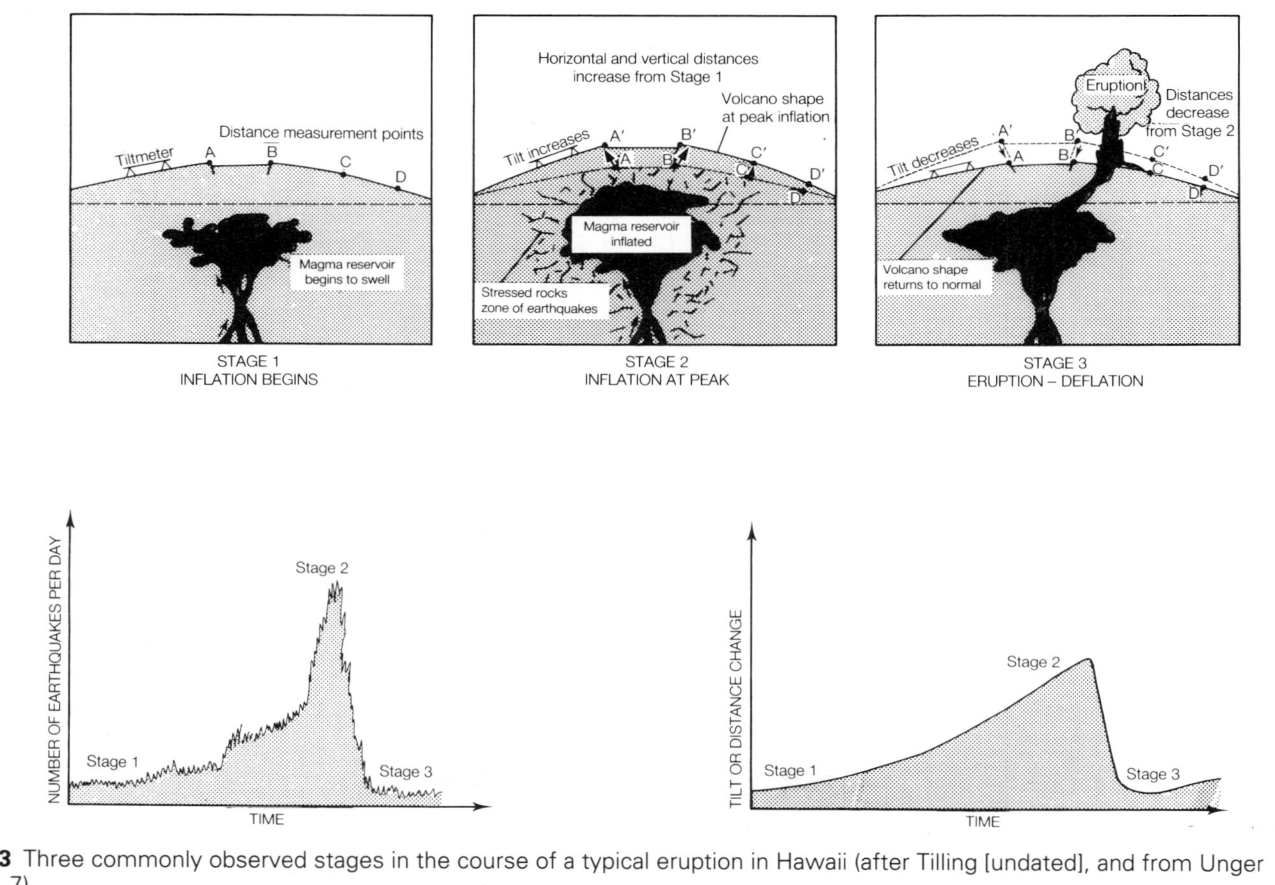

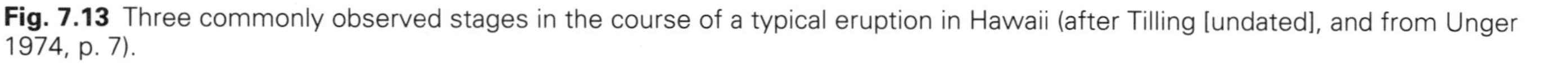

Fig. 7.13 Three commonly observed stages in the course of a typical eruption in Hawaii (after Tilling [undated], and from Unger 1974, p. 7).

Table 7.6: Conclusions of ground-deformation and seismic studies from Mount Etna (based on Chester *et al.* 1985, and several other sources)

Conclusions	Comments
1	During the period of investigation there was no large-scale storage area for magma within the volcanic construct.
2	That some small-scale storage of magma may have occurred in shallow dykes for as long as two years before an eruption.
3	The volume of this stored material is difficult to determine on the basis of ground-deformation studies.
4	If ground deformation is caused by magma pressure, then the amount of storage is small, but if a dyke is emplaced passively because of tectonic dilation of fissures, then much larger amounts of magma could be emplaced in the dyke without there being much vertical deformation.
5	Seismic studies imply that magma is stored at around 20 km depth.

Kilauea) and the relationships between magma movements and eruptions on the one hand and ground inflations and deflations on the other are very complex. A slightly different approach has been used by McGuire *et al.* (1991) and has been concerned not so much with the use of ground deformation as means of predicting primary eruptions, but rather to link deformation with increased slope instability and the generation of failures and debris avalanches.

So far only basaltic volcanoes have been considered but, as Decker and Decker (1989, p. 233) note, when the technique is applied to subduction zone volcanoes 'one problem . . . is their long repose time between eruptions. It could take several hundred years to learn as much about deformation at Vesuvius as has already been learned at Kilauea in the last 30 years'. Nevertheless this has not prevented several research teams attempting to monitor ground deformation and its implications. Two studies have already been mentioned: the cases of Mount St Helens and Usu volcano, Japan; and other examples include Long Valley caldera, California (Decker 1986), Lake Taupo volcano, New Zealand (Otway 1989), the Campi Flegrei volcanic area, Italy (Bianchi *et al.* 1984), and Soufriere of St Vincent (Fiske and Shepherd 1990). However, the majority of these studies are long term and do not at the moment constitute a reliable method of prediction. In the cases of Mount St Helens and Campi Flegrei, they are reactions to emergencies which were already apparent from other precursory signs.

At present studies of ground deformation constitute a very useful predictive device on well-studied volcanoes, which have frequent eruptions and simple relationships between magma movement and eruption. So far this statement applied only to the Hawaiian volcanoes and particularly Kilauea. There has been, nevertheless, some reported success in linking ground deformation and seismic precursory signs. Thatcher (1990) reports research in Japan, in which a satellite with ground-survey sensors (Global Positioning System or GPS) was used in conjunction with recorded seismic activity successfully to monitor a small submarine eruption. For many volcanoes research into ground deformation is still at the preliminary stage and does not at the moment provide a method of prediction which can be used with confidence.

7.3.2.3 Geochemical techniques

Geochemical techniques of specific prediction rely predominantly upon the monitoring of gases which are normally the first phases to reach the surface both before and during eruptions. Monitoring relies on the surveillance of changes in the volume and composition of gases as magma approaches the surface. The theory is simple, but in practice difficulties arise because volcanic gases are often collected in hostile field environments and their contamination is highly likely (Chapter 6, section 6.2). In addition as Gerlach (1983) points out, there is a problem that secondary alteration may occur before gases reach the point of collection. This can include contamination by organic matter and water and reactions with sampling devices. It is now common practice to apply corrections to raw data and produce what are known as 'restored analyses' so that they more faithfully represent original gaseous compositions. Gerlach (1979) reports one extreme cases in which a 'restored analysis' of gases collected on Etna showed only small amounts of H_2, CO and H_2S, whereas the original report implied much higher concentrations. This was due to contamination of the gases by the metallic sampling bottles. Clearly any calibration has to be based on a number of assumptions and this together with variations in the techniques used has meant that, although the geochemical monitoring of fumaroles and hot springs has a high potential as a predictive approach, so far results have been somewhat conflicting (Tazieff 1983b; Chester *et al.* 1985).

In the 1960s and 1970s, research on a limited number of volcanoes gave some indication that general patterns of change in gas species did occur before eruptions. Work by Stoiber on fumaroles located on active volcanoes in Central America, for instance, indicated that many eruptions were presaged by an increase in the ratio of SO_4/Cl (Stoiber and Rose 1969a, 1970; Stoiber *et al.* 1971) and similar results were found by Russian scientists researching the volcanoes of Kamchatka and the Kurile Islands (Menyailov 1975). Later writers have reported a more confused picture. On Mount St Helens in 1980 the CO_2 to SO_2 ratio in the plume dropped from 15 to 1 immediately before eruption (Harris *et al.* 1980), on Kilauea Thomas and Naughton (1979) reported significant changes in the He/CO_2 ratio and during the massive ground inflation in Campi Flegrei, Italy (1983) a very complex picture of gaseous variations was noted (Cioni *et al.* 1989).

Premonitory changes in gas composition often characterize magma movements, whether or not these result in actual eruptions. However, general principles which apply across a range of volcanoes have yet to emerge. One innovation has been the remote sensing of volcanic plumes and several techniques have been developed (Souther *et al.* 1984), the most frequently used being correlation spectrometry or COSPEC. This technique measures ultraviolet absorption and can be employed to monitor SO_2 emissions, while another technique known as MIRAN is concerned with infrared radiation and both are now used on a routine basis to monitor gas emissions before and during eruptions. Examples of the use of COSPEC include: Etna and Stromboli in Italy (Stoiber *et al.* 1978); Mount St Helens, USA (Stoiber *et al.* 1980) and several volcanoes in Japan and Hawaii (Ota *et al.* 1979). One important feature of both approaches, is that they offer the prospect of identifying suspicious changes in gas characteristics on volcanoes in remote and/or poor regions of the world. An excellent review of the potential usefulness of COSPEC is to be found in Stoiber *et al.* (1983).

An exciting approach to the study of volcanic gases may be seen in recent research by Baubron *et al.* (1991). His team has studied gases emanating from soils in the vicinities of several active volcanoes in dissimilar tectonic environments including Italy, the Lesser Antilles and Indonesia. The authors' find correlations between emissions of CO_2 and certain rare gases on the one hand and fumarolic activity and magma degassing at depth on the other. They suggest that this approach may have potential as a method of specific prediction (see also Toutain *et al.* 1992).

7.3.2.4 Thermal monitoring

Volcanoes are localized sources of high geothermal heat flux and as magma moves towards the surface, it might be expected that this would cause a detectable increase in temperature. Early research was not encouraging and after many years spent observing volcanoes in Indonesia, Neumann van Padang (1963a) came to the conclusion that there was not a strong relationship between temperature changes and eruptions. Later Alcarez (1969), working on the temperatures of the crater lake of Taal volcano in the Philippines, produced findings that were equivocal (Fig. 7.14). The September 1965 eruption was preceded by a significant positive anomaly in water temperatures, that of 1966 by

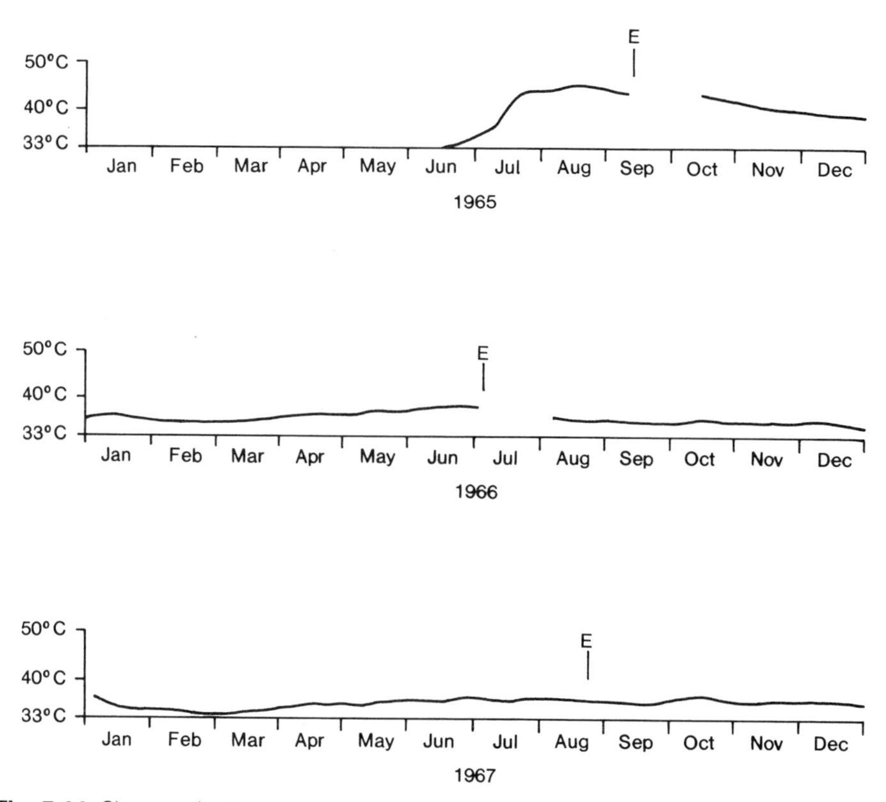

Fig. 7.14 Changes in the temperature of the crater lake on Taal volcano, Philippines, before three eruptions (from Decker 1973, Fig. 16, p. 388; after Alcarez 1969).

a much smaller anomaly, whilst the eruption of 1967 showed no precursory temperature changes. As Decker (1973) notes in reviewing the Alcarez paper, the usefulness of temperature measurements as precursory indications of volcanic activity is complicated by two factors. The first is rainfall (and by extension groundwater), which can obscure the small temperature changes due to increased geothermal heat flux. The second is thermal inertia. Heat conduction, he argues, is very time-dependent and many changes in temperature may either be very shallow in origin or relate to past rather than future eruptions. Also the detailed hydrological conditions in the vicinities of most volcanoes are not known.

Further points are made in a review by Francis (1979). The conductivity of rocks is so low he argues, that a magma body may approach the ground surface at a rate faster than the heat can be conducted from it to produce a measurable rise in temperature. This is more likely with low viscosity basaltic melts which move rapidly, than with slower moving andesitic/rhyolitic magmas. By implication, the direct detection of thermal anomalies associated with basaltic volcanism is difficult if not impossible. Regardless of chemical composition, circulating ground water may act as an effective method of heat transfer and this may well be reflected not only in water temperature changes, but also in the appearance of new fumaroles and increases in temperature at existing ones. The only problem is that fumarole temperatures may also vary because of other factors such as local rainfall and there are instances where eruptions have occurred without any noticeable changes in fumarole temperatures. Finally, he observes that there are relatively few data on the long-term thermal state of volcanoes in periods between eruptions and goes on to suggest that one way round this would be to carry out infrared surveys by remote sensing, so that sequential thermal maps could be compiled. Remote sensing has been carried out on the Island of Vulcano in the Aeolian Islands (Italy) and Brivio *et al.* (1989) report a number of thermal measurements made since 1970. The culmination was an airborne infrared survey over the main cone of La Fossa volcano. Results are not encouraging and highlight the great difficulties involved in constructing temperature maps from aerial thermographs. Brivio and his colleagues found that discrepancies occurred between temperature values obtained from the air and from parallel ground surveys, implying that far more research will have to be carried out before airborne temperatures surveys become reliable methods of specific prediction (see also Mouginis-Mark *et al.* 1989; Alexander 1991; Oppenheimer and Rothery 1991).

Clearly changes in temperature do precede some eruptions and there are well-documented cases of this being observed (e.g. Tokati volcano, Japan; and Fig. 7.15), but many volcanoes have erupted without any thermal changes being detected. At present rather vague rules of thumb have to be used and any major changes in the temperatures of fumaroles, ground water, crater lakes or in remotely sensed geothermal heat fluxes should be considered possible warnings that an eruption may occur in the near future.

7.3.2.5 Other techniques

Gravimetric, geoelectrical and geomagnetic changes are all likely to accompany the movement of magma into shallow depths within the Earth's crust. Monitoring these changes provides a potentially valuable set of new approaches to specific prediction (see Brown *et al.* 1991). Several authors (e.g. Tazieff 1983b; Decker and Decker 1989) have pointed out that, whilst these look promising areas of

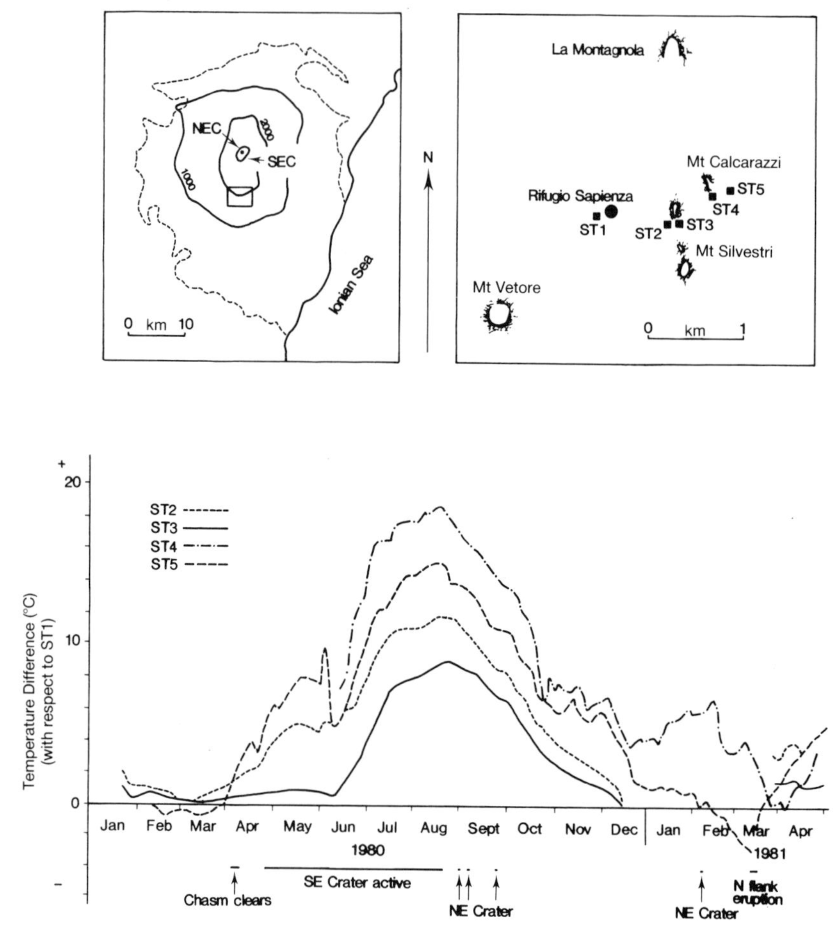

Fig. 7.15 Changes between January 1980 and April 1981 in the relative ground temperature (120 cm depth) measured on Etna's south flank at about 1900 m asl. Note the situation is complex, temperature increases do coincide with southeast (SE) crater activity, but precursory signals (in early April) are not very striking. However, some of the variation does seem to be related to eruptive activity, with heat being transferred from new magma to the monitoring area by the convention of ground water and possibly escaping juvenile gases (based on data from Archambault *et al.* undated; and from Chester *et al.* 1985, Fig. 8.4, p. 304).

research, they are relatively new and even today the number of published studies is minimal and of restricted scope. A full review is outside the scope of this volume, but those readers who are interested should refer to excellent summaries by Halwacks (1983), Robach (1983) and McKee *et al.* (1989). Some progress has been made at Poas volcano, Costa Rica. Over several years a team from the Open University in the United Kingdom has been monitoring gravity. Between 1979 and 1985 no significant changes occurred, but from 1985 to 1989 measurements were interpreted as indicating an increased mass of magma. An eruption occurred in 1989 (Rymer 1992).

A good example of lateral thinking about the problem of specific prediction and a new technique is to be found in a challenging paper by Voight (1988a). He argues that mathematical relationships related to the failure of materials can be applied, by analogy, to the precursory behaviour of a volcano before it erupts. The idea that rock failure accompanies eruptions is a truism, but Voight extends this by arguing that the process of eruption is merely a special case of material failure and can, therefore, be modelled using the same physical laws. By means of a detailed mathematical treatment, Voight derives several equations which incorporate precursory signs of activity—such as ground deformation, seismicity and gaseous emissions—and uses these to 'predict' a number of past eruptions with success. In reviewing Voight's paper Robert Tilling (1988) makes the important point that, whilst this approach to specific prediction is very valuable, it still requires testing more widely across a range of eruptions. Only then can its reliability and usefulness be evaluated fully.

7.4 Concluding remarks

In this chapter a case for the prediction of volcanic hazards has been made, the nature of the hazards outlined and the general and specific approaches to prediction reviewed. Over the last two decades considerable progress has been made, but advance has not occurred at the same rate in all areas. On the positive side, there is now a good knowledge of the hazards communities face in volcanic regions and there are techniques of mapping which allow the advantages of location in volcanic hazards to be maximized, yet at the same time highlight those areas which are likely to suffer damage in the event of an eruption. Specific prediction has been notable for its early optimism which has been tempered by the realization that different volcanoes operate in different ways and that no one approach, or set of techniques, can be applied universally. It is in the field of specific prediction that most research is currently focused and this concentration is likely to increase. It is also probable that the impact of satellite-based remote sensing will increasingly be felt (Alexander 1991) and the Earth Observation System (EOS) is due to be launched before the end of the 1990s. During its 11-year mission it is scheduled to carry out ultraviolet, visible, infrared and microwave sensing. In addition an Orbiting Volcano Observatory (OVO) will be able to chart gas emissions, changes in surface form and thermal anomalies. Finally, laser surveying via and/or from satellites may greatly assist ground-deformation studies, by means of the Global Positioning System, which has a resolution of around 2 cm (Walter 1990). All this promises much for the future, but a note of caution needs to be sounded. Specific prediction has had many false dawns and it should not be forgotten that even on the most intensively studied volcanoes, prediction remains a very inexact science. Only time will tell if my caution is justified, or else whether information technology will in fact answer many of the questions raised in this chapter.

8

Human responses and adjustments to volcanic hazards: a framework for analysis

Natural hazards only exist in relation to human activity aspirations and needs . . . (J. Whittow 1987, p. 311).

8.1 Introduction

In the introductory chapter of his important book, *The Human Impact: Man's Role in Environmental Change*, Andrew Goudie argues that 'in the history of Western thought three basic questions have been posed concerning the relationship of man (*sic*) to the habitable earth. The first of these is whether the earth ... is a purposefully made creation .... The second question is whether the climates, relief and configuration of the continents have influenced both the moral, and social nature of individuals and the character and nature of human activities. The third question ... seeks to find out whether, and to what degree, man (*sic*) has during his long tenure of the earth changed it from its hypothetical pristine condition' (Goudie 1981b, p. 1). The study of natural hazards involves aspects of the second and third questions. With regards to the second, the ways in which different societies and cultures respond and adjust to actual and perceived natural hazards is variable and will clearly have implications for the formulation of policies aimed at reducing losses. The third question is also relevant for, as argued in Chapter 7, over time the world has become more hazardous as populations have placed themselves at risk through both informed and uninformed decisions to settle in areas where extreme events of nature may cause losses.

Studying the interface between the natural environment and human activities has not proved easy and much of the history of geography has consisted of successful and unsuccessful attempts to devise frameworks in which the theme may be brought into focus (Johnston 1979, 1989; Stoddart 1987). As far as research into natural hazards is concerned authors have adopted two approaches: the *dominant*, and the *radical* (Hewitt 1983a; Whittow 1987; Palm 1990). These approaches are not dichotomous, but complementary and strands from both may be seen in the work of most authors.

8.2 The dominant approach

As its name implies this is the most commonly used way of studying natural hazards. The dominant approach has been reviewed extensively and it is not the purpose of this chapter to repeat this, but rather to draw together the essential features as they apply to volcanic hazards. Further information may be gleaned from the following works: White (1973); Burton *et al.* (1978); Warrick (1979); Blong (1984); Chester *et al.* (1985); Whittow (1987); Palm (1990); Bryant (1991); and K. Smith (1992).

The dominant approach was developed from ideas first discussed by H. H. Barrows in 1923 and which he called 'human ecology'. Human ecology is concerned with people 'reacting and adjusting to environments while at the same time attempting to adjust the environment to (their) own needs' (Johnston 1979, p. 39) and Barrows' views were later taken up and applied to natural hazards by Gilbert White, when he considered a range of adjustments which could be innovated to reduce the losses caused by floods in the USA (White 1942). Following the Second World War, White was joined by like-minded colleagues who included Robert Kates, Ian Burton and Kenneth Hewitt. They developed their research into a set of general techniques not only for the study of flooding, but also for all natural hazards including those produced by volcanic eruptions (see Warrick 1979). Social scientists using these approaches were soon referred to as the 'Chicago School', because several worked with Gilbert White at the University of Chicago during the 1950s and 1960s. Some idea of just how dominant the approach became may be judged from the large number of journals which proselytized in its favour and by the fact that many governments established agencies to coordinate planning in disaster-prone regions using many of its precepts. Indeed, in 1972 the United Nations instituted a Disaster Relief Office and in December 1987 designated the 1990s the International Decade for Natural Disaster Reduction (Lechat 1990).

The dominant approach has at its heart five objectives, which are summarized in Table 8.1. As Whittow (1987, p. 308) remarks there is probably a sixth, though it is implicit in the other five, this being to: 'evaluate the hazard's dimensions in order to predict the degree of impact and spatial dimension of the risk zone' (see Chapter 7). The extent and nature of human occupancy (Objective 1) is the foundation upon which the approach is based. It recognizes that not only

Table 8.1: The research objectives of dominant paradigm (based on White 1974, p. 4, and with further information from other sources, especially Mitchell 1989, pp. 175–6)

Objectives
1) To estimate the extent and nature of human occupancy in areas subject to extreme natural events.
2) To determine the range of possible adjustments by social groups to these extreme events.
3) To examine how people perceive extreme events and the hazards resulting from them.
4) To examine the processes by which damage-reducing adjustments are chosen.
5) To estimate the effects of varying public policy upon the set of responses.

Table 8.2: The theoretical range of adjustments to hazards from lava flows (based on Burton *et al.* 1968; and from Chester *et al.* 1985, Table, 9.1, pp. 340–1)

Affect the cause	Modify the hazard	Modify the loss potential	Adjust to losses		
			Spread the losses	Plan for losses	Bear the losses
Types of adjustment					
No known way of altering the eruptive mechanism	1) Protect high value installations	1) Introduce warning systems	1) Public relief from national and local government	Individual family or company insurance	Individual family, company or community loss-sharing
	2) Alter lava flow direction	2) Prepare for a disaster through civil-defence measures	2) Government sponsored and supported insurance schemes		
	3) Arrest forward motion	3) Introduce land-planning measures to control future development in particularly hazard-prone areas	3) International relief from agencies such as the United Nations Disaster Relief Office		
Examples and notes					
	(a) Use of explosives and bombing to divert flows; has been tried in Hawaii and on Etna	(b) Warning systems only available on certain well-monitored volcanoes, in technologically	(c) Public relief available in most countries; the most comprehensive schemes are in the technologically most	(d) Possible to a certain extent in more developed countries, but even in the USA it is limited by the discretion of	This is the traditional form of adjustment and is still widely practised in many volcanic areas

	Emergency barriers tried in Hawaii, Japan and Etna	advanced countries, e.g. USA (Hawaii and volcanoes showing signs of activity in the continental USA), Japan, Iceland and Italy	developed countries, e.g. Canada, USA, Japan and New Zealand	individual companies
	Barriers to divert future flows from inhabited areas; have been suggested for the town of Hilo (Hawaii)	Emergency evacuation plans have been formulated in several countries, e.g. USA, Japan and Soufrìere de Guadeloupe	Government sponsored insurance schemes available in several countries, e.g. former USSR and New Zealand	
	Control forward advance by watering the flow margin; limited success in Hawaii and Heimaey (Iceland)	Land-planning policies are in operation in some areas where 'general prediction' and hazard mapping have been carried out (see Chapter 7). See references for details	UN Disaster Relief Office established only in 1972. May be of great benefit to developing countries in the face of major losses in the future; international relief given by many developed countries in the past, e.g. Parícutin eruption, Mexico	

Based on Burton *et al.* (1968); and Chester *et al.* (1985); with information from:
(a) Mason and Foster 1953; Macdonald 1962, 1972; US Corps of Engineers 1966; Macdonald and Abbot 1970; Grove 1973; Williams and McBirney 1979.
(b) Hawaiian Civil Defense, Department of Defense 1971; Macdonald 1972; UNESCO 1972, 1974; Nakano *et al.* 1974; United Nations 1977; Booth 1977, 1979; Fournier d'Albe 1979; Warrick 1979; Sorenson and Gersmehl 1980.
(c) Gerasimov and Zvonkova 1974; O'Riordan 1974; Burton *et al.* 1978; Warrick 1979.
(d) White 1974; Warrick 1979.

do people create hazards by their presence, but also through their development of the resources of a region (Fig. 8.1). Without people there can be no natural hazards; they are an artifice of the interaction between people and nature. In general, the greater the population density and the higher the level of economic development, then the more severe will be the losses once an eruption occurs. An appreciation of the human geography, economics and sociology of hazard-prone regions and monitoring of changes in them are vital first stages in the dominant approach.

All hazardous events produce responses. These may be individualistic, or involve a single cultural group or a whole society, but their number is not infinite and depends critically on the range of adjustments available at the time. During most of history the principal response to all hazards has been to do nothing (i.e. involuntary loss-bearing) and this is still evident in many economically less developed parts of the world, but over the years an increasing number of alternatives have become available. At any given time the range of adjustments will be finite, wider than in the past but probably narrower than in the future, and will depend upon the technology available to and the institutional structures of the society in question. Objective 2 of the dominant approach is, therefore, to catalogue and evaluate the range of adjustments available for different classes of hazard. For volcanic eruptions adjustments are less extensive than for virtually any other type of hazard and the limited range currently available for hazards posed by lava flows is shown in Table 8.2. It should be noted when reading Table 8.2 that alternatives to loss-bearing which involve technology—such as modifying hazards and their loss potential, and/or highly developed social and administrative systems, including spreading losses and planning for them—(Chester *et al.*

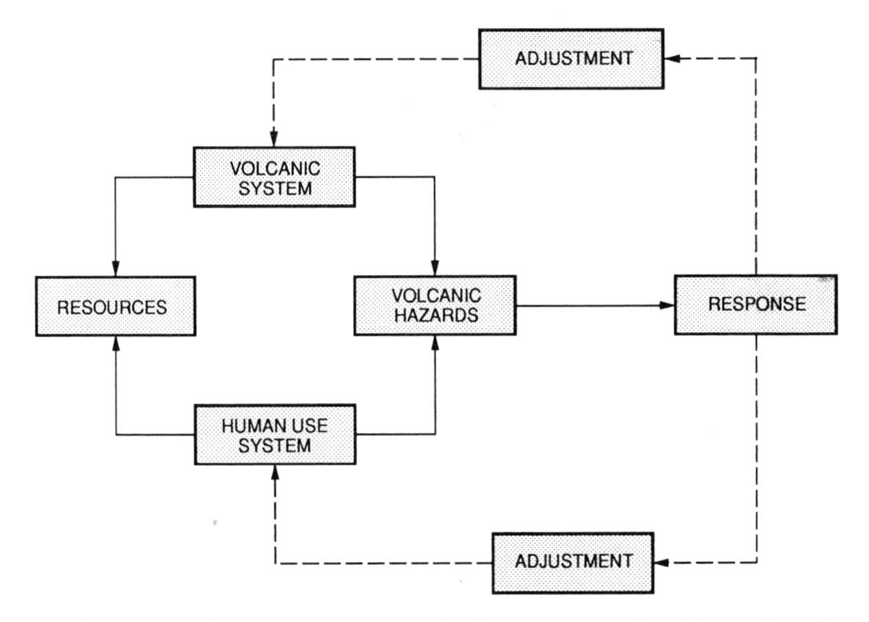

Fig. 8.1 The nature of human responses and adjustments to volcanic hazards under the dominant paradigm (from Chester *et al.* 1985, Fig. 9.1, p. 338; and on a modification of Burton *et al.* 1978; and Warrick 1979).

1985) are not available in all countries of the world. They relate to economic development and even in advanced countries like Japan and the USA they have only been available for a short time. It is a truism of hazard analysis, that the reason why potentially hazardous regions are settled is because long-term benefits outweigh short-term risks and any adjustment cannot adversely affect the economic viability of an area otherwise it is unlikely to receive support. It may appear from Table 8.2 that modifying loss potential particularly by using general prediction (Chapter 7, section 7.3.1), to divide a volcano into zones of relative risk so enabling new development projects to be steered into areas which are relatively safe, is an ideal type of adjustment, but to be effective it requires a close knowledge of its effects on the ratio of benefits to costs.

Table 8.2 also lists measures to spread losses, including public relief, international aid and insurance. These forms of adjustment are usually justified on economic, social and ethical grounds but, paradoxically, may make hazards more acute. It has been noted on many occasions that measures designed to spread losses can cause individuals and societies to be indemnified for their own folly and to persist with hazardous practices (Burton *et al.* 1978). For instance, in New Zealand, O'Riordan (1974) found that government-supported hazard insurance had the effect of actually encouraging new building on shifting foundations. To be effective loss-spreading has to be linked to enforceable land-use zoning policies.

Objectives 3 and 4 (Table 8.1) represented at the time a major innovation, for White and his colleagues recognized that the processes by which individuals, cultural groups and societies choose adjustments when faced by hazards are extremely complex. It is clear from many studies that people do not always act in an economically rational manner. They do not behave 'as omniscient beings who are able to calculate the expected values arising from a range of probabilistic events and their associated consequences' (Warrick 1979, p. 182). An example of this may be seen in the Puna District of Hawaii (USA) where more elderly members of the native Hawaiian cultural group have been observed to take no action when faced with losses caused by incursions of lava. This is because of their beliefs, which hold that the Goddess Pele controls fate and that intervention is neither necessary nor advisable (Murton and Shimabukuro 1974; Warrick 1979). Because people do not act rationally, a model of *bounded rationality* is often invoked by researchers. The model, derived from the work of Simon (1957, 1959), argues that 'choice among a range of alternatives in dealing with a hazard is based on the individual's perception of them, which is conditioned by environmental, social and psychological factors. In effect, a model of *bounded rationality* leaves room for the influence of cultural traits, social norms, cognitive limitations, and personality differences on the individuals' perception of the state of nature and alternative actions, and on his (*sic*) evaluative processes' (Warrick 1979, p. 183, my emphasis). A list of some of the factors involved in individual decision-making in volcanic areas is presented in Table 8.3, but it should be noted that this simplifies the situation because the factors interact one with another. A poor, rural agriculturalist in an economically less developed country is more likely to make decisions on the basis of traditional beliefs and superstitions, than is a highly capitalized, highly educated factory farmer in an economically more developed country.

During the past half century social scientists working within the dominant paradigm have evolved more sophisticated models of decision-making which apply to whole societies rather than to individuals. A model introduced by Burton *et al.* (1978), for example, has had a strong influence. Burton and his colleagues pointed out that any society can cope with extreme natural events by either *adapting* or *adjusting* to them. For volcanic hazards *adjustment* is the response most frequently encountered, but *adaptation* may be important in certain societies. For instance, Russell Blong (1982, 1984, p. 348) quotes the example of a group of highlanders in Papua New Guinea, who have adapted their activities to frequent pyroclastic falls. So favourable to crop growth do they consider these falls to be, that they have developed elaborate rituals to encourage their recurrence. *Adjustments* are subdivided into *purposeful* and *incidental* (Fig. 8.2), with *purposeful* adjustments including most of those listed in Table 8.2. *Incidental* adjustments are, in contrast, policy measures which are not related

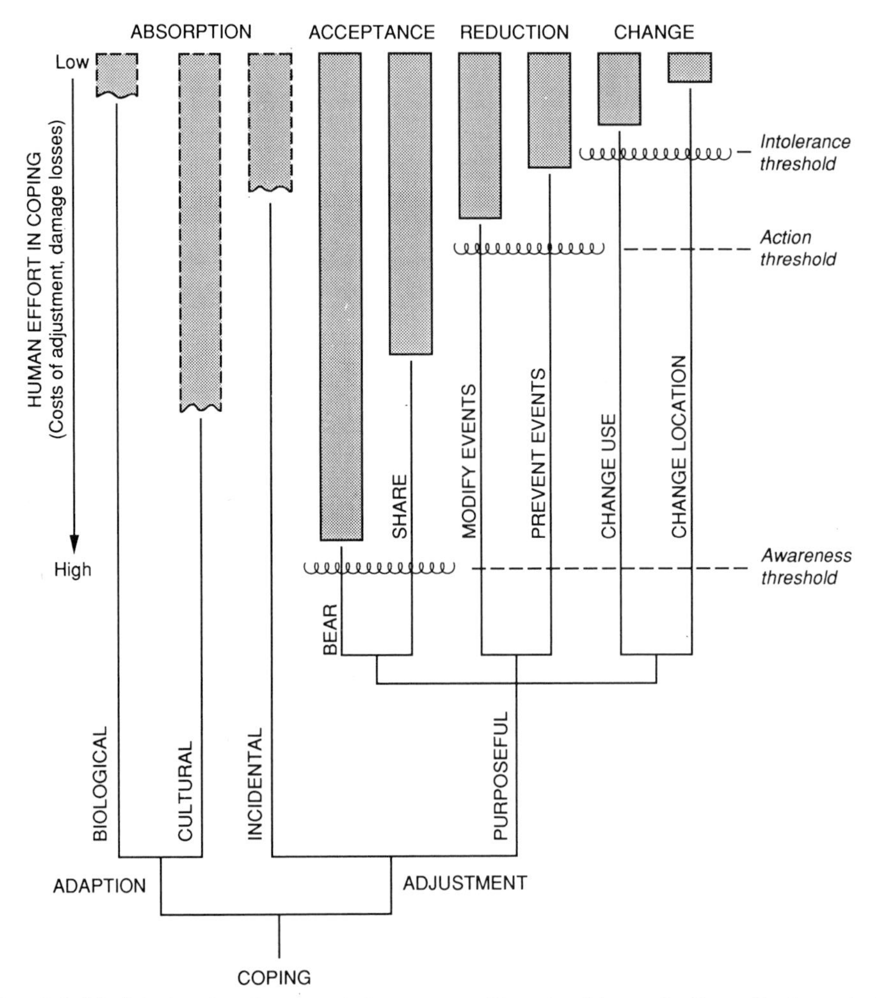

Fig. 8.2 Model showing how societies cope with natural hazards (from Burton *et al.* 1978, Fig. 8.1, p.205).

Table 8.3: Remarks on, and examples of, the factors controlling decision-making and responses to volcanic hazards (based on Chester *et al.* 1985, p. 344)

Factors affecting responses to volcanic hazards	Remarks and examples
Material wealth	Material wealth correlates strongly with other socio-economic traits like education, mobility of residence, employment and scientific awareness. In a given region a volcanic eruption may have initial effects which are broadly egalitarian, but wealth is associated with the ability to take risks and survive (Hass *et al.* 1977). Research by Nolan (1979) of the effects of the Parícutin eruption (Mexico, 1943–52) indicated that richer families could liquidate their assets, whereas poorer Indian communities had no choice but to stay. Considerations of personal wealth may be complicated by the adoption of measures which reduce invidual losses. Individuals, cultural groups and societies at high levels of economic development are more likely to have access to insurance and support from national governments in the event of a disaster. This was certainly the case following the eruptions of Mount St Helens (USA) in the early 1980s. When external aid is involved, as is the case with many economically less developed countries, the perceptions of aid donors may be important in conditioning responses. Following the eruption on the South Atlantic island of Tristan da Cunha in 1961, the reaction of the colonial power (Great Britain) was to evacuate the population into an unaccustomed environment of diseases and high-technology living. Eventually all but four returned to the island (Blair 1964). This example indicates that to be successful 'wealth transfers' through disaster relief, require the perceptions of suitable adjustments by donors and recipients to coincide.
Experience	Both individual and group responses to hazards depend on experience. Generally the longer the experience and the shorter the recurrence interval between eruptions, the more likely is it that future threats will be assessed accurately. In Hawaii Hodge *et al.* (1979) found that newcomers had only a vague impression of true risk. Hawaii has several volcanoes that erupt frequently, but when volcanoes have long intervals between high-magnitude events, then experience may have the opposite effect. In 1975 certain slopes of Mt Baker in the Cascade Range (USA) were closed to holiday-makers, because an eruption was possible. Holiday-makers who had visited the volcano many times and without incident were found to resent the closure far more than those who were first-time visitors and had no previous experience (Hodge *et al.* 1979).
Urban and rural dwellers	Urban and rural dwellers often show differing responses. It has been observed that because urban dwellers live in a built environment they may be less aware than rural inhabitants of the true risks they run. In the Puna District of Hawaii, farmers have used potentially dangerous land for many years, but have not built houses on it. In contrast Hawaiian town and city dwellers have been found to have a less accurate perception

Table 8.3 – *cont.*	
Factors affecting responses to volcanic hazards	Remarks and examples
	of true risk, even though scientific information is readily available (Murton and Shimabukuro 1974; Hodge *et al.* 1979).
Systems of belief	These can complicate both responses and choices of adjustment. An individual's personality traits, such as varying degrees of inner control, perceived and actual leadership aptitudes, the ability to cope, survive and improvise may be crucial in any successful response. Many of these traits correlate with age, intelligence, education and wealth. A further feature is the accuracy with which a person can estimate risk. This depends in part upon the provision of accurate information, but the *gambler's fallacy* has often been observed. This is expressed in a belief that the occurrence of an event at a specific place in a particular year makes it less likely to happen again in the near future.

directly to hazard reduction, but reflect rather the 'normal' processes of economic and social development. These include such measures as better telecommunications—which will assist in the early warning of extreme events—and improved methods of transport—which will enable emergency aid to be moved to a disaster-affected region more rapidly. A major innovation of this model is that it introduces four characteristic modes of coping with natural hazards, which are separated by thresholds (Fig. 8.2; note the modes are on the horizontal axis, the thresholds on the vertical).

Loss absorption is the first mode and it is argued that many societies can absorb the effects of extreme natural events without taking any action. For volcanic hazards this is comparatively rare, although coping with the agriculture effects of the persistent volcanic plume on the eastern flank of Mount Etna, Sicily, is one example (Durbin and Henderson-Sellers 1981; see Chapter 6, section 6.4.2.1). *Loss absorption* may involve *adaption* or *incidental adjustment.* Clearly most volcanic eruptions demand a more positive response and societies affected will cross the *threshold of awareness* and choose either to share or to bear losses (Fig. 8.2 and Table 8.2). If the magnitude of an eruption and/or its effects on people are more severe, then the *action threshold* will be crossed and more positive, interventionist measures will be taken, involving such initiatives as action to protect high-value installations and the introduction of warning systems (Table 8.2). With extremely damaging events, the *threshold of intolerance* will be crossed and drastic action will follow, being exemplified by land-use changes, the relocation of settlements and the introduction of planning measures based on general prediction (Chapter 7, section 7.3.1).

The model of Burton, Kates and White (Burton *et al.* 1978) can distinguish between contrasting forms of adjustment in societies at different levels of economic development (Whittow 1987). This is a development of earlier ideas (e.g. White 1973) and relates to the fifth and final objective listed in Table 8.1: of assessing the effects of public policy upon responses. A *folk* or *pre-industrial* society is more willing to accept modifications of behaviour and land uses and to harmonize with nature, rather than to manage and control it. As Table 8.4 shows,

Table 8.4: Varying responses to natural hazards in different societies (from Chester *et al.* 1985, Table 9.2, p. 346; and modified from White 1973)

Folk or pre-industrial society	Modern technological or industrial society	Comprehensive or post-industrial society
1) Wide range of adjustments. 2) Action by individuals or small groups. 3) Emphasizes harmonization with nature, rather than using technology to control it. 4) Low capital requirements. 5) Responses vary over short distances. 6) Flexible responses and easily abandoned if unsuccessful.	1) Relatively narrow range of adjustments. 2) Requires co-ordinated action by societies. 3) Emphasizes technological control over nature rather than harmonization. 4) High in capital requirements. 5) Responses tend to be uniform. 6) Responses inflexible and difficult to change.	Combines features of the two types of response and, therefore, incorporates a greater range of adjustments than is found in the technological response and is tailored to the needs of the particular society and hazard; this response involves a number of capital and organizational requirements.

this response is marked by low capital requirements, variability over short distances and flexibility. It is only effective 'in the face of low-magnitude hazards; a high magnitude event will probably require massive overseas assistance' (Whittow 1987, p. 313). The responses of *modern technological* or *industrial* societies will be different and emphasis will be placed on shifting the burden from the individual to the society, nation and international community, by loss sharing, insurance, relief aid and technology. Responses emphasize control over nature, a narrow range of adjustments, high capital inputs and, consequently, inflexibility (Table 8.4). Finally, there is an approach typified by *post-industrial* societies. This is the 'ideal' from a planning perspective because it incorporates the best elements of the other two forms of response. 'Its implementation requires an approach to land planning which, while recognizing the true character of the physical threat, at the same time attempts to harmonize technology to both the natural environment and the perception of the individuals [*and social groups*] faced by the danger' (Chester *et al.* 1985, p. 345, my emphasis). For volcanic hazards it is debatable whether any society has yet reached the post-industrial stage, although Japan, Iceland and the USA with their integration of prediction and public policy come nearest to it (Shimozuru 1983; Sigvaldason 1983—see Chapter 9). Even in these countries cooperation between volcanologists on the one hand, and policy-makers on the other can still cause difficulties (e.g. see Peterson 1988). The stages from folk to post-industrial should not be viewed sequentially, since characteristics of more than one type of response may be seen amongst different groups within a society at the same time (White 1973).

8.3 The radical critique

The dominant approach appears to accord closely with reality. Although it accepts that no hazard can exist without human intervention, it argues that physical processes—in particular their magnitude and frequency—are first-order determinants of a disaster and that differences between societies are at a lower, albeit still significant, level of importance. This emphasis can be seen clearly in the translation of the dominant approach into priorities for national and international policy. Two examples illustrate this. The first is the United States Volcanic Studies Program which 'can be classified under three broad headings: fundamental studies of volcanic processes, volcano hazards assessments, and volcano monitoring' (Filson 1987, p. 294). Attitudes of scientists and policy-makers about the nature of the International Decade for Natural Disaster Reduction is the second example. These are summarized by Lechat (1990, p. 1, my emphasis) as:

— 'to improve the capacity of each country to *mitigate* the effects of natural disasters ... paying special attention to assisting developing countries in the assessment of disaster damage potential and in the establishment of *early warning systems* and *disaster-resistant* structures when and where needed;
— to devise appropriate guidelines and strategies for applying *existing scientific and technical knowledge*, taking into account the cultural and economic diversity amongst nations;
— to foster *scientific and engineering endeavours* aimed at closing critical gaps in knowledge in order to reduce loss of life and property;
— to disseminate existing and new *technical information* related to measures for the *assessment*, *prediction* and *mitigation* of natural disasters;
— to develop measures for the *assessment*, *prediction*, *prevention* and *mitigation* of natural disasters through programmes of *technical assistance* and *technology transfer*, demonstration projects and education and training ...'

Clearly this is a physically deterministic brief and, as the emphasized sections indicate, one that is heavily focused on scientific and technological solutions. It would be unfair and pejorative, to claim that the dominant paradigm was deterministic in the sense that the 'nature of human activity was controlled by the parameters of the natural world within which it was set' (Johnston 1979, p. 32), but there is, nonetheless, an underlying deterministic undercurrent.

Whether this deterministic undercurrent and the focus on technological transfers from economically developed to economically less developed countries should be the universal model of human adjustments and responses to natural hazards, is the basis of the radical critique. The most influential critic has been Kenneth Hewitt who in his early career was, paradoxically, associated with the 'Chicago School'. First in a review of a book by Ian Burton, Robert Kates and Gilbert White (*The Environment as Hazard*, Burton *et al.* 1978) and later as the editor of what has become the seminal work of the radical school—*Interpretations of Calamity* (Hewitt 1983a)—Hewitt sets out his disquiet with the dominant paradigm, especially as it is being applied to the relief of natural disasters in economically less developed countries. Hewitt starts by quoting a passage from the philosopher Bertrand Russell, that science (in its broadest sense), will be used in the future to promote the power of dominant groups and then goes on to argue that the term *dominant* may be applied aptly to members of the 'Chicago School'. He argues that the 'dominant view constitutes a technological approach [and] that means, the work subordinates other modes and bases of understanding or action to those using technical procedures. More precisely, technocracy gives precedence in support and prestige to bureaucratically organized institutions, centrally controlled and staffed by or allocating funds to specialized professionals' (Hewitt 1983b, pp. 7–8). In contrast Hewitt believes that hazards should be viewed as normal aspects of societies and goes on to suggest that it is not possible to explain fully the effects of extreme natural events purely in physical terms. Hazards have to be related to the pressures, goals and changes in societies which are often unrelated to the relationships between people and environment. He argues that it is not a coincidence that the areas of the world where major disasters occur are also ones undergoing major social, economic and environmental changes. Quoting the case of the frequently drought-ridden Sahel area of Africa, he poses the question 'what is more characteristic . . . and to be expected by its long time inhabitants: recurrent droughts or the recent history of political, economic and social change' (Hewitt *op. cit.*, p. 26).

Other authors contributing to *Interpretations of Calamity* develop these ideas more fully through the use of case studies, none more so than Susman *et al.* (1983). In their chapter the authors propose a new model of disaster occurrence, which they term a *theory of marginalization.* This they claim has particular relevance to hazards in economically less developed countries. Their starting point is poverty, their creed Marxist, and they argue that underdevelopment is linked to the control and exploitation of resources by governing groups and outside interests, such as multinational corporations. To maintain his or her lifestyle and survive a subsistence agriculturalist is forced into areas which are less productive and where hazards are more likely to occur. 'Disaster proneness is a result of the interface of a population undergoing underdevelopment and a deteriorating physical environment' (Susman *et al.* 1983, pp. 278–9). Furthermore, relief aid serves to reinforce the *status quo* leading to further marginalization and, ultimately, to greater future vulnerability to extreme natural events. The model of marginalization proposed by Susman *et al.* (1983) is reproduced as Fig. 8.3 and the authors' comments on it are summarized in Table 8.5. Some of these comments have already become part of current thinking and have been being accepted by international agencies, but others, such as the political viewpoints expressed in Implications 5 and 6 (Table 8.5), remain highly contentious.

Table 8.5: The implications of marginalization theory (from Susman *et al.* 1983, pp. 279–80)

Implication	Remarks
1	In economically less developed countries, 'disasters will increase as socio-economic conditions and the physical environment deteriorates'.
2	'The poorest classes continue to suffer most losses'.
3	'Relief aid reflects dominant interests, prevents political upheavals and generally works against the classes who have suffered most in a disaster'.
4	'Disaster mitigation based on high technology reinforces underdevelopment and increases marginalization'.
5	'The only way to reduce vulnerability is to concentrate disaster planning within development planning, and that the development planning context must be, broadly speaking socialist'.
6	'Because of the continued forms of exploitation, especially in underdeveloped countries, the only models for successful disaster mitigation are those conceived in the struggle against exploitation'.

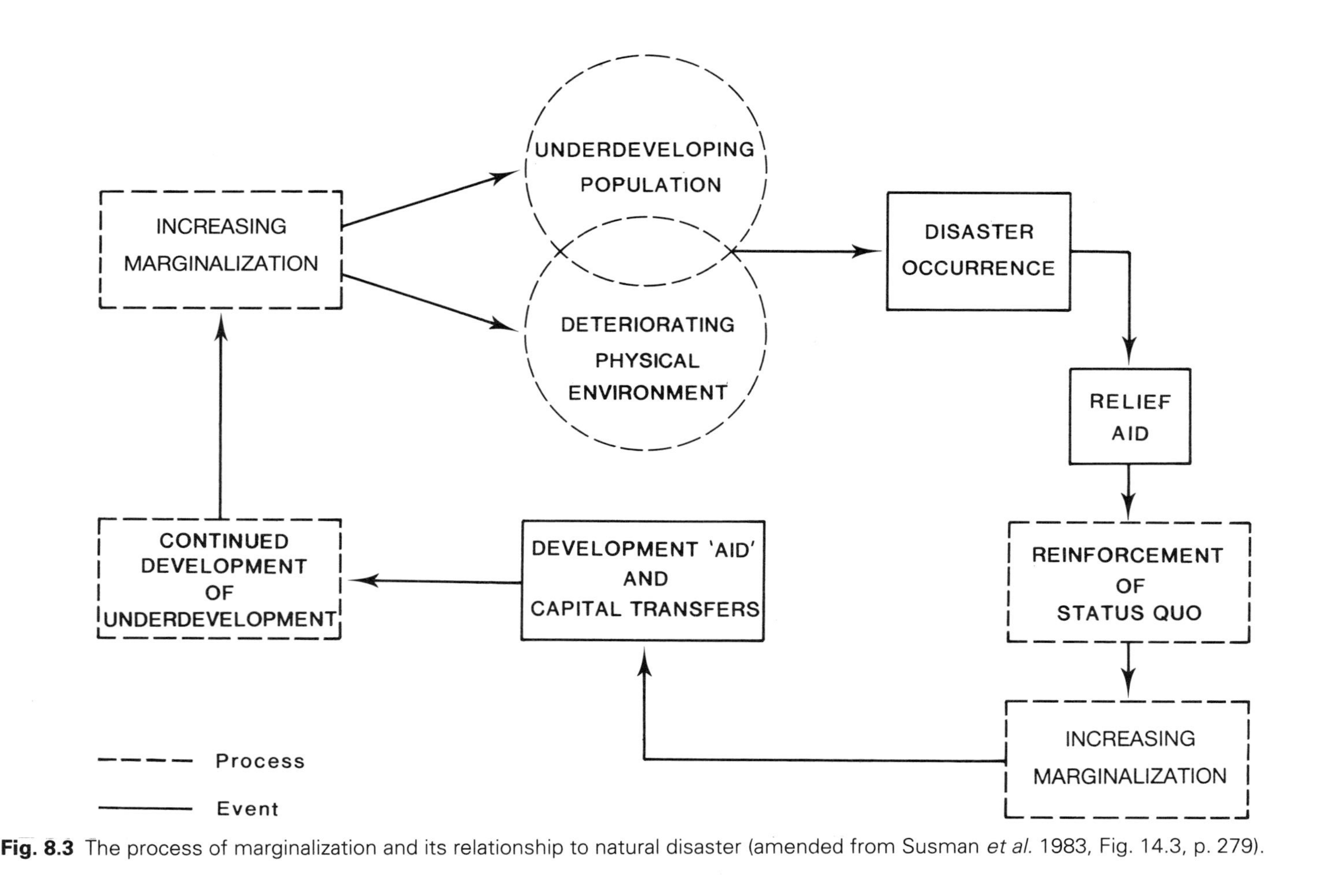

Fig. 8.3 The process of marginalization and its relationship to natural disaster (amended from Susman *et al.* 1983, Fig. 14.3, p. 279).

The authors of *Interpretations of Calamity* focus on the effects of extreme natural events on countries which are economically less developed and, indeed, this is true of the bulk of all published works which are critical of the dominant paradigm. It is not a universal emphasis and, in a thoughtful account of the aftermath of the 1976 Friuli earthquake in Italy, Robert Geipal (1982) argues that folk beliefs and religious dogma deserve more attention from hazard analysts than has normally been the case. This is he claims because most research on disasters and much international policy developed from it, reflects anglophone American experience which is atypical of the situation in even economically developed parts of the world.

Placing volcanic hazards within the context of the radical critique has not been attempted explicitly. Blong (1984), however, recognizes implicitly some aspects of long-term adjustment to volcanic hazards in economically less developed countries which accord closely to the general tenor of the arguments outlined above. For instance, in reviewing the social effects of four eruptions—Mt Lamington, Papua New Guinea, 1951; Parícutin, Mexico, 1943/52; Tristan da Cunha, South Atlantic, 1961/62; and Niuafo'ou, Tonga, 1946—Blong concludes that outcomes are frequently determined by the socio-economic evolution of these societies and that eruptions hasten, rather than alter, existing trends in development. In spite of a lack of published material the radical critique clearly has relevance especially for the study of volcanic eruptions in societies which are at low levels of economic development, since it is in these countries that most of the eruption-related deaths have occurred and will occur in the future. It comes as no surprise to find that in a table produced by the United States Advisory Committee on the International Decade for Natural Disaster Reduction (1987, p. 7) showing the death tolls from *major* natural disasters since 1900, nine volcanic eruptions are identified and all these have occurred in economically less developed countries. Countries such as these are ones in which economic losses, although smaller than the total value of losses that would occur in an economically developed country with an equal population density following an eruption of similar magnitude, will be relatively more serious because of the high proportion these losses represent in terms of total wealth. The cost of the Mount St Helens eruption (USA, 1980) is estimated at US$860 million (Blong 1984, p. 356), or around 0.03% of Gross National Product (GNP), whereas property losses alone during the Nevado del Ruiz eruption (Colombia, 1985) are estimated at US$300 million, or around 1% of GNP (Sigurdsson and Carey 1986). Since property losses in major eruptions comprise only about 10% of the total (e.g. Hunt and MacCready 1980), it may be assumed that total losses from the Ruiz eruption were many times greater and exacted a considerable 'tax' on the Colombian economy. In fact for a selection of natural hazards in economically less developed countries, Harriss *et al.* (1985) and Kates (1987) estimate that costs may be as high as 15–40% of GNP.

The vast majority of examples used by radical writers are concerned with hazards which occur not only in less developed countries, but are also characterized by long onset, long duration, widely distributed effects and impacts upon people who are already marginal to the economy of a state or region. Climatic hazards—in particular drought, floods and crop failures—are, therefore, well represented in the literature. Volcanic hazards do not possess these attributes and are notable for their rapid onset, relatively short duration, areal limitation of their major effects and the fact that—initially at least—they have an influence upon

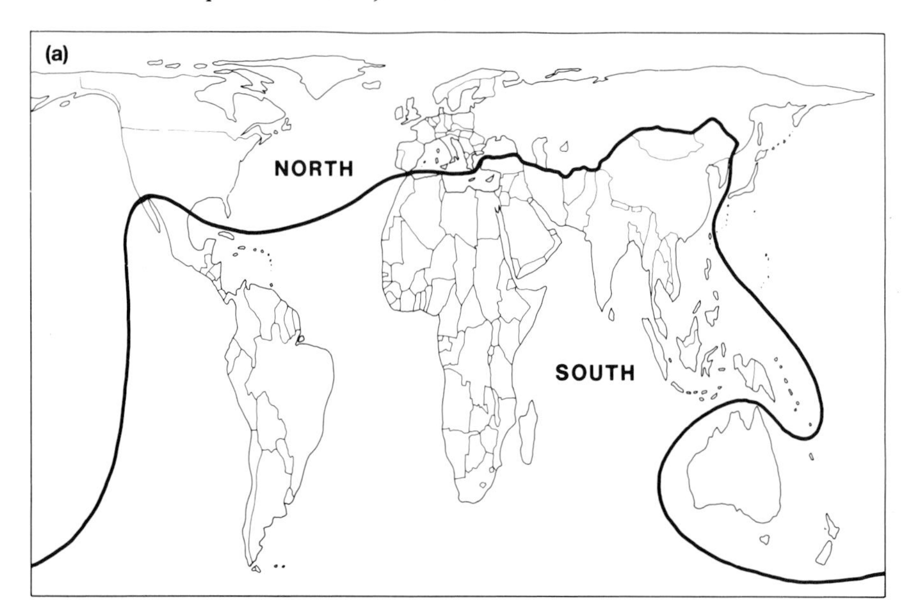

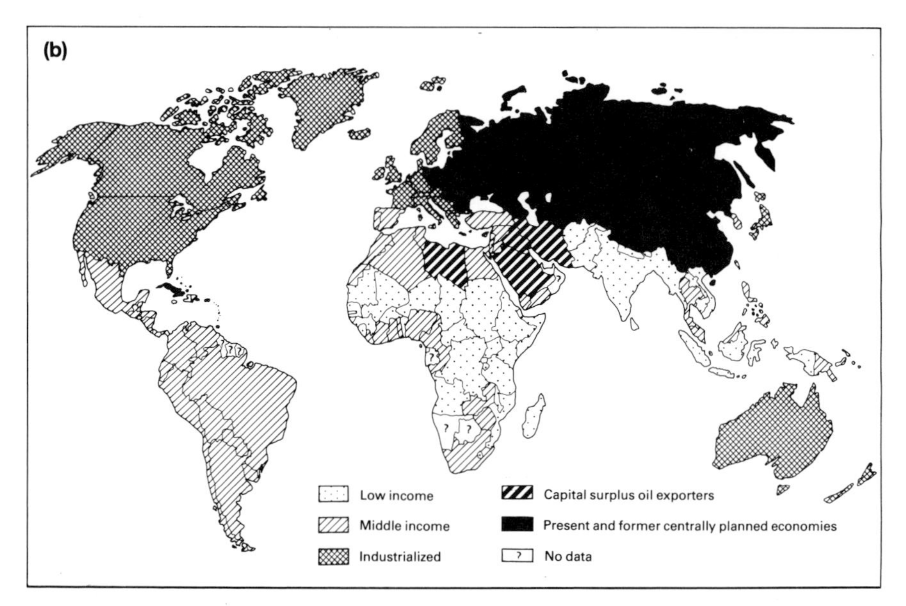

Fig. 8.4 World maps showing:
(a) A classification of countries into a rich 'north' and a poor 'south' (after Dickenson *et al.* 1983, Fig. 1.2, p. 2; data from Brandt Commission 1980). The map uses the Peters' Projection and its surface distortions are distributed at the poles and equator, so that the more densely populated regions are in approximate proportion to each other.
(b) Economic and political groupings of countries (after Dickenson *et al.* 1983, Fig. 1.3, p. 3, with some modifications). Note: The 'centrally planned economies' are those of 1983.

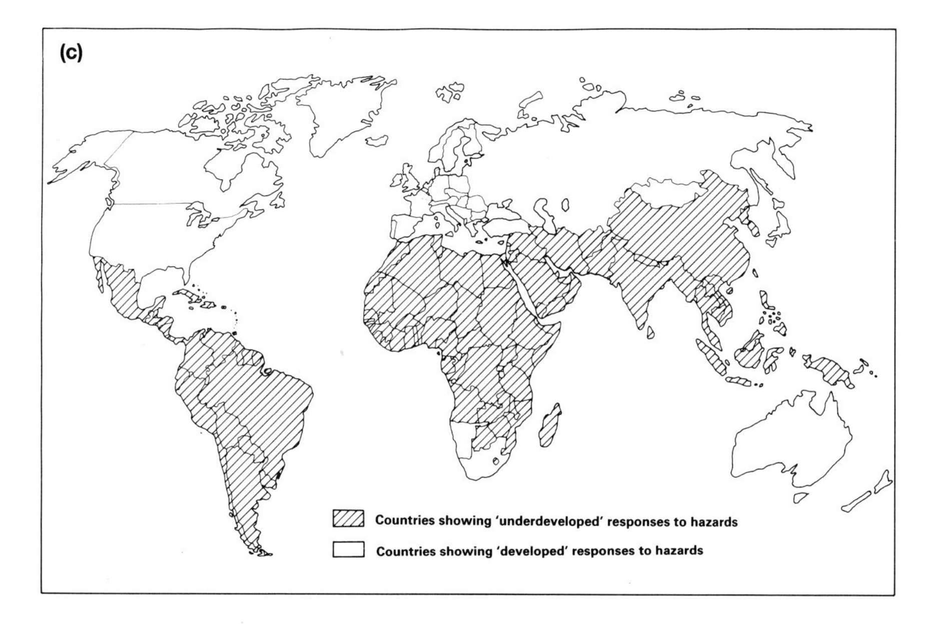

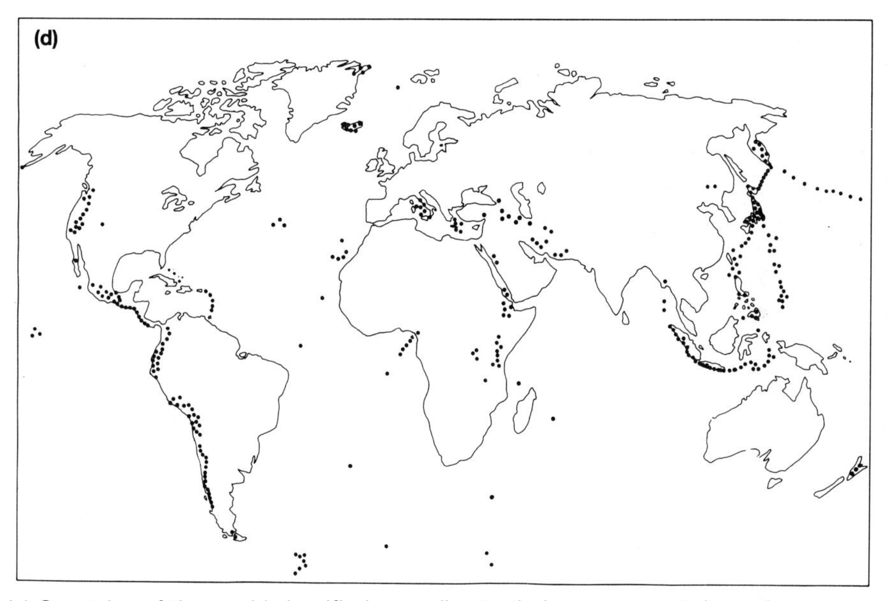

(c) Countries of the world classified according to their responses to hazards.
(d) Active volcanoes of the world (after Ollier 1988, Fig. 12.4, p. 188).

rich and poor alike. In this situation the dominant paradigm has much to offer with its emphasis on prediction (both general and specific) and the fact that many of the adjustments do not involve advanced technology and, when they do, are not normally associated with the marginalization of poor members of a society. Indeed the action taken during the 1983 eruption of Mount Etna, to divert lava flows, had a more immediate beneficial effect on peasant farmers than it did on the owners of tourist facilities, since the latter could spread their losses across a spectrum of business interests and may have carried insurance cover. In fact Hewitt (1983b, p. 25) recognizes this when he writes that it would be 'indefensible to argue that the disruptions occasioned by disaster produce no distinctive, even unique crises phenomena. There are particular aspects of hazard that can be helped by improved geophysical forecasting. Nor are any foreseeable human actions, going to remove the need to bring emergency assistance to ill-equipped victims'. His arguments and those of other radical writers are about the focus and emphasis of hazard research, especially as it applies to poor countries. As Whittow (1987) observes, the radical critique holds the potential danger of replacing physical determinism with social determinism.

8.4 A suggested framework

The present volume places the study of human responses to volcanic hazards within a modified dominant paradigm. It is necessary to guard against: the potential dangers of environmental determinism; the overweening power of dominant groups; the issue of marginalization, particularly as it affects economically less developed countries; and of agency-based approaches, involving uncritical transfers of technology. The major determinants of how a society will cope with volcanic eruptions and respond to them will be conditioned by the nature, frequency and magnitude of a given eruption and the characteristics of that society and its members. The responses that may be chosen by a society from the theoretical range (see Table 8.2) will be predetermined largely by its level of development.

Defining development is no easy task as can be seen from the number of terms and euphemisms currently in use. These include *north* and *south*, *First*, *Second* and *Third Worlds*, *developed* and *underdeveloped* and *developed* and *developing*, but all are really euphemisms for a broad division of the world into rich and poor countries (Fig. 8.4a and b; Brandt Commission 1980; J. P. Dickenson *et al.* 1983). No one set of terms is entirely satisfactory and some are pejorative, implying that a low economic status equates with a position of social and/or cultural inferiority. For this reason a classification of *economically more* and *economically less developed* countries has been used so far in this chapter, but even this is not without its difficulties for many aspects of development have little to do with economics in its strictest sense. Growth to a given stage of development does not involve increases in material wealth alone though this is important. Countries of the Middle East may be 'oil rich' (Fig. 8.4b), but still have to depend on non-monetary aid from abroad. Growth means more than this and involves: increases in human capital (education, training, scientific expertise, knowledge and information); increases in non-human capital (through investment in industry, commerce and communications); and improvements in economic and social organization (an efficient civil service and entrepreneurial skill) (O'Riordan

1976). Adopting this broader definition of growth makes it possible to classify the countries of the world into two groups: those with a *developed response* and those with an *underdeveloped response* to hazards (Fig. 8.4c).

As Fig. 8.4d shows, the majority of the world's active volcanoes occur in countries having underdeveloped responses and between 1600 AD and 1982 these accounted for 86% of all volcano-related deaths (derived from Blong 1984, p. 72). Of the volcanoes which have been active in the Holocene some 82% are found in poor countries (data from Simkin *et al.* 1981). The criteria used to classify the responses of countries to volcanic eruptions are listed in Table 8.6. An extreme case of an underdeveloped response would include all features on the negative side of the table, but clearly this is an over-simplification because most states have some socio-economic and political features which are more positive. Medical provision, literacy, civil administration all vary in their quality and effectiveness and GNP per capita varies from below US$500 in the case of Indonesia to over US$1400 in the case of Costa Rica (Anon. 1988). The classification of a country as having an underdeveloped response is based on it having a majority of negative features. So far only development has been considered and any framework has to take the characteristics of the eruption into account. In the USA there are large differences in the effectiveness of responses to high-magnitude—though infrequent—eruptions in the Cascades and low-magnitude—though frequent—eruptions in Hawaii.

By synthesizing it is possible to derive a four-fold classification of responses to eruptions (Fig. 8.5). This is not wholly satisfactory, though at first sight it appears reasonable and has in fact been used explicitly and implicitly by many writers. There are several reasons for this. First, the theoretical range of adjustments to hazards not only increases over time, but the country in which the volcano is located also undergoes development so that more comprehensive responses may be chosen than in the past. Japan has a history of devastating volcanic eruptions which include: the 1783 Asama pyroclastic flows and lahars (1300 deaths); the Bandai san eruption of 1888 (460 deaths and several villages destroyed); events at Sakura-zima in 1476 and 1779 which killed many people; an avalanche and tsunami at Unzen in 1792 (14,000 deaths); and the seven major eruptions of Usu volcano, which since 1663 have exacted a high death toll. Yet if any of these events were to be repeated in the future then it is likely that the type of adjustment and response would be of an industrial or post-industrial type (Table 8.4). Material losses would be higher but casualties much lower. As Decker and Decker (1989, p. 260) note, the Sakura-zima eruption of 1914 was already notable for the use of seismic monitoring which kept the death toll very low, even though the magnitude of the eruption was high. It is also highly improbable that a repeat of the Laki fissure eruption in Iceland (1783) would lead to a famine causing the death of 10,000 people. Today both Japan and Iceland are wealthy countries, in GNP per capita terms respectively the ninth and the tenth richest in the world, with high levels of literacy, established scientific expertise, efficient civil administrations and, in the case of Japan, one of the strongest economies in the world (Anon. 1988). Hence comparisons between countries and societies can only be made validly over short time periods. Historical examples of responses are, nevertheless, of value because they highlight pre-industrial responses (Table 8.4) and these may hold important lessons for the future. Responses are based on harmonization with nature, have low capital requirements and may have relevance for many economically less developed countries.

Table 8.6: Positive and negative features, conditioning the response to volcanic hazards in a given country. Some of these reflect the development level reached by a country, others the physical characteristics of the eruption. Magnitude and frequency of eruptions are the most important physical characteristics.

Positive features	Negative features
A high level of economic development will involve:	A low level of economic development will involve:
1) wealth to cushion the impact of an eruption and to allow losses to be spread across a society;	1) a lack of wealth and a predominance of loss-bearing. Aid from abroad may bring with it problems of inappropriate technical transfers and the marginalization of poorer sections of the population;
2) scientific expertise to monitor volcanoes using techniques of general and specific prediction;	2) few trained scientists and a lack of monitoring, prediction and advice to policy-makers;
3) the availability of insurance;	3) no insurance, except for high-value foreign-owned installations;
4) policies of land-use planning linked to general prediction;	4) few, if any, land-use planning policies—poor enforcement;
5) efficient civil administration and civil defence to administer disaster relief;	5) an inefficient—often corrupt—civil administration;
6) a high-quality infrastructure of transport and telecommunications;	6) poor communications;
7) a preplanned medical response;	7) few medical resources to devote to disasters;
8) a well educated and informed population;	8) a low level of educational attainment—population ill-informed;
9) the ability to recover quickly from an eruption; and	9) slow recovery from an eruption; and
10) high material losses and a small loss of life, but a low ratio of monetary losses to total national wealth.	10) low material losses and a high loss of life, but a high ratio of monetary losses to total national wealth.
The following physical features of an eruption will minimize losses:	The following physical features of an eruption will maximize losses:
1) A low-magnitude event affecting a small area.	1) A high-magnitude event affecting a wide area.
2) A high ratio of lava flows to pyroclastic products.	2) A high ratio of pyroclastic products to lava flows.
3) A short interval between events. This will make it more likely that planning measures will have been devised and that the population will be aware that they are living in a potentially dangerous place.	3) A long interval between events. This will mean that it is unlikely that any planning measures will be in place.
4) Interaction between eruptions and the external environment, especially water, are highly unlikely.	4) Interactions between eruptions and the external environment are highly likely and could involve hydrovolcanic activity, slope failure and the generation of lahars.

A second reason why the classification used in Fig. 8.5 is unsatisfactory is because the examples chosen show 'typical' responses for the country and hazard in question. There are instances where the reactions to a disaster in a country normally having an underdeveloped response may show aspects of a developed response and *vice versa*. In compiling Fig. 8.5, the author was tempted to include eruptions of Piton de la Fournaise (Réunion Island, Indian Ocean) as exemplifying typical underdeveloped responses to low-magnitude/high-frequency eruptions. Réunion is an overseas Department of France and during the 1977 eruption evacuation was ordered to limit casualties; an action more typical of a developed response (Blong 1984, p. 144). It is also clear that today pre-industrial responses (Table 8.4) do not exist in their pure form, since any major natural disaster will attract international aid. Fig. 8.5 applies to whole countries, but regional differences may be important. When discussing Mount Etna, Chester *et al.* (1985) showed that in spite of Italy being a developed country many pre-1983 responses to eruptions were marked by elements more characteristic of less developed states. The reason for this was that Sicily remains a very poor region within Italy and peripheral to the mainstream of Italian life, economic wealth, state administration and national consciousness.

Finally, although magnitude and frequency are useful ways to classify eruptions and correlate closely with casualty figures and damage, all eruptions are unique in that they do not simply repeat activity which has happened before. The eruption of Nevado del Ruiz (Columbia) was small scale and similar to many others in South America in terms of its magnitude and frequency, yet it was the interaction between this eruption and the covering of snow and ice at the summit which generated the lahars which killed more than 25,000 people (Decker and Decker 1989).

'DEVELOPMENT' LEVEL

'PHYSICAL' CHARACTERISTICS	DEVELOPED RESPONSE	UNDERDEVELOPED RESPONSE
High magnitude/ low frequency	Examples include: Mount St Helens, USA (1980) Ruapehu volcano, New Zealand (1953)	Examples include: El Chichón, Mexico (1982) Agung, Indonesia (1963-64) Arenal, Costa Rica (1968)
Low magnitude/ high frequency	Heimaey, Iceland (1973) Mount Etna, Sicily (1983) Kilauea, Hawaii (many)	Nyiragongo, Zaire (1977) Karthala, Comores, Indian Ocean (1972)

Fig. 8.5 Classification of responses to volcanic eruptions based on physical characteristics of the eruption and level of development attained by the country. Examples are given of eruptions which fall into the four subdivisions of the classification.

SECTION 9.2	SECTION 9.3	SECTION 9.4
Developed responses High-magnitude/low-frequency events (Section 9.2.1)	Underdeveloped responses High-magnitude/low-frequency events (Section 9.3.1)	
Developed responses Low-magnitude/high-frequency events (Section 9.2.2)	Underdeveloped responses Low-magnitude/high-frequency events (Section 9.3.2)	Particular reponses
Developed responses Volcano-related events (Section 9.2.3)	Underdeveloped responses Volcano-related events (Section 9.3.3)	

Fig. 8.6 A suggested classification of responses and adjustments to volcanic hazards, based on the characteristics of societies in which they occur and the nature of the extreme volcanic events. Section numbers (e.g. 9.2.1, etc.) refer to the locations in Chapter 9 of examples of these responses and adjustments.

A framework has to reflect both the nature of and the variations within hazardous phenomena and the complexities of the countries and societies in which they occur. A revised classification is proposed in Fig 8.6. From political, economic, social and cultural perspectives three categories emerge. The first comprises those countries and societies which show a clear predominance of positive features (Table 8.6) and may be said to have developed responses. The underdeveloped category is the converse and includes those countries and societies which show a preponderance of negative features. Thirdly, there are a group of countries and societies where responses and choices of adjustment are not dependent principally on development, but rather on some other overriding characteristic, such as an inefficient or corrupt administration. These countries and societies are said to have *particular* responses. Developed and underdeveloped categories are further subdivided. There is the straightforward distinction between high-magnitude/low-frequency eruptions on the one hand and low-magnitude/high-frequency eruptions on the other. There are also those eruptions which are a result of interactions with other environmental factors and produce damage only indirectly related to volcanic activity. The latter include tsunamis, lahars, hydrovolcanic activity, earthquakes and slope failures. Finally, an historical dimension is vital, for this may reveal traditional (i.e. pre-industrial) forms of coping which may offer much when comprehensive strategies for the future are being discussed.

9

Responses and adjustments to volcanic hazards in different societies

Most natural disasters, or most damages in them, are characteristic rather than accidental features of the places and societies where they occur (Kenneth Hewitt 1983b, p. 25).

9.1 Introduction

Chapter 9 develops the framework outlined in Fig. 8.6, to show the ways in which different societies cope with various types of volcanic hazard. Examples are selective, but have been chosen to represent a wide range of adjustments and responses.

9.2 Developed responses

As Table 9.1 shows Iceland, Japan, New Zealand and the USA have much in common. They are wealthy, have well-educated and well-nourished populations, are notable for the quality of their volcanological expertise and the effectiveness of their civil administrations. Socially and culturally the four countries are very different and it is this diversity, as well as the variety of volcanic hazards they face, which has conditioned their responses to extreme natural events.

Iceland and Japan are notable for the degree of commitment they show to the alleviation of volcanic hazards and this is shared by all sections of their societies. Iceland is a small country with a small population (Table 9.1) and a history dominated by extreme volcanic events. In 1783 the Laki fissure eruption caused the death of much of the island's livestock through fluorine poisoning, leading to the starvation about one-fifth of the population and threatening national survival (Decker and Decker 1989). It comes as no surprise to read in an article in the *New Yorker* magazine that 'almost every scientist of any description sooner or later turns into a volcanologist' and the opinion of one volcanologist that there 'is something that the environment does to Icelanders. Physicists, astronomers come home—they have no cyclotron, no observatory. The land just calls them, and they go into geophysics' (McPhee 1988, p. 48). It is also interesting to note that following the eruption of Heimaey in 1973 the sum required to fund the recovery

Table 9.1: Examples of countries having a 'developed' response to volcanic eruptions. The columns showing eruptions known to have affected people and their activities, the number of volcanoes active in the Holocene and in historical time should be treated with caution. Estimates vary between sources; individual volcanoes within volcanic fields create problems of enumeration and the length of the historic period differs between authors. For full discussion reference should be made to Simkin *et al.* 1981. Figures should be treated as estimates only. The column showing total fertility rates represents the average number of children born to women during their reproductive years. ODA refers to overseas development assistance, negative values indicating a net outflow of funds

Country	Volcanoes[1]			GNP per capita[2] (US$ 1986)	Area (million km^2)	Population 1989* (millions)	Population fertility[3]	ODA as % of GNP (1984–86)[4]	Secondary school enrolment (% of age group 1983–86)[5]		Daily food energy supply as % of requirement, 1983–85[6]
	Eruptions known to have affected people and their activities	Number active in Holocene	Active in historical time (% of those active in Holocene time)						Male	Female	
Iceland	42	68	27(40)	13,370	0.10	0.25	2.04	ND	80	80	114
Japan	>100	89	55(62)	12,850	0.37	123.00	1.83	−0.30	95	97	120
New Zealand	17	33	6(18)	7110	0.27	3.44	1.86	−0.26	84	86	129
USA	59	135	28(21)	17,500	9.36	248.73	1.91	−0.23	99	98	138

[1] Based on Simkin *et al.* (1981) and other sources.
[2,3,4,5,6] From Anon. (1988).
* From various sources.
ND No data.

programme amounted to over 2% of GNP, or about 10% of average family income in extra taxes (Clapperton 1973a–d).

A similar deep-rooted cultural awareness of volcanic hazards may be seen in Japan, though it is more than three times larger in area than Iceland and has a population about 500 times greater. Natural disasters including volcanic eruptions are important elements in Japanese history and culture, and from the time of the Yamato Court (fourth to mid-seventh centuries AD) volcano deities were given a high status within the Shintô pantheon (Aramaki 1983, p. 194). Writing in the 1970s David Jones observed 'that it is small wonder . . . that an awareness of physical factors looms large in Japanese planning, although continued exposure to risks does produce a somewhat passive acceptance of danger and the feeling that any losses must be considered as an almost inevitable "natural tax"' (Jones 1974, p. 188).

In the United States and New Zealand the situation is different. New Zealand was first visited by Europeans in 1642 and it was only from 1800 that settlement started in earnest. The recorded history of volcanic eruptions is, therefore, very short and not representative of the range of activity which has occurred in the Holocene (Dibble 1983; and Table 9.1). The context of extreme natural events is one where direct experience is limited, so that reliance has to be placed on investigations by earth scientists and the subsequent integration of results into planning frameworks. Similar comments apply to the United States where only 21% of volcanoes which have been active in the Holocene have erupted since the country was settled by literate observers (Table 9.1). In both New Zealand and the USA volcanic hazards are regional problems to a much greater degree, than is the case in Iceland and Japan. The USA is of continental size and volcanoes are confined to three broad regions—Hawaii, Alaska and the Cascade Mountains—none of which is at the centre of national life or wealth, while in New Zealand the volcanoes which have been active in the last 2000 years are confined to four districts of North Island: The Bay of Islands; Auckland; Egmont; and Taupo. For the average inhabitant of both countries volcanic eruptions are a poorly understood and remote danger, and this perception has a strong influence on public policies, particularly in the provision of funds for hazard-research programmes. In 1987 the United States Volcano Studies Program, carried out principally under the aegis of the United States Geological Survey (USGS), had a seemingly generous budget of US$11 million (Filson 1987). However, this is modest when compared to the US$34 million devoted to earthquake research (USGS 1987). Historically far more people have been killed by earthquakes, which have affected a greater area of the country much of it, like California, of crucial importance to the creation of national wealth.

In the sections which follow, responses and adjustments to different types of volcanic event in Iceland, Japan, New Zealand and the USA will be reviewed. Because of historical, cultural and social differences it cannot be emphasized strongly enough that, although these responses appear superficially to have much in common, at a deeper level they show important dissimilarities. It is the interaction of a common corpus of scientific knowledge and the particular characteristics of a country and its peoples which defines the nature of responses to volcanic hazards.

9.2.1 Developed responses to high-magnitude/low-frequency events

High-magnitude/low-frequency events such as pyroclastic flows and surges, caldera collapses, rock slides, directed blasts and large-scale fall deposition are rare events in comparison with human life spans. Although there are many examples in the Holocene of high-magnitude/low-frequency events, when this span is reduced the list becomes much smaller. Historically it is not difficult to compile a list of major losses occasioned by high-magnitude volcanic events, such as the 14,000 people killed by Unzen, Japan, in 1792 and the villages buried during the major explosions at Bandai san, Japan, in 1888, but such a list is of marginal relevance since responses to these eruptions were pre-industrial and so provide few insights into how Japan would cope today. In the last 30 years developed countries have suffered little damage from high-magnitude/low-frequency eruptions and the study of adjustments and responses has involved reviews of policy effectiveness and preparedness against theoretical estimates of future eruptions and their effects. The United States and Japan are exceptions to the generalization, because the Mount St Helens eruptions of 1980 and to lesser extent those of Mount Unzen in 1991 were severe tests of the disaster preparedness of both countries and their abilities in handling major volcano-releated emergencies.

Several years before Mount St Helens erupted, action had already been taken to reduce potential losses. Action was stimulated by a brief given to the USGS by Congress under the Disaster Relief Act (1974), namely 'to reduce loss of life, property and natural resources that can result from volcanic eruptions and related consequences' (Bailey *et al.* 1983, p. 14). In fact following the cataclysmic events of 18 May 1980, many of the subsequent adjustments and responses were predetermined by work which had already been carried out. Specifically, personnel employed by the survey had published a hazard map using techniques of general prediction, together with detailed instructions for identifying the warning signs of eruption and procedures for informing government agencies of what action they should take (Crandell and Mullineaux 1978). This exercise and subsequent specific prediction carried out from early 1980 achieved considerable success by restricting access to the volcano and controlling evacuation, so that the eventual death toll was only 57. It is difficult to argue against the opinion expressed by Miller *et al.* (1981, p. 789) 'that forecasts were generally accurate, although the magnitude of the catastrophic and unprecedented landslides and lateral blast of May 18 greatly exceeded our expectation', or to dissent from the view that success was tempered by a failure adequately to warn the people whose property was inundated by vast quantities of ash (Saarinen and Sell 1985). With the exception of certain views to the contrary—including those of Haroun Tazieff (1988) the eminent French volcanologist—the overwhelming opinion of earth scientists was that the work of USGS and other agencies was an overall success. This is also the opinion of the author of this volume. What is more controversial is the manner in which predictions were handled by the geological survey, the media, government agencies and the general public.

Problems started virtually from the day the initial forecasts of Crandell and Mullineaux (1978) were published in a bulletin of the USGS entitled: *Potential Hazards from Future Eruptions of Mount St Helens Volcano, Washington.* Known as the 'Blue Book' (Saarinen and Sell 1985), the bulletin caused consternation amongst the public, since it did not accord with prevailing perceptions of risk and engendered much misunderstanding about what could be achieved by prediction once an eruption started. Consternation, which was not confined to the public but also affected journalists, politicians and even some earth scientists, was due to two factors. First, the Cascades volcanoes had not erupted for many years, the only previous instance this century being at Lassen Peak (California) between 1914–21 and the last activity at Mount St Helens was in 1857. Initial reactions to the 'Blue Book', therefore, ranged from the disinterested to the indignant; 'the idea of dangerous volcanic activity in the serene and verdant region was simply not credible to many people' (Peterson 1988, p. 4163). Secondly, officers of the survey had no experience in: predicting the course of a large eruption; handling the logistics of interagency cooperation; running a large-scale monitoring operation; and briefing the public, the press and politicians. When a volcano capable of high-magnitude events shows signs of reawakening, the course the eruption will take is almost always uncertain (Swanson *et al.* 1985) and rarely can earth scientists be precise (see Chapter 7). Generally they have a range of possibilities from which to choose, since knowledge of past behaviour is never perfect and specific prediction is severely constrained by many factors, including limited equipment, personnel, time and financial resources, and by the probability that previous eruptions may not be fully representative of the total range of volcanic activity (Peterson 1986). Additionally, false forecasts may undermine public confidence. In the case of Mount St Helens this was particularly germane, because in the Cascade Range swarms of earthquakes had never been followed by eruptions and some years earlier publicity had been given to seismic activity in the vicinity of Mount Baker and no eruption had followed (Saarinen and Sell 1985, p. 40). There is always the danger that communicating scientific doubt and qualification to journalists, other government agencies and politicians who seek unambiguous answers may cause severe difficulties. The USA prides itself on ease of public and press access to politicians and officials, and has complex arrangements governing the rights and duties of Federal and State governments and their agencies. Many rights are, moreover, written into legislation and safeguarded by the Constitution. Volcanologists are scientists and not experts in public relations, media communications and political briefings. The rift between two cultures—one grounded in the doubt and circumspection of applied science, the other within a *milieu* dedicated to clear, unambiguous communication—caused severe problems which had to be overcome before responses could be successful.

Looking at interactions between 'experts'—USGS scientists, employees of specialized, state and federal agencies—on the one hand and journalists, other agency employees, private corporations, decision-makers and the general public on the other, makes it possible to chart both successful and unsuccessful aspects of the responses to and adjustments made before, during and immediately after the major eruption of 18 May 1980. The 'balance sheet' which follows is based on detailed research carried out by the social scientists Thomas Saarinen, James Sell (Saarinen and Sell 1985) and other sources where acknowledged. On the positive side was the prompt action and monitoring introduced by the USGS from the time of the initial earthquakes on 20 March and the fact that a hazard map was not only available, but was also widely disseminated. Volcanologists were able to impress their opinion that a major eruption was imminent upon other public officials and an often sceptical public. The United States Forest Service

(USFS) acted closely with the geological survey and, since the volcano was located on their land, took the role of lead agency in coordinating initial responses and organizing press briefings. They introduced restricted zones, encouraged the Federal Aviation Administration to control airspace and organized an early meeting (26 March) to develop a contingency plan. This plan, as vital in conditioning responses as the 'Blue Book', was an adaption of a logistical scheme used for fighting forest fires. Within its jurisdiction the forestry service closed land to the public and managed to secure the agreement of private owners of the summit region to do likewise. On 27 March a set of 'Red' (no public access) and 'Blue' (restricted public access) zones were in place, which were modified as more information became available after the major eruption of 18 May (Fig. 9.1). Successful cooperation with county governments ensured that roads were closed and the reactions of certain local officials on the western side of the volcano, where the effects of eruption according to the 'Blue Book' were to be concentrated, were commendable (see Fig. 7.7). There was a major effort to warn inhabitants and plans were made to ensure evacuation, especially from river valleys with headwaters in the summit region. These plans were subsequently acted upon. The Federal Emergency Management Agency (FEMA) must also take some credit. Although not involved until after 18 May when President Carter declared the State of Washington a major disaster area, the FEMA introduced further measures including the administration of disaster assistance and the control of search and rescue. Additionally, as Saarinen and Sell (1985, p. 152) note, 'a major innovation was the creation of the technical information network, a group of experts from federal, state and private organizations who worked together to develop a set of Technical Information Bulletins on all aspects of the volcanic effects'.

Unsuccessful aspects of the response may be placed at the doors of most of the agencies involved. Media and government officials must also share some blame. Research by sociologists such as Warrick *et al.* (1981) and Sorensen and Mileti (1987) suggests that to be fully effective warnings must be specific, consistent, clear and certain and for the reasons already discussed this was not the case with respect to either the 'Blue Book' or the specific predictions made in the lead up to the events of 18 May. For instance, although the difficulties likely to be encountered by people living to the east of the volcano because of pyroclastic falls were described in the 'Blue Book', these were not emphasized sufficiently in the days and weeks leading up to 18 May. It comes as no surprise to find that because of this lack of warning many counties to the east of the volcano had few contingency plans. This deficiency probably resulted from volcanologists of the USGS playing a subsidiary role to officials of the USFS in the provision of information and Saarinen and Sell (1985, p. 58) suggest that in the future 'the USGS should include more specialized personnel in direct communication'. Like the geological survey the forest service handled the emergency successfully overall, though their response highlighted some weaknesses. The logistical framework developed from their experience of forest fires worked reasonably well, but did not fit the scale of the emergency and difficulties emerged in coordinating numerous agencies over a large area. Criticism can also be made of the State of Washington, which in spite of having several potentially dangerous volcanoes within its boundaries and the possibility of other forms of natural hazard, had a Department of Emergency Services which was 'a neglected, underfunded, ill-equipped agency not fully prepared to deal with a hazard event

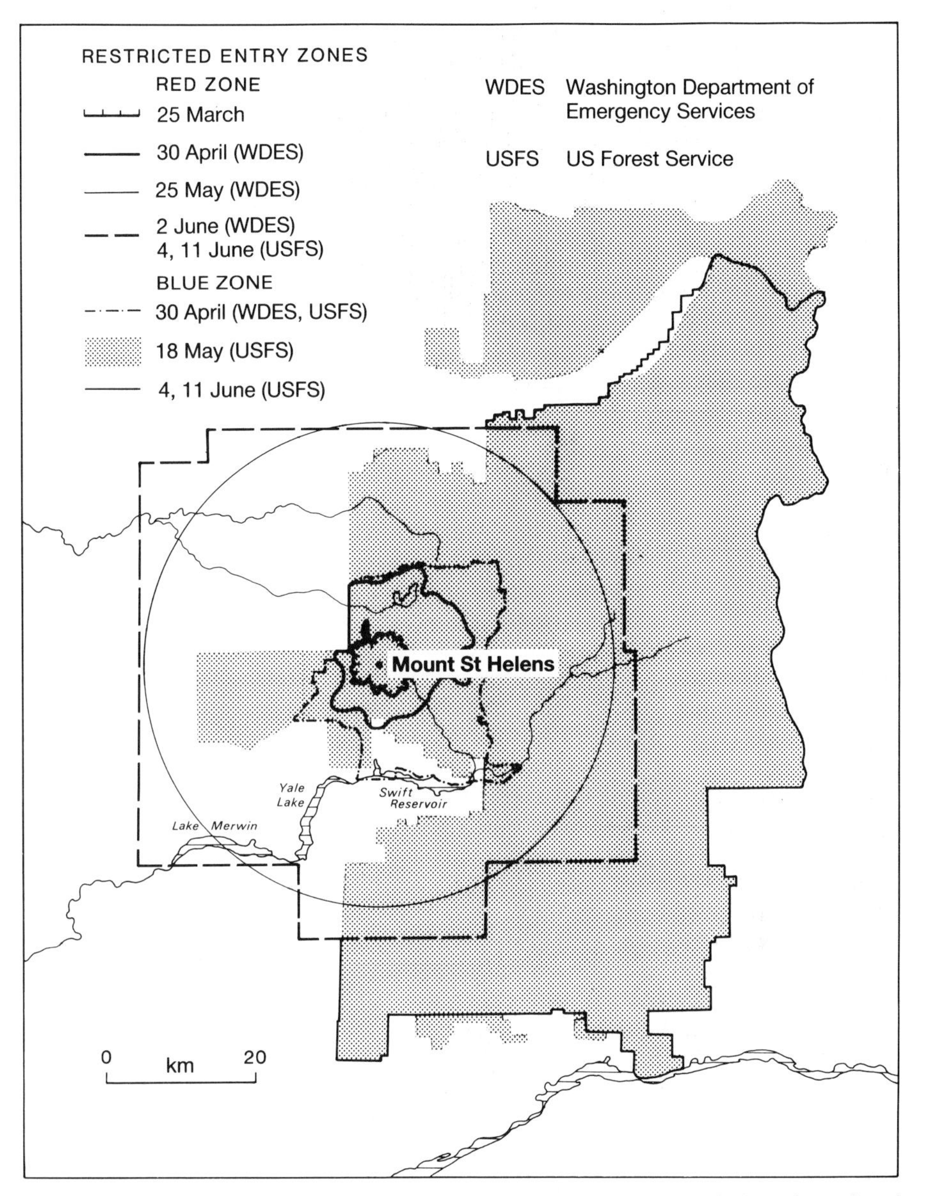

Fig. 9.1 Map of Mount St Helens and its vicinity showing the boundaries of restricted-entry zones established by the State of Washington and the United States Forest Service, between 25 March and 27 June 1980. Note the initial zone boundaries reflect estimates of volcanic risk, while the later boundaries also reflect property rights and political factors (after Foxworthy and Hill 1982, Fig. 13, p. 24; with additional information from Saarinen and Sell 1985).

of the magnitude of Mount St. Helens' (Saarinen and Sell 1985, p. 87). Deficiencies of personnel were in fact more serious than those of equipment and leadership rather than geological expertise was lacking. Following the events of 18 May major changes were made. Finally, at the national level the principal criticism was one of timing. The FEMA only became involved once a major disaster had been declared by the President. Policy was, therefore, reactive. Some commentators have argued that to be more effective federal disaster policy has to be pro-active and use the immense resources of the FEMA in an effective coordinating role before the event.

In any hazardous situation realistic assessments of danger are required so that individual citizens may choose appropriate adjustments and responses. The role of the mass media is, therefore, of critical importance. Studies carried out by Green *et al.* (1981) before the eruption confirmed that perceptions of risk were similar to those of a variety of natural hazards and in particular they found that for 'infrequently occurring hazards the public tends to perceive risk as very low, and cultural adaptation processes operate to encourage discounting or ignoring the true nature of risk' (p. 53). Once the first signs of reawakening were evident, then research showed that the media were a vital element in enhancing awareness, but could have been more effective if authentic notions of risk had been communicated well in advance of the eruption and, once it had begun, more emphasis given to the actions citizens could take to protect themselves.

Once an eruption starts then there is the possibility that friction between the media and scientists can develop so that the quality of information communicated to the public declines. In fact, although better than in some eruptions (see section 9.4), friction between scientists and journalists was not entirely absent from the Mount St Helens eruption (Peterson 1988), but cooperation between all groups improved as personal contacts and relationships developed during its course (Saarinen and Sell 1985). An interesting retrospective study dealing with many social and political aspects of the Mount St Helens eruption is to be found in Pallister *et al.* (1992).

In Japan the only recent example of a major high-magnitude/low-frequency volcanic event has been the eruption of Mount Unzen in 1991, but over the past few decades the country has been devastated by a number of natural disasters. These have included earthquakes in 1964 and 1978, typhoons in 1975 and 1990 and floods in 1975 and 1982. Together these have caused more than 10,000 casualties (Nakabayashi 1984). In addition an earthquake-related tsunami in 1983 killed 107 (Bolt 1988), whilst a combination of rapid tectonic uplift, fluvial erosion and phases of intense and prolonged precipitation means that major landslides are frequent and damaging (Jones 1974). As Fig. 9.2a shows unlike the USA and New Zealand there are few areas of Japan that are not without the risk of natural disaster and most parts of the Empire have been affected at some time during their history; many often and within living memory. The Japanese people and their government are well aware that successful responses require action before an eruption begins. Population densities are not only high in Japan (Fig. 9.2b and Table 9.1), but major concentrations also occur within the vicinities of active volcanoes. Given this history and geography, it is not surprising that for many years significant resources—both scientific and financial—have been devoted to the general and specific prediction. As discussed in Chapter 7, Japan was a pioneer in developing both sets of techniques and today has a comprehensive system of surveillance. Many observatories are run by the

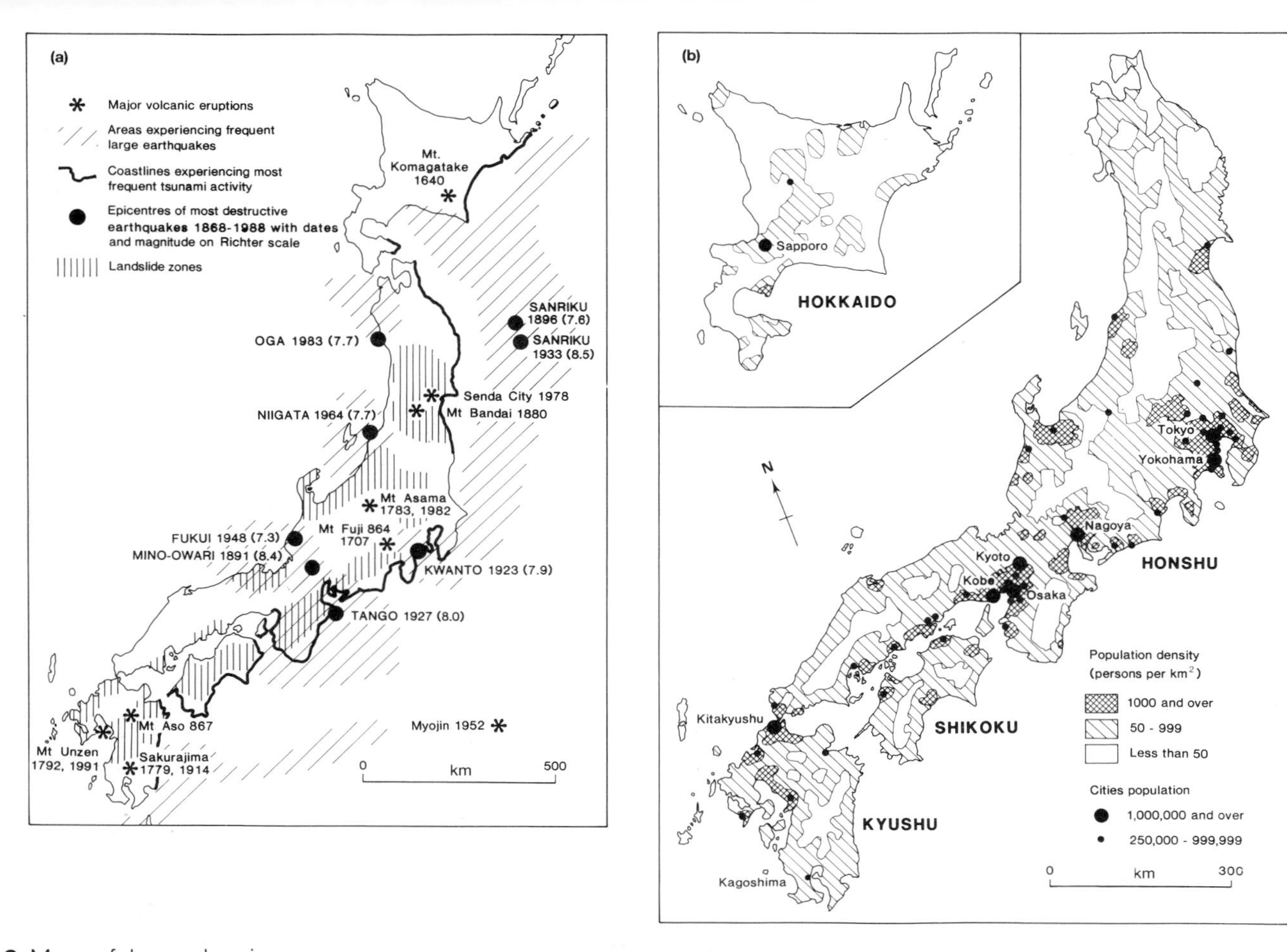

Fig. 9.2 Maps of Japan showing:
(a) Areas experiencing different types of extreme natural events. To this must be added meteorological hazards such as typhoons (amended after Jones 1974, p. 186).
(b) Population distribution (based on Coates 1974, p. 172, and later materials).

Japan Meteorological Agency, others by university groups, and in all, more than 15 of the most potentially dangerous volcanoes are observed (Shimozura 1983a) and for several hazard mapping has been carried out (Shimozuru 1983b). Effort is focused through the Japanese National Program of Predicting Volcanic Eruptions, which started after an eruption of Sakura-zima in 1974, its aims being 'to achieve a state of preparedness and the modernization of volcano surveillance in order to promote research on the prediction of volcanic eruptions and on the mitigation of disasters' (Shimozuru 1983a, p. 182). The programme operates through a series of five-year plans and has government sponsorship and support. Not only scientists but also administrators, the media and local residents are involved and much experience has been gained through research carried out at minor eruptions such as that at Usu volcano in 1977.

A Japanese proverb states that 'natural calamities strike at the time when one forgets their terror' (Shimozuru 1983b, p. 15) and this has not been forgotten by policy-makers. Diligence, attention to detail and efficient administration—aspects of Japanese industry and commerce—may also be seen clearly in the response of the State to the management of all natural hazards, including eruptions. In addition to the 'National Program' outlined above, there is also a comprehensive framework of law enshrined in such measures as the Basic Disasters Prevention Law (1962) and the Active Volcanoes Special Measures Law (1978), with an elaborate warning network and trained emergency services (Jones 1974; Jimbo 1983). Some idea of the success of Japanese policy towards natural hazards may be judged by its high status in international comparisons (Cullen-Tenaka 1986). Deaths commonly account for only ~3% of total casualties (Nakabayshi 1984) and the management of evacuation and resettlement following low-magnitude/high-frequency eruptions (such as those at Oshima volcano in 1986) have been very impressive.

Japan, perhaps more than any other country, has gone furthest along the road towards an industrial response (White 1973; Chester *et al.* 1985; Cullen-Tenaka 1986; and see Table 8.4). Emphasis has been placed on specific prediction, on emergency planning and logistics and less attention has been paid to land-use zoning (Jones 1974), or to integrating economic, land-use planning and disaster planning. It should not be forgotten that the 1970s did not see one major natural disaster, yet the average death toll still averaged over 250 and annual damages of between US$1.7 and 5.1 billion (Jimbo 1983). Also between the last major high-magnitude/low-frequency volcanic eruptions and the Unzen eruption in 1991 the whole nature of Japanese society had changed, the effect being to increase loss potential significantly. The eighteenth-century eruption of Asama which generated pyroclastic flows and lahars (Decker and Decker 1989) would, if repeated today, severely tax the abilities of the authorities to keep the death toll to the 1783 figure of 1300. Economic losses would be hundreds if not thousands of times greater. Earlier this century responses to volcanic eruptions, even severe ones like the Sakura Jima eruption of 1914, still showed a mixture of industrial and pre-industrial elements. Earthquake activity alerted the authorities and the public to the danger and there was both spontaneous and planned evacuation of the island, which lies off the coast of the city of Kagoshima (Fig. 9.2), and later of the city itself (Koto 1916; Jaggar 1924, 1956). Although accounts vary (see Blong 1984, pp. 147–8), there seems little doubt that individual actions were important in the evacuation and that families could not only cope, but also recover quickly. To quote Jaggar (1956, p. 105) 'the migration of more than 50,000 people with packs on their backs and with handcarts bearing household goods demonstrated how easily the Japanese people took to a nomad existence'. Recovery of the city was rapid, most of the migrants being fishermen and agriculturalists. Today Japan is quite different. It is an urbanized country, whose population has more than trebled during the last century and a half, it has one of the strongest manufacturing-based economics in the world and a high material standard of living (see Coates 1974; and Table 9.1). A consequence has been that the fine adjustment of people and environment which grew over thousands of years has been virtually eliminated as pressure on space, particularly on the limited supply of flat land, has become acute. Also for the last 40 years economic growth has taken priority over environmental concerns. It is relevant to ask whether Japan with its responses so strongly based on technology could actually cope with the mass evacuation of, say, the crucial economic region around Tokyo, which would be required if Mount Fuji erupted on even the modest scale of 1707 (see Shimozuru 1983b). Despite the civil discipline customarily shown by the Japanese, in view of the large numbers involved the answer probably ranges from no, if there was little advanced warning, to possibly if warnings were issued and heeded. In Japan there is no doubt that the benefits of development over the last few decades have more than compensated for the economic losses that have occurred from extreme natural events. It is likely that Japan could cope well with any low-magnitude volcanic event, but the problem is of coping with the infrequent and the severe. Developing an argument of Burton *et al.* (1978, pp. 16–127), because the total value of losses from natural hazards increases as economic growth proceeds, then there is statistical probability of a high-magnitude earthquake, a tsunami and a typhoon occurring in any one year. The question is whether even the Japanese economy could cope with multiple catastrophes on this scale. The effect on the world economy would be dire indeed (Hadfield 1991a).

Although not a major eruption in terms of many that have affected Japan in the past, the events of summer 1991 at Mount Fugen—part of the Mount Unzen volcanic complex on the island of Kyushu and close to the major city of Nagasaki (Fig. 9.2a)—provided a severe a test of the disaster preparedness of volcanologists and civil authorities alike. Mount Unzen is a volcano which has erupted five times between 860 AD and 1991, including the instance in 1792 when some 14,000 people were killed in a rock slide and subsequent tsunami. Whether this rock slide was triggered by an eruption or by an earthquake is debated (Decker and Decker 1989). Because of its history and location Unzen has been the subject of detailed volcanological research for many years and from the first precursory signs of activity in November 1989, the government and its agencies prepared themselves for a possible eruption (Global Volcanism Bulletin 1991a; Hadfield 1991b).

The earliest precursory activity was seismic and increased until 17 November 1990 when ash began to be erupted from two new vents. This quickly ceased, but the emission of steam continued and seismic activity remained at very high levels. On 12 February 1991 a second eruption occurred from a new vent and covered a large area with pyroclastic fall materials. Juvenile volcanic material was first detected on 12 May and some days later, heavy rain generated lahars which destroyed two bridges and led to the evacuation of 1300 people. Between 20–23 May dacite lava began to be erupted and further laharic activity caused a further 1100 people to be evacuated. The dome formed of dacite continued to grow and on 24 May a large explosion presaged dome collapse and the generation of a small pyroclastic flow. Further evacuations were ordered because of the hazard posed by lahars but, following action by the authorities to dredge material from a dam on the River Mizunashi—constructed in 1990 to reduce the danger from lahars—this order was rescinded. The dome continued to grow and further small-scale pyroclastic activity occurred on 26 May, together with laharic events which were again responded to by temporary evacuation. On 3 June an explosion was heard and a large pyroclastic flow emerged from the dome and travelled down the River Mizunashi at high speed. A detached pyroclastic surge entered a zone from which the general public had been evacuated and killed 41 people including 15 journalists, four taxi drivers, a few local residents who had not vacated the area, some members of the emergency services and three foreign volcanologists. Flows continued over the next few days and on 7 June the exclusion zone was increased in area and a further 1500 people were asked to leave, bringing the total to over 7200. Later, following additional large pyroclastic flows on 8 and 24 June, the evacuation zone was widened again. Many houses were destroyed and by 10 June nearly 10,000 people had left their homes.

The Unzen eruption of 1991 shows that with careful planning, good coordination, a high level of volcanological expertise and political will, even a moderate high-magnitude/low-frequency event can be handled successfully so that loss of life is minimized. Some additional indication of the care and diligence shown by the Japanese authorities is evident from the following statement by Daisuke Shimozuru (Chairman of the Coordinating Committee for the Prediction of Volcanic Eruptions). 'The Japan Meteorological Agency dispatched an observation team in mid October [1990] to intensify seismic observation. . . . Early in November, volcanic tremors were observed. We were worried about an impending eruption, and asked the Ministry for financial aid for observations by university scientists. . . . The university team set up seismic and deformation nets, in cooperation with the Shimabara Volcano Observatory' (quoted in Global Volcanism Bulletin 1991a). Hence, when the eruptions started, the civil authorities not only had their wide experience of handling natural disasters to draw on, but also useful baseline studies of normal and anomalous activity. There seems little doubt that in handling this event, the Japanese performed in a manner well up to their proven status as a major centre of volcanological expertise and a society able to mobilize its financial and logistical resources so as to minimize losses.

In the 1980 eruption of Mount St Helens David Johnston a volcanologist with the United States Geological Survey lost his life. The three volcanologists killed during the 1991 Unzen eruption were the highly distinguished Harry Glicken from the USA and Katia and Maurice Krafft from France. It is sobering to reflect that the two principal high-magnitude/low-frequency eruptions that have occurred in economically developed countries in the last two decades have caused the deaths of experts actually observing them and emphasizes once again that the detailed course of such eruptions cannot be predicted with any certainty even by the most eminent researchers. Prudent evacuation is still the only real option available to civil authorities and decision-makers.

The issues raised by high-magnitude/low-frequency events in New Zealand, although serious, are not so severe as in Japan. As Table 9.1 shows, population density is low and volcanoes are confined to just one area of the country: North Island (Dibble 1983; Johnson 1989). For many years it has been recognized that the volcanoes of North Island are potentially dangerous and detailed geological studies, together with some research on general and specific prediction, have been features of this country's adjustment to the threat posed by eruptions. As early as 1964, Searle estimated a probability of 2% per century for eruptions from certain basaltic centres within the Auckland District. More potentially dangerous eruptions have occurred in other districts since the time of European settlement. These have included more than 50 explosive eruptions of Ngauruhoe, some of which have been associated with small pyroclastic flows (Decker and Decker 1989). With the exception of 151 people killed in 1953 when a lahar from Ruapehu swept away a railway bridge and destroyed the Wellington to Auckland express, recent losses have been small; not least because of the remote location of many volcanoes, low population densities and the lack of really large eruptions during the last 150 years.

The Mount St Helens eruption was crucially important in the New Zealand context for it heightened awareness of high-magnitude/low-frequency volcanic events and provided valuable lessons to government about the technology and preplanning that are required to ensure successful responses (Dibble *et al.* 1985). In October 1980 a National Civil Defence Planning Committee on Volcanic Hazards was set up with terms of reference expressed in the form of ten objectives. These were applied to areas in and adjacent to the volcanic districts and centres which had been active during the last 4000 years (Fig. 9.3), with the caution that 'it cannot be guaranteed that future eruptions will not occur outside these areas' (Dibble *et al.* 1985, p. 370). Dibble and his colleagues estimate that eruption probabilities are variable and range from around 20% to below 5% per century and give detailed estimates of the likely damage, costs, population and livestock at risk, together with the measures that would have to be taken into account should an eruption of the type specified occur. Considering that it is so recent, the New Zealand response is impressive and shows what can be achieved in a short time when good science is placed within the context of a clear and committed national brief. It is chilling to observe that in the unlikely event of a large pyroclastic flow being generated from the Taupo Volcanic Centre (as happened 1800 years ago; Walker 1980), the population at risk would be 250,000–650,000 (about 7–19% of the New Zealand total), property valued at up to ~9% of GNP would be threatened and that mitigation strategies would be limited indeed. Fortunately the probability of such an event is estimated at only 0.8–2% per century, but if it did occur it would be one of the world's greatest human tragedies even accepting the low population densities involved.

9.2.2 Developed responses to low-magnitude/high-frequency events

Lava flows, small-scale and areally limited pyroclastic falls and other manifestations of Hawaiian and Strombolian volcanic activity are examples of low-magnitude/high-frequency volcanic events. Of the countries listed in Table 9.1, both Japan and New Zealand have volcanoes which are capable of producing low-magnitude events very frequently, but it is in the USA and Iceland that the best examples are to be found. Much has been written already about the Hawaiian volcanoes (Chapter 7) and accounts of human responses to them are available in many texts (Macdonald 1972; Williams and McBirney 1979; Macdonald and Hubbard 1982; Macdonald *et al.* 1983; Heliker *et al.* 1986;

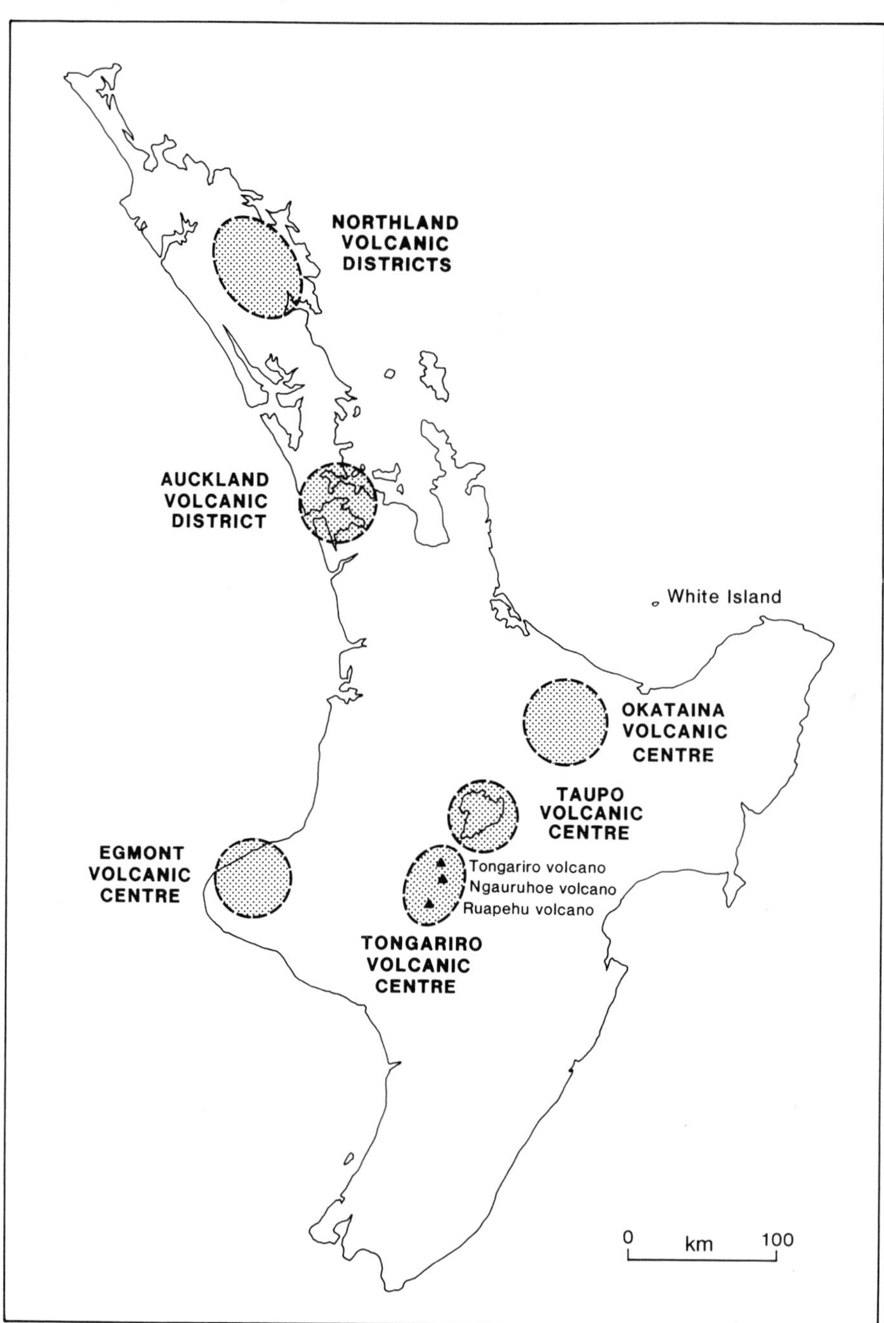

Fig. 9.3 Map showing the active volcanic districts and centres of North Island, New Zealand, in which eruptions have occurred during the last 4000 years (from Dibble 1985, Fig. 1, p. 371).

Decker and Decker 1989). For these reasons this section will be illustrated by examples taken from Iceland.

Iceland is no stranger to volcanic eruptions. Legend, saga and history attest to the effects of volcanic eruptions, the geology of the country is wholly volcanic and, on average, an eruption has occurred once every fifth year since records began around 1100 years ago (Sigvaldason 1983). Most Icelandic eruptions produce typical basaltic products (i.e. lava flows and falls) associated with ocean-ridge volcanism (Chapters 2 and 3) and do not threaten life directly. In the past when Iceland had a poor self-contained subsistence economy loss of life occurred indirectly and on a massive scale, as happened following the Laki fissure eruption of 1783, but today the major threat is to standards of living. In this century and especially since the Second World War, Iceland has achieved the status of a developed country. Partial independence from Denmark was gained in 1874, full independence in 1944 and today Iceland has a GNP per capita ahead of Japan (Table 9.1), a small but highly educated population and an economy based on the export of fish, fish products and smelted aluminium. In spite of a large external debt, a history of chronic inflation and some decline in fishing, Iceland is a wealthy country and one which has innovated a sophisticated response to volcanic hazards, involving several distinctive forms of adjustment (Sigvaldason 1983; Imsland 1989; Banks 1990).

In view of the frequency of eruptions, it seems surprising that losses have not been greater. The country is still sparsely populated (Table 9.1), but changes in Icelandic society have in recent years increased hazardousness significantly. Population has grown from only 143,000 in 1950 to over 250,000 today and changes in its density and distribution have been profound. Today over 36% reside in the capital Reykjavik, an even higher percentage if the surrounding area is taken into account, and most of the property, industrial plants, communications links and power supplies are located within a central northeast/southwest belt which is also the focus of active rifting and fissure volcanism (Fig. 9.4). It is evident, moreover, that the development of energy sources has increased vulnerability. Geothermal energy for both the heating of buildings and the generation of electricity has been developed and, although a feature of several other countries like New Zealand and Italy, in Iceland it now accounts for some 31% of total energy needs; second only to hydroelectric power (37%) and ahead of oil (29%) (Scudder 1990). Power plants are located within the same central belt as present-day volcanism and such areas of natural benevolence are very vulnerable in the event of an eruption. Many historic eruptions if repeated today would cause far higher economic losses than they did at the time.

Responses of the Icelandic authorities and people involve two classes of adjustment. First, there are long-term scientific and technological initiatives aimed at understanding the behaviour of active volcanoes so that general and specific prediction may be carried out and, second, there are measures to 'adjust to losses' through many of the measures listed in Table 8.2. The latter adjustments have been innovated through experience gained in eruptions over the last 20 years and many valuable lessons have been learnt both by government and individuals.

Understanding the volcanoes of Iceland has not proved easy because activity is 'unique', in the sense that the country is in an exposed part of the oceanic rift system and models of general prediction which have been developed for other regions of basaltic volcanism—such as Hawaii (Mullineaux and Peterson 1974)—

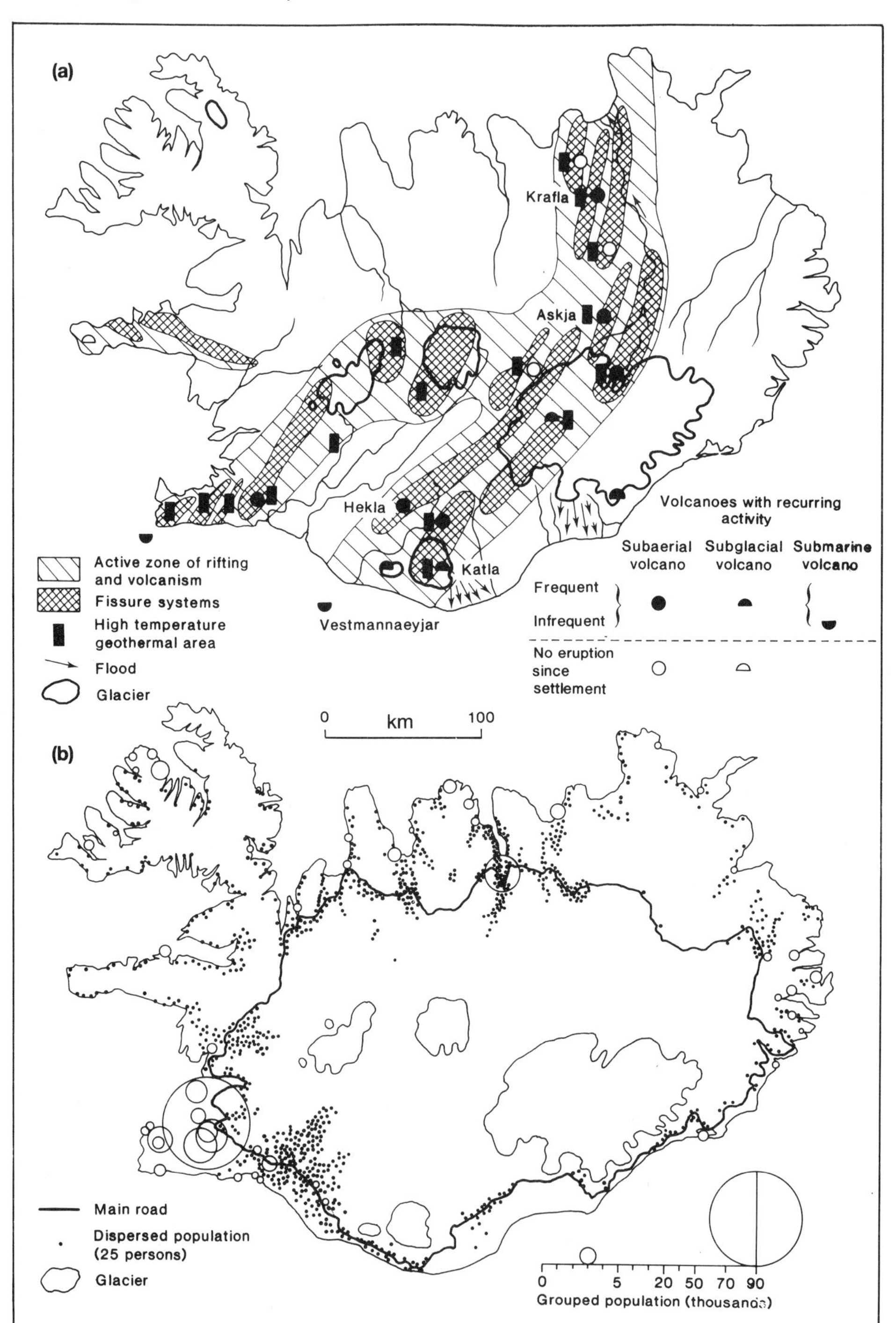

Fig. 9.4 Aspects of volcanic hazard in Iceland showing: (a) active volcanic areas in zones of rifting, with fissure volcanism superimposed; (b) distribution and density of population (based on Sigvaldason 1983, Figs 15.1 and 15.2, p. 194).

are not easily applied. Research using analogies from other areas has been produced, however, as well as studies of individual Icelandic volcanic centres (Thorarinsson 1979; Sigvaldason 1983; Imsland 1989), though by 1989 no hazard maps had been published and only one was being prepared. Problems faced by Icelandic volcanologists are not only scientific but also practical and, despite the interest in and support for applied research shown by all sections of society, the shear scale of the task and its cost are daunting. Around 30% of the country's area coincides with the active volcanic zones shown on Fig. 9.4 and although GNP per capita is high, the population is small, so that total national wealth and the number of trained volcanologists is limited. By way of comparison in 1986 1% of GNP equalled only around US$33 million, whereas in Japan the figure was US$16 billion and in the USA US$42 billion (Anon. 1988). Progress made by Iceland in the field of hazard assessment is, thus, commendable and represents a considerable national achievement; this being even more evident when specific prediction through monitoring and surveillance is taken into account (Sigvaldason 1983). A summary of the progress that has been made in hazard prediction in recent years is given in Table 9.2.

Table 9.2: Summary of volcanic hazards in Iceland: (A) low-magnitude/high-frequency eruptions; and (B) high-magnitude/low-frequency eruptions (compiled from information in Thorarinsson 1979; and Imsland 1989)

Frequency[1]	Size	Remarks[2]
(A)		
Decade	less than 1 km^3	Small-volume eruptions of chemically evolved basalts from volcanic edifices such as Hekla and Katla. Products include fall materials and lava flows. On a local scale falls are dangerous to farmland and vegetation. Lava flows cause semi-permanent sterilization of the land.
	1–10 km^3	Large-volume eruptions of chemically evolved basalts from fissures, without predictable eruption sites, e.g. Krafla area. The hazard is mostly from lava flows.
Centuries	less than 1 km^3	Moderately explosive eruptions from central volcanoes such as Hekla. Hazard mostly from fall deposits.
(B)		
	10–20 km^3	Very large-volume fissure eruptions of basalt from eruption sites which are difficult to predict, e.g. Laki fissure eruption 1783. The effects are likely to be catastrophic.
Millennia		Large-volume Plinian eruptions from central volcanoes.

[1] In addition there are low-magnitude/high-frequency events due to the interaction of volcanic activity and the external environment. These include bursts of glacial water (jökulhlaup). There are also hazards which have a frequency of less than decades and include ground fracturing, which is extremely common in rift zones and can affect roads and buildings, and gas releases which are rarely hazardous.

[2] Distinction is normally drawn between tholeitic and more evolved basalts which are in some cases more silica-rich.

Over the last 30 years responses to losses have improved with each succeeding event. Today Iceland has more recent experience of successfully responding to low-magnitude/high-frequency events involving pyroclastic falls and lava flows than any other developed country. One of the principal eruptions of the last 30 years was the formation of the Island of Surtsey (1963–67), but since this involved hydrovolcanic as well as magmatic activity and interaction of magma with the external environment, it is not relevant to consider it here (see section 9.2.3). Eruptions at Askja (1961) and Hekla (1970) were significant in their impacts, but the best illustration of how Iceland copes with eruptions is the disaster which occurred on the Island of Heimaey in 1973. Like Surtsey, the eruption on Heimaey occurred in the Vestmannaeyjar volcanic archipelago (Fig. 9.5) and was the most destructive of recent years, largely because it was located so close to the principal settlement of the island, Vestmannaeyjar, which before the eruption was the main fishing port and processing centre of southern Iceland accounting for between 17 and 20% of the country's fish exports. On the eve of the eruption Heimaey had a population of 5300 and a prosperous agricultural economy (Clapperton 1973b). The course of the eruption is summarized in Table 9.3 and on Fig. 9.5 and the effects on Vestmannaeyjar, the surrounding country and the very configuration of the island were dire.

The principal initiatives taken by the authorities are summarized in Table 9.4 and, although this table is self-explanatory in the main, a few aspects require elaboration. As already mentioned the exemplary response of the Icelandic authorities reflects a common heritage of fighting the effects of volcanic eruptions and an acceptance throughout the country that the burden has to be shared. It was not only a financial burden but for many a personal one as well. It is notable (Table 9.4) that most of the relief workers were volunteers and that the evacuees

Table 9.3: The chronology and effects of the eruption of Heimaey, 1973 (based on information in Clapperton 1973a, 1973b; and R.S. Williams and Moore 1983)

Dates (approximate)	Events
23–31 January	A 2 km long north/northeast trending fissure opened on the eastern side of Heimaey. The initial eruption was marked by a *fire fountain*, but was soon concentrated about 800 m to the northeast of Helgafell (Fig. 9.5). During the first three days submarine activity occurred offshore at the north and south ends of the fissure. Within two days a cinder cone—Eldfell—rose more than 100 m above sea level. Output was estimated at 170 m^3s^{-1}. Strong winds meant that tephra fell on the town of Vestmannaeyjar burying many buildings.
1 February–10 July	In early February the tephra fall diminished, but a large lava flow approached the eastern side of Vestmannaeyjar and threatened to fill the harbour of this important fishing port. Submarine activity cut electrical power lines and water pipes linking the island to the mainland. By the end of February, Eldfell was over 180 m high and its central crater fed an aa lava flow, which moved to the north, northeast and east threatening Vestmannaeyjar. By the middle of March the effusion rate of the lava fell to 17 m^3s^{-1} and by the end of April to about 9 m^3s^{-1}.

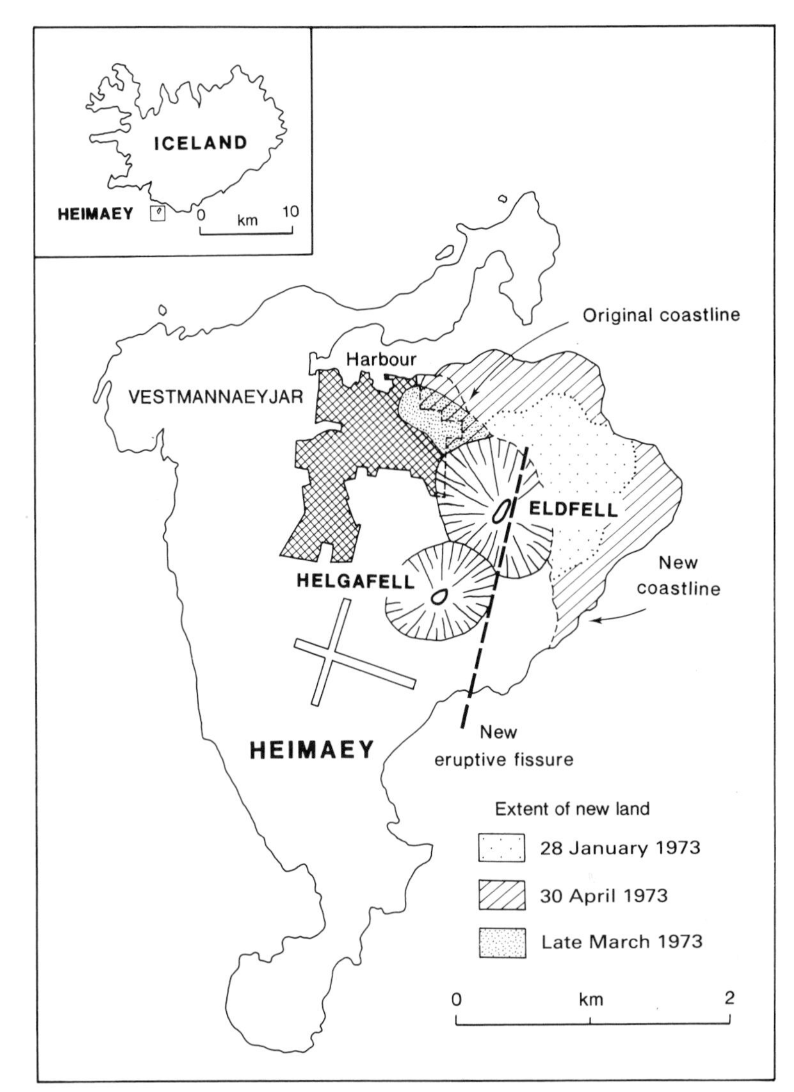

Fig. 9.5 Maps showing the location and extent of the Heimaey eruption, Iceland, 1973 (based on Williams and Moore 1983, p. 6).

readily found temporary accommodation amongst families on the mainland. Coping as a community by using a wide range of adjustments, whilst at the same time investing heavily in aid and recovery programmes and using high technology are characteristics not of an industrial, but of a post-industrial approach to coping with extreme natural events (Table 8.4). In addition and in the long term, the fact that the hospital on the island is now heated with steam from the cooling lava (Decker and Decker 1989) is a further example of how Icelanders harmonize with nature rather than fight against it.

A second aspect of responses and adjustments to the eruption was long term. For many life was never the same again. As Sigvaldason (1983) notes, the

Table 9.4: The principal initiatives (adjustments and responses) taken during the eruption of Heimaey, 1973 (based on Clapperton 1973a, 1973b, 1973c, 1973d; and McPhee 1988)

Stages in the eruption	Responses and adjustment
Initial stage	Within a few hours of the start of the eruption (01:50) hours local time, 23 January), the population of Heimaey had been evacuated. The fishing fleet was moblilized and, fortunately, only a day earlier the Public Safety Committee had completed plans for such an evacuation. Airlifts were also organized. The only casualty was a boy with a broken finger. During the rest of the first week, stress was placed on salvaging the contents of the town and at the same time fire fighters from the island, Reykjavik and Keflavik, tried to save as much of the town as possible. Up to 400 volunteers cleared fall ashes from roofs to prevent collapse.
Main stage	By early February it was evident that something had to be done to arrest the advance of the lava which was threatening the harbour (see Fig. 9.5). On 17 February a pipeline was completed and water sprayed on the lava flow front to cool it, increase its viscosity and reduce its rate of advance. Although many policy-makers and scientists were sceptical, this was reasonably successful. The volunteer force was temporarily evacuated from close to the eruption site because there was concern that the gases were both poisonous and asphyxiating. Much concern was felt whan a visiting European volcanologist claimed that, because of the amount of gas being emitted, an explosive outbreak of lava and pyroclastic material could occur anywhere on the island. Islandic scientists did not agree with this assessment and encouraged the volunteers to continue. Because of the density of the gas (90% CO_2) it rarely rose above 50 cm and only caused problems in closed spaces. By the end of February an all-out attempt was made to save the harbour by spraying water, but in the third week of March lava started to flow rapidly into the town; 25% of which was already destroyed. Pumping equipment was rushed in from the rest of Iceland and the USA and up to 60 m^3s^{-1} of water was sprayed on to the flow front.
Aftermath	A complex programme of recovery was put into place by the Icelandic government. In February it was estimated that the eruption would cost the nation US$ 14 million in loss of revenue from fishing and through the damage caused. This initial figure was soon exceeded. Parliament passed legislation to set up a Relief Fund and impose a special tax on the whole nation. They also requested aid from the Nordic Council and coordinated aid from abroad. In total funds exceeding US$ 36 million were available to enable the recovery to take place. Funds were used for: 1) income compensation for individuals; 2) paying for property destroyed at fire-insurance value; 3) transport costs of moving people and possessions back to Heimaey; and 4) paying the wages of relief workers and the costs of materials and equipment required for reconstruction. Sometime later grass was reseeded on the tephra and grazing re-established.

population of the area in 1979 was only some 88% of its pre-eruption total and a decade earlier Clapperton (1973d, p. 89), reviewing the results of a questionnaire survey of evacuees, reported that some 10% showed no desire to return to Haimaey. He notes that 'older people have been profoundly affected by the changes and, finding the services and comforts of the mainland very attractive, have little real incentive to go back'. The eruption shows that even with a sophisticated comprehensive response, long-term recovery can be slow and depends critically on younger and more adaptable members of a community.

Finally, the Haimaey eruption demonstrates that luck and decisions made on the balance of probabilities rather than hard scientific evidence, are important even in a country as prepared for eruptions as Iceland. As Table 9.4 shows, the authorities were prepared to experiment with the water-cooling of lava flows even though it was a largely untried technique. Also they were able quickly to decide that the experience and expertise of their own scientists and carefully considered pre-eruption plans had to be relied on rather than the opinions of foreign visiting volcanologists, even those of considerable eminence. Volcanic eruptions and their course once started remain difficult to predict and any successful response has within it an element of good fortune. The fishing fleet was in harbour at the time of the eruption so that evacuation could proceed rapidly, the Public Safety Commission had just completed contingency plans and the winds during the eruption reduced falls on critical installations (Clapperton 1973b). Even in the most well-regulated society, good fortune can still be important in reducing losses.

9.2.3 Developed responses to volcano-related events

A volcanic eruption can produce major losses which are not conditioned purely by the nature of the ascending and/or erupting magma, but rather by interactions with elements of the external environment. For example, volcanic earthquakes produced by fracturing of surface rocks by ascending magma caused a magnitude 7.2 earthquake at Kilauea volcano (Hawaii) in 1975, while in the six months leading up to the eruption of Usu (Japan) in 1944 some areas were raised by up to 50 m (Blong 1984). Earthquakes and ground deformation are not the only volcano-related phenomena which have caused casualties and economic losses in developed countries. Neall (1976) notes that in historical times lahars have been responsible for a death toll of over 11,500 in Japan alone, while tsunamis have caused significant loss of life in those countries bordering the Pacific Ocean (Blong 1984).

Both earthquakes and tsunamis can be, and indeed usually are, caused by tectonic processes and consequently responses to them are normally placed within earthquake reduction programmes (see Bolt 1988). In the United States, New Zealand and Japan the drawing of zoning maps (general prediction) and research aimed at specific prediction is being pursued actively and contingency plans are at a high state of readiness because of the size of the potential death toll and projected magnitude of economic losses.

Just as it is difficult to be certain whether an earthquake in a volcanic region is due to magma movement or to tectonic processes, so it is also a problem to decide whether historical tsunamis were caused by volcanic activity or by submarine earthquakes. In an extensive review, Latter (1981) compiled a list of 92 tsunamis caused by volcanic activity and found that they had been caused by a number of mechanisms. Some 22% stemmed from volcanic earthquakes (e.g. Sakura-zima, Japan, 1914), 27% represented the movement of pyroclastic flows and surges into water (e.g. Augustine, Alaska, 1885) and 19% were produced following submarine eruptions. As Blong (1984) notes, although prominence in the volcanological literature is normally given to tsunamis caused by caldera collapse and subsidence, only some 9% of historical cases could be ascribed to this cause. The remainder, about one-quarter of the total, were due to no determined cause. The vast majority of tsunamis are not caused by volcanic eruptions, but by submarine fault movements and in most developed countries responses and adjustments are again considered within earthquake-reduction programmes. At the present time several countries bordering the Pacific Ocean have national warning centres (USA, the former USSR, Japan and French Polynesia) and these centres make predictions based on data provided by seismic and tsunami sensors at sea, land-based seismographic stations and satellites. Some coordination of information and warnings for the whole Pacific basin is produced by the United States (Dohler 1988).

Lahars (using the term as in Chapter 5 to include all volcanic debris flows) are normally dealt with in a manner similar to other high-magnitude/low-frequency events in developed countries. Here the experience of the United States is typical and for the Cascade Range, lahars are invariably included in volcanic hazard mapping (see Crandell *et al.* 1979). In addition responses and adjustments are determined by the same planning contexts as those which apply in the event of high-magnitude, yet infrequent, eruptions (see Saarinen and Sell 1985).

Perhaps the greatest challenge to applied volcanologists and policy-makers in recent years has been the increased recognition of the hazards associated with hydrovolcanism. As discussed in Chapter 5 (section 5.3.6.2), although the effects of interactions between water and magma have been hinted at for many years, it is only recently that their importance has been recognized fully. As Fisher and Schmincke (1984, p. 231) note this has 'opened a Pandora's box of new problems' particularly for the hazard analyst. In the first place, recognition has defined many new situations in which explosive eruptions can occur and many of these are not solely, or even principally, determined by magma composition. Secondly, new knowledge has clarified understanding of surge deposits (see Chapter 5, section 5.6.3; and Table 7.2) and shown that as well as being associated with pyroclastic flows (hot, dry surges), they also occur as products of hydrovolcanic eruptions (cold, wet surges) (Crandell *et al.* 1984a). Thirdly, new knowledge has called for a reassessment of the hazardousness of many volcanic regions. One example of the latter is research carried out on explosive eruptions in Hawaii by Decker and Christiansen (1984). This has shown that, although Hawaiian eruptions are reputed to be benign and non-explosive, some 1% of prehistoric and historic eruptions have been explosive including the 1924 eruption of the Halemaumau crater of Kilauea volcano. As the authors remark, this eruption was 'more curious than dangerous' (p. 131), but the geological record indicates that major and potentially much more dangerous hydrovolcanic events have occurred, the last being only 200 years ago. They conclude with the comment that volcanologists have to be careful that familiarity with low-magnitude spectacular yet harmless eruptions does not breed familiarity and contempt for the possibility of more dangerous events. Because hydrovolcanism is a recent addition to the volcanological canon, it is highly likely that the reinterpretation of the volcanic histories of many volcanoes will reveal past episodes of hydrovolcanic activity and that this will mean that many hazard maps and many contingency plans will have to be modified. In short, the applied volcanologist will have to be careful that hazard zoning and risk assessment keeps pace with advances in pure research.

There is finally a mixed group of hazards that represent the interaction of volcanoes with their external environments, which are normally considered within either existing volcanic hazard assessments or as part of other hazard programmes. One example is landslides, which in the United States are both included on volcanic hazard maps and also as part of a specific programme which covers landslides due to a variety of causes. In 1987 this programme was funded to the tune of US$2 million (Filson 1987). Another example are Jökulhlaups (floods caused by the bursting of a glacial lake), that in Iceland have caused significant casualties in past centuries and which are considered in current programmes of hazard reduction (Imsland 1989).

9.3 Underdeveloped responses

Chile, Colombia, the Comoro Islands, Indonesia, Mexico, Papua New Guinea, the Philippines and Zaire (Table 9.5) are chosen to exemplify underdeveloped responses and adjustments to volcanic hazards. In comparison with countries showing developed responses (Table 9.1), those listed in Table 9.5 are much more varied. Some of this variety is economic and demographic and is easily captured in figures showing differences in living standards, population numbers, educational attainments, food sufficiency and economic dependency (Table 9.5), but deep-seated historical, cultural and social characteristics are also of importance.

At first sight Indonesia would not be expected to cope well with the effects of eruptions. Not only does it have a large number of active volcanoes many of which produce very dangerous pyroclastic flows and lahars, but over time hazardousness has increased in what remains one of the poorest countries in the world. A high proportion of the active volcanoes and of the 175,000 deaths that have resulted from volcanic activity over the past 200 years have occurred on the Island of Java, yet its population density has increased from around 400 per km^2 in 1955 to a projected 1700 per km^2 by the year 2000 (Zen 1983). Despite the poverty of the country the Indonesian response to eruptions is sophisticated, because of the high priority given to it by the authorities and the important place volcanoes hold in the national cultural consciousness. It is claimed that the eruption of Merapi volcano in 1006 AD caused many deaths and destruction so widespread that the Hindu Rajah left Java for Bali and the island fell under the influence of Islam (Decker and Decker 1989). This has had a profound effect on the subsequent religious development and social traditions of the country. Indonesia also gains much from its volcanoes and the possibilities for agriculture presented by the weathering of volcanic rocks to form fertile tropical soils is one reason why the country has been able to sustain such high population densities (Macdonald 1972). Indeed Mohr (1945) went so far as to suggest that there was a strong and simple relationship between population density and soil type. More recent research by Ugolini and Zasoski (1979) has shown that the relationship between volcanic soils and rural population density is much more complex and less physically determined. Reporting work on tephra-derived soils—much of it originally carried out by Tan (1964)—they assert that volcanic andosols are not

Table 9.5: Examples of countries having 'underdeveloped' responses to volcanic eruptions (for additional details and explanations see Table 9.1)

Country	Volcanoes[1]			GNP per capita[2] (US$ 1986)	Area (million km^2)	Population 1989* (millions)	Population fertility[3]	ODA as % of GNP (1984–86)[4]	Secondary school enrolment (% of age group 1983–86)[5]		Daily food energy supply as % of requirement, 1983–85[6]
	Eruptions known to have affected people and their activities	Number active in Holocene	Active in historical time (% of those active in Holocene time)						Male	Female	
Chile	19	79	30(38)	1320	0.76	12.80	2.50	<1	63	69	106
Colombia	8	13	7(54)	1230	1.14	31.72	3.58	<1	77	50	111
Comoro Islands	4	2	1(50)	280	(1860 km^2)	0.43	6.13	35	ND	ND	89
Indonesia	155	~135	82(60)	500	1.92	177.96	3.48	1	45	34	135
Mexico	5	34	14(41)	1850	1.97	84.46	3.98	<1	77	56	116
Papua New Guinea	20	37	15(41)	690	0.46	3.72	5.28	12	15	8	79
Philippines	~40	~40	~14(~35)	570	0.30	59.7	3.91	2	63	66	102
Zaire	5	6	3(50)	160	2.35	39.62	6.09	8	81	33	97

[1] Based on Simkin *et al.* (1981) and other sources.
[2,3,4,5,6] From Anon. (1988).
* From a variety of sources.
ND No data.

Table 9.6: Principal economic, social and political problems facing Chile, Colombia and Mexico in the 1980s and early 1990s (based on Banks 1990, and many other sources)

Country	Principal economic, social and political issues
Chile	Despite having a high proportion of its population employed in agriculture, since 1945 food imports have increased and contributed to record balance of payments problems in the early 1980s. Generally poor and fluctuating prices for export commodities (i.e. copper, gold, silver, coal, iron and nitrates) have also caused difficulties. The application of monetarist economic policies involving programmes of austerity, privatization of state enterprises and lax import restrictions improved some economic indicators in the late 1980s, but at the expense of high unemployment and an external debt of US$ 20 billion. Following the overthrow of the democratically elected Marxist President Allende in 1973, politics were dominated by a military junta under General Pinochet. His presidency ended in 1990.
Colombia	Colombia's economy depends heavily on agricultural exports, particularly coffee. In recent years, however, illegal cultivation, processing and export of cocaine and marijuana have been equally important. Wealth is unevenly distributed, unemployment is high and the country has become a battleground between the government, left-wing guerrillas, drug traffickers and right-wing death squads. In 1989 the Medellin Cartel of drug producers was estimated to have a business worth US$ 6 billion. In the late 1980s drug money was widely held to have infiltrated most organs of the state. Under a new government the situation improved in the early 1990s, but was uncertain by 1992.
Mexico	Some one-third of the population is engaged in agriculture. Increased urbanization—particularly the growth of the capital, Mexico City—has caused many problems which include the provision of adequate housing, health care and pollution. In the last 50 years, industrialization has been rapid, but the benefits have not been distributed widely and the rural poor are virtually unaffected. The 1980s was a period of economic recession, widespread unemployment, the export of capital and the growth of a large international debt which proved impossible to service. These problems were made worse by the earthquake of 1985 and a further collapse in oil prices. Since the 1920s politics has been dominated by the Partico Revolucionario Institucional. According to one reformist member of the party, the party is riddled with corruption, autocracy and bureaucratism.

naturally, but only potentially, fertile. Over the centuries farmers have had to manage soils very carefully, through additions of phosphorus and organic matter to counteract high levels of aluminium.

Indonesia can rely on the quality and expertise its volcanologists. Research is of high quality and expertise of international standing, reflecting the continuation of a tradition which stretches back nearly a century. Before and after the Second World War the volcanoes of Indonesia were studied in considerable detail by R. W. van Bemmelen, Neumann van Padang and other officers serving the Dutch colonial government. Since independence in 1949 this tradition has been maintained and enhanced by the Volcanological Division of the Geological Survey of Indonesia (Kusamadinata 1984). It remains the case, however, that in spite of the

Fig. 9.6 As well as 'high technology' adjustments to volcanic hazards, such as those used by the Volcanological Division of the Geological Survey, the government and people of Indonesia use a variety of more traditional approaches. The photograph shows a simple warning device in a village on the slopes of Semeru volcano, Eastern Java. Natural hazards (i.e. lahars) are announced by four bangs on the gong (photograph reproduced by courtesy of Professor H. Th. Verstappen).

priority given to volcanic-hazard abatement the pattern of losses still remains more typical of an underdeveloped rather than a developed country, both in terms of death tolls and absolute economic losses which fall disproportionately on poorer sections of the community (Fig. 9.6).

Papua New Guinea, like Indonesia, has a colonial heritage having gained independence from Australia as recently as 1975. It is also one of the poorest countries in the world, has low levels of educational attainment and high population-growth rates. It is nonetheless socially, economically and culturally quite distinct. The indigenous population comprise over 1000 groups who speak more than 700 languages. Most practice pantheistic religions, illiteracy remains high—there being no written historical tradition until the era of European settlement—and an estimated 70% of the population relies at least in part on subsistence agriculture and hunting (Banks 1990). The country depends heavily on external aid especially from the former colonial power, although this has decreased of late (see Table 9.5). In spite of low overall population densities ($\sim$8 per km^2), food supply is below basic nutritional requirements for most of the population. Some diversification of the economy has occurred in recent years involving mining, oil exploration and extraction. These activities offer promise for the future, although the effects of budget deficits, low commodity prices and a lack of private investment all presented major difficulties during the 1980s (Banks 1990).

Though population densities are low in Papua New Guinea, volcanoes are extremely active and in the time that has elapsed since the first European settlement little more than a hundred years ago, some 80 eruptions have been reported from 17 volcanoes (Mori *et al.* 1989). Included in this list is the 1951 eruption of Mount Lamington which devastated more than 200 km^2 and killed ~3000 people (Decker and Decker 1989). It is only from the time of colonial settlement that written information on eruptions becomes available, but before this indigenous tribes developed elaborate myths, legends and religious practices which tried to make sense of them. The oral tradition and the cultural milieu associated with it still influence responses to eruptions. As mentioned in Chapter 8, Blong (1982) makes note of a tribal group in the Highland Region which has adjusted its agriculture to frequent tephra falls, while the reactions of different tribes to the Mount Lamington eruption showed great complexity and many characteristic elements of pre-industrial responses to loss bearing (Blong 1984, p. 178; Table 8.4; and section 9.3.4). Unlike Indonesia, Papua New Guinea does not have such a strong tradition of volcanological research and still relies heavily on the efforts of expatriates and peripatetic foreign 'experts'. In this respect Papua New Guinea is more typical of underdeveloped countries than Indonesia and more dependent on responses and adjustments involving technological transfers.

The countries of Central and South America emerge from Table 9.6 as a distinct group. In GNP per capita terms they are not so poor as Indonesia and Papua New Guinea and share many common features in their histories and cultures. Their populations are ethnically mixed (European and Indian), they are all Spanish-speaking and have similar colonial legacies, legal systems, administrative structures, histories of political instability, dominance of powerful groups and religious adherence to Roman Catholicism. Chile, Columbia and Mexico display strong economic pluralism with so-called islands of development and wealth within seas of underdevelopment and poverty. Long-term responses and adjustments to natural disasters highlight many elements of the economic marginalization of the poor which were discussed in Chapter 8 (section 8.3) and by Hewitt (1983a). The three countries have a plethora of economic, social and political problems, which have served to reduce the priority given by governments to the management of risks from active volcanoes and some of these are listed in Table 9.6. Priority is also reduced because volcanic eruptions are often neither the most frequent nor the most serious natural hazards these countries have to face. In addition to crop failures and extreme meteorological events, the toll from earthquakes has been a recurring historical theme. In this century some 15 serious earthquakes have occurred which have included the Colombian earthquake of 1906, earthquakes in Chile in 1939 and 1960 (combined death toll ~36,000) and the Mexico City earthquake of 1985 (9500 killed, 30,000 injured and damage estimated at over US$3 billion; Eiby 1980; Bolt 1988). Scientists from Chile, Colombia and Mexico have made many valuable contributions to volcanology, but the resources made available for pure and applied volcanology are severely limited. Much research has, consequently, involved aid programmes and scientists from abroad. The opinion of Francis (1970, p. 798) expressed with reference to the Andes applies more widely, that despite their importance 'volcanoes . . . are relatively little known geologically. Systematic mapping . . . is not well advanced and such maps as have been made have an economic bias with scant attention paid to the volcanic areas'.

The Comoro Islands located in the Indian Ocean to the north of Madagascar differ greatly from the other countries in Table 9.5. They have only been independent from France since 1975 and remain a poor micro-state with a lingering dependence on the former colonial power. Post-colonial history has been dominated by serious problems including the fact that one of the islands, Mayotte, has showed no desire to secede from French rule. In addition political instability has been endemic, with more than 14 coup attempts since 1975. Economic problems are severe. Earnings from the principal export—vanilla—have fallen, by the late 1980s debt servicing had reached 50% of total export value and serious problems occurred over domestic food supply. Foreign aid particularly from France remains the country's lifeline. The volcanoes of the Comoro Islands are well known geologically (Strong and Jacguot 1970; Upton *et al.* 1974) and in the Holocene, activity has been confined to just two on the most northerly island Njazidja, formerly Grande Comore. In historical times only one of these, Karthala, has been active and has erupted at least 22 times since 1828.

The final country listed in Table 9.5 is the Philippines. This densely populated and poor country has suffered major problems of political instability in the last few decades which have been grafted to underlying issues of social and ethnic complexity (Banks 1990). The inhabitants of the 7000 islands which make up the Philippines are predominantly of Malay ancestry, but the country was claimed by the Spanish Crown in 1521 and more than 350 years later lost to the Empire during the Spanish-American War. In 1898 the Philippines became a self-governing commonwealth of the United States and full independence was granted in 1946. The colonial legacy is still strong and is expressed in: linguistic diversity (English, Spanish and Filipino—based on the Tagalog language common in the area around the capital Manila—being official languages); separatist tendencies amongst the Muslim minority concentrated in the south; continued problems over United States neo-colonial influence; and an economy dominated by the export of primary raw materials to more industrialized nations, with all this implies in terms of economic dependency. Recently the economy has become more diversified. Textile and food-processing industries are now of significance, but the economy still depends upon exports of wood, sugar, coconut products and mined raw materials. Intrinsic economic weakness was exacerbated in the 1970s and early 1980s by financial mismanagement, corruption and armed political rebellion associated with the regime of President Ferdinand Marcos. During the later period of his rule (1980–86) the economy contracted, foreign debt and the balance of payments became major issues and frequent devaluations of the currency took place. Following the demise of the Marcos regime in 1986 and the democratic election of President Corazon Aquino, the economy has improved and by 1988 economic growth had reached 6.8% and inflation fallen to 8.7% (Banks 1990). Nevertheless the economy remains weak, democracy fragile—with several failed coup attempts since 1986—and the issue of United States military presence was not finally resolved until after the Mount Pinatubo eruption in 1991, although talks between the two governments in 1990 led to a consensus that the principal question was not if, but when, American forces would leave.

The Philippines has a long and in recent years distinguished history of dealing with volcanic eruptions (Wilkie 1991; Table 9.5). Despite its material poverty, general standards of education amongst men and, notably for poor countries, women as well, are high. In the central Philippines Mount Mayon has erupted more than 40 times since 1616 and in this densely populated region more than eight eruptions have caused fatalities, while Taal—a stratovolcano with a huge crater lake—has caused deaths in 1716, 1749, 1754, 1911 and 1965 (Decker and Decker 1989). Work on general prediction has been carried out by the Philippine

Commission on Volcanology (COMVOL), which has also drawn up plans for restricting public access and evacuating large numbers of people during volcano-related emergencies (Pena and Newhall 1984). In some respects the Philippines is rather like Indonesia in that its volcanological expertise is well above what might be expected from its economic status as a poor country. Strong links with the United States are of importance.

In section 9.2 it was argued that for countries showing developed responses, historical examples of adjustments were of only marginal relevance in any contemporary discussion of responses. Economic development means that forms of coping which prevailed in the past are not likely to be repeated if similar eruptions occurred today. In poor countries this is not the case and there are many examples where elements of pre-industrial forms of adjustment and response (see Table 8.4) continue to be relevant. This has been recognized by social scientists who have carried out investigations into long-term social and economic adjustments, particularly in Mexico and Papua New Guinea (see Schwimmer 1977; Nolan 1979; and section 9.3.4).

9.3.1 Underdeveloped responses to high-magnitude/low-frequency events

Despite the fact that most countries showing underdeveloped responses to volcanic eruptions are in tectonic situations which make them prone to high-magnitude explosive magmatic eruptions (see Chapters 2 and 3), in any period of 20 years the chances of more than a handful of events affecting the populations of the countries listed in Table 9.5 are remote indeed. This is certainly true of the period since 1970. During this time there have been many eruptions in which volcano-related phenomena such as lahars have caused high death tolls and major economic dislocation (see section 9.3.3), but primarily magmatic eruptions have not exacted such a burden. In this section varying responses and adjustments will be illustrated by means of three examples: the eruption and major loss of life caused by El Chichón in Mexico (1982); the highly successful responses of the Indonesian authorities to the Una Una (also known as Colo) eruption of 1983 (Sudradjat-Suratman 1986); and events at Mount Pinatubo, Philippines, in 1991.

Eruption of El Chichón volcano (Mexico)—also known as Chichonal—in 1982 was a significant event for a number of reasons. First, the effects of the eruption were devastating for the local inhabitants with several villages destroyed and many people killed by pyroclastic flows. Estimates of the death toll, modest in the immediate aftermath of the eruption (tens to hundreds), are now pitched at between 2–3000 (Decker and Decker 1989). Secondly, the impact of this eruption on world climate was profound (see Chapter 6) and, thirdly, all this occurred at a volcano which, apart from some solfataric activity, was assumed to be dormant if not extinct (Simkin *et al.* 1981). The eruption was not predicted and no hazard zoning had been carried out. As the Mount St Helens example shows (see section 9.2.1), even when much is known about the geological history of a volcano, a hazard map has been drawn up and much time, effort and resources have been devoted to specific prediction, the detailed outcome of a high-magnitude/low-magmatic eruption is still difficult to estimate with certainty. In a remote region of a poor country the situation is far worse. Although known to local inhabitants, the fact that a volcano was actually present in the remote, densely forested and isolated Sierra de Chiapas region of southern Mexico was only apparent to scientists when in 1928 it was discovered during a reconnaissance geological survey (Mullerried 1932). Interestingly, Friedrich Mullerried assumed that El Chichón was capable of renewed activity and commented not only on its solfataras, but also on seismic activity in the vicinity of the edifice (Duffield *et al.* 1984). Between the 1930s and the eruption of 1982 little additional research was published so it is not surprising that the scale of the eruption came as a shock to people both within Mexico and to the international volcanological community. The timing of the eruption was unfortunate, as just a few months before the cataclysmic events of March and April 1982 a report on the geothermal potential of the region concluded that a major eruption had in fact occurred ~1200 years ago—involving *inter alia* the generation of pyroclastic flows. Indeed an episode of fall deposition, which occurred 130 years ago, was still part of the collective folk memory of the inhabitants. The authors of the report believed that there was a high risk of renewed activity (quoted by Duffield *et al.* 1984). If this report had been published a few years earlier and been followed by more detailed research, it is likely that the Mexican authorities would have been in a better position than they were to minimize losses. In fact in the months leading up to eruption some seismographs were installed following an increase in earthquake frequency and

Table 9.7: Two stages in the eruption of El Chichón, Mexico, in 1982 (based on information in Tilling 1982; Duffield *et al.* 1984; and Medina 1986)

Stage	Timing	Volcanic activity	Effects on area
1	28 March	Large explosive eruption producing an eruptive column more than 18,000 m high and fall materials covered the area. Near to the volcano these were incandescent and more than 10 cm in size. Possibly a high hydrovolcanic component to this phase.	Falls caused roofs to collapse and structures to burn.
2	3 and 4 April	After weaker explosions on 30 and 31 March and 2 April, high-magnitude explosions occurred, a Plinian column was generated and pyroclastic flows (associated with surges) swept down radial drainage valleys. These flows were caused by column collapse. The eruption column reached a height of 25,000 m. The eruption ended with weaker explosions on 5, 6, 8 and 9 April. Stages 1 and 2 of the eruption removed virtually all the pre-1982 summit dome and replaced it with a 300 m deep crater.	Several villages—including Francisco León—were totally destroyed within a 6.5 m radius. Two months after the eruption, the pyroclastic flow deposits were still too hot to touch. Flows dammed rivers and streams and when these burst, further destruction was caused by hot floods and lahars.

magnitude and there was some spontaneous evacuation of the population, but the absence of a permanent, comprehensive network prevented any worthwhile predictions being made. In addition, and despite an increase in sulphurous odours, no investigation of the summit region was undertaken. Both the inhabitants and the authorities were quite unprepared for the events which began on the night of 28 March at 23:30 hours (local time) (Allard 1983; Medina 1986).

The course of the eruption and its affects are summarized in Table 9.7 and on Fig. 9.7. Destruction of arable land was estimated at over 300 km^2 and, in addition to the high death toll, hundreds of people were injured and rendered homeless (Figs 9.8 and 9.9).

It is easy to be critical of the Mexican authorities in their approach to volcanic hazards and their reaction to the El Chichón emergency. Mexico has 14 volcanoes within its borders which have been active in historical times (Table 9.5) yet in 1982 the government had no programme of general and specific prediction, no contingency plans to deal with evacuation, shelter and resettlement and no administrative structures to ensure coordination of civil-defence measures with the recommendation of scientists. 'In the turmoil and confusion of the El Chichòn eruption, an estimated 2000 people perished—the worst volcanic disaster in Mexico's recorded history' (Souther *et al.* 1984, p. 15). Also there were no protocols governing the relationships between Mexican officials and volcanologists on the one hand and organizations and governments offering help from

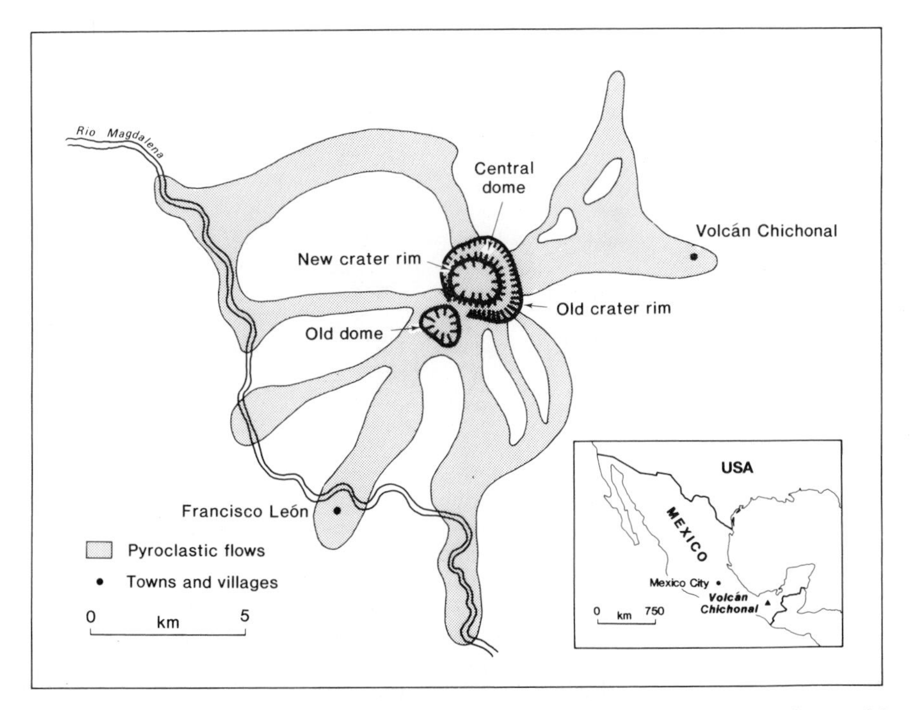

Fig. 9.7 Map showing the location of El Chichón volcano, its summit region and its major pyroclastic flows (from Tilling 1982, p. 172).

Fig. 9.8 The El Chichón eruption generated pyroclastic flows and surges which destroyed the village of Francisco León. The photograph shows reinforcing rods in a partially destroyed wall which have been bent in the same direction as the flow (i.e. from right to left). Photograph reproduced by courtesy of the United States Geological Survey, ref. Earthquake Information Bulletin no. 465.

abroad on the other. Offers of help from the United States Geological Survey were first rejected and then accepted only after appeals by Mexican scientists to their own government. Training of scientists monitoring seismic activity in the vicinity of the volcano in the weeks before the eruption was also claimed to be

Fig. 9.9 Photograph showing all that is left of a tree after the passage of a hot pyroclastic surge during the El Chichón eruption, Mexico, in 1982. The surge moved from left to right, causing the left side of the stump to be abraded and charred, whereas the other side was left unscathed (photograph reproduced with permission of the United States Geological Survey, ref. Earthquake Information Bulletin no. 487).

inadequate. 'Local technicians, unaware that the mountain posed any danger, continued to routinely change, file and send the seismic records to Mexico City without the careful examination that the situation demanded' (Souther *et al.* 1984, p. 17). In short, it is easy to conclude that losses could have been reduced with only minimal extra expenditure, better training and some pre-disaster planning.

The above critique is severe and in some respects unfair. Most poor countries have other and more pressing social, economic and political problems to deal with and Mexico is no exception. In 1982 in addition to endemic poverty and high population-growth rates, Mexico saw the start of a deep recession, an increasing foreign debt, high levels of unemployment and rampant inflation (sec

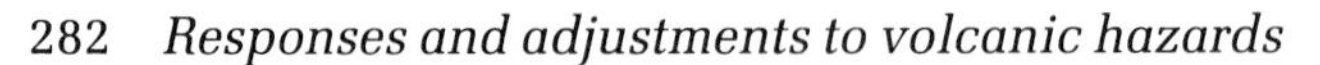

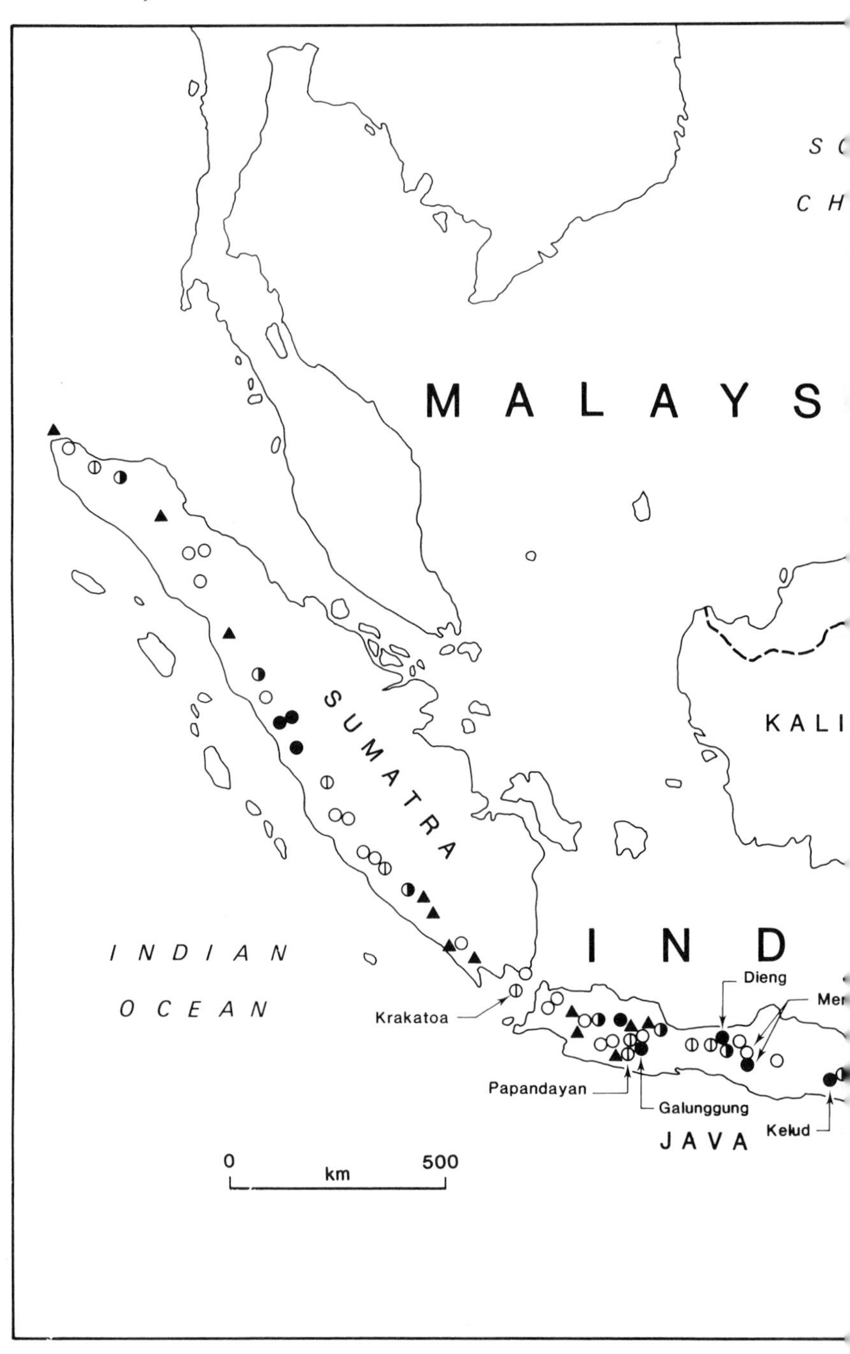

Fig. 9.10 Active volcanoes in Indonesia. The map indicates levels of actual and planned surveillance in 1979 (after Zen 1983, Fig. 17.1, p. 221).

Active volcanoes under constant surveillance
Active volcanoes which will be under constant surveillance in the future
Volcanoes not under constant surveillance but classified as active
Dormant volcanoes (without magmatic activities since 1600) and not under constant surveillance
Solfataric or fumarolic fields
SULAWESI SEA
Una Una (Colo)
SULAWESI
Agung
Tombora
Semeru

Table 9.6). Technical aid programmes unless well thought out in advance, can appear patronizing by casting aspersions on the competence of indigenous scientists, administrators and politicians. The experience of many economically developed countries is that disaster relief is a mixed blessing, particularly if it involves inappropriate technological transfers (see Chapter 8) and the diversion of funds from other development needs (see Burton *et al.* 1978, p. 196). Choosing appropriate methods to assist poor countries is one of the themes dealt with at the end of the chapter.

Eruption of Colo volcano on the Indonesian Island of Una Una in 1983 was a major event even by world standards. Not only did the eruptive plume have a significant climatic effect (Sawada 1989), but some 75% of the island's surface area was affected adversely by pyroclastic flows (Sudradjat-Suratman 1986). As mentioned before, Indonesia is a country that despite its material poverty has a long tradition of coping with the effects of volcanic eruptions. It gives a high priority to both the scientific investigation of active constructs and to integrating these findings with civil-defence measures. In the case of Colo the reputation of both the government and the Volcanological Survey for excellence was more than vindicated and timely evacuation saved at least 7000 lives. In 1979 the Indonesian volcanologist Mudaham Zen published a map showing centres of volcanic activity (Fig. 9.10) and, as the map indicates, Colo, though classified as active, was not under constant and detailed surveillance. This was reasonable because the last eruption was in 1898 (Simkin *et al.* 1981), yet when in early July 1983 the volcano started to display seismic activity monitoring was quickly instituted. Assessment of seismic records showed that earthquakes were increasing in number and magnitude from an initial ten per day, to 40 per day on 14 July and over 110 per day (Richter magnitude 4.6 or greater) on 19 July (Sudradjat-Suratman 1986). On 18 July a solfatara field on the north of the island became active and minor explosions followed. On the basis of this evidence the Volcanological Survey recommended the immediate evacuation of the island and the local administration—the government of Central Sulawest (Celebes)—concurred.

The evacuation of the 7000 inhabitants to the main island of Sulwesi (Fig. 9.10) was completed successfully some 24 hours before the climax of the eruption on 23 July when a plume reached a height of more than 14,000 m and extensive, pyroclastic flows covered most of the island. A further episode of explosive activity occurred on 25 July and more pyroclastic flows were generated. In September activity decreased and from October no further magmatic activity was evident. The economy of the island was almost totally destroyed with losses to housing, livestock and coconut plantations being estimated at nearly 100%. Prompt action by the authorities ensured that the death toll was zero (Sudradjat-Suratman 1986; Decker and Decker 1989).

Like the Icelandic government during the Heimaey eruption in 1973 (section 9.2.2), so ten years later the Indonesian authorities were assisted by the fortuitous circumstance that Colo is located on an island (Una Una) from which seaborne evacuation to the safety of the mainland was possible. It was also extremely fortunate that enough boats were available and could be mustered at short notice. Neither of these factors can detract, however, from the obvious success of the Indonesian authorities and the state volcanological survey in dealing with what could have been a disaster of major proportions. Decker and Decker (1989) make the important point that despite the success of the evacuation, recovery from the effects of the Colo eruption will take many years and in this respect Colo is typical of the situation in many poor countries following major natural disasters.

Mount Pinatubo on the Island of Luzon (Philippines) is a mountain 1745 m in height, which lies some 100 km northwest of the capital Manila and only 15 km to the west of the Clark Air Force Base; in 1991 the largest United States overseas military installation with an establishment of over 15,000 service personnel and families. For many years Pinatubo was considered extinct and as recently as 1981 the standard reference, *Volcanoes of the World* (Simkin *et al.* 1981), did not recognize it as having been active in the Holocene. A more recent report by Wolfe and Self (1982) indicated that the last activity was in the fourteenth century AD, when a large pyroclastic flow was generated.

The eruption which reached its climax on 15 and 16 June 1991 was presaged by more than two months of precursory activity including increased seismicity, some ground deformation and, immediately before the eruption, the emission of small plumes. The activity of 15 and 16 June continued for more than 15 hours, injected ash into the stratosphere (ten days later it covered a wide equatorial zone stretching from Indonesia to Central Africa), generated a large pyroclastic flow and left a small caldera in the summit region.

Although more than 320 people were killed by the eruption and by volcano-related hazards initiated by a typhoon which occurred at the same time, casualties could have been far higher without the prompt intervention of the Philippine Institute of Volcanology and Seismology (PHILVOLCS) (Global Volcanism Bulletin 1991b). The institute not only carried out detailed monitoring of seismic and other precursory signs from early April, but also in early June issued a timely alert. This was a serious warning (Alert Level 3) indicating the probable eruption of a pyroclastic flow within two weeks. With increased precursory activity the level of alert was increased to 4 (eruption likely within 24 hours) and over 12,000 residents were evacuated. In fact the eruption did not happen as soon as was predicted and it was only on 10 June that the United States Air Force decided to move 14,500 service personnel and families to the safety of the Subic Bay Naval Base 30 km away. Between 12 and 15 June strong explosive activity was preceded by over 12 hours of ground tremor, often a sign of impending eruption (Chapter 7), and the later eruption of pyroclastic flows into river valleys on the north, northwest and south flanks of the volcano meant the evacuation of more military personnel. Many local residents fled from the nearby town of Angeles and all inhabitants within a 20 km radius of the volcano were advised to leave. On the eve of the main phase of the eruption, 14 June, the authorities had achieved a great deal. In all 79,000 people had been evacuated and only four people had been killed, 24 injured and four missing.

The culmination of the eruption on 15 and 16 June generated areally extensive pyroclastic flows which extended more than 8 km from the summit. Fortunately before this happened, PHILVOLCS extended the official danger zone to a 40 km radius bringing the total number of evacuees to around 200,000. Several important military organizations left the Philippines for the United States. The typhoon Yunya first reached the area on 14 June and strong storms occurred in the Pinatubo area at the same time as the eruption. The weight of tephra and water caused many roofs to fail and this, rather than the eruption alone, was responsible for many of the casualties (Global Volcanism Bulletin 1991b; Pinatubo Volcano Observatory Team 1991).

Considering the poverty of the Philippines and the recent history of political instability, the civil authorities and PHILVOLCS did very well to keep casualties low and to ensure the removal of much of the population. Unlike the eruptions of Heimaey (Iceland) in 1973 and the Colo (Indonesia) in 1983, the authorities in the Philippines had little good fortune and the occurrence of a typhoon at the same time as the eruption gives a misleading impression of the effectiveness of the scientific and logistical response. Casualty figures in hundreds rather than thousands are commendable in a situation such as this. It is interesting that volcanological expertise together with preplanning can save lives even in a poor country, but features only too typical of underdeveloped responses were the conditions under which the evacuation took place and where evacuees had to live. Journalists drew the obvious comparison between the situation of American service personnel for whom evacuation was something of an inconvenience (the Clark Base subsequently closed) and many indigenous families for whom the evacuation was a disaster. Many had to abandon everything they had with little chance of compensation and no possibility of their lives ever being the same again (McCarty 1991). Laharic activity continued into 1992.

9.3.2 Underdeveloped responses to low-magnitude/high-frequency events

Because of the location of underdeveloped countries with respect to global patterns of tectonic plates and plate boundaries (see Chapter 2), there are relatively few examples of effusive eruptions which have affected people and their activities. Many of the examples are either not well reported in the literature (e.g. Erta Alè, Ethiopia and Nyamuragira, Zaire)—itself an illustration of the manner in which volcanological interest is skewed towards volcanoes located in economically more developed countries—or else occur within countries which are either still colonies or closely related to metropolitan powers (e.g. Piton de la Founaise, Réunion Island, is a Département of France). The example chosen to illustrate underdeveloped responses to frequent lava flows and small-scale limited falls is Karthala volcano in the Indian Ocean. This volcano lies on the island of Njazidja (formerly Grand Comore) and when the first eruption in recent years occurred in 1972 the Comoro Islands were still a French colony and responses reflected this link. Great panic was caused when, following a small-scale eruption of lava in the summit region, wealthy inhabitants—mostly French settlers but including some indigenous islanders—put pressure on the Colonial Governor to evacuate them by the first available flight (Tazieff 1974, 1983c). In the absence of hazard maps, detailed surveillance and any civil-defence strategy, the Governor acted in a manner typical of colonial officials by contacting the eminent French volcanologist Haroun Tazieff in Paris and asking him for a report. The fear was that the eruption would destroy the capital of the island (Moroni, population ~17,000) and its principal airfield, so making rapid evacuation impossible. As Fig. 9.11 shows, Moroni lies at least 10 km from the eruptive vent (2300 m) and Tazieff was able quickly to conclude that the eruption posed no real threat to life and property, since the effusion rate was too low for the lava to travel far (see Chapter 4, section 4.3.1.2) and that, provided the eruption did not take an unexpected course, such as an increase in effusion rate and/or a migration of the eruptive vent downslope, then the inhabitants had little to fear. Such advice acted as a salve and nobody left the island. The eruption declined in intensity and ended on 5 October, just three weeks after it started.

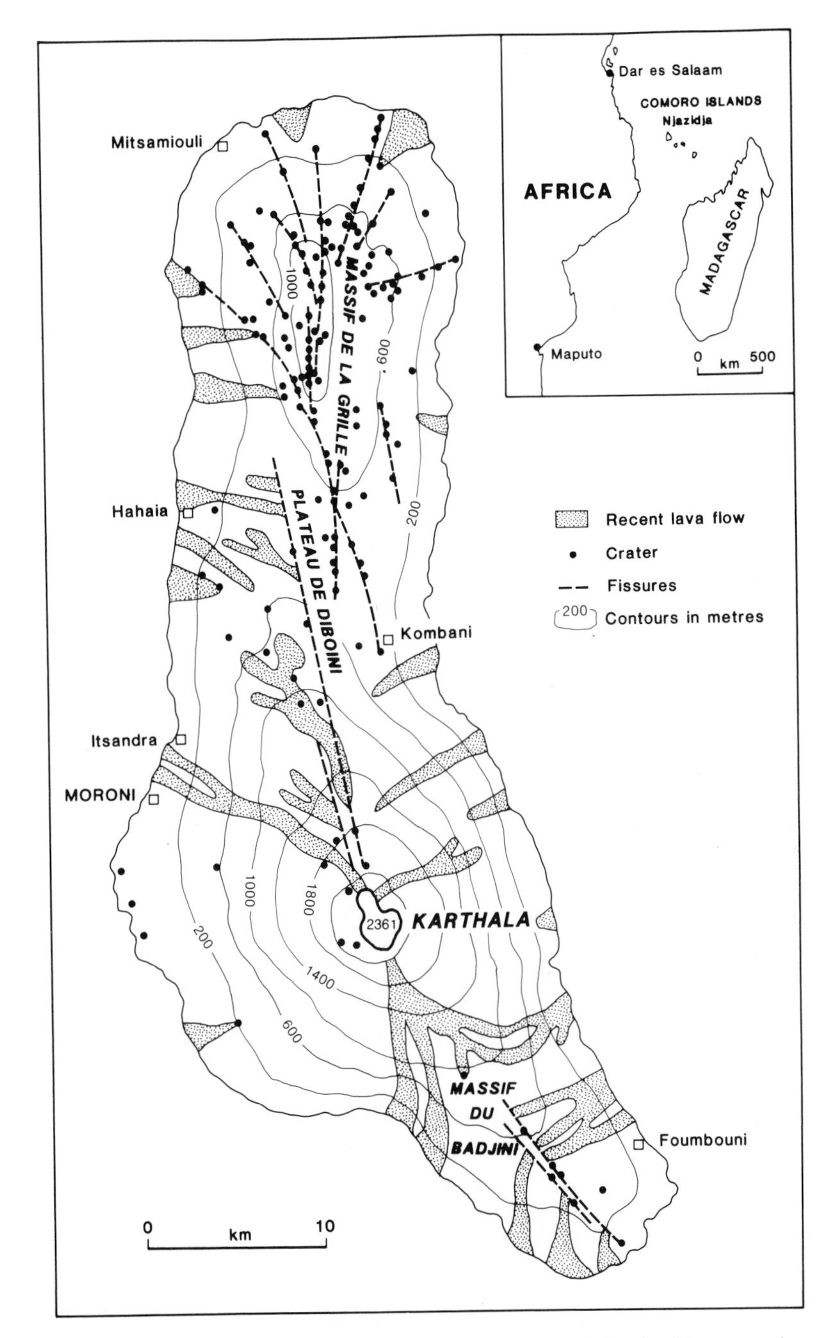

Fig. 9.11 The island of Njazidja (Grande Comore), Comoro Islands. The map shows the principal historic lava flows, craters and fissures in early 1972. The inset map shows the location of the Comoro Islands within the Indian Ocean (based on a number of sources including information in Nuemann van Padang 1963b).

The wealthy inhabitants of Njazidja had every cause to be concerned. Although the volcanoes of the Comoro Islands have been studied closely (e.g. Strong and Jacquot 1970; Upton *et al.* 1974), these works have been concerned with pure research and virtually nothing has been published on either general or specific prediction. In the absence of such investigations and the knowledge that lava flows have caused destruction of cultivated and inhabited coastal regions several times since 1857, it was not surprising that people panicked. In 1860 the area in and around Moroni was particularly badly affected and a large village was destroyed (Neumann van Padang 1963b; Newitt 1984; see Fig. 9.11). The action of the Colonial Governor was not only successful in confirming that no action was necessary but, more importantly, acted to calm fears and ensure that influential members of the community remained on the island so that morale amongst the population as a whole was not affected adversely. All politicians represent and react quickly to the wishes of their actual and perceived constituents and the fact that the Governor acted to propitiate the French settler community is typical of responses to natural hazards in countries under colonial administration.

Two years after independence a much more serious eruption occurred. On the morning of 3 April 1977 ground tremors were felt on Njazidja and at midday lava flowed from a fissure on the southwest flank of Karthala (Fig. 9.11). By 6 April flows had reached the coast (Saint Ours and Beuchamp 1979) and three villages were either partially or totally destroyed. The 1977 eruption occurred when the newly independent state was in severe difficulties. During the year President Ali Soilih had encountered much resistance in trying to establish a programme to abolish semi-feudalism using a mixture of Islamic, Chinese-style Marxism and traditional tribal ideologies. Only six days after the destruction of the villages, central government machinery was disbanded and some 3500 civil servants were dismissed. They were replaced by a National Peoples' Committee composed of recent secondary school graduates (Newitt 1984; Banks 1990). At such a time of political instability, it comes as no surprise to find that 2000 people were rendered homeless. What is more surprising is that the initial response was of a high standard and in fact 4000 people were evacuated successfully from the danger area (Saint Ours and Beuchamp 1979; Blong 1984).

Both the colonial administration in 1972 and the authorities in 1977 were, in their different ways, fortunate in the manner in which the eruptions developed. In 1972 the Governor was lucky that the eruption proved to be small scale and that the delay caused by calling in an expert from Paris did not prove to be disastrous. It was a response based on administrative caution and a reaction to influential lobbying. During 1977 events happened so quickly that evacuation was the only reasonable option available. Both examples demonstrate two points about responses to natural hazards in poor countries. First, in the absence of hazard predictions and contingency planning, responses have to be reactive and *ad hoc*. Secondly, in both 1972 and 1977 poorer sections of the community were marginalized. In 1972 they were marginal to the process of decision-making, while in 1977 they were marginal to the vestigial economy and were rendered homeless. As mentioned before, the Comoro Islands like many of the poorest countries in the world has a plethora of serious social, economic and political problems and the lack of attention given to hazard preparedness is but one symptom of its underdevelopment. In 1991 Karthala erupted again and at the time of writing (late 1992) no comprehensive review had appeared in the scientific literature; itself a testament to a lack of western interest in such countries.

9.3.3 Underdeveloped responses to volcano-related events

According to tables of losses compiled by Russell Blong (1984, p. 72–132) and updated to 1990, more than 240,000 people have been killed by volcanic eruptions. Of these a high proportion have died as a result of volcano-related events in poor countries. Some 16% of these deaths occurred as a result of just two tsunamis generated by nineteenth-century eruptions in Indonesia—Tamboro (1815) and Krakatau (1883)—but in the present century some of the greatest death tolls have been caused by lahars. In poor countries nearly 57,000 deaths have resulted from this cause and one of the highest death tolls was the estimated 25,000 fatalities occasioned by the 1985 eruption of Nevado del Ruiz, Colombia. Lahars and the reactions of authorities to them will be used to illustrate responses to volcano-related emergencies in underdeveloped countries.

For several of the countries listed in Table 9.5 the effects of lahars over the last 30 years have been serious. For instance, Calbuco volcano in southern Chile has erupted nine times since 1837 and in 1961 large lahars caused the destruction of much arable land. Ten years later Villarrica volcano in central Chile melted large volumes of ice and lahars caused the deaths of 15 people, destroyed the town of Conaripe and damaged many roads severely (Blong 1984; Decker and Decker 1989). Two eruptions are particularly instructive in illustrating the nature of underdeveloped responses to lahars: Galunggung (Indonesia) in 1982/83; and Nevado del Ruiz (Mexico) in 1985.

As discussed in section 9.3.1, the history of monitoring active volcanoes and innovating successful responses to eruptions is one field in which Indonesian scientists and government may be justly proud. The scale of the Galunggung eruption in Java (Fig. 9.10), however, severely tested the effectiveness of both pre-disaster planning and subsequent responses. The region in which the eruption took place is both densely populated and important for the Indonesian economy. Some idea of this may be judged from the fact that the eruption affected, both directly and indirectly, some 500,000 people and caused widespread damage to a thriving fish-farming industry and rich agricultural region. Reconstruction of the history of the volcano by Escher (1925) revealed that some thousands of years ago Galunggung collapsed on its southeastern flank to produce a large debris avalanche known, because of its topography, as the 'ten thousand hills'. It is upon this hummocky 'apron' that most of the population of the region is concentrated today and where the principal city of the area, Tasikmalaya (population ~1 million), is located (Fig. 9.12). From the time of the prehistoric collapse until the nineteenth century little is known about Galunggung. In 1822 an eruption generated both pyroclastic flows and lahars, causing more than 4000 casualties, while later eruptions, including one in 1918, were relatively minor and caused no loss of life. In 1982 'only a few people living on or near Galunggung still remembered the 1918 eruption, and the deadly 1822 eruption was virtually forgotten' (Sudradjat and Tilling 1984, p. 14). Activity in 1982/83 falls into three phases (Gourgaud *et al.* 1989). In Phase I (5 April to mid-May) an eruptive plume more than 12 km high was generated together with some pyroclastic flows. Thunderstorms during explosions caused some lahars to be generated. In Phase II

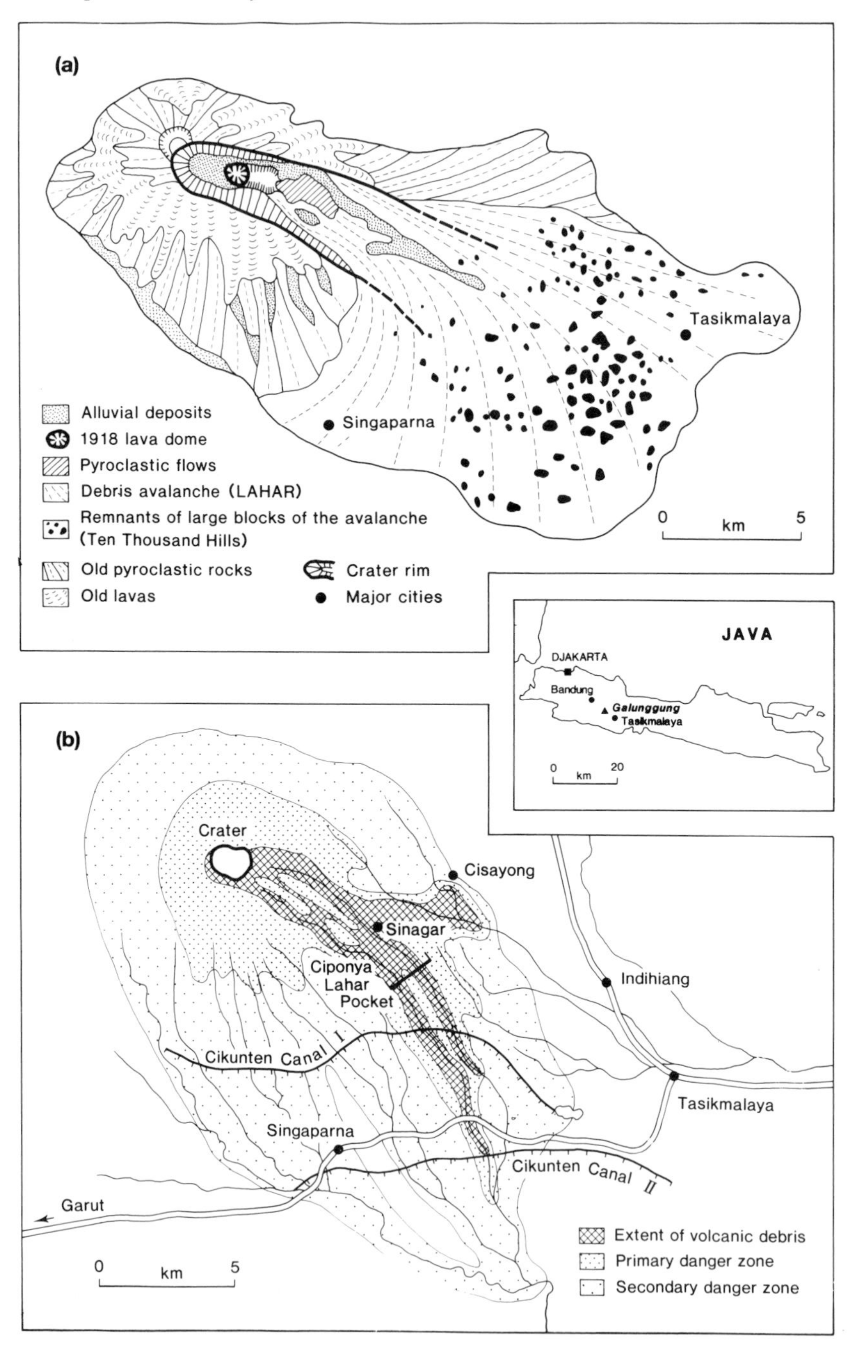

Fig. 9.12 Maps showing the 1982/83 activity of Galunggung, Indonesia.
(a) Sketch showing the large debris avalanche that forms the 'Ten Thousand Hills' (from Sudradjat and Tilling 1984, Fig. 3, p. 14).
(b) Map showing the volcano in November 1982, together with aspects of the hazard assessment (from Blong 1984, Fig. 4.7, p. 169).

(mid-May to the end of November) explosions increased in intensity, became increasingly violent and produced small-scale pyroclastic flows, fall deposits and, following rainfall, substantial lahars within river channels. From December 1982 (Phase III) activity declined, became more typically Strombolian in style and the effects were confined to the crater region (Fig. 9.12). Final activity started on 1 January 1985, continued until 8 January and involved the eruption of some small lava flows.

At the end of the eruption losses were considerable and have been documented in detail by Sudradjat and Tilling (1984). More than 80,000 people had been either evacuated from the region or left of their own accord, many hundreds of homes had been demolished, 35,000 people were homeless, over 170 schools were damaged and/or destroyed, systems of transport were disrupted and the agricultural economy virtually wiped out. Volcanic debris, notably laharic, had destroyed ~94,000 hectares of land in crop—with a further 87,000 hectares being at risk from secondary lahars—and starvation threatened more than 300,000 people. Many of the ponds used by fish farmers had been filled in with losses estimated at US$5 million and most livestock annihilated. To add to these problems, volcanic ashes had contaminated supplies of fresh water. Total financial losses were estimated at over US$100 million.

In contrast to eruptions of Colo volcano (section 9.3.1) the events at Galunggung were of a much larger scale and, hence, involved a far larger and more all-embracing response on the part of the Volcanological Survey and Indonesian civil authorities. A notable aspect of the response and one typical of many large-scale disasters in poor countries was the international dimension which included not only aid agencies (Blong 1984), but also foreign volcanological experts. A common feature of underdeveloped responses is that cooperation with aid agencies from abroad is often at best strained and at worst hostile. With some exceptions this was not the case during or after this eruption.

Although Galunggung had not erupted for 64 years the Volcanological Survey approached the task of disaster management with its customary diligence. As Fig. 9.10 shows, many volcanoes in Indonesia are considered to be more potentially dangerous than Galunggung, yet from the 1950s periodic field visits were undertaken to the summit and from 1975 these were supplemented by an annual three-month seismic-monitoring programme. Unfortunately the eruption occurred with little warning and, despite the fact that seismic monitoring was in operation only two days after the start of the eruption, the Volcanological Survey has no 'baseline' seismic information to compare with the data which were collected as the eruption progressed. General prediction was much better and a hazard map of the volcano was available showing the likely routes of pyroclastic flows and lahars (Sudradjat and Tilling 1984). For a volcano not considered to be in the highest category of risk the initial response of the Volcanological Survey was commendable and shows clearly the benefit of pre-disaster planning. Once the eruption had started seismic monitoring was supplemented by ground-deformation studies, analysis of the plume, petrological investigations, visual observations and cooperation with foreign research groups (such as one from France examining magmatic field changes).

The lack of 'baseline' studies became acute once the eruption started and specific predictions had to be made. Without baseline studies carried out before an eruption it is difficult to know whether, say, seismic signals represent deviations from the norm. The situation became better once the eruption had been under way and towards its end individual eruptions were being predicted with accuracy, but in the initial months this was not the case and caused major difficulties. The eruption directly and indirectly affected millions of people in western Java including the inhabitants and government in the capital, Jakarta. A dearth of clear predictions from the Volcanological Survey allowed a 'vacuum' to develop in which 'independent scientists and other self-proclaimed experts added to the confusion by making baseless observations, misleading interpretations, and irresponsible statements about the possible future behaviour of the volcano. The mounting confusion increased considerably when a rumour began to take hold among the general public, that the eruption could culminate in a huge caldera-forming cataclysm, even though the VSI (Volcanological Survey of Indonesia), based on its monitoring data, convinced the government that such an event was highly unlikely' (Sudradjat and Tilling 1984, p. 16).

Government acted quickly to resolve the confusion and in late September convened a meeting in Bandung (Fig. 9.12) to draw up a contingency plan. Participants included not only Indonesian volcanologists and government officials, but also bankers, journalists, financiers, representative from relief agencies and earth scientists from abroad. Countries represented included: Japan, USA, France and Australia, with additional participation from the United Nations. The principal aim of the meeting was to confirm the veracity of the opinion that a major caldera-forming event was unlikely. This they did by concluding that such an outcome was the most unlikely of a number of scenarios. The meeting also agreed guidelines for the distribution of aid and suggested that additional funds should be provided for future surveillance and hazard mapping.

The sheer complexity of aid provision has been described by Blong (1984, pp. 168–70) and is summarized in Table 9.8. Whether the response will be successful in the long term will only become evident in the future, but a review after one decade suggests that it was an overall success. Indonesia is a poor country, yet more than compensates for this by strong indigenous volcanological expertise, a highly effective government involvement in mitigation strategies and a willingness to both accept and use aid in an effective manner.

It is difficult to conceive of a more stark contrast than that between the successful handling of the Galunggung eruption and the events which took place at Nevado del Ruiz (Colombia) before, during and after the eruptions of November 1985. Because of the scale of the disaster (~25,000 were killed making it at the time the world's fourth largest historical catastrophe), the accessibility of the volcano to scientists from the United States and the fact that many deaths could have been prevented, it is not surprising that this eruption has generated a vast and often critical literature. Amongst the most useful works on the effects of the eruption and the responses of scientists and the authorities to it are: Ellsworth-Jones (1985); Fishlock and Matthews (1985); Clapperton (1986); Espinola (1986); Gueri and Perez (1986); Naranjo *et al.* (1986); Sigurdsson and Carey (1986); Voight (1988b); Barberi *et al.* (1990); Baxter (1990); Hall (1990); Parra and Capeda (1990); Voight (1990); and Verstappen (1992). It is upon these that the following summary is based.

According to archives and historical documents, before the events of 1985 only two previous eruptions of Nevado del Ruiz had been recorded in the comparatively short time this region of Colombia has been settled by literate observers. In 1595 lahars caused the deaths of over 600 people and in 1845 the same process killed nearly 1000, but from 1845 to 1985 the only activity was emission of

Table 9.8: Aid donors and the use made of monetary and non-monetary assistance following the Galunggung eruption of 1982/83 (based on Blong 1984, pp. 168–70, with additional information from Anon. 1982, and other sources)

Principal aid donors	Aid received and action taken
1) United Nations agencies a) Disaster Relief Organisation b) Development Program c) Childrens' Fund d) World Health Organisation e) World Food Program f) International Labour Organisation 2) Bilateral Aid provided by: Australia, Canada, Japan, Netherlands, New Zealand, Sweden, Switzerland and the USA 3) International agencies a) Association of South East Asian Nations b) Catholic Relief Services c) Church World Service	Monetary aid and non-monetary assistance (e.g. medicine, sanitation facilities, food, clothing, emergency shelters and technical assistance) was valued at over US$ 4.4 million. The money was spent on: 1) 18 emergency shelters at Garut and 334 at Tasikmalaya (see Fig. 9.12). 2) Equipping, clothing and feeding refugees and evacuees. A rice ration was the initial response to hunger and this was followed by field kitchens and a coupon system to discourage the sale of food. 3) Following a survey of refugees and evacuees, it was found that 45% (nearly 8000) could be resettled in other villages in Java, while over 9000 (mostly poor and unskilled) were encouraged to migrate to other Indonesian islands. Clearing the shelters meant there was room in the event of secondary lahars.

fumarolic gases and the occurrence of thermal springs. The town of Armero, which took the brunt of the 1985 lahars, was constructed on the debris produced by earlier historical events. By 13 November 1985 Armero had grown into the principal agricultural centre of the volcano and contained a population of around 29,000 (Fig. 9.13). 'Shortly after 21:00 local time, a relatively small magmatic eruption ... produced a series of flows and surges that ... melted part of the summit's snow and ice cap and sent torrents of meltwater, slush, ice and pyroclastic debris in a plexus of sheet and channelled cascades over the volcano flanks.... In lower channels the flows coalesced and entrained debris, vegetation and ponded water to form lahars' (Voight 1990, pp. 152–3). On the western flanks of the volcano over 1800 fatalities occurred and more than 200 homes were obliterated, but shortly after midnight several waves of lahars swept down the Rio Lagunillas (Fig. 9.13) burying Armero in mud and entombing more than 20,000 people (Fig. 9.14). The truly catastrophic losses caused by the eruption are summarized in Table 9.9.

There is a consensus amongst writers reviewing this eruption that many, perhaps most, of the casualties could have been prevented with better planning both before and during the eruption. The eruption did not occur 'out of the blue', but was presaged by fumarolic, seismic and other anomalies for almost a year. In fact the emergency really started in November 1984 when the first volcanic earthquake was felt and when climbers near to the summit reported increased fumarolic activity. Anomalous activity increased and geologists from the Central Hidro Eléctrica de Caldas (CHEC) recommended that programmes of specific prediction should be implemented and involve both geophysical and geochemical

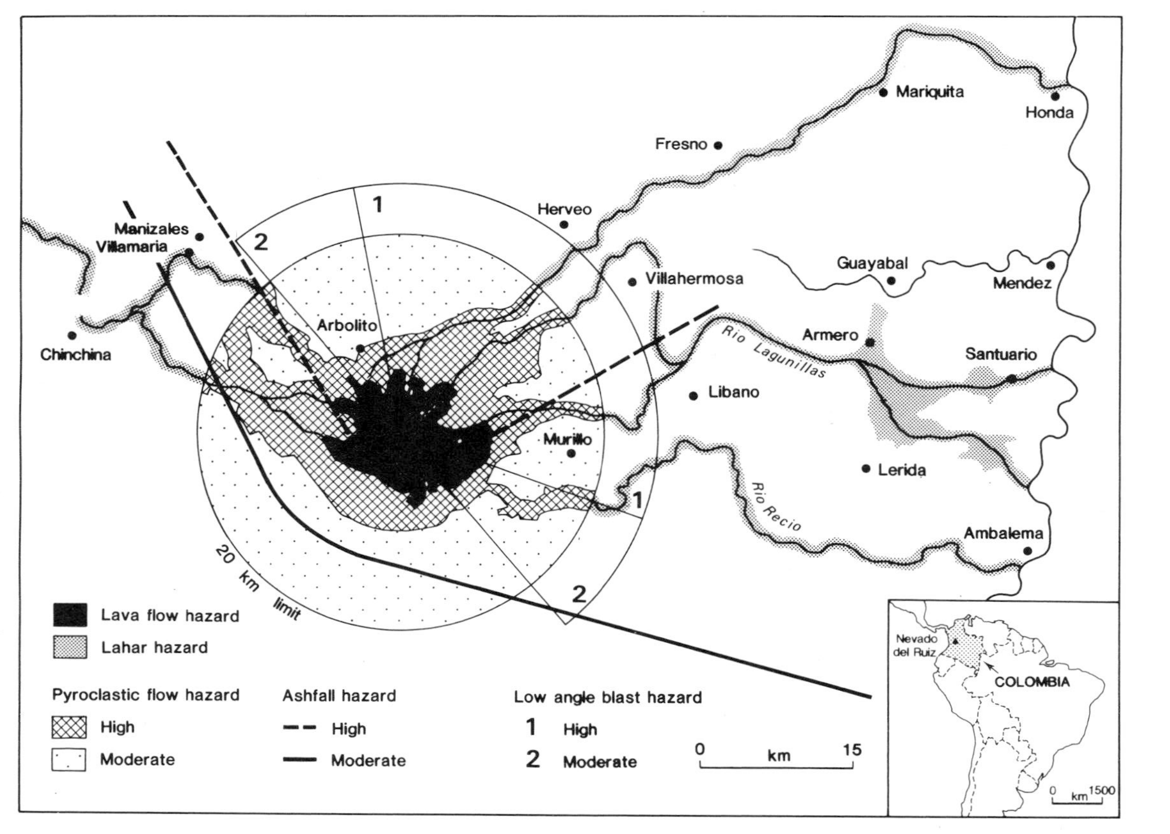

Fig. 9.13 Map showing volcanic hazards on Nevado del Ruiz, Colombia, compiled in November 1985. Inset map shows the location of the volcano within South America (after Parra and Cepeda 1990, Fig. 1, p. 118; and Voight 1990, Fig. 4, p. 163).

Fig. 9.14 Devastation caused to Armero by lahars from the 1985 eruption of Nevado del Ruiz (photograph reproduced by courtesy of Professor H. Th. Verstappen).

Table 9.9: Damage caused by the Nevado del Ruiz eruption in 1985 (based on information in Voight 1988b; and several other sources)

Category of loss	Details
Deaths and injuries	Nearly 70% of the population of Armero was killed (~20,000) and a further 5000 (17%) were injured. On the western flanks of the volcano overbank flooding by lahars caused more than 1800 fatalities.
Agricultural production	60% of the region's livestock, 30% of the sorghum and rice crop and 500,000 bags of coffee were destroyed. Over 3400 ha of agricultural land was lost from production.
Communications	In the vicinity of the volcano virtually all roads, bridges, telephone lines, and power supplies were destroyed. The whole region was isolated.
Industrial, commercial and civic buildings	50 schools, two hospitals, 58 industrial plants, 343 commercial establishments and the National Coffee Research Centre were badly damaged or destroyed.
Housing	Most of the housing was destroyed and nearly 8000 were rendered homeless.
Monetary	The cost of the eruption to the economy of Colombia was estimated at US$ 7.7 billion. This represented about 20% of the country's GNP for the year in question.

monitoring (Hall 1990). The authorities responded quickly and formed a committee of local government officials who consulted the National Geology and Mines Bureau (INGEOMINES). Geologists from CHEC and INGEOMINES visited the volcano several times in February 1985, an expatriate seismologist from the United Nations Disaster Relief Organisation and two Swiss colleagues became involved following an invitation from the national Defensa Civil de Colombia. It was recommended that a hazard map should be prepared, that monitoring equipment—in particular seismographs—must be installed and that emergency evacuation plans should be developed. Between March and November international scientific aid was sought and provided by several countries, there was some emergency planning and a preliminary hazard map using general prediction was drawn up (Fig. 9.13). In fact the final version of the map was due to be presented for approval one day after the eruption started (Parra and Capeda 1990).

On the basis of the last paragraph it would be easy to conclude that the civil authorities acted responsibly, but were victims of bad fortune because plans were not quite ready in time. However, a more detailed review of responses of the scientists and the authorities to this crisis reveals serious weaknesses. As mentioned earlier in this chapter sociological studies such as those of Sorenson and Mileti (1987) indicate that to be effective communications between scientists and government must be consistent, specific, accurate, certain and clear. On all counts this was not the case in the months leading up to the eruption. The lack of 'baseline' studies before the period of crisis to allow comparison with later geophysical anomalies meant that, although the eruption was predicted, forecasts were neither precise enough nor sufficiently reliable to allow the government to risk the economic and political costs of an early evacuation and/or a false alarm. In addition because of the sheer number of Colombian and overseas teams of earth scientists, there was delay in devising plans and controversy about the action to be taken. All contributed to the subsequent disaster. As Voight (1990, p. 151) concludes 'the catastrophe was not caused by technological ineffectiveness or defectiveness, not by an overwhelming eruption, or by an improbable run of bad luck, but rather by cumulative human error—by misjudgment, indecision and bureaucratic shortsightedness'.

In many ways the Nevado del Ruiz eruption is only too typical of responses to natural disasters in poor countries. This example, more so than any other in the chapter, illustrates the truism that disasters are essential rather than accidental features of countries such as Colombia. Colombia simply does not have the resources fully to monitor all its volcanoes in the manner required to develop usable 'baseline' studies, neither does it have the administrative expertise to handle a large multi-agency and multi-national scientific programme. In early 1985 the government of Colombia had many more serious and immediate problems to deal with than an eruption which might take place at some unspecified time in the future. Central to these were an economic crisis, chronic political instability and a long-term fight against narcotics cartels. These external concerns were only too evident just one week before the eruption, when on 6 November President Betancur precipitated a political crisis by sending troops against guerillas who had captured the Palace of Justice in the capital leaving 100 dead, including ten senior judges.

9.3.4 Underdeveloped responses: long-term perspectives

Much attention has been devoted by social scientists to the issue of long-term adjustments and responses to natural disasters in poor countries. Virtually all interest has been focused on non-volcanic disasters, especially droughts and earthquakes (Hewitt 1983a), and published research on the long-term recovery from the effects of volcanic eruptions is comparatively rare. Some studies, however, do exist and perhaps the most comprehensive of these are of Parícutin volcano in Mexico and Mount Lamington in Papua New Guinea.

In the case of Parícutin three decades of sociological investigation have revealed many interesting features (Nolan 1979; Nolan and Nolan 1979; Rees 1979). A basaltic eruption began in 1943 and did not end until 1952. It occurred at a new eruption site within an area where old cinder cones marked former outbursts of lava. Five villages were affected and many subsistence farmers, most of whom owned their lands in small plots but were prevented by tradition from selling without the permission of village leaders, were ruined. The five villages were superficially very similar not only agriculturally but also socially, ethnically and culturally, being isolated from the mainstream of the Mexican economy and life and having Tarascan Indian traditions. It might be expected that the long-term fate of these communities would have been similar and that by the late 1970s recovery would have affected all to the same degree. In fact each of the five villages showed a unique response.

In 1943 Parícutin village was very traditional and still maintained an everyday use of the Tarascan language, but was more open to outside influences than the other 'Indian' village, Angahuan. Parícutin was totally destroyed in 1944 and the people were evacuated by the Mexican government to a new settlement some 20 km away called Caltzontin (Fig. 9.15). Despite receiving more aid than other settlements, 30 years later Caltzontin was a very poor village but one showing the emergence of distinct social stratification between the families of the evacuees who, due to increased population pressure, had insufficient agricultural land to secure even a modest lifestyle and a group of newcomers—mostly urbanites—who worked in the growing town of Uruapan (Fig. 9.15).

In the village of Angahuan the eve of the eruption saw a community which had changed little for many decades. All the inhabitants spoke the Tarascan language—some were monoglot—few had ventured beyond the village and its agricultural lands. Marriage within the community was the norm. The village was not totally destroyed and remained in its original location. Thirty years later little had changed. Despite contacts with the world outside during the eruption and growth of tourism subsequently, the inhabitants avoided anything but strictly commercial contacts and by the late 1970s were showing distinct signs of both social and economic marginalization. To quote Nolan and Nolan (1979, pp. 343–4), 'Angahuan fiercely preserved Indian traditions in dress, language and housing, and the town has been adopted by outsiders who view such qualities with cultural nostalgia. . . . The emergent sale of tradition in order to preserve it . . . is edging the people towards a perpetually marginal existence as "professional Tarascans" in a form of self-imposed reservation'.

In contrast to the traditional villages of Parícutin and Angahuan, Zacan was more ethnically mixed. Tarascan was spoken by only 50% of the population and many of these were elderly; in the early 1940s the village was changing and children were being educated. When disaster struck great changes occurred.

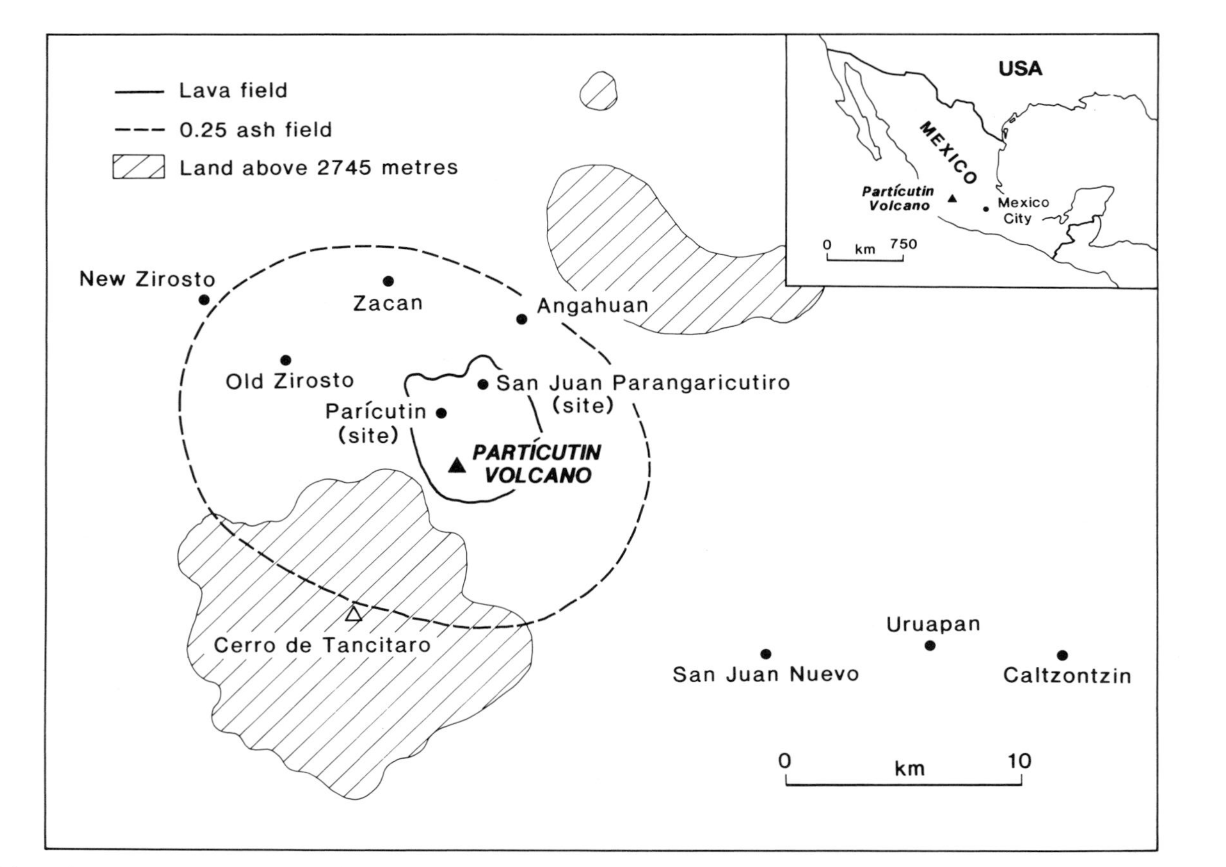

Fig. 9.15 Location map showing the settlements affected by the 1943–52 eruption of Parícutin volcano, Mexico (based on Nolan and Nolan 1979, p. 342).

Agricultural land was back in production quickly, the village retaining its old site (Fig. 9.15), but socially, economically and culturally the community was almost unrecognizable 30 years later. Many families left the village for other parts of Mexico, some of the men went to work in the USA and fluency in the traditional language declined. The village became one of old people living on remittances from sons and daughters living in cities; in this respect only too typical of thousands of villages in Mexico.

Also ethnically mixed was the village of Zirosto in which Tarascan was spoken by only 20% of the inhabitants in 1940. At the time of the eruption there were great differences in wealth between families. Some were still traditional subsistence farmers like those in Parícutin and Angahuan, whilst others had gained modest wealth through activities carried on outside the village, including mule-owning, mule-driving and money-lending. The latter had also purchased the better land. This latent communal fragmentation was exacerbated by the eruption and the fact that the villagers were split between three sites: Old Zirosto, New Zirosto and Miguel Silva, causing much acrimony about what was perceived as being the death of community consciousness.

The most important settlement of the area, San Juan Parangaricutiro, was the centre of local administration (Fig. 9.15). In 1943 it was the settlement most affected by modernization; being unique in having electricity and a single telephone. Some 32% of the inhabitants spoke Tarascan and the town had many outside contacts due in no small measure to its church which contained an ancient crucifix called 'The Lord of Miracles', which attracted pilgrims. Although the town was destroyed it was rebuilt on a new site (San Juan Nuevo Parangaricutiro). New San Juan has been a success and by the 1970s had managed considerable modernizations without sacrificing its traditions (Nolan 1979), with pilgrims still visiting 'The Lord of Miracles' from the rest of Mexico and even the USA. The success of the new town was due largely to the fact that its site was chosen as a communal decision and was built through the efforts of the people themselves. In this case the eruption caused the community to come together and not fragment as was the case in Zirosto.

The Parícutin case study is important for it illustrates that long-term adjustments and responses are very complex. Despite being superficially very similar, the five communities responded to the emergency in ways that exaggerated aspects of community culture which were long standing.

The eruption of the long inactive Mount Lamington, Papua New Guinea, in 1951 produced devastating pyroclastic flows which laid waste over 200 km^2 and killed around 3000 people (Decker and Decker 1989). The eruption occurred in a poor region of a poor country (Table 9.5), which at the time was under the colonial administration of Australia. In 1953 about 3000 evacuees were allowed back into the area to farmland they had occupied previously. Some new larger villages were established under government stimulus (Blong 1984), so that services like schools and health care could be more easily provided and with time these became more innovative than the more traditional settlements, not only because of government encouragement of such enterprises as the cash-cropping of coffee, but also because they were closer to sources of alternative employment, had climates more suited to coffee growing and were located nearer to local markets (Schwimmer 1977). Receptiveness to innovation was in reality much more complex and involved long-standing contrasts amongst tribal groups and the ways they were affected by the eruption. This complexity has been discussed

Table 9.10: The complex long-term responses and adjustments of three tribal groups to the 1951 eruption of Mount Lamington, Papua New Guinea (based on Ingleby 1966; Schwimmer 1977; and Blong 1984)

Tribal group	Effects of eruption	Long-term effects
Sangara	Fled in small groups and most of the older leaders were killed. Survivors were not instructed in traditional religious beliefs.	Innovation not constrained or obstructed by the attitudes of the older generation.
Sasembata	Fled from the 1951 eruption in discreet tribal groups under the leadership of village constables. This decreased the power of traditional village elders. Constables also enhanced their authority by interceding with the authorities over building supplies and food relief.	Altered the power structure of the tribal group.
Orokaivan	Alienation from the land meant separation from ancestors, and that religious rites for infants and the dead could not be carried out since these had to be performed on taro fields associated with dead ancestors. Also there was concern that tribal land would be taken over by other groups and this was considered a major danger to the survival of the tribe. Because of this the Orokaivans had to accept the principle of sole tenure of land in 1966–67, even though this could have had severe consequences for the organization of the tribe. In accepting the idea of sole tenure, the tribe was conscious of the advent of cash cropping and the spread of people from other groups over their traditional land. Later when cattle grazing became viable, the Orokaivans demanded the return of 'their' land.	Clearly the eruption had great effects on tribal life. By the mid-1960s, however, few children had any knowledge of the eruption or fear of the volcano. It 'lived' on in legends, songs and lore.

by a number of writers (Ingleby 1966; Schwimmer 1977; Blong 1984) and is summarized in Table 9.10.

9.4 Particular responses

At first sight the countries listed in Table 9.11 as having particular responses to volcanic eruptions appear to have little in common. Italy is a developed country, a

Table 9.11: Examples of countries having 'particular' responses to volcanic eruptions (for additional details and explanations see Table 9.1)

Country	Volcanoes			GNP per capita[2] (US$ 1986)	Area (million km^2)	Population 1989* (millions)	Population fertility[3]	ODA as % of GNP (1984–86)[4]	Secondary school enrolment (% of age group 1983–86)[5]		Daily food energy supply as % of requirement, 1983–85[6]
	Eruptions known to have affected people and their activities	Number active in Holocene	Active in historical time (% of those active in Holocene time)						Male	Female	
Italy	32	17	12 (71)	8570	0.30	57.55	1.60	−0.33	74	73	138
St Vincent	5	1	1(100)	~630+	(387 km^2)	0.12	~3.0	ND	ND	ND	88
Guadeloupe	2	2	2(100)	4340+	(1780 km^2)	0.34	2.07	ND	ND	ND	124

[1] Based on Simkin (1981) and other sources.
[2,3,4,5,6] From Anon. (1988).
* From a variety of sources.
ND No data.

major industrial power and might be expected to have responses to volcanic eruptions which are similar to those of countries listed in Table 9.1. In like manner the Caribbean countries of St Vincent—a British colony until independence in October 1979—and the French overseas Département—Guadeloupe—might be expected to approximate in their responses to the countries listed in Table 9.5. However, it is a notable feature of Italy, St Vincent and Guadeloupe that their responses to volcanic eruptions are impossible to classify on a development axis and that cultural, administrative and other factors intervene to make such responses individualistic.

In the case of Italy much has been written about responses to a variety of extreme natural events including earthquakes, landslides and volcanic eruptions. Many authors have reiterated the point that 'whilst financial provision and political will are adequate, effective relief is severely hampered by poor administration, logistical difficulties in ensuring that aid actually reaches victims, and the lengthy period it takes for regions to recover following a catastrophe' (Chester *et al.* 1985, p. 359). In southern Italy, where all currently active volcanoes are located and where many high-magnitude earthquakes have occurred, corruption within local administrations is endemic and the activities of organized crime syndicates—the Mafia in Sicily and the Camorra in the Naples Region—means that disaster relief is often illegally expropriated (Littlewood 1985). Over the past century responses to natural disasters have continued to suffer from maladministration and corruption, even though the country has improved its international economic position and is today a rich developed state. For instance, in 1924 Benito Mussolini visited Messina in Sicily and was shocked to find that many victims of the 1908 earthquake were still housed in makeshift dwellings (Admiralty 1945). More than 50 years later in 1982 the Pope made a pastoral visit to western Sicily and the villages affected by an earthquake in 1968. He was horrified by the lack of recovery and the level of endemic poverty (Chester *et al.* 1985).

In the case of volcanic eruptions and threats of eruption, the situation has been very similar. In 1970 a seismic crisis in the Phlegraean Fields volcanic region to the northwest of Naples (Fig. 9.16) was a classic example of bureaucratic muddling and has been widely criticized. Despite being an area of only 80 km^2, the Phlegraean Fields contains nearly 400,000 people, many important industries and several military establishments. It has been a centre of eruption for at least 50,000 years and only 34,000 years ago more than 250×10^6 tonnes of pyroclastic material was erupted, covering 1.5×10^6 km^2. The last (small) eruption occurred as recently as 1538 and formed Monte Nuovo (Kilburn 1986; see Fig. 9.16). To many volcanologists worldwide the Phlegraean Fields are known as an area of slow vertical land movements (bradyseisms) and many buildings from the classical age are now below sea level (Lyell 1847), but before the eruption of 1538 some *upward* vertical movement was noted (Barberi *et al.* 1984). When on 2 March 1970 uplift at Pozzuoli (Fig. 9.16) had reached 80 cm and much of the old town had already been damaged by volcanic earthquakes, the authorities decided to evacuate 3000 people. Scientific advice was conflicting and in fact no eruption occurred. Some ten years of controversy then ensued about whether or not an eruption had been imminent (Tazieff 1971; Imbo 1980) and about the appropriateness of the response. As Barberi *et al.* (1984, p. 183) note 'confusion was total, two commissions of enquiry were appointed with conflicting terms of reference and conflict [between them] rapidly arose reaching

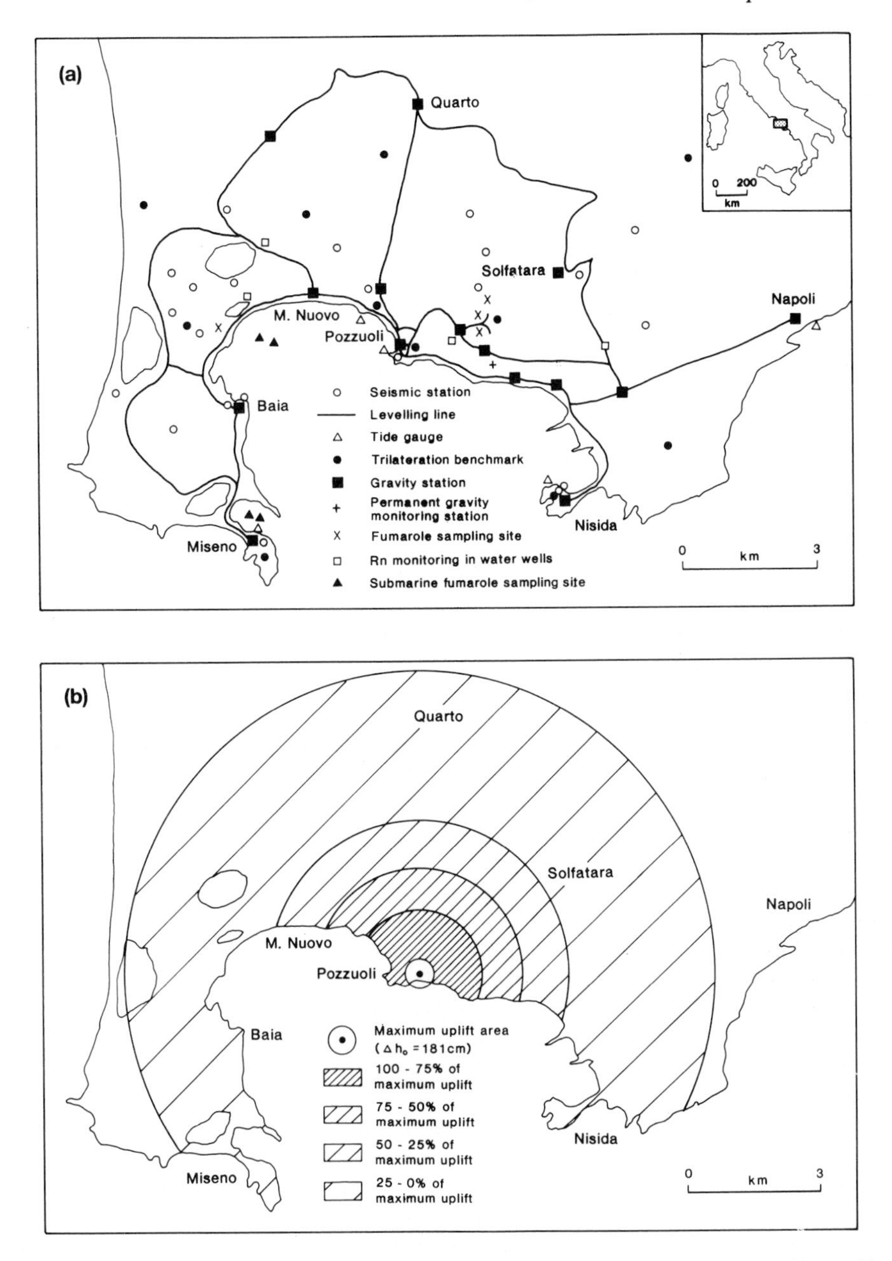

Fig. 9.16 Phlegraean Fields, Italy, showing: (a) surveillance network; (b) area affected by uplift, January 1982–October 1984 (based on Barberi *et al.* 1984, Figs 2 and 3, pp. 177 and 178).

violent polemic levels'. Much of the town remained damaged for many years.

The Italian authorities learnt much from this débâcle and even more in the aftermath of the 1980 Irpinian (or Avellino) earthquake. In the former case a proper system of specific prediction was instituted in the Phlegraean Fields and hazard evaluation (specifically concerned with advice to the government) was made the responsibility of a newly created National Group for Volcanology (GNV), while in the latter the sheer incompetence of the earthquake response and the corruption associated with the distribution of disaster relief—despite funds of around US$1.3 billion being made available (Anon. 1981)—led to the establishment of a Ministry of Civil Protection.

There is little doubt that much has changed since the events at Pozzuoli and Avellino and this may be seen clearly by looking at responses to three more recent volcanic crises. The 1983 eruption of Etna is interesting for it shows the enterprising involvement of the authorities in dealing with a potential disaster. This was the first time in Italy during recent centuries in which an attempt was made to divert a lava flow. The eruption started on 28 March at around 2400 m on the south flank of Etna and in less than 30 days had caused damage estimated at over US$30 million to tourist facilities on the mountain (Chester *et al.* 1985). Residents living further down the mountain were fearful for their livelihoods and homes, putting pressure on the authorities to act. Action was swift. First, an ancient law preventing lava-flow diversion was suspended and, secondly, the Minister for Civil Protection called a meeting of scientific advisors under Professor Franco Barberi of the University of Pisa. The committee came up with the recommendation that earthen barriers should be constructed to divert the lava away from high-value installations and that, as a secondary action, the walls of the lava flow should be breached by explosives to start a second flow front and so reduce the forward movement of the main flow. It was a measure designed to give the authorities time to build the barriers. The consensus of scientific opinion was that this was a highly successful response (Lockwood 1983) and in fact many valuable tourist installations were saved, but perhaps expecting a similar litany of incompetence many journalists either missed, or chose to misinterpret, the whole character of the response. Many news reports were of a hubristic tenor, emphasizing the breaching of the flow, rather than the secondary nature of this initiative, others placed the whole exercise within the context of a battle for votes in the June election and still others argued—after the event—that settlements further down the mountain had not been in immediate danger. To the present author's knowledge no press report mentioned the major organizational change that had occurred since the 1970 events at the Phlegraean Fields and especially since the Irpinian earthquake only three years before.

The second example relates to the events which occurred in the Phlegraean Fields between January 1982 and January 1985. Again uplift was concentrated around Pozzuoli (Fig. 9.16), but this time was accompanied by the monitoring of such phenomena as gas composition and seismic activity which have been shown elsewhere to be good precursory signs of volcanic activity and useful in specific prediction (Fig. 9.16; and Chapter 7, section 7.3.2). One problem faced by scientists was that the data were capable of being interpreted in several ways (Barberi *et al.* 1984). Observations of uplift, seismic activity and geochemical changes seemed to point to an increase in the volume of a shallow magma chamber and an imminent eruption, but as time passed there was no systematic reduction in the depths of focus of earthquakes, fumarolic gases remained constant in composition and no changes in microgravity were noted. It was concluded that if an eruption did occur, then its magnitude would be small and its style similar to that of 1538. However, prudent contingency plans and hazard maps were compiled for more serious scenarios (Rosi and Santacroce 1984).

Fig. 9.17 Destruction caused by volcanic earthquakes in Pozzuoli, Naples Bay Region, Italy. The town was badly damaged in 1970 and between January 1982 and January 1985 (author's photograph).

Gradually it became clear that an eruption was not about to happen and that the major hazard faced by the inhabitants was through damage caused by volcanic earthquakes and the danger resulting from collapsed buildings (Fig. 9.17). Some 40,000 people were evacuated from Pozzuoli and further plans were in place should the crisis develop in unexpected ways. In January 1985 the crisis was over. To the great consternation of media critics the Italian government was able to build a new settlement, Monteruscello, within one year, with a planned initial intake of 20,000 people (Kilburn 1986).

In December 1991 an eruption began on Mount Etna and by the following April lava had reached the outskirts of Zafferana, an eastern flank village, destroyed some property and devastated agricultural land (McKenzie 1992). According to the world's press the reactions of villagers ranged from the resigned to the angry—many pouring scorn on the lack of prediction. The government did not do more because its attention was diverted by an impending election and generally the response was ineffective. Controversy was also evident amongst volcanologists both within and outside Italy. Much of the criticism was unfair, ill-informed and incorrect. In fact the efforts of the Italian Ministry of Civil Protection were impressive and included attempts to divert the lava through the use of barriers, explosives and, most spectacularly, helicopters from the United States Navy dropping huge rafts of boulders into a hole in the flow hopefully to spread the lava and arrest its progress. Monitoring the volcano has been continuous for many years and increased as Italian and foreign research teams moved in after the eruption had started. Transport was ready in case of civilian evacuation and in the person of the Ministry of Civil Protection's chief volcanologist, Franco Barberi, the authorities had on hand one of the world's foremost and most experienced experts in handling volcano-related emergencies. In view of the current state of the art in volcanic hazard prediction and lava-flow manipulation it is difficult to see what more the authorities could have done.

There is little doubt that much has changed in the Italian response to natural disasters since 1980. Today government is more interventionist in its approach and more typical of a developed country. In 1985 the Ministry of Civil Protection carried out a trial evacuation in a mountainous area of northern Tuscany in order to test the logistical effectiveness of its pre-disaster planning. This was judged a great success, but highlighted problems of traffic congestion and caused many protests to be made by displaced people (Alexander 1987). It is also clear that Italy is now capitalizing on its strengths in pure and applied volcanology (e.g. Luongo and Scandone 1991) and this may be seen in the responses to the eruptions of Mount Etna in 1983, 1989 and 1992 (Gruppo Nazionale per il Volcanologia [GNV] 1990) and the launch in 1991 of a new journal, *Acta Vulcanologica*. On the other side of the coin there have been no *major* disasters or threatened disasters since the early 1980s and only time will tell whether the effort put into contingency planning is adequate to deal with a real catastrophe, or whether Italy will again show its traditional propensity for political muddle and bureaucratic incompetence during emergencies, and the long recovery period, corruption and criminality which follow.

The eruptions of La Soufrière on the island of Guadeloupe in 1976 and of Soufrière volcano, St Vincent, in 1979 illustrate how responses can vary significantly between two countries which are superficially similar. Richard Fiske of the Smithsonian Institution (Washington DC) has made a close study of these eruptions and shows how relationships between scientists, public officials and the news media may be crucial in conditioning the nature and success of responses.

In the case of La Soufrière the situation was one in which the absence of long-term studies of activity and an initial mistake by scientists were compounded by conflicting advice to the authorities and its misrepresentation by the international media. The mistake was made on 12 August, approximately one month after large quantities of steam and ash had started to be erupted. A group of scientists claimed that the fall contained a significant quantity of fresh volcanic glass, implying that new magma had risen to high levels within the volcano and that a very dangerous phase of explosive activity was about to begin. Acting on this advice the local *préfet* ordered the evacuation of the 72,000 people living on the volcano (some 25% of the island's total) causing hardship to the evacuees and a high cost to the French treasury both in lost tax revenues and in relief aid (Fiske 1984). When this was reported by the world's press, tourism, a mainstay of the economy, collapsed virtually overnight. In fact 'the identification of fresh volcanic glass . . . was a mistake . . . and there was no clear evidence that magma was about to erupt to the surface' (Fiske 1984, p. 171).

Later the eruption became a media event and the initial mistake was compounded. On 30 August four scientists received minor-impact injuries near to the summit and the reporting of the incident became exaggerated and garbled. Front-page headlines appeared in newspapers in the United States claiming that a major eruption was imminent and a group of Miami-based journalists chartered an aircraft and flew to the island, only to be told by Richard Fiske that the explosion that had caused injuries to the scientists was trivial. Further problems were caused by the fact that by August there were four separate research teams on

Guadeloupe: two from France, one from the United States and one from Trinidad, and differences between them were exploited by the press. Indeed at times more attention was paid to these disagreements than to the actual events the journalists were supposed to be reporting. One group of French scientists continued to support the evacuation, the other argued that there was no immediate danger and the civil authorities were bewildered. The evacuation quickly degenerated into farce. In the end so bitter did the controversy become, that the French authorities in Paris had no option but to convene a committee involving foreign volcanologists to evaluate the claims of the rival French teams. A report was issued stating that the volcano, whilst still requiring close monitoring, appeared to pose less of a threat than initially thought. Within hours the minister responsible announced that the evacuation was at an end.

The eruption of Soufrière, St Vincent, was without doubt a much more potentially dangerous event than that on Guadeloupe three years earlier (Fiske 1984). From 13 to 26 April 1979 a series of powerful explosions burst through the crater lake, a plume rich in fall materials reached an altitude of up to 20 km and small pyroclastic flows and lahars were triggered. At the end of April explosions ended and a dome of basaltic andesite began to be built on the crater floor. Despite being a much poorer country than Guadeloupe (Table 9.11), careful monitoring had been carried out for several years and in 1979 there were baseline measurements of seismic activity, ground elevation and tilt, and the temperature of water in the crater lake. These measurements provided precursory indications that the volcano was about to erupt. When the eruption began the evacuation of people was unplanned, but the authorities soon took over and imposed order. Dissemination of information from scientists to the authorities and to journalists was facilitated by a single small team of fewer than eight from the University of the West Indies, augumented at times by additional personnel from the USA and the United Kingdom. A consensus was maintained by the team and speculation by individual members of it kept purposefully to a minimum. The authorities exercised strict control over the press through formal briefings in the capital, Kingstown, and journalists were prevented from entering the danger zone. No direct access to the volcanological team was granted.

These studies by Fiske (1984) illustrate that the success of a response is not solely dependent on development, efficient administration, the provision of relief aid and scientific expertise. In the case of Guadeloupe all these factors were positive. The scientific teams were well qualified (many of their members having international reputations for excellence), the evacuation programme until it degenerated later in the eruption was efficiently carried out, and in Caribbean terms Guadeloupe was relatively wealthy and relief aid was forthcoming from France. The problems were ones of management both before and during the eruption. First, no 'baseline' studies were available against which to test anomalous activity during the crisis. Secondly, there was no coordination and checking of the scientific forecasts made by the largely independent research teams. Finally, there was no proper system of press briefing. In contrast the eruption on St Vincent shows just how restrictive it can be to work within an administrative context defined by the secretive norms of the British Civil Service. As Fiske (1984, p. 176) concludes the ideal administrative structure lies in the middle ground between these two extremes. In the author's opinion there is little doubt that the *dirigiste* management style of the authorities on St Vincent was preferable to the chaotic approach of the government of Guadeloupe.

9.5 Conclusion

Perhaps the most important lesson to emerge from a review of responses and adjustments to volcanic hazards in different societies is that they are both variable and complex. Whilst it is only too evident that the physical characteristics of extreme volcanic events—in particular their magnitudes and frequencies—are important in determining losses, it is fundamental—indeed it is almost a truism—that in most cases they are of secondary significance. As the quotation which begins this chapter states 'most natural disasters ... are characteristic rather than accidental features of the places and societies where they occur' (Hewitt 1983b, p. 25). In Chapter 8 attention was paid to the so-called *radical* critique of much of what has been traditionally carried out under the umbrella of natural hazards research, especially of its implicitly deterministic agenda and focus on uncritical technological transfers (see Chapter 8, sections 8.2 and 8.3). It was also pointed out that this thinking had found its way into many of the initial reports prepared in connection with the International Decade for Natural Disaster Reduction (IDNDR) (Lechat 1990; see also Degg 1992), although valuable work by the United Nations in the mid-1980s did avoid many of the pitfalls of the dominant approach (Chapter 8, section 8.2), yet at the same time capitalized on its evident strength. For instance, in a publication entitled *Volcanic Emergency Management*, the United Nations provided a very valuable flow chart and checklist to illustrate the requirements—both administrative and scientific—that any country needs to consider in drawing up an emergency plan (UNDRO 1985).

It is encouraging to report that so far the volcanological community has eschewed the dominant approach in its pure form and in a recently published document the International Association of Volcanology and Chemistry of the Earth's Interior (IAVCEI) has spelt out a balanced and affordable programme for poor countries, which shows considerable awareness of the cultural, political and social differences between places, yet at the same time attempts to innovate best international scientific practice (IAVCEI 1990). It is perhaps not coincidental that six out of the 11 members of the task group which compiled the IAVCEI document were drawn from economically less developed countries. The principal initiatives of the package of measures are summarized in Table 9.12. All the initiatives would cost US$10 million per year for the decade, which considering what is attempted and what could be achieved is a modest figure indeed, especially when viewed against the scale of potential losses. Whether this programme can be achieved by the year 2000 is unlikely to be determined by the willingness of scientists to be cooperative or even by cost, but rather by social and political considerations. Already some actions have been taken which have included: an education programme at the University of Hawaii to enable economically less developed nations to attain self-sufficiency in applied volcanology (Anderson and Decker 1992); the establishment of a corps of volunteers in the Philippines to observe volcanoes and increase public awareness of risk (Delos-Reyes 1992); and the creation of interdisciplinary consultancy groups to provide educational courses in and advise on the mitigation of a range of environmental hazards (see S. C. Scott 1992).

Table 9.12: Principal initiatives envisaged by the International Association of Volcanology and Chemistry of the Earth's Interior (IAVCEI) during the international Decade of Natural Disaster Reduction (after IAVCEI 1990)

Initiative	Details
Hazard and risk mapping	Reconnaissance mapping of hazards (general prediction) and risks at previously unmapped volcanoes.
Volcano surveillance	Baseline monitoring and minimum surveillance (specific prediction) at volcanoes that are not presently monitored and which threaten people. Emphasis on affordable procedures and the involvement of local people.
Public education	Improved education about volcanic hazards (e.g. printed materials, community talks, films, videos, community field trips, observatory open days, workshops in schools, involvement of local volunteers in volcano surveillance, symposia for public officials and decision-makers, simulations of evaucations and establishment of twin-city relationships between two cities with similar volcano threats).
Dialogue with public officials	Scientists helping civil-defence officials and community leaders on questions of emergency planning, land-use planning and other risk-reducing activities.
'Decade Volcano' demonstration projects	It is proposed that about ten volcanoes are selected from around the world and be the subject of intense, integrated, multinational and multi-disciplinary cooperation. It is intended that this will demonstrate the range of activities required for hazard mitigation.
IAVNET	Electronic mail and long-distance communication links between volcanologists to allow coordination and advise both before and during eruptions.
Reference materials	An initiative designed to build up archive data sets on individual volcanoes and similar volcanoes elsewhere.
Volcano training	Training for new scientists/technicians and for in-service retraining of existing staff, including scientists, planners and civil-defence officials.
Low-cost equipment	For poor countries new equipment is required which is low in cost, reliable and easy to repair.
Satellite monitoring	Innovation of satellites to collect: multispectral images to map geological features, SO_2 plumes, ash in plumes, etc., data from ground sensors and relay this to observatories and facilitate voice communications between scientists 'on the ground' during an eruption and those in other parts of the world.
Crisis assistance	1) Supplement the personnel available locally during a volcanic crisis; 2) build up pools of regional expertise; 3) help national scientists and civil-defence officials to prepare for eruptions during periods of volcanic calm; and 4) help national scientists towards self-sufficiency.
Seed money	In poor countries volcanologists often find it difficult to gain sufficient money for rudimentary programmes of hazard mitigation. This initiative aims to provide external funds if these can be matched by funds generated locally.
Publication	Publication and the wide dissemination of the lessons learnt from volcanic eruptions.

References

Abich, H. 1841: *Geologische Beobachtungen über die vulkanischen Erscheinungen und Bildungen in Unter und Mittel Italien*. Braunschweig.

Achache, J. 1987: Plate tectonics: a framework for understanding our living planet. *Impact of Science on Society* 145, 5–19.

Acharya, H. 1989: Estimation of tsunami hazard from volcanic activity—suggested methodology with Augustine volcano as an example. *Natural Hazards* 1, 341–8.

Acharya, J. 1981: Volcanism and aseismic slip in subduction zones. *Journal of Geophysical Research* 86 (B), 335–44.

Acharya, J. 1982: Volcanic activity and large earthquakes. *Journal of Volcanology and Geothermal Research* 13, 373–8.

Admiralty 1945: *Italy*. (BR517c, vol IV). London Naval Intelligence Division.

Affronti, F. 1967: Straordinario gradiente pluviometrico del Monte Etna. *Schweiz Met. Zentralanstalt Veröffentlichungen* 4, 115–23.

Affronti, F. 1969a: Inversioni Termiche estive sul Monte Etna. *La Meteorologie* 10–11, 109–17.

Affronti, F. 1969b: Polveri da esplosioni dell 'Etna e nucleazioni. *Rivista di. Meteorologia* 21, 41–55.

Alcarez, A. 1969: Is Taal volcano drawing close to another eruptive activity? *Comvel Letter, Philippine Commission on Volcanology* 5, 1–10.

Alexander, D. 1987: Italy: land of disasters. *Geographical Magazine* 54(5), 226–32.

Alexander, D. 1991: Information technology in real-time for monitoring and managing natural disasters. *Progress in Human Geography* 15(3), 238–60.

Alison, M.L. and Clifford, S.M. 1987: Ice covered water volcanism on Ganymede. *Journal of Geophysical Research* 92B, 7865–76.

Allard, P. 1983: Facing hazards from the reawakening of dormant explosive volcanoes: the examples of Mt St Helens, El Chichón and Galunggung in 1980–1982. In Tazieff, H. and Sabroux, J.C. (eds), *Forecasting Volcanic Events*, IAVCEI Proceedings in Volcanology 1. Berlin: Springer-Verlag, 561–619.

Allard, P., Carbonnelle, J., Dajlevic, D., Le Bronec, J., Morel, P., Robe, M.C., Maurenas, J.M., Faivre-Pierret, R., Martin, D., Sabroux, J.C. and Zettwoog, P. 1991: Eruptive and diffuse emissions of CO_2 from Mount Etna. *Nature* 351, 387–91.

Allen, C.C. 1980: Icelandic subglacial volcanism: Thermal and physical effects. *Journal of Geology* 88, 108–17.

Allen, J.R.L. 1982: *Developments in sedimentology. Vols 30A and 30B: Sedimentary structures: Their character and physical basis*. Amsterdam: Elsevier.

Anderson, A.T. 1974: Before eruption H_2O content of some high alumina magmas. *Bulletin Volcanologique* 37, 530–52.

Anderson, D.L. 1974: Earthquakes and the rotation of the earth. *Science* 186, 49–50.

Anderson, D.L. 1981: Plate tectonics on Venus. *Geophysical Research Letters* 8, 309–11.

Anderson, D.L. 1982: Hotspots, polar wander, Mesozoic convection and the geoid. *Nature* 297, 391–3.

Anderson, J.L. and Decker, R.W. 1992: Volcanic risk mitigation through training. In McCall, G.J.H., Laming, D.J.C. and Scott, S.C. (eds), *Geohazards: Natural and Man-made*. London: Chapman and Hall, 7–12.

Anderson, S.W. and Fink, J.H. 1990: The development and distribution of surface textures at the Mount St Helens dome. In Fink, J.H. (ed.), *Lava flows and domes*, IAVCEI Proceedings in Volcanology 2. Berlin: Springer-Verlag, 25–46.

Anderson, T. and Fleet, J.S. 1903: Report on the eruptions of the Soufrière in St Vincent, and a visit to Montagne Pelée in Martinique, Part 1. *Philosophical Transactions of the Royal Society of London* 200A, 353–553.

Angell, J.K. and Korshover, J. 1985: Surface temperature changes following six major volcanic episodes between 1780 and 1980. *Journal of Climate and Applied Meteorology* 24, 937–51.

Anon. 1978: Overview and recommendations. In *Geophysical Prediction*, Studies in Geophysics, Geophysics Study Committee. Washington DC: National Academy of Sciences, 1–15.

Anon. 1981: The earthquake in Southern Italy on 23 November 1980 (Part 1). *Italy: Documents and Notes*, New Series 16, 31–59.

Anon. 1982: The earthquake in Southern Italy on 23 November 1980 (Part 2). *Italy: Documents and Notes*, Third Series 1, 9–41.

Anon. 1983: New sulfuric acid clouds from El Chichón spotted. *Bulletin of the American Meteorological Society* 64(4), 393–4.

Anon. 1988: *World Resources 1988–89: An Assessment of the Resource Base that Supports the Global Economy.* World Resources Institute and the International Institute for Environment and Development in Collaboration with the United Nations Environment Programme. New York: Basic Books.

Aoki, K., Ishiwaki, K. and Kanisawa, S. 1981: Flourine geochemistry of basaltic rocks from continental and oceanic regions and petrogenetic application. *Contributions to Mineralogy and Petrology* 78, 53–9.

Aramaki, S. 1983: Volcanoes. In *Kodansha Encyclopedia of Japan*, vol 8. Tokyo: Kodansha, 193–5.

Archambault, C., Scarpinati, G., Stoschek, J. and Tanguy, J.C. (undated): Analyse et surveillance thermiques de volcans actifs. *Bulletin PIRPSEV*, 66, CNRS-INAG, Paris.

ARCYANA 1975: Transform fault and rift valley from bathyscaph and diving saucer. *Science* 190, 108–16.

Arndt, N.T. and Nisbet, E.G. 1982: *Komatiites.* London: Allen and Unwin.

Arvidson, R.E., Greeley, R., Malin, M.C., Saunders, S., Izenberg, N., Plaut, J.J., Stofan, E.R. and Shepard, M.K. 1992: Surface modification of Venus as inferred from Magellan observations of plains. *Journal of Geophysical Research* 97E, 13, 303–18.

Associated Press 1977: Dispatch from Hong Kong, distributed to newspapers in USA, 5 January 1977.

Auer, V. 1956: The Pleistocene of Fuego-Patagonia. Part I: The ice age and interglacial ages. *Suomalaisen Tiedeaktemian Toimituksia Helsinki Sarja A, III, Geologica-Geographica* 45, 226.

Auer, V. 1958: The Pleistocene of Fuego-Patagonia. Part II: The history of the flora and vegetation. *Suomalaisen Tiedeaktemian Toimitiksia Helsinki Sarja* A, III. *Geologica-Geographica* 50, 239.

Auer, V. 1959: The Pleistocene of Fuego-Patagonia. Part III: Shoreline displacements. *Suomalaisen Tiedeaktemian Toimitiksia Helsinki Sarja* A, III, 247.

Auer, V. 1965: The Pleistocene of Fuego-Patagonia. Part IV: Bog profiles. *Suomalaisen Tiedeaktemian Toimitiksia Helsinki Sarja* A, III, 160.

Bailey, R.A. and Hill, D.P. 1990: Magmatic unrest at Long Valley caldera, California, 1980–1990. *Geoscience Canada* 17(3), 175–9.

Bailey, R.A., Dalrymple, G.B. and Lanphere, M.A. 1976: Volcanism, structure and geochronology of the Long Valley caldera, Mono County, California. *Journal of Geophysical Research* 81, 725–44.

Bailey, R.A., Beauchemin, P.R., Kapinos, F.P. and Klick, D.W. 1983: *The Volcanic Hazards Program: Objectives and Long-range Plans.* Washington DC: United States Geological Survey, Open File Report 83/400.

Baker, B.H., Mohr, P.A. and Williams, L.A.J. 1972: *Geology of the Eastern Rift System of Africa.* Geological Society of America Special Paper, 136.

Baker, P.E. 1979: Geological aspects of volcano prediction. *Journal of the Geological Society (London)* 136, 341–5.

Baker, P.E. 1985: Volcanic hazards on St Kitts and Montserrat, West Indies. *Journal of the Geological Society (London)* 142, 279–95.

Baker, V.R., Komatsu, G., Parker, T.J., Gulick, V.C., Kargel, J.S. and Lewis, J.S. 1992: Channels and valleys on Venus: Preliminary analysis of Magellan data. *Journal of Volcanology and Geothermal Research* 97E, 13, 421–13, 444.

Ballard, R.D. and van Andel, T.H. 1977: Morphology and tectonics of the inner rift valley at lat. 36°50′ on the Mid-Atlantic Ridge. *Geological Society of America Bulletin* 88, 507–30.

Ballard, R.D., Hekinian, R. and Francheteau, J. 1984: Geological setting of hydrovolcanic activity at 12°50′N on the East Pacific Rise—a submersible study. *Earth and Planetary Science Letters* 69, 176–86.

Banks, A.S. 1990: *Political Handbook of the World: 1989.* New York, State University of New York: CSA Publications.

Banks, N.G. and Hoblitt, R.P. 1981: Summary of temperature studies of 1980 deposits. In Lipman, P.W. and Mullineaux, D.R. (eds), *The 1980 eruptions of Mount St Helens*, United States Geological Survey Paper 1250, 295–313.

Barberi, F. and Gasparini, P. 1976: Volcanic hazards. *Bulletin International Association of Engineer-*

ing Geology 14, 217–32.
Barberi, F., Innocenti, F., Lirer, L., Munno, R., Pescatore, T. and Santacroce, R. 1978: The Campanian ignimbrite: a major prehistoric eruption in the Neopolitan area (Italy). *Bulletin Volcanologique* 41(1), 10–31.
Barberi, F., Rosi, M., Santocroce, R. and Sheridan, M.F. 1983: Volcanic hazard zonation: Mt Vesuvius. In Tazieff, H. and Sabroux, J.C. (eds), *Forecasting Volcanic Events*, IAVCEI Developments in Volcanology 1. Amsterdam: Elsevier, 149–61.
Barberi, F., Corrado, G., Innocenti, F. and Luongo, G. 1984: Phlegraean Fields 1982–1984: Brief chronicle of a volcanic emergency in a densely populated area. *Bulletin Volcanologique* 47(2), 175–85.
Barberi, F., Martini, M. and Rosi, M. 1990: Nevado del Ruiz volcano (Colombia): pre-eruption observations and the November 13 1985 catastrophic event. *Journal of Volcanology and Geothermal Research* 42, 1–12.
Bardintzeff, J.M. 1985: Calc-alkaline nuéss ardentes: A new classification. *Journal of Geodynamics* 3, 303–25.
Barrows, H.H. 1923: Geography as human ecology. *Annals Association of American Geographers* 13, 1–14.
Basaltic Volcanism Study Project (BVSP) 1981: *Basaltic Volcanism on the Terrestrial Planets*. New York: Pergamon Press.
Baubron, J.C., Allard, P., Sabroux, J.C., Tedesco, D. and Toutain, J.P. 1991: Soil gas emanations as precursory indicators of volcanic eruptions. *Journal of the Geological Society (London)* 148(3), 571–6.
Baxter, P.J. 1990: Medical aspects of volcanic eruptions: 1. Main causes of death and injury. *Bulletin of Volcanology* 52, 532–44.
Baxter, P.J., Bernstein, R.S., Falk, H., French, J. and Ing, R. 1982: Medical aspects of volcanic disasters: An outline of the hazards and emergency response measures. *Disasters* 6(4), 268–76.
Beane, J.E., Turner, C.A., Hooper, P.R., Subbarao, K.V. and Walsh, J.N. 1986: Stratigraphy, composition and form of the Deccan Basalts, western Ghats, India. *Bulletin of Volcanology* 48, 61–83.
Begoyavlenskaya, G.E., Braitseva, O.A., Melekestev, V., Kiriyanov, V.Y. and Miller, C.D. 1985: Catastrophic eruptions of the directed-blast type at Mount St Helens, Bezymianny and Shiveluch volcanoes. *Journal of Geodynamics* 3, 189–218.
Bernard, A., Demaiffe, D., Mattielli, N. and Punongbayan, R.S. 1991: Anhydrite-bearing pumices from Mount Pinatubo: Further evidence for the existence of sulphur-rich silicic magmas. *Nature* 354, 139–40.
Best, M.G. 1982: *Igneous and metamorphic petrology*. San Francisco.
Beverage, J.P. and Culbertson, J.K. 1964: Hyperconcentrations of suspended sediment. *Journal of Hydraulics: Division American Society of Civil Engineering Proceedings* 90 (HY6), 117–28.
Bianchi, R., Coradini, A., Federico, C., Gilberti, G., Sartoris, G. and Scandone, R. 1984: Modelling of surface ground deformation in the Phelgraean Fields volcanic area, Italy. *Bulletin Volcanologique* 47(2), 321–30.
Bickle, M.J. 1978: Heat loss from the earth: A constraint on Archaean tectonics from the relation between geothermal gradients and the rate of plate production. *Earth and Planetary Science Letters* 40, 301–15.
Biju-Duval, B., Dercourt, J. and Le Pichon, X. 1977: From Tethys Ocean to the Mediterranean Seas: a plate tectonics model of the evolution of the Western Alpine System. In Biju-Duval, B. and Montadent, L. (eds), *Structural History of the Mediterranean Basin*. Paris: Edition Technip, 143–64.
Binzel, R.P. 1990: Pluto. *Scientific American* 262(6), 26–33.
Blackburn, E.A., Wilson, L. and Sparks, R.S.J. 1976: Mechanisms and dynamics of strombolian activity. *Journal of the Geological Society (London)* 132, 429–40.
Blair, J.P. 1964: Home to Tristan da Cunha. *National Geographic* 125, 60–81.
Blake, S. 1981a: Volcanism and the dynamics of open magma chambers. *Nature* 289, 783–5.
Blake, S. 1981b: Eruption from zoned magma chambers. *Journal of the Geological Society (London)* 138, 281–7.
Blake, S. 1990: Viscoplastic models of lava domes. In Fink, J.H. (ed.), *Lava Flows and Domes*. IAVCEI Proceedings in Volcanology. Berlin: Springer-Verlag, 88–126.
Blake, S., Wilson, C.J.N., Smith, I.E.M. and Walker, G.P.L. 1992: Petrology and dynamics of the Waimhia mixed magma eruption, Taupo volcano, New Zealand. *Journal of the Geological Society (London)* 149(2), 193–207.

Blick, G.H., Otway, P.M. and Scott, B.J. 1989: Deformation monitoring of Mt Erebus, Antarctica 1980–1985. In Latter, J.H. (ed.), *Volcanic Hazards: Assessment and Monitoring*, IAVCEI Proceeding in Volcanology. Berlin: Springer-Verlag, 554–61.

Blong, R.J. 1982: *The Time of Darkness, Local legends and Volcanic Reality in Papua New Guinea.* Seattle: University of Washington Press; and Canberra: Australian National University Press.

Blong, R.J. 1984: *Volcanic Hazards.* Australia Academic Press.

Blot, C. 1965: Relations entre les séismes profonds et les éruptions volcaniques au Japon. *Bulletin Volcanologique* 28, 25–64.

Blot, C. 1981: Earthquakes at depth beneath volcanoes, forerunners of their activities, application to White Island, New Zealand. *Journal of Volcanology and Geothermal Research* 9, 277–91.

Blot, C. and Priam, R. 1964: Une nouvelle éruption du Volcan de Lopevi (Nouvelles Hebrides) et son analogie séismic avec les éruptions précédentes. *Bulletin Volcanologique* 27, 341–5.

Bolt, B.A. 1978: *Earthquakes: A Primer.* San Francisco: W.H. Freeman.

Bolt, B.A. 1982: *Inside the Earth.* San Francisco: W.H. Freeman.

Bolt, B.A. 1988: *Earthquakes: A Primer.* San Francisco: W.H. Freeman (second edition).

Bolt, B.A., Horn, W.L., Macdonald, G.A. and Scott, R.F. 1975: *Geological Hazards.* Berlin: Springer-Verlag.

Bonatti, E. 1987: The rifting of continents. *Scientific American* 256(3), 74–81.

Bond, A. and Sparks, R.S.J. 1976: The Minoan eruption of Santorini, Greece. *Journal of the Geological Society (London)* 132, 1–16.

Booth, B. 1973: The Granadilla Pumice deposit of southern Tenerife, Canary Islands. *Proceedings of the Geologists' Assocation* 84(3), 353–70.

Booth, B. 1977: Mapping volcanic risk. *New Scientist*, 22 September, 743–5.

Booth, B. 1979: Assessing volcanic risk. *Journal of the Geological Society (London)* 136, 331–40.

Booth, B. 1984: Tenerife. In Crandell, D.R., Booth, B., Kusumadinata, K., Shimozuru, D., Walker, G.P.L. and Westercamp, D. (eds), *Source-Book for Volcanic-Hazards Zonation.* Paris: UNESCO, 67–72.

Booth, B., Croasdale, R. and Walker, G.P.L. 1983: Volcanic hazard on Sao Miguel, Azores. In Tazieff, H. and Sabroux, J.C. (eds), *Forecasting Volcanic Events.* Developments in Volcanology 1. Amsterdam: Elsevier, 99–108.

Boss, A.P. 1983: Convection. *Reviews of Geophysics and Space Physics* 21, 1511–20.

Boudal, C. and Robin, C. 1989: Volcan Popocatepetl: Recent eruptive history, and potential hazards and risks in future eruptions. In Latter, J.H. (ed.), *Volcanic Hazards: Assessment and Monitoring*, IAVCEI Proceedings in Volcanology 1. Berlin: Springer-Verlag, 110–29.

Bowen, N.L. 1922: The reaction principle in petrogenesis. *Journal of Geology* 30, 177–98.

Bowen, N.L. 1928: *The Evolution of Igneous Rocks.* Princeton New Jersey: Princeton University Press.

Bradley, R.S. 1988: The explosive eruption signal in Northern Hemisphere continental temperature records. *Climatic Change* 12, 221–43.

Brandt Commission 1980: *North-South: A Programme for Survival.* London: Pan Books.

Bray, J.R. 1974: Volcanism and glaciation during the past 40 millennia. *Nature* 252, 679–80.

Bray, J.R. 1977: Volcanic dust veils and North Atlantic climatic change. *Nature* 268, 616–17.

Brazier, S.A., Davis, A.N., Sigurdsson, H. and Sparks, R.S.J. 1982: Fall-out and deposition of volcanic ash during the 1979 explosive eruption of the Soufriere of St Vincent. *Journal of Volcanology and Geothermal Research* 14, 335–59.

Brinkley, S.R., Kirkwood, J.F., Lampson, C.W., Revelle, R. and Smith, S.B. 1950: Shock from underwater and underground blasts. In *The Effects of Atomic Weapons.* Washington DC: US Government Printing Office, 83–113.

Brivio, P.A., Lo Giudice, E. and Zilioli, E. 1989: Thermal infrared surveys at Volcano Island: An experimental approach to the thermal monitoring of volcanoes. In Latter, J.H. (ed.), *Volcanic Hazards: Assessment and Monitoring*, IAVCEI, Proceedings in Volcanology 1. Berlin: Springer-Verlag, 357–72.

Brophy, J.G. and Marsh, B.D. 1986: On the origin of high-alumina arc basalt and the mechanics of melt extraction. *Journal of Petrology* 27, 763–89.

Brown, G.C. and Mussett, A.E. 1981: *The Inaccessible Earth.* London: Allen and Unwin.

Brown, G.C., Rymer, H. and Stevenson, D. 1991: Volcano monitoring by microgravity and energy budget analysis. *Journal of the Geological Society (London)* 148(3), 585–93.

Bryant, E.A. 1991: *Natural Hazards.* Cambridge: Cambridge University Press.

Bryson, R.A. and Goodman, B.M. 1982: The climatic effect of explosive volcanic activity: Analysis of the historical data. In Deepak, A. (ed.), *Atmospheric Effects and Potential Climatic Impact of the*

1980 Eruptions of Mount St Helens. Washington DC: NASA (NASA-CP-2240), 191–202.
Buchanan, D.J. 1974: A model for fuel-coolant interactions. *Journal of Physics D: Applied Physics* 7, 1441–57.
Bucher, W.H. 1950: The crust of the Earth. *Scientific American* 182, 32–41.
Buck, W.R., Martinez, F., Steckler, M.S. and Cochran, J.R. 1988: Thermal consequences of lithospheric extension: pure and simple. *Tectonics* 7, 213–34.
Budyko, M.I. 1968: On the causes of climatic variations. *Sveriges Meteorologiska Och Hydrologiska Institut Medd. ser B*, 28, 6–13.
Bulau, J.R., Waff, H.S. and Tyburczy, J.A. 1979: Mechanical and thermodynamical constraints of fluid distribution in partial melts. *Journal of Geophysical Research* 84(B), 6102–8.
Burke, K. and Dewey, J.F. 1973: Plume-generated triple junctions: Key indicators in applying plate tectonics to old rocks. *Journal of Geology* 81, 406–33.
Burke, K.C. and Wilson, J.T. 1976: Hot spots on the Earth's surface. *Scientific American* (August), 46–57.
Burke, K. and Francis, P. 1985: Climatic effects of volcanic eruptions. *Nature* 314, 136.
Bursik, M.I. and Woods, A.W. 1991: Buoyant, superbuoyant and collapsing eruption columns. *Journal of Volcanology and Geothermal Research* 45, 347–50.
Burton, I., Kates, R.W. and White, G. 1978: *The Environment as Hazard*. New York: Oxford University Press.
Caldeira, K. and Rampino, M.R. 1990: Carbon dioxide from Deccan volcanism at a K/T boundary greenhouse effect. *Geophysical Research Letters* 17(9), 1299–302.
Campbell, D.B. and Burns, B.A. 1980: Earth-based radar imagery of Venus. *Journal of Geophysical Research* 85, 8271–81.
Carey, S.N. and Sigurdsson, H. 1982: Influence of particle aggregation on deposition of distal tephra from the May 18, 1980, eruption of Mount St Helens Volcano. *Journal of Geophysical Research* 87B, 7061–72.
Carey, S.N. and Sparks, R.S.J. 1986: Quantitative models of the fallout and dispersal of tephra from volcanic columns. *Bulletin of Volcanology* 48, 109–25.
Carey, S.N. and Sigurdsson, H. 1987: Temporal variations of column height and magma discharge rate during the 79 AD eruption of Vesuvius. *Geological Society of America Bulletin* 99, 303–14.
Carey, S.N. and Sigurdsson, H. 1989: The intensity of Plinian eruptions. *Bulletin of Volcanology* 51, 28–40.
Carey, S.W. 1988: *Theories of the Earth and Universe*. Standford: Stanford University Press.
Carlson, R.W. 1987: Geochemical evolution of the crust and mantle. *Reviews of Geophysics* 25, 1011–20.
Carmichael, I.S.E., Turner, F.J. and Verhoogen, J. 1974: *Igneous Petrology*. New York: McGraw-Hill.
Carr, M.H. 1973: Volcanism on Mars. *Journal of Geophysical Research* 78, 4049–62.
Carr, M.H. 1974: The role of lava erosion in the formation of lunar rilles and martian channels. *Icarus* 22, 1–23.
Carr, M.H. 1981: *The Surface of Mars*. New Haven and London: Yale University Press.
Carr, M.H. 1984: Earth. In Carr, M.H. (ed.), *The Geology of the Terrestrial Planets*. Washington DC: National Aeronautics and Space Administration, 79–105.
Carr, M.H. 1987: Volcanic processes in the Solar System. *Earthquakes and Volcanoes* 19(4), 128–37.
Carr, M.J. and Stoiber, R.E. 1973: Intermediate depth earthquakes and volcanic eruptions in Central America, 1861–1972. *Bulletin Volcanologique* 37, 326–38.
Carta, S., Figari, R., Sartoris, A., Sassi, E. and Scandone, R. 1981: A statistical model for Vesuvius and its volcanological significance. *Bulletin Volcanologique* 44, 117–51.
Cas, R.A.F. 1978: Silicic lavas in Palaeozoic flysch-like deposits in New South Wales, Australia: Behaviour of deep subaqueous silicic flows. *Geological Society of America Bulletin* 89, 1708–14.
Cas, R.A.F. and Wright, J.V. 1987: *Volcanic Successions: Modern and Ancient*. London: Allen and Unwin.
Cas, R.A.F. and Wright, J.V. 1991: Subaqueous pyroclastic flows and ignimbrites: An assessment. *Bulletin of Volcanology* 53, 357–80.
Chappel, J. 1973: Astronomical theory of climatic change. *Quaternary Research* 3, 221–36.
Chappel, J. 1975: On possible relationships between upper Quaternary glaciations, geomagnetism, and vulcanism. *Earth and Planetary Science Letters* 26, 370–6.
Chayes, F. 1966: Alkaline and sub-alkaline basalts. *American Journal of Science* 264, 128–45.
Chester, D.K. 1986: Comments on 'A statistical analysis of flank eruptions on Etna Volcano' by F. Mulgaria, S. Tinti and E. Boschi. *Journal of Volcanological and Geothermal Research* 28, 385–95.

Chester, D.K. and Duncan, A.M. 1979: Interrelationships between volcanic and alluvial sequences in the evolution of the Simeto River Valley, Mount Etna, Sicily. *Catena* 6, 293–315.
Chester, D.K. and Duncan, A.M. 1982: The interaction of volcanic activity in Quaternary times upon the evolution of the Alcantara and Simeto Rivers, Mount Etna, Sicily. *Catena* 9, 319–42.
Chester, D.K., Duncan, A.M., Guest, J.E. and Kilburn, C.R.J. 1985: *Mount Etna: The Anatomy of a Volcano*. London: Chapman and Hall.
Chester, D.K., Duncan, A.M. and Guest, J.E. 1987: The pyroclastic deposits of Mount Etna, Sicily. *Geological Journal* 22, 225–43.
Chorley, R.J. and Kennedy, B.A. 1971: *Physical Geography: A Systems Approach*. London: Prentice-Hall.
Christiansen, R.L. and Petersen, D.W. 1981: Chronology of the 1980 eruptive activity. In Lipman, P.W. and Mullineaux, D.R. (eds), *The 1980 Eruptions of Mount St Helens, Washington*. United States Geological Survey, Professional Paper 1250, 17–31.
Christie-Blick, N. 1982: Pre-Pleistocene glaciations on the earth. Implications for the climatic history of Mars. *Icarus* 50, 421–43.
Cintala, M.J., Wood, C.A. and Head, J.W. 1977: The effects of target characteristics on fresh crater morphology: preliminary results from the Moon and Mercury. *Proceedings of the 8th Lunar Science Conference*, 3409–25.
Cioni, R., Corazza, E., Fratta, M., Guidi, M., Magro, G. and Marini, L. 1989: Geochemical precursors at Solfatara Volcano, Pozzuoli (Italy). In Latter, J.H. (ed.), *Volcanic Hazards: Assessment and Monitoring*, IAVCEI Proceedings in Volcanology 1. Berlin: Springer-Verlag, 384–99.
Clapperton, C.M. 1973a: Eruption on Helgafell. *Geographical Magazine* 45 (April), 481–6.
Clapperton, C.M. 1973b: Thrice threatened Heimaey. *Geographical Magazine* 45 (April), 495–500.
Clapperton, C.M. 1973c: Dying fires of a new volcano. *Geographical Magazine* 45 (June), 623–33.
Clapperton, C.M. 1973d: Back home to work on a volcanic island. *Geographical Magazine* 46 (November), 83–96.
Clapperton, C.M. 1977: Volcanoes in space and time. *Progress in Physical Geography* 1(3), 375–411.
Clapperton, C.M. 1986: The danger of tropical volcanoes. *Geographical Magazine* 58 (February), 74–80.
Clapperton, C.M. 1990: Glacial and volcanic geomorphology of the Chimborazo-Carihuairazo Massif, Ecuadorian Andes. *Transactions of the Royal Society of Edinburgh: Earth Sciences* 81, 91–116.
Clapperton, C.M. and Smyth, M.-A. 1986: Late Quaternary debris avalanche at Chimborazo, Equador. *Rev. Ciaf. (Bogota)* 11, 1–12.
Coates, B.R. 1974: Giant of the East. *Geographical Magazine* 47 (December), 170–4.
Coates, D.R. 1977: Landslide perspectives. *Geological Society America: Review of Engineering Geology* 3, 3–28.
Cogley, J.G. and Henderson-Sellers, A. 1984: The origin and earliest state of the Earth's hydrosphere. *Review of Geophysics and Space Physics* 22(2), 131–75.
Cole, J.W. 1979: Structure, petrology and genesis of Cenozoic volcanism, Taupo volcanic zone, New Zealand—a review. *New Zealand Geology and Geophysics* 22, 631–57.
Cole, J.W. 1984: Taupo-Rotorua depression—an ensialic marginal basin of North Island, New Zealand. In Kokelaar, B.P. and Howells, M.F. (eds), *Marginal Basin Geology: Volcanic and Associated Sedimentary and Tectonic Processes in Modern and Ancient Marginal Basins*. Geological Society Special Publication 16, 109–20.
Cole, P.D. 1991: Migration direction of sand-wave structures in pyroclastic surge deposits: Implications for depositional processes. *Geology* 19, 1108–11.
Colgate, S.A. and Sigurgeirsson, T. 1973: Dynamic mixing of magma and water. *Nature* 244, 552–4.
Condie, K.C. 1981: *Archaen Greenstone Belts*. Amsterdam: Elsevier.
Condie, K.C. 1982: *Plate Tectonics and Crustal Evolution*. New York: Pergamon (second edition).
Condie, K.C. 1989: *Plate Tectonics and Crustal Evolution*. New York: John Wiley (third edition).
Cornell, W., Carey, S. and Sigurdsson, H. 1983: Computer simulation of the transport and deposition of the Campanian Y-5 ash. *Journal of Volcanology and Geothermal Research* 17, 89–109.
Corradini, M.L. 1981: Phenomological modelling of the triggering phase of small-scale steam explosion experiments. *Nuclear Science Engineering* 78, 154–70.
Cotton, C.A. 1944: *Volcanoes as Landscape Forms*. Christchurch: Whitcombe and Tombs.
Cox, A. and Hart, R.B.H. 1986: *Plate Tectonics: How it Works*. Palo Alto and Oxford: Blackwell.
Cox, K.G. 1988: The Karoo Province. In Macdouggall, J.D. (ed.), *Continental Flood Basalts*. Dordrecht: Kluwer Academic Publishers, 239–73.
Cox, K.G. 1989: The role of mantle plumes in the development of continental drainage patterns.

Nature 342, 873–7.
Cox, K.G. 1991: A superplume in the mantle. *Nature* 352, 564–5.
Cox, K.G., Bell, J.D. and Pankhurst, R.J. 1979: *The Interpretation of Igneous Rocks*. London: Allen and Unwin.
Crandell, D.R. 1971: Postglacial lahars from Mount Rainier volcano, Washington. *United States Geological Survey Professional Paper* 677, 1–75.
Crandell, D.R. 1973: *Potential Hazards from Future Eruptions of Mount Rainier, Washington*. United States Geological Survey, Miscellaneous Geologic Investigations Map, I–826.
Crandell, D.R. 1975: *Assessment of Volcanic Risk on the Island of Oahu, Hawaii*. United States Geological Survey Open-File Report, 75–287.
Crandell, D.R. 1976: *Preliminary Assessment of the Potential Hazards from Future Volcanic Eruptions in Washington*. United States Geological Survey, Miscellaneous Field Studies Map, MF-774.
Crandell, D.R. 1980: *Recent Activity of Mount Hood, Oregon and Potential Hazards from Future Eruptions*. United States Geological Survey Bulletin, 1942.
Crandell, D.R. and Waldron, H.H. 1956: A recent volcanic mudflow of exceptional dimensions from Mount Rainier, Washington. *American Journal of Science* 254, 349–62.
Crandell, D.R. 1984: United States. In Crandell, D.R., Booth, B., Kusumadinata, K., Shimozuru, D., Walker, G.P.L. and Westerkamp, D. (eds), *Source Book for Volcanic-Hazards Zonation*. Paris: UNESCO, 72–91.
Crandell, D.R. and Mullineaux, D.R. 1978: *Potential Hazards from Future Eruptions of Mount St Helens Volcano, Washington*. United States Geological Survey Bulletin, 1383–C.
Crandell, D.R. and Hoblitt, R.P. 1986: Lateral blasts at Mount St Helens and hazard zonation. *Bulletin of Volcanology* 48, 27–37.
Crandell, D.R., Mullineaux, D.R. and Rubin, M. 1975: Mount St Helens Volcano: Recent and future behaviour. *Science* 187, 438–51.
Crandell, D.R., Mullineaux, D.R. and Miller, C.D. 1979: Volcanic-hazards studies in the Cascade Range of the Western United States. In Sheets, P.D. and Grayson, D.K. (eds), *Volcanic Activity and Human Ecology*. New York: Academic Press, 195–218.
Crandell, D.R., Booth, B., Kusamadinata, K., Shimozuru, D., Walker, G.P.L. and Westercamp, D. 1984a: *Source—Book for Volcanic-Hazards Zonation*. Paris UNESCO.
Crandell, D.R., Miller, C.D., Glicken, H.X., Christiansen, R.L. and Newhall, C.G. 1984b: Catastrophic debris avalanche from ancestral Mount Shasta volcano, California. *Geology* 12, 143–6.
Crisp, J.A. 1984: Rates of magma emplacement and volcanic output. *Journal of Volcanology and Geothermal Research* 20, 177–211.
Cristofolini, R. and Romano, R. 1980: *Pericoli da Attivita Volcanica Nell' Area Etnea Catania*. Catania: Comitato di consulenza Tecnico-Scientifico per il volcano Etna.
Cronan, D. 1985: A wealth of sea-floor minerals. *New Scientist* (6 June), 34–8.
Cronin, J.F., 1971: Recent volcanism and the stratosphere. *Science* 172, 847–9.
Cruz-Reyna, S. De la. 1991: Poisson-distributed patterns of explosive eruptive activity. *Bulletin of Volcanology* 54, 57–67.
Cullen-Tenaka, J.M. 1986: Where (or what) in the world is volcanic hazard management. In Keller, S.A.C. (ed.), *Mount St Helens: Five Years Later*. Cheney, Washington: Eastern Washington University Press, 395–9.
Curtis, G.H. 1968: The stratigraphy of the ejecta from the 1912 eruption of Mount Katmai and Novarupta, Alaska. In Coats, R.R., Hay, R.L. and Anderson, C.A. (eds), *Studies in Volcanology (Howell Williams Volume)*. Geological Society of America Memoir 16, 153–210.
Cutts, J.A., Thompson, T.W. and Lewis, B.H. 1981: Origin of bright ring-shaped craters in radar images of Venus. *Icarus* 48, 428–52.
Dalrymple, G.B., Silver, E.A. and Jackson, E.D. 1973: Origin of the Hawaiian Islands. *American Scientist* 61, 294–308.
Dana, J.D. 1890: *Characteristics of Volcanoes*. New York: Dodd, Mead and Co.
Davidson, J.P., McMillan, N.J., Moorbath, S., Wörner, G., Harmon, R.S. and Lopez-Escobar, L. 1990: The Nevados de Payachata volcanic region, II: Evidence for widespread crustal involvement in Andean magmatism. *Contributions to Mineralogy and Petrology* 105, 412–32.
Davies, P.A. and Runcorn, S.K. (eds) 1980: *Mechanisms of Continental Drift and Plate Tectonics*. London: Academic Press.
Davis, J.C. 1973: *Statistics and Data Analysis in Geology*. New York: Wiley.
Decker, R.W. 1973: State of the art in volcano forecasting. *Bulletin Volcanologique* 37, 372–93.
Decker, R.W. 1978: State of the art in volcano forecasting. In *Studies in Geophysics*. Washington DC: National Academy of Sciences, 47–57.

Decker, R.W. 1986: Forecasting volcanic eruptions. *Annual Review of Earth and Planetary Science* 14, 267–91.
Decker, R. and Christiansen, R.L. 1984: Explosive eruptions of Kilauea volcano, Hawaii. In *Explosive Volcanism: Inception, Evolution and Hazards.* Washington DC: National Academy Press, 122–33.
Decker, R.W. and Decker, B. 1981a: *Volcanoes.* San Francisco: W.H., Freeman.
Decker, R.W. and Decker, B. 1981b: The eruptions of Mount St Helens. *Scientific American* 24(3), 68–80.
Decker, R.W. and Decker, B. 1988: Volcanology in Hawaii. *Earthquakes and Volcanoes* 20(1), 4–30.
Decker, R.W. and Decker, B. 1989: *Volcanoes.* New York: W.H. Freeman.
Decker, R.W. and Decker, B. 1991: *Mountains of Fire.* Cambridge: Cambridge University Press.
Deepak, A. (ed.) 1982: *Atmospheric Effects and Potential Climatic Impact of the 1980 Eruptions of Mount St Helens.* Washington DC: NASA (NASA-CP-2240).
Degg, M. 1992: Natural disasters: Recent trends and future prospects. *Geography* 77(3), 198–209.
Deirmendjian, D. 1973: On volcanic and other particulate turbidity anomalies. *Advances in Geophysics* 16, 267–96.
Delos-Reyes, P.J. 1992: Volunteer observers program: a tool for monitoring volcanic and seismic events in the Philippines. In McCall, G.J.H., Laming, D.J.C. and Scott, S.C. (eds), *Geohazards: Natural and Man-Made.* London: Chapman and Hall, 13–24.
Denlinger, R.P. 1990: A model for some eruptions at Mount St Helens, Washington based on subcritical crack growth. In Fink, J.H. (ed), *Lava Flows and Domes*, IAVCEI Proceedings in Volcanology. Berlin: Springer-Verlag, 70–87.
de Silva, S.L. 1989: The Antiplano-Puna volcanic complex of the central Andes. *Geology* 17, 1102–6.
de Silva, S.L. and Francis, P.W. 1991: *Volcanoes of the Central Andes.* Berlin: Springer-Verlag.
Dewey, J.F., Pitman, W.C., Ryan, W.B.F., and Bonnin, J. 1973: Plate tectonics and the evolution of the Alpine System. *Geological Society of America Bulletin* 84, 3137–80.
Dibble, R.R. 1983: Recent volcanic activity and volcanological surveillance in New Zealand. In Tazieff, H. and Sabroux, J.C. (eds), *Forecasting Volcanic Events*, IAVCEI Developments in Volcanology 1. Amsterdam: Elsevier, 237–57.
Dibble, R.R., Nairn, I.A. and Neall, V.E. 1985: Volcanic hazards of North Island, New Zealand—Overview. *Journal of Geodynamics* 3, 369–96.
Dickenson, J.P., Clarke, C.G., Gould, W.T.S. Hodgkiss, A.G., Siddle, D.J., Smith, C.T. and Thomas-Hope, E.M. 1983: *A Geography of the Third World.* London: Methuen.
Dickinson, W.R. 1973: Widths of modern arc-trench gaps proportional to past duration of igneous activity in associated magmatic arcs. *Journal of Geophysical Research* 78, 3376–89.
Di Paola, G.M. 1972: The Ethiopian rift valley (between 7°00′ and 8°40′ (lat. North)). *Bulletin Volcanologique* 36, 517–60.
Dieterich, J.H. and Decker, R.W. 1975: Finite element modelling of the surface deformation associated with volcanism. *Journal of Geophysical Research* 80, 4094–102.
Dobran, F., Barberi, F. and Casarosa, C. 1990: *Modelling of Volcanological Processes and Simulation of Volcanic Eruptions.* Giardini, Rome: GNR, Gruppo Nazionale per La Vulcanologia Italy, Report VSG90–01.
Doelter, C. 1876: Die Vulkangruppe der Poninischen Inseln. *Denkschriften der Kaiserlichen Akademie der Wissenschaften in Wien Mathematisch-Naturwissenschafliche Klasse* 36/2, 141–84.
Dohler, G.C. 1988: A general outline of the ITSU Master Plan for the tsunami warning system in the Pacific. *Natural Hazards* 1, 295–302.
Donahue, T.M., Hoffman, J.H., Hodges, R.R. and Watson, A.J. 1982: Venus was wet: A measurement of the ratio of deuterium to hydrogen. *Science* 216, 630–3.
Dragoni, M., Bonafede, M. and Boschi, E. 1986: Downslope flow models of a Bingham liquid: Implications for lava flows. *Journal of Volcanology and Geothermal Research* 30, 305–25.
Druitt, T. 1991: Terminology and process in the flow-surge debate. *IAVCEI Commission on Explosive Volcanism—Newsletter* 20 (no page numbers)
Druitt, T. and Sparks, R.S.J. 1982: A proximal ignimbrite beccia facies on Santorini volcano. *Journal of Volcanology and Geothermal Research* 13, 147–71.
Druitt, T.H. and Sparks, R.S.J. 1984: On the formation of calderas during ignimbrite eruptions. *Nature* 310, 679–81.
Duffield, W.B., Tilling, R.I. and Canul, R. 1984: Geology of El Chichón volcano, Chiapas, Mexico. *Journal of Volcanology Geothermal Research* 20, 117–32.
Duncan, A.M. 1976: Pyroclastic flow deposits in the Adrano area of Mount Etna, Sicily. *Geological Magazine* 113, 357–63.
Duncan, A.M., Chester, D.K. and Guest, J.E. 1981: Mount Etna Volcano: Environmental impact and

problems of volcanic prediction. *Geographical Journal* 147, 164–79.
Duncan, A.M., Chester, D.K. and Guest, J.E. 1984: The Quaternary stratigraphy of Mount Etna, Sicily: The effects of differing palaeoenvironments on styles of volcanism. *Bulletin Volcanologique* 41(3), 497–516.
Duncan, A.M., Chester, D.K. and Guest, J.E. 1986: Volcanism in Italy: The hazard implications of hydromagmatic eruptions. *Interdisciplinary Science Reviews* 11(4), 377–86.
Durbin, C.S. 1981: *The Climatology of a Volcano: Mount Etna, Sicily.* Unpublished B.Sc. thesis, University of Liverpool.
Durbin, C.S. and Henderson-Sellers, A. 1981: Meteorological importance of the volcanic activity of Mount Etna. *Weather* 36, 284–91.
Du Toit, A.L. 1920: The Karoo dolerites. *Transactions of the Geological Society of South Africa* 33, 1–42.
Dzurisin, D. 1978: The tectonic and volcanic history of Mercury as inferred from studies of scarps, ridges, troughs and other lineaments. *Journal of Geophysical Research* 83B, 4883–906.
Eaton, G.P. 1982: The Basin and Range Province: Origin and tectonic significance. *Review of Earth and Planetary Science* 10, 409–40.
Eaton, G.P. 1984: The Miocene Great Basin of western North America as an extending back-arc region. *Tectonophysics* 102, 275–95.
Eaton, J.P. and Murata, K.J. 1960: How volcanoes grow. *Science* 132, 925–39.
Edmonds, J.M. and von Damm, K. 1983: Hot springs on the ocean floor. *Scientific American* 248(4), 70–85.
Eggers, A., Krausse, J., Rush, H. and Ward, J. 1976: Gravity changes accompanying volcanic activity at Paceya Volcano, Guatemala. *Journal of Volcanology and Geothermal Research* 1, 229–36.
Eggler, D.H. 1972: Water-saturated and under-saturated melting relations in a Paricutin andesite and an estimate of water content in the natural magma. *Contributions to Mineralogy and Petrology* 34, 261–71.
Eiby, G.A. 1980: *Earthquakes.* Auckland, New Zealand: Heinemann.
Einarsson, T. 1949: The eruption of Hekla 1947–1948; IV.3. The flowing lava. Studies of its main physical and chemical properties. *Soc. Scientiarum Islandica, Reykjavik*, p. 70.
Ellsaesser, H.W. 1983: *Isolating the Climatogenic Effects of Volcanoes.* Proc. USS/USSR Bilateral Meeting of Experts on Anthropogenic Climatic Change, Leningrad, p. 31 (Lawrence Livermore National Laboratory, Rep. UCRL-89161).
Ellsaesser, H.W. 1986: Comments on 'Surface temperature changes following the six major volcanic episodes between 1780 and 1980'. *Journal of Climate and Applied Meteorology* 25, 1184–5.
Ellsaesser, H.W., MacCracken, M.C., Walton, J.J. and Grotch, S.L. 1986: Global climatic trends as revealed by recorded data. *Reviews of Geophysics* 24(4), 745–92.
Ellsworth-Jones, W. 1985: A town of 25,000 is gone ... there is no Armero. *Sunday Times*, 17 November.
Embleton, C. and Thornes, J. 1979: *Process in Geomorphology.* London: Edward Arnold.
Endo, E.T. and Murray, T. 1991: Real-time seismic amplitude measurement (RSAM): A volcano monitoring and prédiction tool. *Bulletin of Volcanology* 53, 533–45.
Escher, B.G. 1925: L'emboulement préhistorique de Tasikmalaja et le volcan Galounggoung. *Leidsche Geologische Mededeelingen* 1, 8–21.
Espinolo, A. 1986: Notas sobre la Actividad Historica del Volcan Nevado del Ruiz. *INGEOMINAS* (Colombia), 1–17.
Fenner, C.N. 1920: The Katmai region, Alaska, and the great eruption of 1912. *Journal of Geology* 28, 569–606.
Filson, J.R. 1987: Geological hazards: Programs and research in the USA. *Episodes* 10(4), 292–5.
Finch, R.H. 1930: Rainfalls accompanying explosive eruptions of volcanoes. *American Journal of Science* 19, 147–50.
Fink, J.H. 1990 (ed.): *Lava Flows and Domes: Emplacement Mechanisms and Hazard Implications*, IAVCEI Proceedings in Volcanology 2. Berlin: Springer-Verlag.
Fisher, R.V. 1961: Proposed classification of volcaniclastic sediments and rocks. *Geological Society America Bulletin* 72, 1409–14.
Fisher, R.V. 1977: Erosion by volcanic base-surge density currents: U-shaped channels. *Geological Society America Bulletin* 88, 1287–97.
Fisher, R.V. 1979: Models for pyroclastic surges and pyroclastic flows. *Journal of Volcanology and Geothermal Research* 6, 305–18.
Fisher, R.V. 1986: Systems of transport and deposition within pyroclastic surges: Evidence from Mount St Helens, Washington. *EOS Transactions American Geophysical Union* 67, 1246.
Fisher, R.V. and Schmincke, H.U. 1984: *Pyroclastic Rocks.* Berlin: Springer-Verlag.

Fishlock, T. and Matthews, G. 1985: Death toll officially put at 20,000. *The Times*, 16 November.
Fiske, R.S. 1963: Subaqueous pyroclastic flows in the Ohanapecosh Formation, Washington. *Geological Society of America Bulletin* 74, 391–406.
Fiske, R. 1981: Scientists and the news media: Contrasting interaction during the two volcanic crises in the eastern Caribbean. *International Association of Volcanology and Chemistry of the Earth's Interior*, IAVCEI Symposium 'Arc Volcanism' (abstract).
Fiske, R. 1984: Volcanologists, journalists, and the concerned public: A tale of two crises in the Eastern Carbibbean. In *Explosive Volcanism: Inception, Evolution and Hazards*. Washington DC: National Academy Press, 170–6.
Fiske, R.S. and Matsuda, T. 1964: Submarine equivalents of ash flows in the Tokiwa Formation, Japan. *American Journal of Science* 262, 76–106.
Fiske, R. and Kinoshita, W.T. 1969: Inflation of Kilauea volcano prior to its 1967–1968 eruption. *Science* 165, 341–9.
Fiske, R.S. and Shepherd, J.B. 1990: Twelve years of ground-tilt measurements on the Soufriere of St Vincent. *Bulletin of Volcanology* 52, 227–41.
Folk, R.L. 1974: *Petrology of Sedimentary Rocks*. Austin, Texas: Hemphill Publishing.
Folk, R.L. and Ward, W. 1957: Brazos River bar: A study in the significance of grain size parameters. *Journal of Sedimentary Petrology* 27, 3–26.
Forgione, G., Luongo, G. and Romano, R. 1989: Mount Etna, Sicily: Volcanic hazard assessment. In Latter, J.H. (ed.), *Volcanic Hazards Assessment and monitoring*, IAVCEI, Proceedings in Volcanology. Berlin: Springer-Verlag 137–50.
Foster, H.L. and Mason, A.C. 1955: 1950 and 1951 eruptions of Mihara yama, Oshima volcano, Japan. *Geological Society of America Bulletin* 66, 731–62.
Fournier d'Albe, E.M. 1979: Objectives of volcanic monitoring and prediction. *Journal of the Geological Society (London)* 136, 321–6.
Foxworthy, B.L. and Hill, M. 1982: Volcanic eruptions of 1980 at Mount St Helens: The first 100 days. *United States Geological Survey Professional Paper*, 1249.
Francheteau, J., Needham, H.D., Choukroune, P., Juteau, T., Seguret, M., Ballard, R.D., Fox, P.J., Normark, W., Carranza, A., Cordoba, D., Guerrero, J., Rangin, C., Bougault, H., Cambon, P. and Hekinian, R. 1979: Massive deep-sea sulphide ore deposits discovered on the East Pacific Rise. *Nature* 277, 523–8.
Francis, P.W. 1970: Mysterious volcanoes of the high Andes. *Geographical Magazine* 42(11), 795–804.
Francis, P.W. 1979: Infra-red techniques for volcano monitoring and prediction—a review. *Journal of the Geological Society (London)* 136, 355–9.
Francis, P. 1983: Giant volcanic calderas. *Scientific American* 248(6), 46–56.
Francis, P. and Self, S. 1987: Collapsing volcanoes. *Scientific American* 256(6), 91–7.
Francis, P., O'Callaghan, L., Kretzchmar, G.A., Thorpe, R.S., Sparks, R.S.J., Page, R.N., de Barrio, R.E., Gillou, G. and Gonzalez, O.E. 1983: The Cerro Galan ignimbrite. *Nature* 301, 51–3.
Francis, P.W., Gardeweg, M., Raminrez, C.F. and Rothery, D.H. 1985: Catastrophic debris avalanche deposit of Socompa, northern Chile. *Geology* 13, 600–3.
Franklin, B. 1789: Meteorological imaginations and conjectures. *Memoirs Manchester Literary and Philosophical Society* 2, 373–7.
Frazzetta, G. and Romano, R. 1978: Approcio de studio per la stesura di una carta del rischio vulcano (Etna—Sicilia). *Memorie Societa Geologia Italiana* 19, 691–7.
Frazzetta, G., Gillot, P.Y., La Volpe, L. and Sheridan, M.F. 1984: Volcanic hazards at Fossa of Vulcano: Data from the last 6000 years. *Bulletin Volcanologique* 47(1), 105–24.
Frey, F.A. and Clague, D.A. 1984: Geochemistry of diverse basalt types from Loihi Seamount, Hawaii—petrologenic implications. *Earth and Planetary Science Letters* 66, 337–55.
Furnes, H., Friedliefsson, I.B. and Atkins, F.B. 1980: Subglacial volcanics—on the formation of acid hyaloclastites. *Journal of Volcanology and Geothermal Research* 8, 95–110.
Galindo, I., Otaola, J.A. and Zenteno, G. 1984: Atmospheric impact of the volcanic eruptions of El Chichón over Mexico. *Geofisica Internacional* 23(2), 373–83.
Garvin, L. 1991: Smoke signals from the deep. *Nature* 351, 699.
Gass, I.G. 1982: Ophiolites. *Scientific American* 247(2), 108–17.
Gass, I.G., Chapman, D.S., Pollack, H.N. and Thorpe, R.S. 1978: Geological and geophysical parameters of mid-plate volcanism. *Philosophical Transactions of the Royal Society of London* 288A, 581–97.
Geipal, R. 1982: *Disaster and Reconstruction*. London: Allen and Unwin.
Gentilli, J. 1948: Present day volcanicity and climatic change. *Geological Magazine* 85, 172–5.
Geophysics Study Group 1984: *Explosive Volcanism: Inception, Evolution and Hazards*. Washington

DC: National Academy Press.
Gerasimov, I.P. and Zvonkova, T.V. 1974: Natural hazards in the territory of the USSR: Study, control and warning. In White, G.F. (ed.), *Natural Hazards: Local, National, Global.* London: Oxford University Press, 243–55.
Gerlach, T.M. 1979: Evaluation and restoration of the 1970 volcanic gas analyses from Mt Etna, Sicily. *Journal of Volcanology and Geothermal Research* 6, 165–78.
Gerlach, T.M. 1983: Intrinsic chemical variations in high temperature volcanic gases from basic lavas. In Tazieff, H. and Sabroux, J.C. (eds), *Forecasting Volcanic Events*, Developments in Volcanology 1. Amsterdam: Elsevier, 323–36.
Gerlach, T. 1991: Etna's greenhouse pump. *Nature* 351, 352.
Gibson, I.L. and Walker, G.P.L. 1963: Some composite rhyolite/basalt lavas and related composite dykes in eastern Iceland. *Proceedings of the Geologists Association* 74, 301–18.
Gill, J.B. 1981: *Orogenic Andesites and Plate Tectonics.* Berlin: Springer-Verlag.
Gill, J.B., Stork, A.L. and Whelan, P.M. 1984: Volcanism, accompanying back-arc development in the southwest Pacific. *Tectonophysics* 102, 207–24.
Gilliland, R.L. 1982: Solar, volcanic and CO_2 forcing of recent climatic change. *Climatic Change* 4, 111–31.
Glass, B. 1982: *Introduction to Planetary Geology.* Cambridge: Cambridge University Press.
Global Volcanism Bulletin (1991a): Reports on Pinatubo (Luzon) and Unzen (Kyushu). *Bulletin of the Global Volcanism Network (Smithsonian Institution)* 16(5), 2–11.
Global Volcanism Bulletin (1991b): Reports on Pinatubo (Luzon) and Unzen (Kyushu). *Bulletin of the Global Volcanism Network (Smithsonian Institution)* 16(6), 2–6.
Goudie, A. 1981a: *Geomorphological Techniques.* London: Allen and Unwin.
Goudie, A. 1981b: *The Human Impact: Man's Role in Environmental Change.* Oxford: Blackwell.
Goudie, A. 1990: *The Human Impact: Man's Role in Environmental Change.* Oxford: Blackwell (second edition).
Gourgaud, G., Camus, G., Gerbe, M.-C., Morel, A., Sudradjat, A. and Vincent, P.M. 1989: The 1982–83 eruption of Galunggung (Indonesia): A case study of volcanic hazards with particular relevance to air navigation. In Latter, J.H. (ed.), *Volcanic Hazards: Assessment and Monitoring* IAVCEI Proceedings in Volcanology 1. Berlin: Springer-Verlag, 151–63.
Gow, A.J. and Williamson, T. 1971: Volcanic ash in the Antarctic ice sheet and its possible climatic implications. *Earth and Planetary Science Letters* 13, 210–18.
Greeley, R. 1977a: Volcanic morphology. In Greeley, R. and King, J.S. (eds), *Volcanism of the Eastern Snake River Plain, Idaho: A Comparative Planetary Geology Guidebook.* Washington DC: NASA, 5–23.
Greeley, R. 1977b: Basaltic plains volcanism. In Greeley, R. and King, J.S. (eds), *Volcanism of the Eastern Snake River Plain, Idaho: A Comparative Planetary Geology Guidebook.* Washington DC: NASA, 45–59.
Greeley, R. 1985: *Planetary Landscapes.* London: Allen and Unwin.
Greeley, R. 1987: *Planetary Landscapes.* London: Allen and Unwin (revised edition).
Greeley, R. and Theilig, E. 1978: Small volcanic constructs in the Chryse Planitia Region of Mars. In *Report Planetary Geology Program—1977–1978*, NASA, TM-79729, 202.
Greeley, R. and Spudis, P.D. 1981: Volcanism on Mars. *Reviews of Geophysics and Space Physics* 19, 12–41.
Greeley, R., Arvidson, R.E., Elachi, C., Geringer, M.A., Plaut, J.J., Saunders, R.S., Schubert, G., Stofan, E.R., Thouvenot, E.J.P., Wall, S.D. and Weitz, C.M. 1992: Aeolian features on Venus: Preliminary Magellan results. *Journal of Geophysical Research* 97E, 13,319–46.
Green, D.H. 1972: Magmatic activity as the major process in the chemical evolution of the Earth's crust and mantle. *Tectonophysics* 13, 47–71.
Green, D.H. and Ringwood, A.H. 1968: Genesis of calc-alkaline igneous rock suites. *Contributions to Mineralogy and Petrology* 8, 277–88.
Green, D.H., Hibberson, W.O. and Jaques, A.L. 1979: Petrogenesis of mid-ocean ridge basalts. In McElhinny, M.W. (ed.), *The Earth: its Origin, Structure and Evolution.* London: Academic Press, 265–300.
Greene, M., Perry, R. and Lindell, M. 1981: The March 1980 eruptions of Mt St Helens: Citizens perceptions of volcano threat. *Disasters* 5(1), 49–67.
Gregory, K.J. 1985: *The Nature of Physical Geography.* London: Edward Arnold.
Grove, N. 1973: Volcano overwhelms an Icelandic village. *National Geographic* 144 (July), 40–67.
Gruppo Nazionale per il Volcanologia (GNV) 1990: *Mount Etna: The 1989 Eruption.* Pisa: Giardini.
Gudmundsson, A. 1988: Formation of collapse calderas. *Geology* 16, 808–10.
Gueri, M. and Perez, L.J. 1986: Medical aspects of the 'El Ruiz' avalanche disaster; Colombia.

Disasters 10(2), 150–7.
Guest, J.E. and Sanchez, J. 1969: A large dacite lava flow in Northern Chile. *Bulletin Volcanologique* 33, 779–90.
Guest, J.E. and Greeley, R. 1977: *Geology on the Moon*. London: Wykeham.
Guest, J.E. and Murray, J.B. 1979: An analysis of hazard from Mount Etna Volcano. *Journal of the Geological Society of (London)* 136, 347–54.
Guest, J.E. and Duncan, A.M. 1981: Internal plumbing of Mount Etna. *Nature* 290, 584–6.
Guest, J.E., Butterworth, P., Murray, J. and O'Donnell, W. 1979: *Planetary Geology*. New York: Halsted.
Guest, J.E., Underwood, J.R. and Greeley, R. 1980: Role of lava tubes in flows from the Observatory Vent, 1971 eruption on Mount Etna. *Geological Magazine* 117, 601–6.
Guest, J.E., Duncan, A.M. and Chester, D.K. 1988: Monte Vulture Volcano (Basilicata, Italy): An analysis of morphology and volcaniclastic facies. *Bulletin of Volcanology* 50, 244–57.
Guffanti, M. and Weaver, C.S. 1988: Distribution of late Cenozoic volcanic vents in the Cascade range: Volcanic arc segregation and regional tectonic considerations. *Journal of Geophysical Research* 93B, 6513–29.
Gurnis, M. 1988: Large-scale mantle convection and the aggregation and dispersal of supercontinents. *Nature* 332, 695–9.
Haas, J.E., Kates, R.W. and Bowden, M.J. 1977: *Reconstruction following disaster*. Cambridge, Mass: MIT Press.
Hadfield, P. 1991a: The global earthquake. *The Sunday Times*, 28 April, 20–32.
Hadfield, P. 1991b: Telltale 'swarms' warned of Japanese eruption. *New Scientist* (8 June), 13.
Haggett, P. 1965: *Locational Analysis in Human Geography*. London: Edward Arnold.
Hall, M.L. 1990: Chronology of the principal scientific and governmental actions up to the November 13, 1985 eruption of Nevado del Ruiz, Colombia. *Journal of Volcanology and Geothermal Research* 42, 101–15.
Hall, M.L. 1992: The 1985 Nevado del Ruiz eruption: scientific, social and governmental response and interaction before the event. In McCall, G.J.H., Laming, D.J.C. and Scott, S.C. (eds), *Geohazards: Natural and Man-Made*. London: Chapman and Hall, 43–52.
Hallam, A. 1975: Alfred Wegener and the hypothesis of continental drift. *Scientific American* 232(2), 88–97.
Halwacks, M. 1983: Electrical and electromagnetic methods. In Tazieff, H. and Sabroux, J.C. (eds), *Forecasting Volcanic Events*. Developments in Volcanology 1, Amsterdam: Elsevier, 507–28.
Hamilton, Sir William 1776: *Campi Phlegraei. Observations on the Volcanoes of the Two Sicilies*. London.
Hamilton, W.B. 1988: Plate tectonics and island arcs. *Geological Society of America Bulletin* 100, 1503–27.
Hammer, C.U. 1977: Past volcanism revealed by Greenland ice sheet impurities. *Nature* 270, 482–6.
Hammer, C.U. 1980: Acidity of polar ice cores in relation to absolute dating, past volcanism and radio-echoes. *Journal of Glaciology* 25, 359–72.
Hammer, C.U., Clausen, H.B. and Dansgaard, W. 1981: Past volcanism and climate revealed by Greenland ice cores. *Journal of Volcanology and Geothermal Research* 11, 3–10.
Hampton, M.A. 1972: The role of subaqueous debris flow in generating turbidity currents. *Journal of Sedimentary Petrology* 42, 775–93.
Handler, P. 1985: Possible association between the climatic effects of stratospheric aerosols and corn yields in the United States. *Agricultural Meteorology* 35, 205–28.
Handler, P. 1986a: Possible association between the climatic effects of stratospheric aerosols and sea surface temperatures in the eastern tropical Pacific Ocean. *Journal of Climatology* 6, 31–41.
Handler, P. 1986b: Stratospheric aerosols and the Indian Monsoon. *Journal of Geophysical Research* 91(D), 14475–90.
Handler, P. 1989: The effect of volcanic aerosols on global climate. *Journal of Volcanology and Geothermal Research* 37, 233–49.
Hansen, J.E., Wang, W.C., and Lacis A.A. 1978: Mount Agung eruption provides a test of global climatic perturbations. *Science* 199, 1065–8.
Hargreaves, R. and Ayres, L.D. 1979: Morphology of the Archean metabasalt flows, Utik Lake, Manitoba. *Canadian Journal of Earth Science* 16, 1452–66.
Harlow, D.H. 1971: *Volcanic Earthquakes*. Unpublished M.A. thesis, Dartmouth College, Hanover, New Hampshire, USA.
Harris, D.M., Rose, W.I., Bornhorst, T. and Casadevall, T.J. 1980: Variations of SO_2 and CO_2 emission rates at Mount St Helens, July 22 to August 29, 1980. *EOS (Transactions American Geophysical Union)* 61, 1139.

Harriss, R.W., Hohenemser, C. and Kates, R.W. 1985: Human and non-human mortality. In Kates, R.W., Hohenemser, C. and Kasperson, J.X. (eds), *Perilous Progress: Managing the Hazards of Technology*. Boulder: Westview Press, 129–55.

Hartmann, W.K. and Wood, C.A. 1971: Moon: Origin and evolution of multi-ring basins. *The Moon* 3, 2–78.

Hatch, F.H., Wells, A.K. and Wells, M.K. 1972: *Petrology of Igneous Rocks*. London: Allen and Unwin (thirteenth edition).

Haughton, D.R., Roeder, P.L. and Skinner, B.J. 1974: Solubility of sulphur in mafic magmas. *Economic Geology* 69, 451–67.

Hawaiian Civil Defense, Department of Defense 1971: *The State of Hawaii Plan for Emergency Preparedness: Disaster Assistance (Vol. 3)*. Honolulu.

Hays, J.D., Imbrie, J. and Shackleton, N.J. 1976: Variations in the Earth's orbit: pacemaker of the ice ages. *Science* 194, 1121–32.

Head, J.W. 1976: Lunar volcanism in space and time. *Review of Geophysics and Space Physics* 14, 265–300.

Head, J.W. and Gifford, A. 1980: Lunar mare domes: Classification and modes of origin. *Moon and Planets* 22, 235–58.

Head, J.W. and Wilson, L. 1986: Volcanic processes and landforms on Venus: Theory, predictions, and observations. *Journal Geophysical Research* 91B, 9407–46.

Head, J.W., Crumpler, L.S., Aubele, J.C., Guest, J.E. and Saunders, R.S. 1992: Venus volcanism: Classification of volcanic features and structures, associations and global distribution from Magellan data. *Journal of Geophysical Research* 97E, 13,153–98.

Heiken, G. and Wohletz, K. 1985: *Volcanic Ash*. Berkeley California: University of California Press.

Hékinian, R. 1984: Undersea volcanoes. *Scientific American* 251(1), 34–43.

Heliker, C., Takahashi, T.J. and Wright, T.L. (eds) 1986: Volcano monitoring at the US Geological Survey's Hawaiian Volcano Observatory. *Earthquakes and Volcanoes* 18(1), 3–69.

Henyey, T.L. and Lee, T.C. 1976: Heat flow in Lake Tahoe, California-Nevada, and the Sierra Nevada-Basin and Range transition. *Geological Society of America Bulletin* 87, 1179–87.

Hess, H.H. 1962: History of Ocean basins. In Engle, A.E.J. *et al.* (eds), *Petrologic Studies: A Volume in Honor of A.F. Buddington*. Boulder, Colorado: Geological Society of America, 599–620.

Hewitt, K. (ed.) 1983a: *Interpretations of Calamity*. London: Allen and Unwin.

Hewitt, K. 1983b: The idea of calamity in a technocratic age. In Hewitt, K. (ed.), *Interpretations of Calamity*. London: Allen and Unwin, 3–30.

Hildreth, W. 1981: Gradients in silicic magma chambers: Implications for lithospheric magmatism. *Journal of Geophysical Research* 86B, 10153–92.

Hirschboeck, K.K. 1980: A new worldwide chronology of volcanic eruptions. *Palaeogeography, Palaeoclimatology, Palaeoecology* 29, 223–41.

Hobbs, P.V. 1991: Atmospheric effects of smoke from the Kuwaiti oil fires. *Eos* 73, 32.

Hoblitt, R.P. 1986: Observations of the eruptions of July 22 and August 7, 1980 at Mount St Helens, Washington. *United States Geological Survey Professional Paper* 1335, 1–44.

Hoblitt, R.P. and Kellogg, K.S. 1979: Emplacement temperatures of unsorted and unstratified deposits of volcanic rock debris as determined by palaeomagnetic techniques. *Geological Society of America Bulletin* 90, 633–42.

Hodge, D., Sharp, V. and Marts, M. 1979: Contemporary responses to volcanism: Case studies from the Cascades and Hawaii. In Shetts, P.D. and Grayson, D.K. (eds), *Volcanic Activity and Human Ecology*. New York: Academic Press, 221–7.

Hodges, C.A. 1973: Mare ridges and lava lakes. *Apollo 17 Preliminary Science Report*, NASA SP-330, 31.12–31.21.

Hoernes, R. 1893: *Erdbebenkurde*. Leipzig: Verlag Veit.

Holcomb, R.T., Moore, J.G., Lipman, P.W. and Belderson, R.H. 1988: Voluminous submarine lava flows from Hawaiian volcanoes. *Geology* 16, 400–4.

Hole, M.J. 1990: Antarctic volcanoes. *NERC News* (Natural Environmental Research Council, UK) 14 (July 1990), 4–6.

Holland, H.D. 1976: The evolution of sea water. In Windley, B.F. (ed.), *The Early History of the Earth*. New York: Wiley, 559–67.

Holmes, A. 1928: Radioactivity and earth movements. *Transactions of the Geological Society of Glasgow* 18, 559–606.

Holmes, A. 1965: *Principles of Physical Geology*. London: Nelson.

Honnorez, J. and Kirst, P. 1975: Submarine basaltic volcanism: Morphometric parameters for discriminating hyaloclastites and hyalotuffs. *Bulletin Volcanologique* 39, 1–25.

Honza, E. 1983: Evolution of arc volcanism related to marginal sea spreading and subduction at

trenches. In Shimozuru, D. and Yokoyama, I. (eds), *Arc Volcanism: Physics and Tectonics*. Tokyo: Terra Scientific Publishing Company.
Houghton, B.P., Latter, J.H. and Hackett, W.R. 1987: Volcanic hazard assessment for the Ruapehu composite volcano, Taupo Volcanic Zone, New Zealand. *Bulletin of Volcanology* 49, 737–51.
Howard, K.A., Wilhelms, D.E. and Scott, D.H. 1974: Lunar basin formation and highland stratigraphy. *Reviews of Geophysics and Space Physics* 12, 309–27.
Hoyt, D.V. 1978: An explosive volcanic eruption in the Southern Hemisphere in 1928. *Nature* 275, 630–2.
Hsü, K.J. 1975: Catastrophic debris streams (Sturzstroms) generated by rockfalls. *Geological Society America Bulletin* 86, 129–140.
Hsü, K.J. 1978: Albert Heim: observations on landslides and relevance to modern interpretations. In Voight, B. (ed.), *Rock Slides and Avalanches, I. Natural Phenomena*. Amsterdam: Elsevier.
Hulme, G. 1973: Turbulent lava flow and the formation of lunar simuous rills. *Modern Geology* 4, 107–17.
Hulme, G. 1974: The interpretation of lava flow morphology. *Geophysical Journal of the Royal Astronomical Society* 39, 361–83.
Hulme, G. and Fielder, G. 1977: Effusion rates and rheology of lunar lavas. *Philosophical Transactions of the Royal Society London* 285A, 227–34.
Humphreys, W.J. 1940: *Physics of the Air*. New York: MacGraw Hill.
Hunt, C.E. and McCready, J.S. 1980: *The Short-Term Economic Consequences of the Mount St Helens Volcanic Eruptions in May and June 1980*. Seattle: Washington State Department of Commerce and Economic Development.
Huppert, H.E. and Sparks, R.S.J. 1980: Restrictions on the compositions of mid-ocean ridge basalts: A fluid dynamical investigation. *Nature* 286, 46–8.
Huppert, H.E. and Sparks, R.S.J. 1984: Double-diffusive convection due to crystallisation in magmas. *Annual Review Earth and Planetary Science* 12, 11–37.
Huppert, H.E. and Sparks, R.S.J. 1985: Komatiites I: Eruption and flow. *Journal of Petrology* 26, 694–725.
Huppert, H.E., Shepard, J.B., Sigurdsson, H. and Sparks, R.S.J. 1982a: On lava dome growth, with application to the 1979 lava extrusion of the Soufriere of St Vincent. *Journal of Volcanology and Geothermal Research* 14, 199–222.
Huppert, H.E., Sparks, R.S.J. and Turner, J.S. 1982b: Effects of volatiles on mixing in calc-alkaline magma systems. *Nature* 297, 554–7.
Huppert, H.E., Sparks, R.S.J., Turner, J.S. and Arndt, N.T. 1984: Emplacement and cooling of komatiite lavas. *Nature* 309, 19–22.
Hyde, J.H. and Crandell, D.R. 1978: Postglacial volcanic deposits at Mount Baker, Washington and potential hazards from future eruptions. *United States Geological Survey Professional Paper* 1022–C.
IAVCEI 1950–75: *Catalogue of Active Volcanoes and Solfatara Areas of the World*. Rome: International Association of Volcanology and Chemistry of the Earth's Interior.
IAVCEI 1990: Reducing volcanic disasters in the 1990s. *Bulletin Volcanological Society Japan (ser. 2)* 35(1), 80–95.
Imamura, A. 1930: Topographical changes accompanying earthquakes on volcanic eruptions. *Publications Earthquake Investigations in Foreign Languages (Tokyo)* 25, 143.
Imbò, G. 1928: Sistemi eruttivi Etnei. *Bulletin Volcanologique* 15, 89–119.
Imbò, G. 1977: Seismic prediction at Pozzuoli. *Nature* 227, 511.
Imbò, G. 1980: Ancora sulla crisi bradisismica puteolana del 1970. *Accademia Pontaniana* 29, 53–83.
Imsland, P. 1989: Study models for volcanic hazard in Iceland. In Latter, J.H. (ed.), *Volcanic Hazards, Assessment and Monitoring*, IAVCEI Proceedings in Volcanology 1. Berlin: Springer-Verlag, 36–57.
Ingersoll, A.P. 1981: Jupiter and Saturn. *Scientific American* 245(6), 66–80.
Ingleby, I. 1966: Mount Lamington fifteen years later. *Australian Territories* 6, 28–34.
Inman, D.L. 1952: Measures for describing the size distribution of sediments. *Journal of Sedimentary Petrology* 22, 125–45.
Innes, J.L. 1983: Debris flows. *Progress in Physical Geography* 7(4), 469–502.
Iverson, R.M. 1990: Lava domes modelled as brittle shells that enclose pressurized magma, with application to Mount St Helens. In Fink, J.H. (ed.), *Lava Flows and Domes*, IAVCEI Proceedings in Volcanology 2. Berlin: Springer-Verlag, 47–69.
Jaggar, T.A. 1904: The initial stages of the spine on Pelée. *American Journal of Science* (fourth series), 17, 34–40.

Jaggar, T.A. 1917: Volcanological investigations at Kilauea. *American Journal of Science* (fourth series), 44, 161–220.
Jaggar, T.A. 1924: Sakura-jima, Japan's greatest volcanic eruption. *National Geographic* 45, 441–70.
Jaggar, T.A. 1949: *Steam Blast Volcanic Eruptions*. Hawaiian Volcano Observatory, Special Report 4, 1–137.
Jaggar, T.A. 1956: *My Experience with Volcanoes*. Honolulu Hawaiian Volcano Research Association.
Jaggar, T.A. and Finch, R.M. 1929: Tilt records for thirteen years at the Hawaiian Volcano Observatory. *Bulletin of the Seismological Society of America* 19, 38–51.
Jakosky, B.M. 1986: Volcanoes, the stratosphere and climate. *Journal Volcanology Geothermal Research* 28, 247–55.
Jimbo, G. 1983: Natural disasters. In *Kodansha Encyclopedia of Japan*. Tokyo: Kodansha.
Johnson, A.M. 1970: *Physical Processes in Geology*. San Francisco: Freeman Cooper.
Johnston, R.J. 1979: *Geography and Geographers: Anglo-American Human Geography since 1945*. London: Edward Arnold.
Johnston, R.J. 1989: *Environmental Problems: Nature, Economy and State*. London: Bellhaven Press.
Johnson, R.W. (ed.) 1989: *Intraplate Volcanism in Eastern Australia and New Zealand*. Cambridge: Cambridge University Press.
Jones, A.E. 1943: Classification of lava surfaces. *Transactions of the American Geophysical Union* Part 1, 265–8.
Jones, D.K.C. 1974: Japan under strain. *Geographical Magazine* 47(3), 185–92.
Jones, J.G. 1970: Intraglacial volcanoes of the Laugarvatn region, southwest Iceland, II. *Journal of Geology* 78, 127–40.
Judd, J.W. 1875: Contributions to the study of volcanoes, VII. The Ponza Islands. *Geological Magazine* 12, 298–308.
Jurdy, D.M. 1987: Plates and their motions. *Reviews of Geophysics* 25, 1286–92.
Jurewicz, S.R. and Watson, E.B. 1984: Distribution of partial melt in a felsic system: The importance of surface energy. *Contributions to Mineralogy and Petrology* 85, 125–9.
Kantha, L.H. 1981: 'Basalt fingers'—origin of columnar joints. *Geological Magazine* 118, 251–64.
Karig, D.E. 1974: Evolution of arc systems in the western Pacific. *Annual Review of Earth and Planetary Science* 2, 51–75.
Kasting, J.F., Toon, O.B. and Pollack, J.B. 1988: How the climate evolved on the terrestrial planets. *Scientific American* 258(2), 46–53.
Kasumadinata, K. 1984: Indonesia. In Crandell, D.R., Booth, B., Kasumadinata, K., Shimozuru, D., Walker, G.P.L. and Westercamp, D., *Source-Book for Volcanic Hazards Zonation*. Paris: UNESCO, 55–60.
Kates, R.W. 1987: Hazard assessment and management. In McLaren, D.J. and Skinner, B.J. (eds), *Resources and World Development*. Chichester: John Wiley, 741–53.
Keller, G. 1967: Alter und Abfolge des vulkanischen Ereignisse auf den Aolischen Inseln/Sizilien. *Bericht Naturforschenden Gesellschaft Freiburg* 57, 33–67.
Kelly, P.M. and Sear, C.B. 1982: The formulation of Lamb's Dust Veil Index. In Deepak, A. (ed.), *Atmospheric Effects and Potential Climatic Impact of the 1980 Eruptions of Mount St Helens*. Washington DC: NASA (NASA-CP-2240), 293–8.
Kelly, P.M. and Sear, C.B. 1984: Climatic impact of explosive volcanic eruptions. *Nature* 311 (5988), 740–3.
Kennett, J.P. and Thunell, R.C. 1975: Global increase in Quaternary explosive volcanism. *Science* 187, 497–503.
Kent, P.E. 1966: The transport mechanism of catastrophic rock falls. *Journal of Geology* 74, 79–83.
Kerr, R.A. 1981: Mount St Helens and a climate quandary. *Science* 211, 371–4.
Kerr, R.A. 1989: Volcanoes can muddle the greenhouse. *Science* 245, 127–8.
Kieffer, S.W. 1981a: Blast dynamics at Mount St Helens on 18 May 1980. *Nature* 291, 568–70.
Kieffer, S.W. 1981b: Fluid dynamics of the May 18th blast at Mount St Helens. In Lipman, P.W. and Mullineaux, D.R. (eds), *The 1980 Eruptions of Mount St Helens, Washington*. United States Geological Survey Professional Paper 1250, 379–400.
Kieffer, S.W. 1982: Dynamics and thermodynamics of volcanic eruptions: Implications for the plumes on Io. In Morrison, D. (ed.), *Satellites of Jupiter*. Tuscon: University of Arizona Press, 647–723.
Kieffer, S.W. 1984: Factors governing the structure of volcanic jets. In *Explosive Volcanism: Inception, Evolution and Hazards*, Washington DC: National Academy Press, 143–57.
Kienle, J. and Swanson, S.E. 1983: The hazards of Augustine. *The Northern Engineer* 15(3), 10–37.
Kienle, J., Dean, K.G. and Garbeil, H. 1990: Satellite surveillance of volcanic ash plumes, application to aircraft safety. *Eos* 71(7), 265–6.

Kilburn, C.R.J. 1981: Pahoehoe and aa lavas: A discussion and continuation of the model of Peterson and Tilling. *Journal of Volcanology and Geothermal Research* 11, 373–82.

Kilburn, C.R.J. 1986: In the jaws of the volcano. *New Scientist* (6 February), 42–6.

Kilburn, C.R.J. 1990: Surfaces of aa flow-fields on Mount Etna, Sicily: Morphology, rheology, crystallization and scaling phenomena. In Fink, J. (ed.), *Lava Flows and Domes*, IAVCEI Proceedings in Volcanology 2, Berlin: Springer-Verlag, 129–57.

King, R. 1973: *Sicily*. Newton Abbot: David and Charles.

King, R. 1975: Geographical perspectives on the evolution of the Sicilian Mafia. *Tijdschrift Economische Sociale Geographie* 66, 21–34.

Kinoshita, W.T., Swanson, D.A. and Jackson, D.B. 1974: The measurement of crustal deformation related to volcanic activity at Kilauea Volcano, Hawaii. In Civetta, L., Gasparini, P., Luongo, G. and Rapolla, A. (eds), *Physical Volcanology*. Amsterdam: Elsevier, Ch. 4.

Klein, F.W. 1982: Patterns of historical eruptions of Hawaiian volcanoes. *Journal of Volcanology and Geothermal Research* 12, 1–35.

Kokelaar, B.P. 1983: The mechanisms of surtseyan volcanism. *Journal of the Geological Society (London)* 140, 939–44.

Kolelaar, B.P. 1986: Magma-water interactions in subaqueous and emergent basaltic volcanism. *Bulletin of Volcanology* 48, 275–89.

Kokelaar, B.P. and Durant, G.P. 1983: The submarine eruption and erosion of Surtla (Surtsey), Iceland. *Journal of Volcanology and Geothermal Research* 19, 239–46.

Kondo, J. 1988: Volcanic eruptions, cool summers, and famine in the northeastern part of Japan. *Journal of Climate* 1, 775–88.

Koto, B. 1916: The great eruption of Sikura-jima in 1914. *Journal of the College Science, Imperial University of Tokyo*, 38(3), 229.

Kozu, S. 1934: The great activity of Komagatake in 1929. *Tschermak's Mineralogische und Petrographische Mitteilungen* 45, 133–74.

Kubotera, A. 1974: Volcanic tremors at Aso Volcano. In Civetta, L., Gasparini, P., Luongo, G. and Rapolla, A. (eds), *Physical Volcanogy*. Amsterdam: Elsevier, 29–47.

Kusamadinata, K. 1984: Indonesia. In Crandell, D.R., Booth, B., Kusumadinata, K., Shimozuru, D., Walker, G.P.L. and Westercamp, D. (eds), *Source-Book for Volcanic-Hazards Zonation*, Paris: UNESCO, 55–60.

Kushiro, I. 1974: Melting of hydrous upper mantle and a possible generation of andesitic magma: An approach from synthetic systems. *Earth and Planetary Science Letters* 22, 294–9.

Kushiro, I. 1976: Changes in viscosity and structure of melt of $NaAlSi_2O_6$ composition at high pressures. *Journal of Geophysical Research* 81, 6347–50.

Kushiro, I., Yoder, H.S. and Mysen, B.O. 1976: Viscosities of basalt and andesites melts at high pressures. *Journal of Geophysical Research* 81, 6351–6.

Kutzbach, J.E. and Otto-Bliesner, B.L. 1982: The sensitivity of the African-Asia monsoonal climate to orbital parameter changes for 9000 BP in a low-resolution general circulation model. *Journal of Atmospheric Science* 39, 1177.

Kyle, P.R., Jezek, P.A., Mosley-Thompson, E. and Thompson, L.G. 1981: Tephra layers in the Byrd station ice-core and the Dome C ice-core, Antarctica and their climatic importance. *Journal of Volcanology and Geothermal Research* 11, 29–39.

Lamb, H.H. 1970: Volcanic dust in the atmosphere; with a chronology and assessment of its meteorological significance. *Philosophical Transactions of the Royal Society of London* A266, 425–533.

Lamb, H.H. 1972: *Climate Past, Present and Future, Vol. 1*. London: Methuen.

Lamb, H.H. 1977: *Climate Past, Present and Future, Vol. 2. Climate History and the Future*. London: Methuen.

Lamb, H.H. 1982a: Climatic changes in our own times and future trends. *Geography* 67(3), 203–20.

Lamb, H.H. 1982b: *Climatic History and the Modern World*. London: Methuen.

Larson, R.L. 1991: Latest pulse of the Earth: Evidence for a mid-Cretaceous superplume. *Geology* 19, 547–50.

Larsson, W. 1937: Vulkanische Äsche vom Ausbruch des chilenischen Vulkans Quizapu (1932) in Argentinien gesammelt. *Geological Institute Upsala Bulletin* 26, 27–52.

Latter, J.H. 1971: The interdependence of seismic and volcanic phenomena. *Bulletin Volcanologique* 35, 127–42.

Latter, J.H. 1981: Tsunamis of volcanic origin: Summary of causes, with particular reference to Krakatoa, 1883. *Bulletin Volcanologique* 44(3), 467–90.

Latter, J.H. 1989a: Preface. In Latter, J.H. (ed.), *Volcanic Hazards, Assessment and Monitoring*, IAVCEI Proceedings in volcanology 1. Berlin: Springer-Verlag, v–viii.

Latter, J.H. 1989b (ed.): *Volcanic Hazards, Assessment and Monitoring*, IAVCEI Proceedings in Volcanology 1. Berlin: Springer-Verlag.
Leavitt, S.W. 1982: Annual volcanic carbon dioxide emission: An estimate from eruption chronologies. *Environmental Geology* 4, 15–21.
Le Bas, M.J. and Streckeisen, A.L. 1991: The IUGS systematics of igneous rocks. *Journal of the Geological Society (London)* 148(5), 825–35.
Lechat, M.F. 1990: The International Decade for Natural Disaster Reduction: Background and objectives. *Disasters* 14(1), 1–6.
Le Grand, H.E. 1988: *Drifting Continents and Shifting Theories: The Modern Revolution in Geology and Scientific Change*. Cambridge: Cambridge University Press.
Legrand, M. and Delmas, R.J. 1987: A 220 year continuous record of volcanic H_2SO_4 in the Antarctic ice sheet. *Nature* 327, 671–6.
Le Maitre, R.W. 1976: The chemical variability of some common igneous rocks. *Journal of Petrology* 17, 589–637.
Le Maitre, R.W. (ed.), Bateman, P., Dudek, A., Keller, J. *et al.* 1989: *A Classification of Igneous Rocks and Glossary of Terms: Recommendations of the International Union of Geological Sciences Subcommission on the Systematics of Igneous Rocks*. Oxford: Blackwell.
Lipman, P.W. and Mullineaux, D.R. (eds) 1981: *The 1980 Eruptions of Mount St Helens, Washington*. USGS Professional Paper 1250.
Lipman, P.W., Self, S. and Heiken, G. 1984: Introduction to calderas; special issue. *Journal of Geophysical Research* 89B, 8219–21.
Lirer, L., Pescatore, T., Booth, B. and Walker, G.P.L. 1973: Two Plinian pumice fall deposits from Somma-Vesuvius, Italy. *Geological Society of America Bulletin* 84, 759, 772.
Littlewood, P. 1985: Social and political aspects of the south Italian earthquake of 1980. *Disasters* 9(3), 206–12.
Lockwood, J.G. 1983: Diversion of a lava flow at Mount Etna. *Volcano News* 15, 4–6.
Lonsdale, P. and Batiza, R. 1980: Hyaloclastite and lava flows on young seamounts examined with a submersible. *Geological Society of America Bulletin* 91, 545–54.
Lopez, R. and Guest, J.E. 1982: Lava flows on Mount Etna, a morphological study. In Coradini, A. and Fulchignoni, M. (eds), *The Comparative Study of the Planets*. Dordrecht: D. Reidel, 441–59.
Lorenz, V. 1973: On the formation of maars. *Bulletin Volcanologique* 37, 183–204.
Lorenz, V. 1987: Phreatomagmatism and its relevance. *Chemical Geology* 62, 149–56.
Lowman, P.D. 1976: Crustal evolution in silicate planets: Implications for the origin of continents. *Journal of Geology* 84, 1–26.
Lucchitta, B.K. 1976: Mare ridges and related highland scarps—result of vertical tectonism? *Proceedings of the 7th Lunar and Planetary Science Conference*, 2761–82.
Luhr, J.F. 1991: Volcanic shade causes cooling. *Nature* 354, 104–5.
Luongo, G. and Scandone, R. (eds) 1991: Campi Flegrei. *Journal of Volcanology and Geothermal Research* 48(1/2), 1–223.
Lunine, J.I. 1989: Origin and evolution of the outer Solar System atmospheres. *Science* 245, 141–6.
Luyendyk, B.P. and Macdonald, K.C. 1977: Physiography and structure of the inner floor of the Famous rift valley: observations with a deep-towed instrument package. *Geological Society of America Bulletin* 88, 648–63.
Lyell, Sir Charles 1830: *Principles of Geology*. London: J. Murray.
Lyell, Sir Charles 1847: *Principles of Geology*. London: J. Murray (seventh edition).
Lyell, Sir Charles 1858: On the structure of lavas which have consolidated on steep slopes; with remarks on the mode of origin of Mount Etna, and on the theory of 'Craters of Elevation'. *Philosophical Transactions of The Royal Society London* 148, 703–86.
Lyttleton, R.A. 1982: *The Earth and its Mountains*. Chichester: John Wiley.
MacCracken, M.C. and Luther, F.M. 1984: Preliminary estimate of the radiative and climatic effects of the El Chichón eruption. *Geofiscia Internacional* 23, 385.
Macdonald, G.A. 1962: The 1959 and 1960 eruptions of Kilauea volcano, Hawaii and the construction of walls to restrict the spread of the lava flows. *Bulletin Volcanologique* 24, 249–94.
Macdonald, G.A. 1967: Forms and structures of extrusive basaltic rocks. In Hess, H.H. and Poldervaart, A. (eds), *The Poldervaart Treatise on Rocks of Basaltic Composition*. New York: Interscience Publishers, Vol. 1, 1–61.
Macdonald, G.A. 1972: *Volcanoes*. New Jersey: Prentice Hall.
Macdonald, G.A. and Abbott, A.T. 1970: *Volcanoes in the Sea: The Geology of Hawaii*. Honolulu: University of Hawaii Press.
Macdonald, G.A. and Hubbard, D.H. 1982: *Volcanoes of the National Parks in Hawaii*. Honolulu: Hawaii National Parks Assocation.

Macdonald, G.A., Abbott, A.T. and Peterson, F.L. 1983: *Volcanoes in the Sea*. Honolulu: Hawaii National Parks Assocation.

Macdonald, K.C. 1982: Mid-ocean ridges: fine scale tectonic, volcanic and hydrothermal processes within the plate boundary zone. *Review of Earth and Planetary Science* 10, 155–90.

Macdonald, K.C. and Luyendyk, B.P. 1977: Deep tow studies of the structure of the mid-Atlantic ridge crest near lat. 37°N. *Geological Society of America Bulletin* 88, 621–36.

Macdonald, K.C. and Luyendyk, B.P. 1982: The crust of the East Pacific Rice. In Decker, R. and Decker, B. (eds), *Volcanoes and The Earth's Interior*. San Francisco: W.H. Freeman, 17–30.

Macdougall, J.D. (ed.) 1988: *Continental Flood Basalts*. Dordrecht: Kluwer Academic Publishers.

MacKenzie, W.S., Donaldson, C.H. and Guilford, C. 1982: *Atlas of Igneous Rocks and their Textures*. London: Longman.

Macpherson, G.J. 1984: A model for predicting the volumes of vesicles in submarine basalts. *Journal of Geology* 92, 72–82.

Maddox, J. 1984: From Santorini to Armageddon. *Nature* 307, 107.

Mahoney, J.J. 1988: Deccan traps. In Macdougall, J.D. (ed.), *Continental Flood Basalts*. Dordrecht: Kluwer Academic Publishers, 151–95.

Malin, M.C. 1980: Lengths of Hawaiian lava flows. *Geology* 8, 306–8.

Malin, M.C. and Sheridan, M.F. 1982: Computer-assisted mapping of pyroclastic surges. *Science* 217, 637–40.

Malinconico, L.L. 1987: On the variation of SO_2 emission from volcanoes. *Journal of Volcanology and Geothermal Research* 33, 231–7.

Marsh, B.D. 1982: On the mechanisms of igneous diapirism, stoping and zone melting. *American Journal of Science* 282, 808–55.

Marsh, B.D. 1984: Mechanisms and energetics of magma formation and ascension. In *Explosive Volcanism: Inception, Evolution and Hazards*. Washington DC: National Academy Press, 67–83.

Marsh, B.D. 1987: Magmatic processes. *Reviews of Geophysics* 25, 1043–53.

Marsh, B.D. and Kantha, L.H. 1978: On the heat and mass transfer from an ascending magma. *Earth and Planetary Science Letters* 39, 435–43.

Marshall, P. 1935: Acid rocks of the Taupo-Rotorua volcanic district. *Transactions Royal Society New Zealand* 64, 323–66.

Martin, D. 1989: A stirring tale of crystals and currents. *New Scientist* (25 November), 53–9.

Martini, M. 1989: The forecasting significance of chemical indicators in areas of quiescent volcanism: Examples from Vulcano and Phlegrean Fields (Italy). In Latter, J.H. (ed.), *Volcanic Hazards, Assessment and Monitoring*, IAVCEI Proceedings in Volcanology 1. Berlin: Springer-Verlag, 372–83.

Mason, A.C. and Foster, H.L. 1953: Diversion of lava flows at O Shima, Japan. *American Journal of Science* 251, 249–58.

Mass, C. and Schneider, S.H. 1977: Statistical evidence on the influence of sunspots and volcanic dust on long term temperature records. *Journal of Atmospheric Science* 34, 1995–2004.

Masursky, H., Eliason, E., Ford, P.G., McGill, G.E., Pettengill, G.H., Schaber, G.G. and Schubert, G. 1980: Pioneer Venus radar results: Geology from images and altimetry. *Journal of Geophysical Research* 85A, 8232–60.

Matthews, R.K. 1969: Tectonic implications of glacio-eustatic sea level fluctuations. *Earth Planetary Science Letters* 5, 459–62.

McBirney, A.R. 1963: Factors governing the intensity of submarine volcanism. *Bulletin Volcanologique* 26, 455–69.

McBirney, A.R. 1971: Thoughts on some current concepts of orogeny and volcanism. *Comments on Earth Sciences: Geophysics* 2, 69–76.

McBirney, A.R. 1980: Mixing and unmixing of magmas. *Journal of Volcanology and Geothermal Research* 7, 357–71.

McBirney, A.R. and Murase, T. 1971: Factors governing the formation of pyroclastic rocks. *Bulletin Volcanologique* 34, 372–84.

McBirney, A.R. and Murase, T. 1984: Rheological properties of magmas. *Annual Review Earth and Planetary Science* 12, 337–57.

McCall, G.J.H., Laming, D.J.C. and Scott, S.C. 1992: *Geohazards*. London: Chapman and Hall.

McCarthy, T. 1991: Chaos rules as thousands flee volcano. *The Independent*, 17 June, 10.

McClelland, L., Simkin, T., Summers, M., Nielsen, E. and Stein, T. 1989: *Global Volcanism 1975–1985*. Washington DC: Smithsonian Institution.

McEwen, A.S. and Soderblom, L.A. 1983: Two classes of volcanic plumes on Io. *Icarus* 55, 191–217.

McGill, G.E., Warner, J.L., Malin, M.C., Arvidson, E., Eliason, E., Nozette, S. and Reasenberg, R.D. 1983: Topographic geography, surface properties, and tectonic evolution. In Hunten, D.M., Colin,

L., Donahue, T.M. and Moroz, V.I. (eds), *Venus*. Tucson: University of Arizona Press, 69–130.
McGuire, W.J., Murray, J.B., Pullen, A.D. and Saunders, S.J. 1991: Ground deformation monitoring at Mt Etna; evidence for dyke emplacement and slope instability. *Journal of the Geological Society (London)* 148(3), 577–83.
McKee, C., Mori, J. and Talai, B. 1989: Microgravity changes and ground deformation at Rabaul Caldera 1973–1985. In Latter, J.H. (ed.), *Volcanic Hazards, Assessment and Monitoring*, IAVCEI Proceedings in Volcanology 1. Berlin: Springer-Verlag, 399–428.
McKenzie, D. 1984: A possible mechanism for epeirogenic uplift. *Nature* 307, 616–18.
McKenzie, D. 1985: The extraction of magma from the crust and mantle. *Earth and Planetary Science Letters* 74, 81–91.
McKenzie, D. 1992: Pundits pontificate while Etna erupts. *New Scientist* 134(1818), 5.
McKenzie, D., Ford, P.G., Johnson, C., Parsons, B., Sandwell, D., Saunders, S. and Solomon, S.C. 1992: Features on Venus generated by plate boundary processes. *Journal of Geophysical Research* 97E, 13,533–13,544.
McKinnon, W.B. 1991: Icy clues to Triton's origin. *Nature* 354, 431.
McLean, D.L. 1985: Deccan traps mantle degassing in the terminal Cretaceous marine extinctions. *Cretaceous Research* 6, 235–59.
McLean, D.L. 1988: K-T transition into chaos. *Journal of Geological Education* 36, 237–43.
McNutt, S.R. 1989: Some seismic precursors to eruptions at Pavlof Volcano, Alaska, October 1973–April 1986. In Latter, J.H. (ed.), *Volcanic Hazards Assessment and Monitoring*, IAVCEI Proceedings in Volcanology. Berlin: Springer-Verlag, 463–85.
McPhee, J. 1988: The control of nature (volcano-part 1). *New Yorker*, 2 February 1988, 43–78.
Medina, J.P. 1986: The organisation of the group to monitor volcano Nevado del Ruiz on Colombia. *Eos (Transactions of the American Geophysical Union)* 67(16), 403.
Melekestev, I.V., Braitseva, O.A. and Ponomareva, V.V. 1989: Prediction of volcanic hazards on the basis of the study of dynamics of volcanic activity, Kamchatka. In Latter, J.H. (ed.), *Volcanic Hazards, Assessment and Monitoring*, IAVCEI Proceedings in Volcanology. Berlin: Springer-Verlag, 10–35.
Menard, H.W. 1964: *Marine Geology of the Pacific*. New York: McGraw-Hill.
Menyailov, I.A. 1975: Prediction of eruptions using changes in the composition of volcanic gases. *Bulletin Volcanologique* 39, 112–25.
Middlemost, E.A.K. 1985: *Magmas and Magmatic Rocks*. London: Longman.
Miles, M.K. and Gildersleeves, P.B. 1978: A statistical study of the likely influence of some causative factors on the temperature changes since 1665. *Meteorological Magazine* 107, 193–204.
Miller, C.D. 1980: *Potential Hazards from Future Eruptions of Mt Shasta Volcano, Northern California*. United States Geological Survey Bulletin 1503.
Miller, C.D. 1990: Volcanic hazards in the Pacific Northwest. *Geoscience Canada* 17(3), 183–7.
Miller, C.D., Mullineaux, D.R. and Crandell, D.R. 1981: Hazards assessments at Mount St Helens. In Lipman, R.W. and Mullineaux, D.R. (eds), *The 1980 Eruptions of Mount St Helens, Washington*. United States Geological Survey Professional Paper 1250, 789–802.
Miller, C.D., Mullineaux, D.R., Crandell, D.R. and Bailey, R.A. 1982: *Potential Hazards from Volcanic Eruptions in the Long Valley—Mono Lake Area, East-Central California and Southwest Nevada—A Preliminary Assessment*. United States Geological Survey Circular 877.
Minakami, T. 1942: On the distribution of volcanic ejecta, II. The distribution of Mt Asama pumice in 1783. *Tokyo University Earthquake Research Institute Bulletin* 20, 93–106.
Minakami, T. 1950: The explosive activities of volcano Asama in 1935. *Bulletin of the Earthquake Research Institute* 12, 629–44, 790–800.
Minakami, T. 1959a: The study of eruptions and earthquakes originating from volcanoes: Part 1. *Bulletin of the Volcanological Society of Japan* 4, 104–14 (in Japanese).
Minakami, T. 1959b: The study of eruptions of earthquakes originating from volcanoes: Part 2. *Bulletin of the Volcanological Society of Japan* 4, 115–30 (in Japanese).
Minakami, T. 1960: The study of eruptions and earthquakes originating from volcanoes: Part 3. *Bulletin of the Volcanological Society of Japan* 4, 133–51 (in Japanese).
Mitchell, B. 1989: *Geography and Resource Analysis*. Burnt Mill, Harlow: Longman.
Mitchell, J.M. 1970: A preliminary evaluation of atmospheric pollution as a cause of the global temperature fluctuation of the past century. In Singer, S.F. (ed.), *Global Effects of Environmental Pollution*. Berlin: Springer-Verlag, 139–259.
Mitchell, J.M. 1982: El Chichón: Weather-maker of the century? *Weatherwise* 35(6), 252–9.
Mogi, K. 1958: Relations between eruptions of various volcanoes and the deformation of the ground surface around them. *Bulletin of the Earthquake Research Institute* 36, 94–134.
Mohr, E.C.J. 1945: The relationship between soil and population density in the Netherlands Indies. In

Honig, P. and Verdoorn, F. (eds), *Science and Scientists in the Netherlands Indies*. New York, 254–62.
Mohr, P.A. 1983: Ethiopian flood basalt province. *Nature* 303, 577–84.
Mohr, P.A. and Wood, C.A. 1976: Volcano spacing and lithospheric attenuation in the eastern rift of Africa. *Earth and Planetary Science Letters* 33, 126–44.
Moore, H.G. and Schilling, J.G. 1973: Vesicles, water and sulfur in Reykjanes Ridge basalts. *Contributions to Mineralogy and Petrology* 41, 105–18.
Moore, H.J., Arthur, D.W.G. and Schaber, G.G. 1978: Yield strengths of flows on the Earth, Mars and Moon. *Proceedings Lunar Planetary Science Conference, 9th*, 4451–78. Houston: Lunar Planetary Institute.
Moore, H.J., Boyce, J.M., Schaber, G.G. and Scott, D.H. 1980: Lunar remote sensing and measurement. *United States Geological Survey Professional Paper* 1046–B.
Moore, J.G. 1967: Base surge in recent volcanic eruptions. *Bulletin Volcanologique* 30, 337–63.
Moore, J.G. 1970: Water content of basalt erupted on the ocean floor. *Contributions to Mineralogy and Petrology* 28, 272–9.
Moore, J.G. 1975: Mechanism of formation of pillow lava. *American Journal of Science* 63, 269–77.
Moore, J.G. and Fiske, R.S. 1969: Volcanic substructure inferred from dredge samples and ocean-bottom photographs, Hawaii. *Geological Society of America Bulletin* 80, 1191–202.
Moore, J.G. and Melson, W.G. 1969: Nuees ardentes of the 1968 eruption of Mayon Volcano, Philippines. *Bulletin Volcanologique* 33, 600–20.
Moore, J.G. and Simon, T.W. 1981: Deposits and effects of the May 18 pyroclastic surge. In Lipman, P.W. and Mullineaux, D.R. (eds), *The 1980 Eruptions of Mount St Helens*. United States Geological Survey Professional Paper 1250, 421–38.
Moore, J.G., Nakamuru, K. and Alcaraz, A. 1966: The 1965 eruption of Taal volcano. *Science* 151, 955–60.
Moore, J.G., Batchelder, J.N. and Cunningham, C.G. 1977: CO_2-filled vesicles in mid-ocean basalt. *Journal of Volcanology and Geothermal Research* 2, 309–27.
Moore, J.M. and Malin, M.C. 1988: Dome craters on Ganymede. *Geophysical Research Letters* 15(3), 225–8.
Moorhouse, W.W. 1970: *A Comparative Atlas of the Textures of Archaean and Younger Volcanic Rocks*. Geological Society of Canada Special Paper, 8.
Morgan, N. 1991: The fires that cracked a continent. *New Scientist* (8 June), 42–5.
Morgan, W.J. 1972: Deep mantle convection plumes and plate motions. *American Association of Petroleum Geologists Bulletin* 56, 203–13.
Mori, J., McKee, C., Itikarai, P., Lowenstein, P., de Saint Ours, P. and Talai, B. 1989: Earthquakes of the Rabaul seismo-deformational crisis September 1983 to July 1985: Seismicity on a caldera ring fault. In Latter, J.H. (ed.), *Volcanic Hazards: Assessment and Monitoring*, IAVCEI Proceedings in Volcanology 1. Berlin: Springer-Verlag, 429–63.
Morley, C.K., Wescott, W.A., Stone, D.M., Harper, R.M., Wigger, S.T. and Karanja, F.M. 1992: Tectonic evolution of the northern Kenyan rift. *Journal of the Geological Society (London)* 149, 333–48.
Mouginis-Mark, P.J., Pieri, D.C., Francis, P.W., Wilson, L., Self, S., Rose, W.L. and Wood, C.A. 1989: Remote sensing of volcanoes and volcanic terrains. *Eos*, 28 December, 1567 and 1571.
Muir-Wood, R. 1985: *The Dark Side of the Earth*. London: Allen and Unwin.
Mukherjee, B.K., Indira, K. and Dani, K.K. 1987: Low latitude volcanic eruptions and their effects on Sri Lankan rainfall during the north east monsoon. *Journal of Climatology* 7, 145–55.
Mulargia, F., Tinti, S. and Boschi, E. 1985: A statistical analysis of flank eruptions on Etna Volcano. *Journal of Volcanology and Geothermal Research* 23, 263–72.
Mulargia, F., Tinti, S. and Boschi, E. 1986: A statistical analysis of flank eruptions on Etna Volcano. Reply to comments of D.K. Chester. *Journal of Volcanology and Geothermal Research* 28, 389–95.
Mullerried, F.K.G. 1932: 'El Chichón', volcan en actividad, descubierto en el estado de Chiapas. *Rev. Acad. Nac. Ciec. 'Antonio Alzate'*, 53(11–12), 411–16.
Mullineaux, D.R. 1975: *Preliminary Map of Volcanic Hazards in the 48 Conterminous United States*. United States Geological Survey, Miscellaneous Field Investigations Map MF 786.
Mullineaux, D.R. and Peterson, D.W. 1974: *Volcanic Hazards on the Island of Hawaii*. United States Geological Survey Open File Report 74–239.
Murai, I. 1961: A study of the textural characteristics of pyroclastic flow deposits in Japan. *Bulletin of the Earthquake Research Institute* 39, 133–254.
Murase, T. 1962: Viscosity and related properties of volcanic rocks at 800° to 1400°C. *Journal of the Faculty of Science Hokkaido University* ser. 7 1, 487–584.
Murase, T. and McBirney, A.R. 1973: Properties of some common igneous rocks and their melts at

high temperatures. *Geological Society of America Bulletin* 84, 3563–92.
Murphy, J.B. and Nance, R.D. 1991: Supercontinent model for the contrasting character of late Proterozoic orogenic belts. *Geology* 19(5), 469–72.
Murphy, J.B. and Nance, R.D. 1992: Mountain belts and the supercontinent cycle. *Scientific American* 266(4) (April), 34–41.
Murray, B.C., Malin, M.C. and Greeley, R. 1981: *Earthlike Planets*. San Francisco: W.H. Freeman.
Murton, B.J. and Shimabakuro, S. 1974: Human adjustment and volcanic hazard in the Puna District, Hawaii. In White, G.F. (ed.), *Natural Hazards: Local, National, Global*. New York: Oxford University Press, 151–61.
Murray, J.B. and Guest, J.E. 1982: Vertical ground deformation on Mt Etna, 1975–1982. *Geological Society of America Bulletin* 93, 1166–75.
Mutch, T.A., Arvidson, R.E., Head, J.W., Jones, K.L. and Saunders, R.S. 1976: *The Geology of Mars*. Princeton: Princeton University Press.
Mysen, B.O., Virgo, D. and Seifert, F.A. 1982: The structure of silicate melts: Implications for chemical and physical properties of natural magma. *Reviews of Geophysics and Space Physics* 20, 353–83.
Nairn, I.A. 1981: *Some Studies of the Geology, Volcanic History and Geothermal Resources of the Okataina Volcanic Centre, Taupo Zone, New Zealand*. Unpublished Ph.D. thesis, Victoria University.
Nakabayashi, I. 1984: Assessing intensity of damage of natural disasters in Japan. *Ekistics* 308, 432–8.
Nakano, T., Kadomuru, H., Mizutani, T., Okuda, M. and Sekiguchi, T. 1974: Natural hazards: Report from Japan. In White, G.F. (ed.), *Natural Hazards: Local, National, Global*. New York: Oxford University Press, 231–43.
Nance, R.D., Worsley, T.R. and Moody, J.B. 1988: The supercontinent cycle. *Scientific American* 259(1), 72–9.
Naranjo, J.L., Sigurdson, H., Carey, S.N. and Fritz, W. 1986: Eruption of the Nevado del Ruiz Volcano, Colombia, 13 November 1985: Tephra fall and lahars. *Science* 233, 961–3.
Neall, V.E. 1976: Lahars as major geological hazards. *Bulletin of the International Association of Engineering Geology* 14, 233–40.
Neumann van Padang, M. 1934: Haben bei den Ausbruchen des Slametvulkans Eruptionsregen stattgefunden? *Leidsche Geologische Medelingen* 6, 79–97.
Neumann van Padang, M. 1960: Measures taken by the authorities of the Volcanological Survey to safeguard the population from the consequences of volcanic outbursts. *Bulletin Volcanologique* 23, 181–92.
Neumann van Padang, M. 1963a: The temperatures in the crater region of some Indonesian volcanoes before eruption. *Bulletin Volcanologique* 26, 319–36.
Neumann van Padang, M. 1963b: *Catalogue of the Active Volcanoes and Solfatara Fields of Arabia and the Indian Ocean*. Rome: International Association of Volcanology, vol. 16.
Newell, R.E. 1970: Stratospheric temperature change from the Mount Agung volcanic eruption of 1963. *Journal of Atmospheric Science* 27, 977–8.
Newell, R.E. 1981a: Further studies of the atmospheric temperature change produced by the Mt Agung volcanic eruption in 1963. *Journal of Volcanology and Geothermal Research* 11, 61–6.
Newell, R.E. 1981b: Introduction. *Journal of Volcanology and Geothermal Research* 11, 1–2.
Newell, R.E. 1983: Workshop on Mount St Helens eruptions of 1980: Atmospheric effects and potential climatic impact. *Bulletin of the American Meteorological Society* 64(2), 154–6.
Newell, R.E. and Weare, B.C. 1976: Factors governing tropospheric mean temperature. *Science* 194, 1413–14.
Newhall, C.G. 1982: *A Method for Estimating Intermediate and Long-Term Risks from Volcanic Activity, with an Example from Mount St Helens*. Washington DC: United States Geological Survey, Open File Report, 82–396.
Newhall, C.G. 1984: *Semiquantitative Assessment of Changing Volcanic Risk at Mount St Helens, Washington*. United States Geological Survey, Open File Report, 84–272.
Newhall, C.G. and Self, S. 1982: The Volcanic Explosivity Index (VEI): An estimate of the explosive magnitude for historical volcanism. *Journal of Geophysical Research* 87(C), 1231–8.
Newhall, C.G. and Melsom, W.G. 1983: Explosive activity associated with the growth of volcanic domes. *Journal of Volcanology and Geothermal Research* 17, 111–31.
Newitt, M. 1984: *The Comoro Islands: Struggle against dependency in the Indian Ocean*. London: Gower Press.
Ninkovich, D. and Down, W.L. 1976: Explosive Cenozoic volcanism and climatic implications. *Science* 194, 899–906.

Nisbet, E.G. 1984: Modelling mantle temperatures during the Archaean. *Nature* 309, 110.
Nolan, M.L. 1979: Impact of Parícutin on five communities. In Sheets, P.D. and Grayson, D.K. (eds), *Volcanic Activity and Human Ecology*. New York: Academic Press, 293–335.
Nolan, M.L. and Nolan, S. 1979: Five towns of Parícutin. *Geographical Magazine* 51(5), 338–45.
Nozette, S. and Lewis, J.S. 1982: Venus: Chemical weathering of igneous rocks and buffering of atmospheric composition. *Science* 216, 181–3.
Oberbeck, V.R., Quaide, W.L. and Greeley, R. 1969: On the origin of sinuous rilles. *Modern Geology* 1, 75–80.
Officer, C.B. and Drake, C.L. 1985: Terminal Cretaceous environmental events. *Science* 227, 1161–7.
Officer, C.B., Hallam, A., Drake, C.L. and Devine, J.D. 1987: Late Cretaceous and Paroxysmal Cretaceous/Tertiary extinctions. *Nature* 326, 143.
O'Hara, M.J. and Matthews, R.E. 1981: Geochemical evolution in an advancing, periodically replenished, periodically tapped, continuously fractionated magma chamber. *Journal of the Geological Society (London)* 138, 237–7.
Oliver, R.C. 1976: On the response of hemispheric mean temperature to stratospheric dust: An empirical approach. *Journal of Applied Meteorology* 15, 933–50.
Ollier, C.D. 1984: Geomorphology of the South Atlantic volcanic islands, Part 1: The Tristan da Cunha Group. *Zeitscrift für Geomophologie* 28, 367–82.
Ollier, C.D. 1988: *Volcanoes*. Oxford: Blackwell.
Omori, F. 1911 and 1920: The Usu-san eruption and earthquake and elevation phenomena. *Bulletin Imperial Earthquake Investigation Committee* 5, 101–7; and 9, 41–76.
Oppenheimer, C.M.M. and Rothery, D.A. 1991: Infrared monitoring of volcanoes by satellite. *Journal of the Geological Society (London)* 148(3), 563–9.
O'Riordan, T. 1974: The New Zealand natural hazard insurance scheme: Application to North America. In White, G.F. (ed.), *Natural Hazards: Local, National, Global*. New York: Oxford University Press, 217–19.
O'Riordan, T. 1976: *Environmentalism*. London: Pion Books.
Osborn, E.F. 1979: The reaction principle. In Yoder, H.S. (ed.), *The Evolution of Igneous Rocks: Fiftieth Anniversary Perspectives*. Princeton, New Jersey: Princeton University Press, 133–69.
Ota, K., Matsuo, N. and Sugoshi, T. 1979: *Sulfur Dioxide Emissions from Usu Volcano*. Coordinating Committee for prediction of volcanic eruptions. Tokyo: Report 14, Japan Meteorological Agency.
Otway, P.M. 1989: Vertical deformation monitoring by periodic water level observations, Lake Taupo, New Zealand. In Latter, J.H. (ed.), *Volcanic Hazards, Assessment and Monitoring*, IAVCEI Proceedings in Volcanology 1. Berlin: Springer-Verlag, 561–75.
Pallister, J.S., Hoblitt, R.P., Crandell, D.W. and Mullineaux, D.R. 1992: Mount St Helens a decade after the 1980 eruptions: Magmatic models, chemical cycles, and a revised hazards assessment. *Bulletin of Volcanology* 54, 126–46.
Palm, R.I. 1990: *Natural Hazards: An Integrative Framework for Research and Planning*. Baltimore: Johns Hopkins Press.
Pantic, N. and Stefanovic, P. 1982: Complex interaction of cosmic and geological events that affect the variation of earth climate through geological history. In Berger, A.L., Imbrie, J., Hays, J. and Kukla, G. and Saltzman, B. (eds), *Milankovich and Climate*. Dordrecht: Reidel, 251–64.
Parra, E. and Capeda, H. 1990: Volcanic hazards maps of the Nevado del Ruiz volcano, Colombia. *Journal of Volcanology and Geothermal Research* 42, 117–27.
Patterson, C.C. and Settle, D.M. 1987: Magnitude of lead flux to the atmosphere from volcanoes. *Geochimica et Cosmologica Acta* 51, 675–81.
Peale, S.J., Cassen, P. and Reynolds, R.T. 1979: Melting of Io by tidal dissipation. *Science* 203, 892–4.
Pearce, J.A., Lippard, S.J. and Roberts, S. 1984: Characteristics and tectonic significance of supra-subduction zone ophiolites. *Geological Society of London Special Publication* 16, 77–94.
Peckover, R.S., Buchanan, D.J. and Ashby, D. 1973: Fuel-coolant interactions in submarine vulcanism. *Nature* 245, 307–8.
Pena, O. and Newhall, C.G. 1984: Philippines. In Crandell, D.R., Booth, B., Kusumadinata, K., Shimozuru, D., Walker, G.P.L. and Westercamp, D. (eds), *Source-Book for Volcanic-Hazards Zonation*. Paris: UNESCO, 65–7.
Perret, F.A. 1937: The eruption of Mt Pelée 1929–1932. *Carnegie Institution (Washington) Publication* 458, 1–126.
Peterson, D.W. 1986: Volcanoes: Tectonic setting and impact on society. In *Active Tectonics*. Washington DC: National Academy Press, Studies in Geophysics.
Peterson, D.W. 1988: Volcanic hazards and public response. *Journal of Geophysical Research* 93(B), 4161–70.

Peterson, D.W. and Tilling, R.I. 1980: Transition of a basaltic lava from pahoehoe to aa, Kilauea volcano, Hawaii. *Journal of Volcanology and Geothermal Research* 7(3/4), 271–93.
Pichler, H. 1965: Acid hyaloclastites. *Bulletin Volcanologique* 28, 293–310.
Pieri, D.C. and Bologa, S.M. 1986: Eruption rate, area, and length relationships for some Hawaiian lava flows. *Journal of Volcanology and Geothermal Research* 30, 29–45.
Phillips, R.J. and Malin, M.C. 1983: The interior of Venus and tectonic implications. In Hunten, D.M., Colin, L., Donahue, T.M. and Moroz, V.I. (eds), *Venus*. Tucson: University of Arizona Press, 159–214.
Pinatubo Volcano Observatory Team 1991: Lessons from a major eruption: Mt Pinatubo, Philippines. *Eos (Transaction of the American Geophysical Union)* 72(49), 545–54.
Pinkerton, H. and Sparks, R.S.J. 1978: Field measurements of the rheology of lava. *Nature* 276, 383–6.
Plescia, J.B. and Saunders, R.S. 1979: The chronology of Martian volcanoes. *Proceedings of the 10th Lunar Science Conference*, 2841–59.
Pollack, J.B., Toon, O.B., Sagan, C., Summers, A., Baldwin, B. and Van Camp, W. 1976: Volcanic explosions and climatic change: A theoretical assessment. *Journal of Geophysical Research* 81, 1071–83.
Pollard, D.D. 1976: On the form and stability of open hydraulic fractures in the Earth's crust. *Geophysical Research Letters* 3, 513–16.
Popper, K.R. 1959: *The Logic of Scientific Discovery*. London: Hutchinson.
Porter, S.C. 1981: Recent glacier variations and volcanic eruptions. *Nature* 291, 139–42.
Postma, G. 1986: Classification for sediment gravity-flow deposits based on flow conditions during sedimentation. *Geology* 14, 291–4.
Prata, A.J., Barton, I.J., Johnson, R.W., Kamo, K. and Kingwell, J. 1991: Hazard from volcanic ash. *Nature* 354, 25.
Press, F. 1972: The Earth's interior as inferred from a family of models. In Robertson, E.C. (ed.), *The Nature of the Solid Earth*. New York: McGraw-Hill, 147–71.
Press, F. and Siever, R. 1986: *The Earth*. New York: W.H. Freeman (fourth edition).
Pyle, D.M., Ivanovich, M. and Sparks, R.S.J. 1988: Magma-cumulate mixing identified by U-Th disequilibrium dating. *Nature* 331, 157–9.
Ramage, C.S. 1986: El Ninõ. *Scientific American* 254(6), 76–83.
Ramberg, I.B. and Neumann, E.R. (eds) 1978: *Tectonics and Geophysics of Continental Rifts*. Dordrecht: Reidel (Part 2).
Rampino, M.R. and Self, S. 1982: Historic eruptions of Tambora (1815), Krakatau (1883) and Agung (1963), their stratospheric aerosols and climatic impact. *Quaternary Research* 18, 127–43.
Rampino, R.B. and Stothers, R.B. 1985: Climatic effects of volcanic eruptions. *Nature* 311, 272.
Rampino, M.R., Self, S. and Fairbridge, R.W. 1979: Can rapid climatic change cause volcanic eruptions? *Science* 206, 826–9.
Rasmusson, E.M. and Wallace, J.M. 1983: Meteorological aspects of the El Ninõ/Southern Oscillation. *Science* 322, 1195–1202.
Rees, J.D. 1979: Effects of the eruption of Parícutin volcano on landforms, vegetation, and human occupancy. In Sheets, P.D. and Grayson, D.K. (eds), *Volcanic Activity and Human Ecology*. New York: Academic Press, 249–92.
Reiner, M. 1960: *Deformation, Strain and Flow*. London: H.K. Lewis.
Reynolds, M.A., Best, J.G. and Johnson, R.W. 1980: 1953–57 eruption of Tuluman volcano: Rhyolitic volcanic activity in the northern Bismarck Sea. *Geological Survey of Papua New Guinea Memoir* 7.
Riccò, A. 1907: Periodi di riposo dell Etna. *Bolletino Academia Gioenia di Scienze Naturali (Catania)* 94, 2–6.
Ringwood, A.E. 1962: A model for the upper mantle. *Journal of Geophysical Research* 67, 857–66.
Ringwood, A.E. 1969: Composition and evolution of the upper mantle. *American Geophysical Union, Geophysical Monographs* 13, 1–17.
Ringwood, A.E. 1974: The petrological evolution of island arc systems. *Journal of the Geological Society (London)* 130, 183–204.
Ringwood, A.E. 1975: *Composition and Petrology of the Earth's Mantle*. New York: McGraw-Hill.
Ringwood, A.E. and Irifune, T. 1988: Nature of the 650 km seismic discontinuity: Implications for mantle dynamics and differentiation. *Nature* 331, 131–6.
Rittmann, A. 1930: Geologie von Ischia. *Zeitschrift fur Volcanologia Erganzungsband* 6.
Robach, R. 1983: Geomagnetism and volcanology. In Tazieff, H. and Sabroux, J.C. (eds), *Forecasting Volcanic Events: Developments in Volcanology 1*. Amsterdam: Elsevier, 495–506.
Robock, A. 1978: Internally and externally caused climatic change. *Journal of Atmospheric Science* 35, 1111–22.

Robock, A. 1979: The 'Little Ice Age': Northern Hemisphere average observations and model calculations. *Science* 206, 1402–4.
Robock, A. 1981: A latitudinally dependent volcanic dust veil index, and its effect on climate simulation. *Journal of Volcanology and Geothermal Research* 11, 67–80.
Robock, A. 1984: Climate model simulation of the effects of the El Chichón eruption. *Geofisica Internacional* 23(3), 403–14.
Robson, G.R. 1967: Thickness of Etnean lavas. *Nature* 216, 251–2.
Robson, G.R. and Barr, K.G. 1963: The effect of stress on faulting and minor intrusions in the vicinity of a magma body. *Bulletin Volcanologique* 27, 1–16.
Robson, G.R. and Tomblin, J.F. 1966: *Catalogue of the Active Volcanoes of the World: Part XX, West Indies*. Rome: International Association Volcanology, 1–56.
Romano, R. and Sturiale, C. 1982: The historical eruptions of Mt Etna (volcanological data). In Romano, R. (ed.), *Mount Etna Volcano, a Review of the Recent Earth Science Studies*. Memorie Societa Geologica Italiana 23, 75–97.
Roobol, M.J. and Smith, A.L. 1989: Volcanic and associated hazards in the Lesser Antilles. In Latter, J.H. (ed.), *Volcanic Hazards, Assessment and Monitoring*. Berlin: Springer-Verlag, 57–86.
Rose, W.I. 1977: Scavenging of volcanic aerosol by ash: Atmospheric and volcanologic implications. *Geology* 5, 621–4.
Rose, W.I., Chuan, R.L., Giggenbach, W.F., Kyle, P.R. and Symonds, R.B. 1986: Rates of sulfur dioxide and particle emissions from White Island volcano, New Zealand, and an estimate of the total flux of major gaseous species. *Bulletin of Volcanology* 48, 181–8.
Rosi, M. and Santacroce, R. 1984: Volcanic hazard assessment in the Phlegraean Fields: A contribution based on stratigraphic and historical data. *Bulletin Volcanologique* 47(2), 359–70.
Rosi, M. and Sbrana, A. (eds) 1987: *Phlegrean Fields*. Roma Quaderni 'de la Ricerca Scientifica' 114, Consiglio Nazionale delle Ricerche.
Ross, C.S. and Smith, R.L. 1961: Ash-flow tuffs: Their origin, geologic relations and identification. *United States Geological Survey Professional Paper* 366, 1–77.
Rowland, S.K. and Walker, G.P.L. 1987: Toothpaste lava: Characteristics and origin of a lava structural type transitional between pahoehoe and aa. *Bulletin of Volcanology* 49, 631–41.
Rowley, P.D., Kuntz, M.A. and Macleod, N.S. 1981: Pyroclast-flow deposits. In Lipman, P.W. and Mullineaux, D.R. (eds), *The 1980 Eruptions of Mount St Helens, Washington*. Washington DC: United States Geological Survey Professional Paper 1250, 489–512.
Rymer, H. 1992: Hot and bothered: New measures of volcanoes' activity can help prevent disaster. *The Times Higher Education Supplement*, 15 May, 16.
Saarinen, T.F. and Sell, J.L. 1985: *Warning and Response to the Mount St Helens Eruption*. Albany: State University of New York Press.
Saemundson, K. 1978: Fissure swarms and central volcanoes of the nonvolcanic zones of Iceland. *Geological Journal Special Issue* 10, 415–32.
Sagan, C. 1975: The Solar System. *Scientific American* 233(3), 23–31.
Saint-Ours, P. de and Beuchamp, G.E. 1979: Karthala. *Bulletin of Volcanic Eruptions* 17, 13.
Sakuyama, M. and Nesbitt, R.W. 1986: Geochemistry of the Quaternary rocks of the northeast Japan arc. *Journal of Volcanology and Geothermal Research* 29, 413–50.
Saunders, R.S., Spear, A.J., Allin, P.C., Austin, R.S., Berman, A.L., Chandlee, R.C., Clark, J., de Charon, A.V., De Jong, E.M., Griffith, D.G., Gunn, J.M., Hensley, S., Johnson, W.T.K., Kirby, C.E., Leung, K.S., Lyons, D.T., Michaels, G.A., Miller, J., Morris, R.B., Morrison, A.D., Piereson, R.G., Scott, J.F., Shaffer, S.J., Slonski, J.P., Stofan, E.R., Thompson, T.W. and Wall, S.D. 1992: Magellan mission summary. *Journal of Geophysical Research* 97E, 13,067–13,090.
Sawada, Y. 1989: The detection capability of explosive eruptions using GMS imagery, and the behaviour of dispersing eruption clouds. In Latter, J.H. (ed.), *Volcanic Hazards: Assessment and Monitoring*, IAVCEI Proceedings in Volcanology. Berlin: Springer-Verlag, 233–46.
Scandone, R. 1983: Problems related with the evaluation of volcanic risk. In Tazieff, H. and Sabroux, J.C. (eds), *Forecasting Volcanic Events*, Developments in Volcanology 1. Amsterdam: Elsevier, 57–69.
Scandone, R. 1990: Chaotic collapse of calderas. *Journal of Volcanology and Geothermal Research* 42, 285–302.
Schaber, G.G., Strom, R.G., Moore, H.J., Soderblom, L.A., Kirk, R.L., Chadwick, D.J., Dawaon, D.D., Gaddis, L.R., Boyce, J.M. and Russell, J. 1992: Geology and distribution of impact craters on Venus: What are they telling us? *Journal of Geophysical Research* 97E, 13,257–303.
Schick, R. 1981: Source mechanisms of volcanic earthquakes. *Bulletin Volcanologique* 44(3), 491–7.
Schmincke, H.U., Rautenschlein, M., Robinson, P.T. and Mehegan, J.M. 1983: The Troodos

extrusive series of Cyprus: a comparison with oceanic crust. *Geology* 11, 410–12.
Schneider, S.H. 1988: Whatever happened to nuclear winter?—an editorial. *Climatic Change* 12, 215–19.
Schneider, S.H. and Mass, C. 1975a: Volcanic dust, sunspots, and temperature trends. *Science* 190, 744–6.
Schneider, S.H. and Mass, C. 1975b: Volcanic dust, sunspots and long term climatic trends: Theories in search of verification. In *Proceedings of WMO/IAMP Symposium on Long Term Climatic Fluctuations*, Norwich, 18–23 August 1975. Geneva: World Meteorological Organisation Report 421, 365–72.
Scholz, C.H., Barazangi, M. and Sbar, M. 1971: Late Cenozoic evolution of the Great Basin, western United States, as an ensialic interarc basin. *Geological Society of America Bulletin* 82, 2979–90.
Schubert, G. 1979: Solidus convection in the mantles of terrestrial planets. *Annual Review of Earth Planetary Science* 7, 287–343.
Schultz, P.H. and Gault, D.E. 1975: Seismic effects from major basin formation on the Moon and Mercury. *Moon* 12, 159–77.
Schumm, S.A. and Lichty, R.W. 1965: Time, space and causality in geomorphology. *American Journal of Science* 263, 110–19.
Schwimmer, E.G. 1977: What did the eruption mean? In Leiber, M.D. (ed.), *Exiles and Migrants in Oceania*. Honolulu: University of Hawaii Press, 296–341.
Scott, D.H. 1982: Volcanoes and volcanic provinces: Martian western hemisphere. *Journal of Geophysical Research* 87B, 0839–51.
Scott, S.C. 1992: The international decade for natural hazard reduction and the geohazards unit at Polytechnic South West, Plymouth, UK. In McCall, G.J.H., Laming, D.J.C. and Scott, S.C. (eds) *Geohazards: Natural and Man-Made*. London: Chapman and Hall, 217–22.
Scott, W.E. 1990: Patterns of volcanism in the Cascade Arc during the past 15,000 years. *Geoscience Canada* 17(3), 179–83.
Scrope, C.P. 1862: *Considerations on Volcanoes*. London: W. Phillips and G. Yard (second edition).
Scrope, G.J.P. 1825: *Considerations on Volcanoes, the Probable Causes of their Phenomena and their Connection with the Present State and Past History of the Globe; Leading to the Establishment of a New Theory of the Earth*. London: W. Phillips and G. Yard.
Scudder, B. 1990: Energy galore: Tapping Iceland's subterranean heat energy. *Geographical Magazine* 57, 40–6.
SEAN 1975–92: *Bulletin Smithsonian Scientific Event Alert Network*. Washington DC: Smithsonian Institution.
SEAN 1985: Ruiz volcano. *Scientific Event Alert Network Bulletin* 10(5), 5. Washington DC: Smithsonian Institution.
SEAN 1986: Lake Nyos. *Scientific Event Alert Network Bulletin* 12(1), 2–3. Washington DC: Smithsonian Institution.
SEAN 1989: Ruiz (Colombia). *Scientific Event Alert Network Bulletin* 14(8), 2–5. Washington DC: Smithsonian Institution.
Sear, C.B., Kelly, P.M., Jones, P.D. and Goodess, C.M. 1987: Global surface-temperature responses to major volcanic eruptions. *Nature* 330, 365–7.
Searle, E.J. 1964: Volcanic risk in the Auckland Metropolitan District. *New Zealand Journal of Geology and Geophysics* 7, 94–100.
Self, S. 1976: The recent volcanology of Terceira, Azores. *Journal of the Geological Society (London)* 132, 645–66.
Self, S. and Sparks, R.S.J. 1978: Characteristics of pyroclastic deposits formed by the interaction of silicic magma and water. *Bulletin Volcanologique* 41, 196–212.
Self, S. and Sparks, R.S.J. (eds) 1981: *Tephra Studies*. Dordrecht: Reidel.
Self, S. and Wright, J.V. 1983: Large wave-forms from the Fish Canyon tuff, Colarado. *Geology* 11, 443–6.
Self, S., Wilson, L. and Nairn, I.A. 1979: Vulcanian eruption mechanisms. *Nature* 277, 440–3.
Self, S., Kienle, J. and Huot, J.P. 1980: Ukinrek maars, Alaska, II. Deposits and formation of the 1977 craters. *Journal of Volcanology and Geothermal Research* 7, 39–65.
Self, S., Rampino, M.R. and Barbera, J.J. 1981: The possible effects of large 19th and 20th century volcanic eruptions on zonal and hemispheric surface temperatures. *Journal Volcanology Geothermal Research* 11, 41–60.
Self, R., Rampino, M.R., Newton, M.S. and Wolff, J.A. 1984: Volcanological study of the great Tambora eruption of 1815. *Geology* 12, 659–63.
Self, S., Rampino, M.R. and Carr, M.J. 1989: A reappraisal of the 1835 eruption of Cosiguina and its atmospheric impact. *Bulletin Volcanologique* 52, 57–65.

Shand, S.J. 1950: *Eruptive Rocks, their Genesis, Composition, Classification, and their Relation to Ore-Deposits with a Chapter on Meteorites*. London: Murby.
Sharp, A.D.L., Davis, P.M. and Gray, P. 1980: A low-velocity zone beneath Mount Etna and magma storage. *Nature* 287, 587–91.
Shaw, H.R. 1969: Rheology of basalt in the melting range. *Journal of Petrology* 10, 510–35.
Shaw, H.R. 1970: Earth tides, global heat flow and tectonics. *Science* 168, 1084–7.
Shaw, H.R., Peck, D.L. and Okamura, A.R. 1968: The viscosity of basaltic magma: An analysis of field measurements in Makaopuhi lava lake, Hawaii. *American Journal of Science* 266, 255–64.
Shepherd, J.B. 1989: Eruptions, eruption precursors and related phenomena in the Lesser Antilles. In Latter, J.H. (ed.), *Volcanic Hazards, Assessment and Monitoring*, IAVCEI Proceedings in Volcanology 1. Berlin: Springer-Verlag, 292–312.
Sheridan, M.F. 1979: Emplacement of pyroclastic flows: A review. *Geological Society of America Special Paper* 180, 125–36.
Sheridan, M.F. and Updike, R.G. 1975: Sugarloaf Mountain tephra a Pleistocene rhyolitic deposit of base-surge origin. *Geological Society of America Bulletin* 86, 571–81.
Sheridan, M.F. and Regan, D.M. 1976: Compaction of ash-flow tuffs. In Chilinarian, G.V. and Wolf, K.H. (eds), *Developments in Sedimentology*. Amsterdam: Elsevier, vol. 18b, 677–713.
Sheridan, M.F. and Wohletz, K.H. 1981: Hydrovolcanic explosions: The systematics of water-pyroclast equilibrium. *Science* 212, 1387–9.
Sheridan, M.F. and Malin, M.C. 1983: Application of computer-assisted mapping to volcanic hazard evaluation of surge eruptions: Vulcano, Lipari and Vesuvius. *Journal of Volcanology and Geothermal Research* 17, 187–202.
Sheridan, M.F. and Wohletz, K. 1983: Hydrovolcanism: Basic considerations and review. *Journal of Volcanology Geothermal Research* 17, 1–29.
Shervais, J.W. and Kimbrough, D.L. 1985: Geochemical evidence for the tectonic setting of the Coast Range ophiolite—A composite island arc-oceanic crust terrane in western California. *Geology* 13, 35–8.
Shimozuru, D. 1983a: Volcanic surveillance and prediction of eruptions in Japan. In Tazieff, H. and Sabroux, J.C. (eds), *Forecasting Volcanic Events*. Developments in Volcanology 1. Amsterdam: Elsevier, 173–93.
Shimozuru, D. 1983b: Volcanic hazard assessment of Mount Fuji. *Natural Disaster Science* 5(2), 15–31.
Shimozuru, D., Miyazaki, N., Gyoda, N. and Matahelumual, J. 1969: Volcanological survey of Indonesian volcanoes. *Bulletin of the Earthquake Research Institute* 47, 969–90.
Shimozuru, D. and Kagiyama, T. 1989: Some significant features of pre-eruption volcanic earthquakes. In Latter, J.H. (ed.), *Volcanic Hazards, Assessment and Monitoring*, IAVCEI Proceedings in Volcanology 1. Berlin: Springer-Verlag, 504–13.
Siebert, L. 1984: Large volcanic debris avalanches: Characteristics of source areas, deposits and associated eruptions. *Journal of Volcanology and Geothermal Research* 22, 163–97.
Siebert, L. 1992: Threats from debris avalanches. *Nature* 356, 658–9.
Siebert, L., Glicken, H. and Kienle, J. 1989: Debris avalanches and lateral blasts at Mount St Augustine volcano, Alaska. *National Geographic Research* 5(2), 232–49.
Sigurdsson, H. 1982: Volcanic pollution and climate: The 1783 Laki eruption. *Eos* 63, 601–2.
Sigurdsson, H. 1991: *Geologic Observations in the Crater of Soufriere Volcano, St Vincent*. University of West indies, West Indies Seismic Research Unit Special Publication, 1981/1.
Sigurdsson, H. and Sparks, R.J.S. 1978: Rifting episodes in north Iceland in 1874–75 and the eruptions of Askja and Sveinagja. *Bulletin Volcanologique* 41, 1–19.
Sigurdsson, H. and Carey, S. 1986: Volcanic disasters in Latin America and the 13th November 1985 eruption of Nevado del Ruiz volcano in Columbia. *Disasters* 10(3), 205–16.
Sigurdsson, H., Carey, S., Cornell, W. and Pescatore, T. 1985: The eruption of Vesuvius in AD 79. *National Geographic Research* 1(3), 332–87.
Sigvaldason, G.E. 1983: Volcanic prediction in Iceland. In Tazieff, H. and Sabroux, J.C. (eds), *Forecasting Volcanic Events*, Developments in Volcanology 1. Amsterdam: Elsevier, 193–215.
Sigvaldason, G.E. 1989: International conference on Lake Nyos disaster, Yaounde, Cameroon 16–20 March 1987: Conclusions and recommendations. *Journal of Volcanology and Geothermal Research* 39, 97–107.
Silver, E.A. and Beutner, E.C. 1980: Melanges. *Geology* 8, 32–4.
Simkin, T. and Fiske, 1983: *Krakatau 1883: The Volcanic Eruption and its Effects*. Washington DC: Smithsonian Institution.
Simkin, T., Siebert, L., McClelland, L., Bridge, D., Newhall, C. and Latter, J.H. 1981: *Volcanoes of the World*. Washington DC: Smithsonian Institution.

Simon, H.A. 1957: *Administrative Behaviour*. New York: Macmillan.

Simon, H.A. 1959: Theories of decision making in economic and behavioural science. *American Economic Review* 49, 253–83.

Smith, B.A., Soderblom, L.A., Beebe, R., Bliss, D., Boyce, J.M., Briggs, G.A., Brown, R.H., Collins, S.A., Cook, A.F., Croft, S.K., Cuzzi, J.N., Danielson, G.E., Davies, M.E., Dowling, T.E., Godfrey, D., Hansen, C.J., Harris, C., Hunt, G.E., Ingersoll, A.P., Johnson, T.V., Krauss, R.J., Masursky, H., Morrison, D., Owen, T., Plescia, J.B., Pollack, J.B., Porco, C.P., Rages, K., Sagan, C., Shoemaker, E.M., Stromovsky, L.A., Stoker, C., Strom, R.G., Suomi, V.E., Synnott, S.P., Terrile, R.J., Thomas, P., Thompson, R. and Veverka, J. 1986: Voyager 2 in the Uranian system: Imaging science results. *Science* 233, 43–64.

Smith, J.V. 1985: Protection of the human race, against natural hazards (asteroids, comets, volcanoes, earthquakes). *Geology* 13, 675–8.

Smith, K. 1992: *Environmental Hazards*. London: Routledge.

Smith, R.L. 1960a: Ash flows. *Geological Society America Bulletin* 71, 795–842.

Smith, R.L. 1960b: Zones and zonal variations in welded ash flows. *United States Geological Survey Professional Paper* 354F, 149–59.

Smith, R.L. 1979: Ash-flow magmatism. *Geological Society of America Special Paper* 180, 5–27.

Smith, R.L. and Bailey, R.A. 1968: Resurgent cauldrons. *Memoir Geological Society America* 116, 613–62.

Smithsonian Institution (1968–75): *Reports of Volcanic Eruptions*. Washington DC: Smithsonian Institution.

Smyth, M.-A. 1991: *Movement and Emplacement Mechanisms of the Rio Pita Volcanic Debris Avalanche and its Role on the Evolution of Cotopaxi Volcano*. Unpublished Ph.D. thesis, University of Aberdeen (Scotland).

Snyder, G.L. and Fraser, G.D. 1963: Pillowed lavas, I. Intrusive layered lava pods and pillowed lavas, Unalaska Islands, Alaska. *United States Geological Survey Professional Paper* 454 B, 1–23.

Soderblom, L.A. and Johnson, T.V. 1982: The Moons of Saturn. *Scientific American* 246(1), 73–86.

Solomon, S.C. 1976: Some aspects of core formation in Mercury. *Icarus* 28, 509–21.

Solomon, S.C., Smrekar, S.E., Bindschadler, D.L., Grimm, R.E., Kaula, W.M., McGill, G.E., Phillips, R.J., Saunders, R.S., Schubert, G., Squyres, S.W. and Stofan, E.R. 1992: Venus tectonics: An overview of Magellan observations. *Journal of Geophysical Research* 97E, 13,199–256.

Sorensen, J.H. and Gersmehl, P.J. 1980: Volcanic hazard warning system: Persistence and transferability. *Environmental Management* 4, 125–36.

Sorenson, J.H. and Mileti, D.S. 1987: Public warning needs. *United States Geological Survey Open File Report* 87–269, 9–75.

Souther, J.G., Tilling, R.I. and Punongbayan, R.S. 1984: Forecasting eruptions in the Circum-Pacific. *Episodes* 7(4), 10–18.

Sparks, R.S.J. 1975: Stratigraphy and geology of the ignimbrites of Vulsini volcano, Central Italy. *Geologische Rundschau* 64, 497–523.

Sparks, R.S.J. 1976: Grain size variations in ignimbrites and implications for the transport of pyroclastic flows. *Sedimentology* 23, 147–88.

Sparks, R.S.J. 1978: Gas release rates from pyroclastic flows: an assessment of the role of fluidisation in their emplacement. *Bulletin Volcanologique* 41, 1–9.

Sparks, R.S.J. and Walker, G.P.L. 1973: The ground surge deposit: A third type of pyroclastic rock. *Nature* 241, 62–4.

Sparks, R.S.J. and Wilson, L. 1976: A model for the formation of ignimbrite by gravitational column collapse. *Journal of the Geological Society (London)* 132, 441–51.

Sparks, R.S.J. and Walker, G.P.L. 1977: The significance of vitric-enriched air-fall ashes associated with crystal-enriched ignimbrites. *Journal of Volcanology and Geothermal Research* 2, 329–41.

Sparks, R.S.J. and Marshall, L.A. 1986: Thermal and mechanical constraints on mixing between mafic and silicic magmas. *Journal of Volcanology and Geothermal Research* 29, 99–124.

Sparks, R.S.J., Self, S. and Walker, G.P.L. 1973: Products of ignimbrite eruption. *Geology* 1, 115–18.

Sparks, R.S.J., Sigurdsson, H. and Wilson, L. 1977: Magma mixing: A mechanism for triggering acid explosive eruptions. *Nature* 267, 315–18.

Sparks, R.S.J., Wilson, L. and Hulme, G. 1978: Theoretical modelling of the generation, movement and emplacement of pyroclastic flows by column collapse. *Journal of Geophysical Research* 83B, 1727–39.

Sparks, R.S.J., Meyer, P. and Sigurdsson, H. 1980: Density variation amongst mid-ocean ridge basalts: Implications for magma mixing and the scarcity of primative basalts. *Earth and Planetary Science Letters* 46, 419–30.

Sparks, R.S.J., Wilson, L. and Sigurdsson, H. 1981: The pyroclastic deposits of the 1875 Askja

eruption, Iceland. *Philosophical Transactions Royal Society London* 29, 241–73.
Sparks, S.R.J., Moore, J.G. and Rice, C.J. 1986: The initial giant umbrella cloud of the May 18th, 1980, explosive eruption of Mount St Helens. *Journal of Volcanology and Geothermal Research* 28, 257–74.
Spence, D.A. and Turcotte, D.L. 1985: Magma driven propagation of cracks. *Journal of Geophysical Research* 90B, 575–80.
Spera, F.J. 1980: Aspects of magma transport. In Hargraves, R.B. (ed.), *Physics of Magmatic Processes*. Princeton: Princeton University Press, 265–324.
Spera, F.J., Yuen, D.A., Greer, J.C. and Sewell, G. 1986: Dynamics of magma withdrawal from stratified magma chambers. *Geology* 14, 723–6.
Spray, A. 1962: The origin of columnar jointing, particularly in basalt flows. *Journal of the Geological Society of Australia* 8, 191–216.
Statham, I. 1977: *Earth Surface Sediment Transport*. Oxford: Clarendon Press.
Steinberg, G.S. and Lorenz, V. 1983: External ballistics of volcanic explosions. *Bulletin Volcanologique* 46(4), 333–48.
Stoddart, D.R. 1987: To reclaim the high ground: Geography for the end of the century. *Transactions of the Institute of British Geographers* NS 12, 327–36.
Stoiber, R.E. and Rose, W.I. 1969: Recent volcanic and fumarolic activity at Santiaquito volcano, Guatemala. *Bulletin Volcanologique* 33, 577–8.
Stoiber, R.E. and Rose, W.I. 1970: The geochemistry of Central American volcanic gases and condensates. *Geological Society America Bulletin* 81, 2891–2912.
Stoiber, R.E. and Jepsen, E. 1973: Sulfur dioxide contributions to the atmosphere by volcanoes. *Science* 182, 577–8.
Stoiber, R.E., Leggett, D., Jenkins, T., Murrmann, R. and Rose, W.J. 1971: Organic compounds in volcanic gas from Santiaquito volcano, Guatemala. *Geological Society America Bulletin* 82, 2299–2302.
Stoiber, R.E., Malone, G.B. and Bratton, G.P. 1978: Volcanic emission of SO_2 at Italian and Central American volcanoes. *Geological Society America Abstract* 10, 148.
Stoiber, R.E., Williams, S.N. and Malinconico, L.L. 1980: Mount St Helens, Washington, 1980 volcanic eruption: Magmatic gas component during the first 16 days. *Science* 208, 1258–9.
Stoiber, R.E., Malinconico, L.L. and Williams, S.N. 1983: Use of the correlation spectrometer at volcanoes. In Tazieff, H. and Sabroux, J.C. (eds), *Forecasting Volcanic Events*, Development in Volcanology 1. Amsterdam: Elsevier, 425–45.
Stommel, H. and Stommel, E. 1979: The year without a summer. *Scientific American* 240(6), 134–40.
Stommel, H. and Swallow, J.C. 1983: Do late grape harvests follow large volcanic eruptions. *Bulletin of the American Meteorological Society* 64, 794–5.
Stothers, R.B. and Rampino, M.R. 1983a: Historic volcanism, European dry fogs and Greenland acid precipitation, 1500 BC to AD 1500. *Science* 222, 411–13
Stothers, R.B. and Rampino, M.R. 1983b: Volcanic eruptions in the Mediterranean Region before AD 630 from written and archaeological sources. *Journal of Geophysical Research* 88(B), 6357–71.
Stothers, R.B., Wolff, J.A. and Rampino, M.R. 1986: Basaltic fissure eruptions, plume heights and atmospheric aerosols. *Geophysical Research Letters* 13, 725–8.
Stothers, R.B., Rampino, M.R., Self, S. and Wolff, J.A. 1988: Volcanic winter? Climatic effects of large volcanic eruptions. In Latter, J.H. (ed.), *Volcanic Hazards: Assessment and Monitoring*. Berlin: Springer-Verlag, 3–10.
Strom, R.G., Trask, N.J. and Guest, J.E. 1975: Tectonism and volcanism on Mercury. *Journal of Geophysical Research* 80, 2478–2507.
Strong, D.F. and Jacquot, C. 1970: The Karthala, Caldera, Grande Comore. *Bulletin Volcanologique* 34, 663–80.
Stuart, W.D. and Johnston, M.J.S. 1975: Intrusive origin of the Matsushiro earthquake swarm. *Geology* 3, 63–7.
Sudradjat, A. and Tilling, R. 1984: Volcanic hazards in Indonesia: the 1982–1983 eruption of Galunggung. *Episodes* 7(2), 13–19.
Sudradjat-Saratman, A. 1986: Una Una (Colo). *Bulletin of Volcanic Eruptions* 23, 20.
Summerfield, M.A. 1985: Plate tectonics and landscape development on the African continent. In Morisawa, M. and Hack, J.T. (eds), *Tectonic Geomorphology*. London: Allen and Unwin.
Summerfield, M.A. 1991: *Global Geomorphology*. Burnt Hill, Harlow: Longman.
Surkov, Y.A., Moskalyeva, P., Shcheglov, O.P., Kharyukova, V.P., Manvelyan, O.S., Kirichenko, V.S. and Dudin, A.D. 1983: Determination of the elemental composition of rocks on Venus by Venera 13 and Venera 14. *Journal of Geophysical Research* 88 (supplement) A481–493.
Susman, P., O'Keefe, P. and Wisner, B. 1983: Global disasters, a radical interpretation. In Hewitt, K.

(ed.), *Interpretations of Calamity*. London: Allen and Unwin, 263–80.
Swanson, D.W. and Casadevall, T.J. 1983: Volcanology. *Reviews of Geophysics and Space Physics* 21, 1419–35.
Swanson, D.A. and Holcomb, R.T. 1990: Regularities in growth of the Mount St Helens dacite dome 1980–1986. In Fink, J.H. (ed.), *Lava Flows and Domes*, IAVCEI Proceedings in Volcanology. Berlin: Springer-Verlag, 25–46.
Swanson, D.A. and Wright, T.H. 1981: The regional approach to studying the Columbia River basalt group. In Subarao, K.V. and Sukheswala, R.N. (eds), *Deccan Volcanism and Related Basalt Provinces in Other Parts of the World*. Geological Society of India Memoir 3, 58–80.
Swanson, D.A., Casadevall, T.J., Dzurisin, D., Holcomb, R.T., Newhall, C.G., Malone, S.D. and Weaver, C.S. 1985: Forecasts and predictions of the eruptive activity at Mount St Helens, USA, 1975–1984. *Journal of Geodynamics* 3, 397–423.
Symons, G. (ed.) 1888: *The Eruption of Kratatoa and Subsequent Phenomena: Report of the Krakatoa Committee*. London: The Royal Society.
Tan, K.H. 1964: The andosols in Indonesia. In *FAO-UNESCO Meeting on the Classification and Correlation of Soils from Volcanic Ash (Tokyo, Japan)*. World Soil Resources Report 14, 101–10.
Tauber, M.E. and Kirk, D.B. 1976: Impact craters on Venus. *Icarus* 28, 351–7.
Taylor, G.A. 1958: The 1951 eruption of Mount Lamington, Papua. *Australian Bureau Mineral Resources Geological and Geophysical Bulletin* 38, 117.
Tazieff, H. 1967: The menace of extinct volcanoes. *Impact*. UNESCO 2–17, 135–48.
Tazieff, H. 1971: *Petit discours à propos de la panique que eut lieu à Pozzuoli à l'occasion du bradisismo Flegreo*. Stromboli.
Tazieff, H. 1974: Karthala. *Bulletin of Volcanic Eruptions* 12, 12.
Tazieff, H. 1983a: Some general points about volcanism. In Tazieff, H. and Sabroux, J.C. (eds), *Forecasting Volcanic Events*, Developments in Volcanology 1. Amsterdam: Elsevier, 9–25.
Tazieff, H. 1983b: Some general points about volcano monitoring and forecasting. In Tazieff, H. and Sabroux, J.C. (eds), *Forecasting Volcanic Events*, Developments in Volcanology 1. Amsterdam: Elsevier, 165–73.
Tazieff, H. 1983c: Estimating eruptive peril: Some case histories. In Tazieff, H. and Sabroux, J.C. (eds), *Forecasting Volcanic Events*, Developments in Volcanology 1. Amsterdam: Elsevier, 547–61.
Tazieff, H. 1988: Forecasting volcanic eruptive disasters. In El-Sabh, M.I. and Murty, T.S. (eds), *Natural and Man-Made Hazards*. Dordrecht: D. Reidel, 751–72.
Tazieff, H. 1989: Mechanisms of the Nyos carbon dioxide disaster and the so-called phreatic steam eruptions. *Journal of Volcanology and Geothermal Research* 39, 109–16.
Tazieff, H. and Sabroux, J.C. (eds) 1983: *Forecasting Volcanic Events*, Developments in Volcanology 1. Amsterdam: Elsevier.
Thatcher, W. 1990: Precursors to eruption. *Nature* 343, 590–1.
Thomas, D.M. and Naughton, J.J. 1979: Helium/carbon dioxide ratios as premonitors of volcanic activity. *Science* 204, 1195–6.
Thompson, L.G. and Mosley-Thompson, E. 1981: Temporal variability of microparticle properties in Polar ice sheets. *Journal of Volcanology and Geothermal Research* 11, 11–27.
Thompson, R.N. 1984: Dispatches from the basalt front. I. Experiments. *Proceedings Geologists Association* 95, 249–62.
Thompson, S.L. and Schneider, S.H. 1986: Nuclear Winter reappraised. *Foreign Affairs* (Summer 1986), 981–1005.
Thorarinsson, S. 1954: *The Eruption of Hekla 1947–8, II3, The Tephra-Fall from Hekla on March 29th 1947*. Reykjavik: Societas Science Islandica. p. 68.
Thorarinsson, S. 1974: The terms tephra and tephrochronology. In Westgate, J.A. and Gold, C.M. (eds), *World Bibliography and Index of Quaternary Tephrochronology*. University of Alberta, 1–528.
Thorarinsson, S. 1979: On the damage caused by volcanic eruptions with special reference to tephra and gases. In Sheets, P.D. and Grayson, D.K. (eds), *Volcanic Activity and Human Ecology*. New York: Academic Press, 125–61.
Thornes, J.B. and Brunsden, D. 1977: *Geomorphology and Time*. London: Methuen.
Tilling, R.I. undated: *Monitoring Active Volcanoes*. Washington DC: United States Geological Survey (1984–421–618/10001).
Tilling, R.I. 1982: The 1982 eruption of El Chichón, Southeastern Mexico. *Earthquake Information Bulletin* 14(5), 154–72.
Tilling, R.I. 1988: Volcano prediction—lessons from material science. *Nature* 332, 108–9.
Tilling, R.I. and Bailey, R.A. 1984: Volcano hazards program in the USA. *27th International Geological Congress*, Moscow, August 4–14th 1984. Reports Volume 6, 106–18.

Tilling, R.I. and Bailey, R.A. 1985: Volcano hazards, program in the United States. *Journal Geodynamics* 3, 425–46.

Tinkler, K.J. 1985: *A Short History of Geomorphology*. London: Croom-Helm.

Toksöz, M.N. 1980: The subduction of the lithosphere. In *Scientific American, Earthquakes and Volcanoes*. San Francisco: W.H. Freeman, 125–35.

Tomkeieff, S.I. 1940: Basalt lavas of the Giant's Causeway. *Bulletin Volcanologique* 6, 89–143.

Toon, O.B. 1982: Volcanoes and climate. In Deepak, A. (ed.), *Atmospheric Effects and Potential Climatic Impact of the 1980 Eruptions of Mount St Helens*. Washington DC: NASA (NASA CP-2240), 15–24.

Toutain, J.-P., Baubron, J.-C., Le Bronec, J., Allard, P., Briole, P., Marty, B., Miele, G., Tedesco, D. and Luongo, G. 1992: Continuous monitoring of distal gas emanations at Vulcano, Italy. *Bulletin of Volcanology* 54, 147–55.

Trask, N.J. and Guest, J.E. 1975: Preliminary geologic terrain map of Mercury. *Journal of Geophysical Research* 80, 2461–77.

Tricart, J. 1965: *Principles et Méthodes de la Géomorphologie*. Paris: Masson.

Turco, R.P., Toon, O.B., Ackerman, T.P., Pollack, J.B. and Sagan, C. 1983: Nuclear winter: Global consequences of multiple nuclear explosions. *Science* 222, 1283–92.

Turco, R.P., Toon, O.B., Ackerman, T.P., Pollack, J.B. and Sagan, C. 1984: The climatic effects of nuclear war. *Scientific American* 251, 33–43.

Turner, J.S., Huppert, H.E. and Sparks, R.S.J. 1983: An experimental investigation of volatile exsolution in evolving magma chambers. *Journal of Volcanology and Geothermal Research* 16, 263–78.

Ugolini, F.G. and Zasoski, R.J. 1979: Soils derived from tephra. In Sheets, P.D. and Grayson, D.K. (eds), *Volcanic Activity and Human Ecology*. New York: Academic Press, 83–124.

Ui, T. 1983: Volcanic dry avalanche deposits—identification and comparison with nonvolcanic debris stream deposits. *Journal of Volcanology and Geothermal Research* 18, 135–50.

Ui, T. and Glicken, H. 1986: Internal structural variations in a debris-avalanche deposit from ancestral Mount Shasta, California, USA. *Bulletin of Volcanology* 48, 189–94.

Ui. T., Yamamoto, H. and Suzuki-Kamata, K. 1986: Characterization of debris avalanche deposits in Japan. *Journal of Volcanology and Geothermal Research* 29, 231–43.

UNDRO 1985: *Volcanic Emergency Management*. New York: United Nations.

Unger, J.D. 1974: Scientists probe Earth's secrets at the Hawaiian Volcano Observatory. *Earthquake Information Bulletin* 6(4), 3–11.

United Nations, 1977: *Disaster Prevention and Mitigation. Volume 1, Volcanological Aspects*. New York: Office of the United Nations Disaster Relief Co-ordinator.

United States Advisory Committee on the International Decade for Natural Hazard Reduction 1987: *Confronting Natural Disasters: An International Decade for Confronting Natural Hazard Reduction*. Washington DC: National Academy Press.

Upton, B.G.J., Wadsworth, W.J. and Latrille, E. 1974: The 1972 eruption of Karthala volcano, Grande Comore. *Bulletin Volcanologique* 38, 136–48.

US Corps of Engineers 1966: *Review Report on Survey for Lava Flow Control, Island of Hawaii*. Honolulu: Corps of Engineers.

USGS 1987: *United States Geological Survey Yearbook, Fiscal Year 1986*. Washington DC: United States Government Printing Office.

van Andel, T.H. 1984: Plate tectonics at the threshold of middle age. *Geologie en Mijnbouw* 63, 337–41.

van Bemmelen, R.W. 1949: *The Geology of Indonesia*. The Hague Government Printing Office, vol. 1A.

Verstappen, H. Th. 1992: Volcanic hazards in Colombia and Indonesia: Lahars and related phenomena. In McCall, G.J.H., Laming, D.J.C. and Scott, S.C. (eds), *Geohazards: Natural and Man-Made*. London: Chapman and Hall, 33–42.

Viljoen, R.P. and Voljoen, M.J. 1982: Komatiites—an historical review. In Arndt, N.T. and Nisbet, E.G. (eds), *Komatiites*. London: Allen and Unwin, 5–17.

Vine, F.J. and Matthews, D.H. 1963: Magnetic anomalies over ocean ridges. *Nature* 199, 947–9.

Vogel, J.S., Cornell, W., Nelson, D.E. and Southon, J.R. 1990: Vesuvius/Avellino, one possible source of seventeenth century BC climatic disturbances. *Nature* 344, 534–7.

Voight, B. 1988a: A method for prediction of volcanic eruptions. *Nature* 332, 125–30.

Voight, B. 1988b: Countdown to catastrophe. *Earth and Mineral Sciences (Pennsylvania State University)* 57(2), 17–30.

Voight, B. 1990: The 1985 Nevado del Ruiz volcano catastrophe: Anatomy and retrospection. *Journal of Volcanology and Geothermal Research* 42, 151–88.

Voight, B., Glicken, H., Janda, R.J. and Douglass, P.M. 1981: Catastrophic rockslide avalanche of May 18. In Lipman, P.W. and Mullineaux, D.R. (eds), *The 1980 Eruptions of Mount St Helens, Washington*. United States Geological Survey Professional Paper 1250, 69–80.

Volcanological Society of Japan (1961–date): *Bulletin of Volcanic Eruptions*. Tokyo: Volcanological Society of Japan.

Volz, F.E. 1975a: Burden of volcanic dust and nuclear debris after injection into stratosphere at 40°–58°N. *Journal of Geophysical Research* 80, 2649–52.

Volz, F.E. 1975b: Distribution of turbidity after 1912 Katmai eruption in Alaska. *Journal of Geophysical Research* 80, 2643–8.

von Wältershausen, S.W. 1880: *Der Atna*, Vols 1 and 2. Leipzig: Engelman.

Vossler, T., Anderson, D.L., Aras, N.K., Phelen, J.M. and Zoller, W.H. 1981: Trace element composition of the Mount St Helens plume stratospheric samples from May 18th eruption. *Science* 211, 827–30.

Wadge, G. 1978: Effusion rate and the shape of aa lava flow-fields on Mount Etna. *Geology* 6, 503–6.

Wadge, G. 1976: Deformation of Mount Etna, 1971–1974. *Journal of Volcanology and Geothermal Research* 1, 237–63.

Wadge, G. 1985: Avoiding another St Pierre. *Volcanic News* 22, 4.

Wadge, G. and Isaacs, M.C. 1988: Mapping the volcanic hazards from Soufriere Hills volcano, Montserrat, West Indies, using an image processor. *Journal of the Geological Society (London)* 145(4), 541–53.

Waitt, R.B. and Dzurisin, D. 1981: Proximal air-fall deposits from the May 18 Eruption—stratigraphy and field sedimentology. In Lipman, P.W. and Mullineaux, D.R. (eds), *The 1980 Eruptions of Mount St Helens, Washington*. United States Geological Survey Professional Paper 1250, 601–16.

Walker, A.B., Redmayne, D.W. and Browitt, C.W.A. 1992: Seismic monitoring of Lake Nyos, Cameroon, following the gas release disaster of August 1986. In McCall, G.J.H., Laming, D.J.C. and Scott, S.C. (eds), *Geohazards: Natural and Man-Made*. London: Chapman and Hall, 65–79.

Walker, G.P.L. 1967: Thickness and viscosity of Etnean lavas. *Nature* 213, 484–5.

Walker, G.P.L. 1970: Compound and simple lava flows and flood basalts. *Bulletin Volcanologique* 35, 579–90.

Walker, G.P.L. 1971: Grainsize characteristics of pyroclastic deposits. *Journal of Geology* 79, 696–714.

Walker, G.P.L. 1973a: Lengths of lava flows. *Philosophical Transactions of the Royal Society London* 274, 107–18.

Walker, G.P.L. 1973b: Explosive volcanic eruptions—a new classification scheme. *Geologische Rundschau* 62, 431–46.

Walker, G.P.L. 1974: Volcanic hazards and the prediction of volcanic eruptions. In Funnell, B.M. (ed.), *Prediction of Geological Hazards*. Geological Society of London, Miscellaneous Paper 3, 23–41.

Walker, G.P.L. 1980: The Taupo pumice: Product of the most powerful known (ultraplinian) eruption. *Journal of Volcanology and Geothermal Research* 8, 69–94.

Walker, G.P.L. 1981a: Generation and dispersal of fine ash and dust by volcanic eruptions. *Journal of Volcanology and Geothermal Research* 11, 81–92.

Walker, G.P.L. 1981b: Volcanological applications of pyroclastic studies. In Self, S. and Sparks, R.S.J. (eds), *Tephra Studies*. Amsterdam: D. Reidel, 391–403.

Walker, G.P.L. 1981c: Studies on individual volcanoes of tephra layers: New Zealand case histories of pyroclastic studies. In Self, S. and Sparks, R.S.J. (eds), *Tephra Studies*, Dordrecht: Reidel, 317–30.

Walker, G.P.L. 1981d: Plinian eruptions and their products. *Bulletin Volcanologique* 44(2), 221–40.

Walker, G.P.L. 1982: Volcanic hazards. *Interdisciplinary Science Reviews* 7(2), 148–57.

Walker, G.P.L. 1983: Ignimbrite types and ignimbrite problems. *Journal of Volcanology and Geothermal Research* 17, 65–88.

Walker, G.P.L. 1984: Downsag calderas, ring faults, caldera sizes, and incremental caldera growth. *Journal of Geophysical Research* 89B, 8407–16.

Walker, G.P.L. 1985: Origin of coarse lithic breccias near ignimbrite source vents. *Journal of Volcanology and geothermal Research* 25, 157–71.

Walker, G.P.L. 1991: Structure, and origin by injection of lava under surface crust, of tumuli, 'lava rises', 'lava-rise pits', and 'lava-inflation clefts' in Hawaii. *Bulletin of Volcanology* 53, 546–58.

Walker, G.P.L. and Croasdale, R. 1971: Two plinian-type eruptions from the Azores. *Journal of the Geological Society (London)* 127, 303–17.

Walker, G.P.L. and Croasdale, R. 1972: Characteristics of some basaltic pyroclastics. *Bulletin Volcanologique* 35, 303–17.

Walker, G.P.L., Wilson, L. and Bowell, E.L.G. 1971: Explosive volcanic eruptions I. The rate and fall of pyroclasts. *Geophysical Journal of the Royal Astronomical Society* 22, 377–83.
Walker, G.P.L., Heming, R.F. and Wilson, C.J.N. 1980a: Low-aspect ratio ignimbrites. *Nature* 283, 286–7.
Walker, G.P.L., Wilson, L. and Froggatt, P.C. 1980b: Fines depleted ignimbrite in New Zealand—the product of a turbulent pyroclastic flow. *Geology* 8, 245–9.
Walter, L.S. 1990: The uses of satellite technology in disaster management. *Disasters* 14(1), 20–35.
Warrick, R.A. 1979: Volcanoes as hazard: An overview. In Sheets, P.D. and Grayson, D.K. (eds), *Volcanic Activity and Human Ecology*. New York: Academic Press, 161–89.
Warrick, R.A. 1981: *Four Communities Under Ash after Mount St Helens*. Program on Technology, Environment and Man-Monograph 34, University of Colorado, Institute of Behavioural Science.
Waters, A.C. and Fisher, R.V. 1971: Base surges and their deposits: Calelinhos and Taal volcanoes. *Journal of Geophysical Research* 76, 5596–5614.
Watson, E.B. 1982: Melt infiltration and magma evolution. *Geology* 10, 236–40.
Watts, W.A. 1985: A long pollen record from Laghi di Monticchio, southern Italy: A preliminary account. *Journal of the Geological Society (London)* 142, 491–9.
Weertman, J. 1971: Theory of water-filled crevasses in glaciers applied to vertical magma transport beneath oceanic ridges. *Journal of Geophysical Research* 76, 1171–83.
Wegener, A. 1912: Die Entstehung der Kontinenten. *Geologische Rundschau* 3, 276–92.
Wentworth, C.K., Carson, M.H. and Finch, R.H. 1945: Discussion on the viscosity of lava. *Journal of Geology* 53, 94–104.
Wexler, H. 1951a: Spread of Krakatoa volcanic dust cloud as related to the high level circulation. *Bulletin of the American Meteorological Society* 32(2), 48–51.
Wexler, H. 1951b: On the effects of volcanic dust on isolation and weather. *Bulletin of the American Meteorological Society* 32(1), 10–15.
Wexler, H. 1952: Volcanoes and climate. *Scientific American* 186, 74–80.
Wheller, G.E., Varne, R., Foden, J.D. and Abbott, M.J. 1987: Geochemistry of Quaternary volcanism in the Sunda-Banda arc, Indonesia, and three component genesis of island-arc basaltic magmas. *Journal of Volcanology and Geothermal Research* 32, 137–60.
White, G.F. 1942: Human Adjustment to Floods: A Geographical Approach to the Flood Problem in the United States. Research Paper 29, Department of Geography, University of Chicago.
White, G.F. 1973: Natural hazards research. In Chorley, R.J. (ed.), *Directions in Geography*. London: Methuen, 193–212.
White, G.F. 1974: Natural hazards research: Concepts, methods and policy implications. In White, G.F. (ed.), *Natural Hazards: Local, National, Global*. New York: Oxford University Press, 3–16.
White, R.S. and McKenzie, D.P. 1989: Magmatic rift zones: The generation of volcanic continental margins and flood basalts. *Journal of Geophysical Research* 94B, 7685–7730.
Whitford-Stark, J. and Wilson, L. 1976: Atmospheric motions produced by hot lava. *Weather* 32, 25–7.
Whitham, A.G. and Sparks, R.S.J. 1986: Pumice. *Bulletin of Volcanology* 48, 209–23.
Whittow, J. 1987: Hazards—adjustment and mitigation. In Clark, M.J., Gregory, K.J. and Gurnell, A.M. (eds), *Horizons in Physical Geography*. London: Macmillan, 307–19.
Wickman, F.E. 1966a: Repose period patterns of volcanoes. I. Volcanic eruptions regarded as random phenomena. *Arkiv Mineralogi Geologi* 4, 4, 291–301.
Wickman, F.E. 1966b: Repose period patterns of volcanoes, II. Eruption histories of some East Indian volcanoes. *Arkiv Mineralogi Geologi* 4, 303–17.
Wickman, F.E. 1966c: Repose period patterns of volcanoes, III. Eruption histories of some Japanese volcanoes. *Arkiv Mineralogi Geologi* 4, 319–35.
Wickman, F.E. 1966d: Repose period patterns of volcanoes, IV. Eruption histories of some selected volcanoes. *Arkiv Mineralogi Geologi* 4, 337–50.
Wickman, F.E. 1966e: Repose period patterns of volcanoes, V. General discussion and a tentative stochastic model. *Arkiv Mineralogi Geologi* 4, 351–67.
Wickman, F.E. 1976: Markov models of repose-period patterns of volcanoes: In Merriam, D.F. (ed.), *Random Processes in Geology*. Berlin: Springer-Verlag, 135–61.
Wijkman, A. and Timberlake, L. 1984: *Natural Disasters: Acts of God or Acts of Man?* London and New York: Earthscan, International Institute for Environment and Development.
Wilhelms, D.E. 1976: Mercurian volcanism questioned. *Icarus* 28, 551–8.
Wilhelms, D.E. 1980: *Stratigraphy of Part of the Lunar Nearside*. Washington DC, United States Geological Survey Professional Paper 1046–A.
Wilkie, T. 1991: Experts await 'avalanche of fire' in the Philippines. *The Independent*, 13 June, 13.
Williams, H. 1932: The history and character of volcanic domes. *University of California (Berkeley)*

Publication, Bulletin Department of Geological Science 21, 51–146.
Williams, H. 1941: Calderas and their origin. *University of California Publications in Geological Sciences* 25, 239–346.
Williams, H. 1952: The great eruption of Cosiguina, Nicaragua. *University of California Publication Geological Science* 29, 21–46.
Williams, H. and McBirney, A.R. 1979: *Volcanology*. San Francisco: Freeman Cooper.
Williams, H., Turner, F.J. and Gilbert, C.M. 1982: *Petrology: An Introduction to the Study of Rocks in Thin Section*. San Francisco: Freeman Cooper (second edition).
Williams, R.S. and Moore, J.G. 1983: *Man Against Volcano: The Eruption of Heimaey, Vestmannaeyjar, Iceland*. United States Geological Survey.
Williams, S.N. and Self, S. 1983: The October 1902 plinian eruption of Santa Maria volcano, Guatemala. *Journal of Volcanology and Geothermal Research* 16, 33–56.
Williams, S.N., Stoiber, R.E., Nestor-Garcia, P., Adela-London, C.J., Gemmell, J.B., Lowe, D.R. and Connor, C.B. 1986: Eruption of the Nevado del Ruiz volcano, Columbia on 13 November 1985: Gas flux and fluid geochemistry. *Science* 233, 964–7.
Wilson, C.J.N. 1980: The role of fluidisation in the emplacement of pyroclastic flows. *Journal of Volcanology and Geothermal Research* 8, 231–49.
Wilson, C.J.N. 1984: The role of fluidisation in the emplacement of pyroclastic flows, 2: experimental results and their interpretation. *Journal of Volcanology and Geothermal Research* 20, 55–84.
Wilson, C.J.N. 1986: Pyroclastic flows and ignimbrites. *Science Progress (Oxford)* 70, 171–207.
Wilson, C.J.N. and Walker, G.P.L. 1982: Ignimbrite depositional facies: The anatomy of a pyroclastic flow. *Journal of the Geological Society (London)* 139, 581–92.
Wilson, J.T. 1963a: Evidence from islands on the spreading of ocean floors. *Nature* 197, 536–8.
Wilson, J.T. 1963b: A possible origin of the Hawaiian Islands. *Canadian Journal of Physics* 41, 863–70.
Wilson, J.T. 1965: A new class of faults and their bearing on continental drift. *Nature* 207, 343–7.
Wilson, J.T. 1966: Did the Atlantic close and then re-open? *Nature* 211, 676–81.
Wilson, J.T. 1988: Convection tectonics: Some possible effects upon the Earth's surface of flow from the deep mantle. *Canadian Journal of Earth Science* 25, 1199–1208.
Wilson, L. 1972: Explosive volcanic eruptions—II. The atmospheric trajectories of pyroclasts. *Geophysical Journal of the Royal Astronomical Society* 30, 381–92.
Wilson, L. 1976: Explosive volcanic eruptions—III. Plinian eruption columns. *Geophysical Journal of the Royal Astronomical Society* 45, 543–56.
Wilson, L. 1980: Relationships between pressure, volatile content and ejecta velocity in three types and Geothermal Research 8, 297–313.
Wilson, L. and Head, J.W. 1981a: Ascent and eruption of basaltic magma on the Earth and Moon. *Journal of Geophysical Research* 86B, 2971–3001.
Wilson, L. and Head, J.W. 1981b: Lunar sinuous rill formation by thermal erosion: Eruption conditions, rates and durations. *Lunar Planetary Science* XII, 427–9. Houston: Lunar Planetary Institute.
Wilson, L. and Head, J.W. 1983: A comparison of volcanic eruption processes on Earth, Moon, Mars, Io and Venus. *Nature* 302, 663–9.
Wilson, L. and Walker, G.P.L. 1987: Explosive volcanic eruptions IV. Ejecta dispersal in plinian eruptions: The control of eruption conditions of atmospheric properties. *Geophysical Journal of the Royal Astronomical Society* 89, 651–678.
Wilson, L., Sparks, R.S.J., Huang, T.C. and Watkins, N.D. 1978: The control of eruption column heights by eruption energetics and dynamics. *Journal of Geophysical Research* 83B, 1829–36.
Wilson, L., Sparks, R.S.J. and Walker, G.P.L. 1980: Explosive volcanic eruptions, IV. The control of magma properties and conduit geometry on eruption column behavior. *Geophysical Journal of the Royal Astronomical Society* 63, 117–48.
Wilson, L., Pinkerton, H. and Macdonald, R. 1987: Physical processes in volcanic eruptions. *Annual Review Earth Planetary Sciences* 15, 73–95.
Wilson, M. 1989: *Igneous Petrogenesis*. London: Harper-Collins.
Windley, B.F. 1977: *The Evolving Continents*. New York: John Wiley.
Wohletz, K.H. 1983: Mechanisms of hydrovolcanic pyroclastic formation: Grain size, scanning electron microscopy and experimental studies. *Journal of Volcanology and Geothermal Research* 17, 31–63.
Wohletz, K.H. 1986: Explosive magma-water interactions: Thermodynamics, explosion mechanisms, and field studies. *Bulletin of Volcanology* 48, 245–64.
Wohletz, K.H. and Sheridan, M.F. 1979: A model of pyroclastic surge. *Geological Society of America Special Paper* 180, 177–194.

Wohletz, K.H. and Sheridan, M.F. 1983: Hydrovolcanic explosions II. Evolution of basaltic tuff rings and tuff cones. *American Journal of Science* 283, 385–413.

Wohletz, K.H. and McQueen, R.G. 1984: Experimental studies of hydromagmatic volcanism. In *Explosive Volcanism: Inception, Evolution and Hazards.* Washington DC: National Academy Press, 158–69.

Wohletz, K.H. and Valentine, G.A. 1990: Computer simulations of the explosive volcanic eruptions. In Ryan, M.P. (ed.), *Magma Transport and Storage.* New York: John Wiley, 113–35.

Wohletz, K.H., McGetchin, T.R., Sandford, M.T. and Jones, E.M. 1984: Hydrodynamic aspects of caldera-forming eruptions: Numerical models. *Journal of Geophysical Research* 89B, 8269–85.

Wolfe, J.A. and Self, S. 1982: Structural lineaments and Neogene volcanism in south western Luzon. In *The Tectonics and Geologic Evolution of the South-East Asia Seas and Islands.* Geophysical Monograph Series 27, 157–72.

Woodcock, N. 1992: Saddam versus Pinatubo as atmospheric polluters. *Geoscientist* 2(2), 14.

Wood, C.A. 1984: Amazing and portentous summer of 1783. *Eos (Transactions of the American Geophysical Union)* 65, 410.

Wood, C.A. and Kienle, J. (eds) 1990: *Volcanoes of North America.* Cambridge: Cambridge University Press.

Woods, A.W. 1988: The fluid dynamics and thermodynamics of plinian eruption columns. *Bulletin of Volcanology* 50, 169–93.

Woods, A.W. and Bursik, M.I. 1991: Particle fallout, thermal disequilibrium and volcanic plumes. *Bulletin of Volcanology* 53, 559–70.

Wright, J.V. 1979: *Formation, Transport and Deposition of Ignimbrites and Welded Tuffs.* Unpublished Ph.D. thesis, Imperial College, University of London.

Wright, J.V. and Walker, G.P.L. 1977: The ignimbrite source problem: Significance of a co-ignimbrite lag-fall deposit. *Geology* 5, 729–32.

Wright, J.V. and Walker, G.P.L. 1981: Eruption, transport and deposition of ignimbrite: A case study from Mexico. *Journal of Volcanology and Geothermal Research* 9, 111–31.

Wright, J.V., Smith, A.L. and Self, S. 1980: A working terminology of pyroclastic deposits. *Journal of Volcanology and Geothermal Research* 8, 315–36.

Wright, J.V., Smith, A.L. and Self, S. 1981a: A terminology for pyroclastic deposits. In Self, S. and Sparks, R.S.J., *Tephra Studies.* Dordrecht: D. Reidel, 457–63.

Wright, J.V., Smith, A.L. and Fisher, R.V. 1981b: Towards a model for ignimbrite-forming eruptions. In Self, S. and Sparks, R.S.J., *Tephra Studies.* Dordrecht: D. Reidel, 433–9.

Wyllie, P.J. 1971: *The Dynamic Earth: Textbook in Geosciences.* New York: J. Wiley.

Wyrwoll, K.-H. and McConchie, D. 1986: Accelerated plate motion and rates of volcanicity as controls on Archean climates. *Climatic Change* 8, 257–65.

Yoder, H.S. 1976: *Generation of Basaltic Magma.* Washington DC: National Academy of Sciences.

Zen, M.T. 1983: Mitigating volcanic disasters in Indonesia. In Tazieff, H. and Sabroux, J.C., *Forecasting Volcanic Events*, Developments in Volcanology 1. Amsterdam: Elsevier, 219–36.

Zimbelman, J.R. 1985: Estimation of rheological properties for flows on the Martian volcano Ascraeus Mons. *Journal of Geophysical Research* B90, D157–62 (supplement).

Zittel, K.A. von 1901: *History of Geology and Paleontology*, translated by M.M. Ogilvie-Gordon. London: W. Scott.

Index